Power Systems

Electrical power has been the technological foundation of industrial societies for many years. Although the systems designed to provide and apply electrical energy have reached a high degree of maturity, unforeseen problems are constantly encountered, necessitating the design of more efficient and reliable systems based on novel technologies. The book series Power Systems is aimed at providing detailed, accurate and sound technical information about these new developments in electrical power engineering. It includes topics on power generation, storage and transmission as well as electrical machines. The monographs and advanced textbooks in this series address researchers, lecturers, industrial engineers and senior students in electrical engineering. ** Power Systems is indexed in Scopus**

More information about this series at http://www.springer.com/series/4622

Klaus Schon

High Voltage Measurement Techniques

Fundamentals, Measuring Instruments, and Measuring Methods

Springer

Klaus Schon
Formerly with the Physikalisch-Technische
Bundesanstalt Braunschweig und Berlin
Braunschweig, Niedersachsen, Germany

ISSN 1612-1287 ISSN 1860-4676 (electronic)
Power Systems
ISBN 978-3-030-21769-3 ISBN 978-3-030-21770-9 (eBook)
https://doi.org/10.1007/978-3-030-21770-9

This Springer imprint is published by the registered company Springer Nature Switzerland AG
The registered company address is: Gewerbestrasse 11, 6330 Cham, Switzerland

Preface

The present book is the translation of the German textbook "Hochspannungsmesstechnik" published in 2017 and the thoroughly revised and extended version of "High Impulse Voltage and Current Measurement Techniques" published in 2013. The new edition is intended as an actual and comprehensive introduction into high-voltage measurement techniques. It covers the standardized measurement quantities as defined in the standards, the measuring circuits and methods in the fields of direct, alternating, and impulse voltages and currents, electro- and magneto-optical sensors, electric and dielectric quantities, and partial discharges. In most areas of high-voltage measurements, digital techniques and data processing have been successfully introduced, replacing the well-known analogue measuring instruments and techniques of the past. This significant development, which is still ongoing, provides an enlarged variety of new technical solutions combined with the prospect of sophisticated analysis and higher accuracy.

High-voltage measurement technique is generally considered to be very important for reliable transmission of electrical energy at high-voltage potential. Furthermore, it poses high demands on the engineers and technicians entrusted with testing and measurements. In addition to the transmission of electrical energy, high voltages and currents occur in many other areas of physics and engineering, in which they are made use of for various applications. Catchwords for these applications are as follows: plasma physics, particle accelerators, power electronics, medical technology, aluminum production, spot-welding technology, electronic ignition systems for combustion engines, electro-shock devices, and electromagnetic compatibility. Even in these areas, precise high-voltage measurement technique is of special significance, either to prevent the device under test being stressed too high or too low, or to guarantee the quality of the application.

As a general point, international agreements on testing and metrology have taken more and more influence on electrical engineering, not least because of the globalized market economy. This is particularly evident, on the one hand, in national and international test specifications, which lay down (more or less) exact measurement and evaluation procedures, and, on the other, in the worldwide network of accredited testing and calibration laboratories. Accreditation according to internationally accepted rules is a prerequisite for mutual recognition and acceptance of the test and measurement results. The introduction of digital techniques and the significantly improved characteristics of digital recorders and personal computers, achieved during the last two decades, allow sophisticated data processing by applying rather complex software algorithms. Examples are, among others, the objective evaluation of the recorded waveforms, filtering of the measured data, detection of partial discharges, and application of convolution for investigating the dynamic performance of voltage dividers and current sensors.

For understanding the content of this book, basic knowledge of high-voltage engineering is a prerequisite for the reader. While in Europe, measuring systems as well as testing and measurement technology are tailor-made for alternating voltages up to the maximum transmission voltage of 400 kV, more than twice this level is required in other parts of the world for bridging large distances between energy suppliers and consumers. As a result of the enormous economic development in the Asiatic region, voltages higher than 1000 kV for alternating voltage transmission and 800 kV for direct voltage transmission are under discussion. In this context, the experts also discuss the question whether the proven measuring devices and measurement techniques can even be used at these higher voltages.

I would like to express my greatest thanks to Prof. Dr.-Ing. Dr.-Ing. h.c. Dieter Kind, Professor at the Technical University of Braunschweig and Past-President of the Physikalisch-Technische Bundesanstalt Braunschweig und Berlin (PTB). He has strongly influenced and fostered my professional life at the High-Voltage Laboratory of the PTB, supported me on many small and big occasions, and introduced me to the international group of high-voltage experts. My hearty thanks are also due to my colleagues at the PTB and a number of high-voltage experts in other institutes and companies for their valuable support in elaborating the manuscript and providing me with photographs and printing permissions. Finally, I would like to thank the Springer staff for the good and efficient support of the book project.

Braunschweig, Germany Klaus Schon
April 2019

Contents

Abbreviations

3PARD	3-Phase Amplitude Relation Diagram
3PFRD	3-Phase Frequency Relation Diagram
3PTRD	3-Phase Time Relation Diagram
A/D Converter	Analog-to-Digital Converter
AC	Alternating Current
AkkStelleG	German Accreditation Act
BGO	$Bi_4Ge_3O_{12}$ or $Bi_{12}GeO_{20}$
BIPM	Bureau International des Poids et Mesures
CD	Coupling Device
CD-ROM	Compact Disc Read-Only Memory
CIGRE	Conseil International des Grands Réseaux Électriques
CISPRE	Comité International Spécial des Perturbations Radioélectriques
CMC	Calibration and Measurement Capability
D/A Converter	Digital-to-Analog Converter
DAKKS	German Accreditation Body
DATech	German Accreditation Body Technology
DC	Direct Current
DFT	Discrete Fourier Transform
DIN	German Institute for Standardization
DKD	German Calibration Service
DKE	German Electrotechnical Commission
DM	Digital Measuring Instrument
DSP	Digital Signal Processor
EA	European Co-operation for Accreditation
EB	Effective Bit Number
EMP	Electromagnetic Pulse
EMV	Electromagnetic Compatibility
f.s.d	Full-Scale Deflection
FFT	Fast Fourier Transform
FOF	Fluorescent Optical Fiber

FPGA	Field-Programmable Gate Array
FS	Spark Gap
GIL	Gas-Insulated Line
GIS	Gas-Insulated Switchgear
GPS	Global Positioning System
GRIN Lens	Gradient Index Lens
GTEM	Gigahertz Transverse Electromagnetic
GUM	Guide to the Expression of Uncertainty in Measurement
HVDC	High-Voltage Direct Current
IEC	International Electrotechnical Commission
ISH	International Symposium on High Voltage Engineering
ISO	International Organization for Standardization
LD	Laser Diode
LED	Light-Emitting Diode
LI	Lightning Impulse Voltage
LIC	Lightning Impulse Voltage Chopped
LNO	$LiNbO_3$
LSB	Least Significant Bit
LWL	Fiber Optic Cable
MOSFET	Metal–Oxide–Semiconductor Field-Effect Transistor
NEMP	Nuclear Electromagnetic Pulse
OP	Operational Amplifier
PC	Personal Computer
PD	Partial Discharge
PE	Polyethylene
PMF	Polarization-Maintaining Fiber
Proc.	Proceedings
PRPD	Phase-Resolved Partial Discharge
PTB	Physikalisch-Technische Bundesanstalt
SI units	International System of Units
SI	Switching Impulse Voltage
TC	Technical Committee
TDG	Test Data Generator
TEA Laser	Transversely Excited Atmospheric Laser
TEM	Transversal Electromagnetic
THD	Total Harmonic Distortion
UHV	Ultrahigh Voltage
UV	Ultraviolet Light
UVC	Ultraviolet Light of Very Short Wavelength
VIM	International Vocabulary of Metrology
WLAN	Wireless Local Area Network
WTO	World Trade Organization
XLPE	Cross-Linked Polyethylene

Chapter 1
Introduction

Abstract High direct, alternating and impulse voltages as well as the corresponding currents play an important role in the electrical energy supply, but also in many other areas of physics and technology. The high-voltage apparatus and other equipment used for this purpose are subjected to a series of tests prior to commissioning, which allow limited information on their reliability and expected lifetime. The decisive factor is the proper execution of high-voltage insulation, which is based inter alia on the specific knowledge of the solid, liquid and gaseous insulating materials used. All tests require accurate measurement techniques and the use of calibrated measuring systems. In this context are terms and content such as quality assurance, calibration, traceability of the measurements to the SI units, measurement uncertainty, internationally agreed test specifications, accredited testing and calibration laboratories. In the following chapters, the old but still valid fundamental basics and principles of high-voltage measurement techniques are combined with the more recent developments in all the above-mentioned fields under particular consideration of digital measurement technology and data transmission.

The transmission of electrical energy from the energy suppliers to the urban centers takes place predominantly via overhead transmission lines at high voltage to keep the line currents and transmission losses low. In the metropolitan areas themselves, the energy is distributed further via underground high-voltage cables, *gas-insulated switchgear* (GIS) and *gas-insulated lines* (GIL). The energy transmission with three-phase alternating (AC) voltages, which can be transformed with power transformers to the desired voltage levels, is used worldwide. The highest voltage levels for electrical power transmission are, for example, 400 kV in Europe, 750 kV in North America and 1000 kV in Asia. The frequency of the approximately sinusoidal AC voltage is 50 Hz in Europe and in many other countries, while 60 Hz is common in North and Central America and parts of South America. Other frequencies exist, for example, the German railway network is operated with its own electrical power supply with a single-phase AC voltage of 110 kV and a frequency of 16.7 Hz. Furthermore, a large number of local supply networks exist

K. Schon, *High Voltage Measurement Techniques*, Power Systems,
https://doi.org/10.1007/978-3-030-21770-9_1

for regional trains with different voltages and frequencies. High AC voltages are also required in other areas of physics and engineering, including devices and systems for generating high direct (DC) voltages and impulse voltages.

Electrical energy transmission over distances of more than 700 km advantageously takes place at high DC voltage, since in this case lower transmission losses occur than at AC voltage and more power can be transmitted. In *high-voltage DC current* (HVDC) transmission networks, voltages are generally in the range of up to 500 kV. In Asia, networks with a maximum of 800 kV are in operation or planned. On the other hand, short DC transmission lines are also used to interconnect two asynchronous AC voltage networks (*back-to-back HVDC installation*). HVDC systems are also installed with submarine cables or underground cables with a length of up to several 100 km. In Germany, the future energy supply will change significantly after switching off nuclear power plants. For example, the energy increasingly generated by wind turbines in the north of the country is to be transported via HVDC transmission lines to the south in addition to the existing AC grid. Finally, it should be noted that a futuristic planning study examines the technical, economic and political conditions for the case that the solar energy generated in the Sahara with photovoltaic systems is transported via HVDC links to Europe.

High DC voltages are used in a variety of other applications, such as X-ray machines, dust filter systems, coating and painting facilities, aluminum production, etc. Particularly high DC voltages of up to 25 MV, which are free from harmonics, can be generated with *electrostatic band generators* according to *van de Graaff*. However, they can only be loaded with low currents of a few milliamperes and therefore are not suitable for electrical power transmission. They are mainly used in accelerator systems for basic research in nuclear physics.

Transient overvoltages with peak values of more than 1 MV can occur in power supply networks with overhead lines. The impulse voltages are thus greater than the maximum DC and AC operating voltages. Causes of the overvoltages are direct or indirect *lightning strokes* on overhead lines or outdoor switchgear, short circuits or flashovers due to failure of the electrical insulation, switching operations in substations and the response of overvoltage arresters. The rise times of these transient voltages are mainly in the range of microseconds to milliseconds. In special cases, for example, in the event of *disruptive discharges*, the voltage collapse can take place in a very short time of less than 1 μs. Extremely short impulse voltages lasting from a few 100 ns down to 1 ns occur in gas-insulated systems during switching operations and flashovers. Transient voltages of more than 1 kV can also occur in the low-voltage network when switching an electrical household appliance. This may affect or even destroy the appliance itself or neighboring devices.

The apparatus used in power grids is also subjected to high DC, AC and impulse currents. For example, in the case of a short circuit, the AC current may be superimposed by a transient DC component. The resulting short-time current then reaches briefly peak values of up to 200 kA or even more. Direct or remote lightning strokes in the supply grid can cause transient currents with peak values in the range of 100 kA and rise times of 1 μs. When a transmission line is struck by lightning, current impulses propagate on either side of the line and cause high transient voltages at both ends of the line, which are superimposed on the operating voltage. Surge arresters are therefore used to protect the apparatus from overvoltages. When the arrester reacts, the transmission line can be discharged, resulting in an approximately rectangular current impulse of approximately 1 ms duration.

High impulse voltages and currents with rise times in the microsecond or nanosecond range also occur in other fields of physics and engineering or are useful for certain applications, as the following examples show. In plasma physics, extremely high magnetic fields are generated for the short-time magnetic confinement of hot plasma. The impulse currents required for electrical spot-welding have peak values of up to 200 kA. Electronic ignition systems for internal combustion engines generate impulse voltages with a maximum peak value of 30 kV. In power electronics, impulse voltages and currents of several 10 kV and up to 10 kA occur or are required for testing, for example, for solar modules. Electricity meters are tested with impulse currents consisting of a line-frequency sinusoidal half-oscillation with amplitudes of up to several kiloamperes. In medical technology, the destruction of kidney stones and gallstones as well as calcifications in joints is achieved by acoustic shock waves generated by electric impulses. The effect of electro-shock devices is based on high-voltage impulses that temporarily paralyze the muscle functions of the attacked person or animal. Finally, mention should be made of the wide spectrum of impulses used in *electromagnetic compatibility* (EMC) testing of small electronic devices to very complex systems such as aircraft.

The insulation of the apparatus can be severely stressed by the high voltages and currents during operation, which affects the service life. Knowledge of the electrical and dielectric properties of the solid, liquid and gaseous insulation materials used is therefore an important part of high-voltage engineering. If the high-voltage insulation of the apparatus is not perfect, for example, due to gas cavities, *partial discharges* will occur in this region above a certain inception voltage. The long-term influence of partial discharges on the surrounding insulation often leads to a gradual degradation of the electrical strength and possible failure of the system.

The reliability of the electrical energy supply is an important prerequisite for a prosperous economy in every country and for the welfare of the population. Therefore, the apparatus of the energy supply is subjected to a series of acceptance tests prior to its use. The electrical, mechanical and thermal stresses which may occur during practical operation of the apparatus are thus simulated in laboratory tests, possibly also as *on-site tests* at the site of operation. These include, on the one hand, tests with the voltage or current type corresponding to the mains operation of the apparatus, and, on the other hand, tests with impulse voltages or currents. The level of the internationally standardized test voltages depends on the rated voltage of the apparatus. For EMC testing of electronic devices and systems, the electromagnetic test fields between plate- or strip-shaped electrode arrangements are generated with very steeply rising impulse voltages. The effect of the electromagnetic pulse (EMP), which is released by a nuclear explosion at very high altitude, can also be simulated in this way.

In addition to the voltage tests, all high-voltage apparatus are subjected to partial discharge tests in a usually shielded test laboratory or during on-site tests. The phenomenon of partial discharges is very complex and not yet fully understood. However, it has been known for decades of experience that there is a risk of long-term damage to the insulation and premature failure of the apparatus if the partial discharge magnitude exceeds a value individually specified for each apparatus. Therefore, in addition to an initial acceptance test, permanent online monitoring of the partial discharges takes place increasingly in order to detect a possible failure of the apparatus in time. In further measurements, the electrical and dielectric properties of the high-voltage insulation are checked. These include quantities such as insulation resistance, conductivity, capacitance and *dissipation factor* of the test object.

To carry out all necessary tests, thorough knowledge of the measurement technique is required. This is important in order to avoid overloading or underloading of the test object, or because the quality of an application, for example, a medical treatment or electrical spot welding, must be ensured. The measurement of high voltages and currents, partial discharges and dielectric properties of insulating materials has a long tradition. There are now two crucial changes. The mechanical meters and measurement methods used for decades have been largely replaced by electronic measuring instruments and suitable measurement methods several years ago. The introduction of digital measurement technology with numerical data processing is yet another key breakthrough and means the end for most analog measuring circuits and instruments.

All measuring instruments used in the tests must be thoroughly checked with regard to their uncertainty of measurement. In this context, terms and contents such as quality assurance, calibration, traceability of the measurements to the SI units, measurement uncertainty, internationally recognized test specifications, accredited testing and calibration laboratories, etc., are of particular importance.

High-voltage and power engineering are discussed extensively in the relevant literature, especially in the current conference volumes of national and international lectures and proceedings, for example, "International Symposium on High Voltage Engineering" (ISH). Comprehensive representations can be found in several textbooks, but the corresponding measurement techniques are only briefly presented [1–5]. The resourceful reader can also get more or less detailed information on individual topics via the well-known search engines on the internet. The textbooks in references [6–8] deal with high-voltage measurement techniques and related fields in more detail, but were published decades ago or are available only as an unchanged copy of older editions.

Current topics of the high-voltage measurement technique are dealt with in recent publications [9, 10]. The present book is a translation of [10], which is a revised and extended version of [9]. It now comprises the measurement techniques for high DC, AC and impulse voltages and the corresponding currents, partial discharges as well as electrical and dielectric measurement quantities. The book has been written with the intention of combining the old but still valid fundamental basics and principles of high-voltage measurement techniques with recent developments in all the above-mentioned fields. These developments have resulted from improved test and measurement equipment, the introduction of digital methods and numerical data processing, special calibration procedures and uncertainty calculations, as well as from the increasing importance of the relevant test standards.

References

1. Beyer, M., Boeck, W., Möller, K., Zaengl, W.: Hochspannungstechnik. Theoretische und praktische Grundlagen für die Anwendung. Springer, Berlin, Heidelberg, New York (1986)
2. Kind, D., Feser, K.: High-Voltage Test Techniques, 2nd edn. Butterworth Heinemann, Oxford (2001) [German edition: Kind, D., Feser, K.: Hochspannungsversuchstechnik, 5th edn. Friedr. Vieweg & Sohn, Braunschweig/Wiesbaden (1995)]
3. Küchler, A.: High Voltage Engineering. Fundamentals—Technology—Applications. Springer, Heidelberg, Dordrecht, London, New York (2013) [German edition: Küchler, A.: Hochspannungstechnik. Grundlagen – Technologie – Anwendungen, 3rd edn. Springer, Berlin, Heidelberg (2017)]
4. Kuffel, E., Zaengl, W.S., Kuffel, J.: High Voltage Engineering—Fundamentals, 2nd edn. Elsevier Newness, Oxford (2000)
5. Hauschild, W., Lemke, E.: High-Voltage Test and Measuring Techniques. Springer, Heidelberg, New York, Dordrecht, London (2013)
6. Schwab, A.J.: High-Voltage Measurement Techniques. M.I.T Press (1972) [German edition: Schwab, A.J.: Hochspannungsmesstechnik. Messgeräte und Messverfahren, 2nd edn. Springer, Berlin, Heidelberg, New York (1981)]
7. Aŝner, A.M.: Stoßspannungs-Meßtechnik. Springer, Berlin, Heidelberg, New York (1974)
8. Hyltén-Cavallius, N.: The measurement of high impulse voltages and currents. In: Claudi, A., Bergman, A., Berlijn, S., Hällström, J. (eds.) A Review of Seven Decades of Development. SP, Boras (2004)
9. Schon, K.: High Impulse Voltage and Current Measurement Techniques. Springer, Heidelberg, Dordrecht, London, New York (2013) [German edition: Schon, K.: Stoßspannungs- und Stoßstrommesstechnik. Springer, Heidelberg, Dordrecht, London, New York (2010)]
10. Schon, K.: Hochspannungsmesstechnik. Grundlagen – Messgeräte – Messverfahren. Springer, Heidelberg, Dordrecht, London, New York (2017)

Chapter 2
High Alternating Voltages and Currents

Abstract The transmission of electrical energy from the power plant to the consumer takes place predominantly with high alternating (AC) voltages, so that this voltage and thus also the alternating currents have special significance. Each apparatus for the electrical energy supply is therefore tested for reliability prior to commissioning. The test and measurement procedures as well as requirements for test voltages and test currents are specified in national and international test regulations. High AC voltages are also important because they are needed for the generation of high DC and impulse voltages as well as numerous applications in physics and engineering. Furthermore, high alternating voltages are used for tests on insulating material in terms of dielectric properties and partial discharges. The chapter introduces the standardized quantities and measurement methods, briefly describes the basic principles of AC voltage and current generators, and discusses the measuring systems and instruments in more detail. Although analog measurement methods and devices are still in use, including the standard sphere gap, to-day the main focus is on digital measuring systems with computer-aided data processing, allowing online and on-site tests.

High alternating voltages—and thus also high alternating currents—play an important role in the transmission of electrical energy from the power plant to the consumer, so that this type of voltage is particularly important. Each high-voltage device for the power supply is tested for reliability before use, with the test and measurement procedures as well as the test voltage and current requirements specified in national and international test standards. Alternating voltages are also particularly important as they are necessary for the generation of high direct and impulse voltages. Furthermore, high alternating voltages are used for tests on insulating material in terms of dielectric properties and partial discharges. The chapter introduces the standardized measurement quantities and methods, briefly describes the basic principles of voltage and current generators, and deals primarily with the measurement systems and instruments that are widely used today in digital versions with computer-aided data processing.

K. Schon, *High Voltage Measurement Techniques*, Power Systems,
https://doi.org/10.1007/978-3-030-21770-9_2

2.1 Alternating Test Voltages

High *alternating (AC) voltages* are the prerequisite for the economic transmission of electrical energy over medium distances and are necessary for the generation of high direct (DC) and impulse voltages as well as for dielectric tests and partial discharge measurements. The basic test and measurement procedures for the apparatus of the electrical power transmission at high voltage are laid down in a number of national and international test standards [1–5]. The terms and requirements for the generation of AC test voltages are specified in IEC 60060-1 [1], the details for measurements and calibrations in IEC 60060-2 [2]. For on-site tests, the definitions and requirements are given in IEC 60060-3 [3]. The use of sphere gaps and rod-rod gaps is dealt with in IEC 60052 [5]. For the devices in the low-voltage range with rated AC voltages of maximum 1 kV, special test regulations are given in IEC 61180 [6].

The generated AC test voltage shall be approximately sinusoidal with a frequency between 45 and 65 Hz. Other frequencies are possible, e.g. 16.7 Hz for testing the equipment of the German railway. For specific tests, frequencies well below or above this range are recommended, e.g. 1 Hz or even 0.1 Hz for on-site cable testing.

The *peak value of the AC test voltage* is defined as the mean of $\hat{u}_+$ and $\hat{u}_-$:

$$\hat{u} = \frac{\hat{u}_+ + \hat{u}_-}{2}, \tag{2.1}$$

where $\hat{u}_+$ and $\hat{u}_-$ are the positive and negative maximum values, respectively (Fig. 2.1). The difference between the positive and negative maximum values shall be less than 2%. The peak value $\hat{u}$, divided by $\sqrt{2}$, gives the *value of the AC test voltage* that characterizes the AC voltage and to which the requirements in the test standards refer:

$$\boxed{U = \frac{\hat{u}}{\sqrt{2}} = \frac{\hat{u}_+ + \hat{u}_-}{2\sqrt{2}}}. \tag{2.2}$$

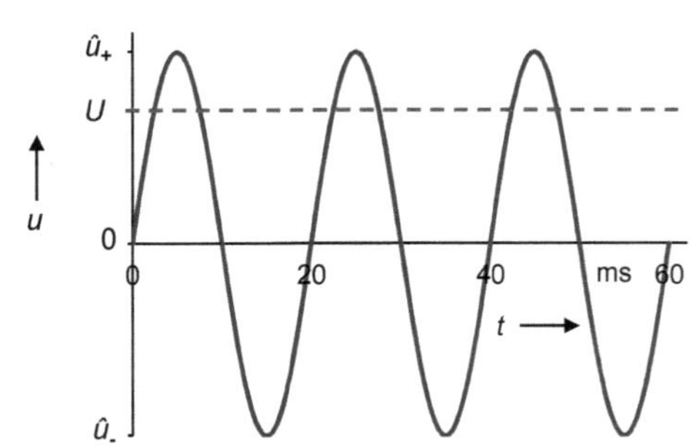

Fig. 2.1 Example of a sinusoidal AC voltage with the maximum values $\hat{u}_+$ and $\hat{u}_-$, the peak value $\hat{u} = (\hat{u}_+ + \hat{u}_-)/2$ and the value of the AC test voltage $U = \hat{u}/\sqrt{2}$

Older analog peak voltmeters often only measure the maximum value of one polarity. If the measured values $\hat{u}_+$ and $\hat{u}_-$ differ by less than 2%, the maximum value displayed is accepted as the peak value $\hat{u}$ according to Eq. (2.1). The test voltage value relevant for a high-voltage test object is specified in the test standards. For test durations not exceeding 1 min, the test voltage shall be kept constant within ±1% of the specified value (±3% for longer test duration).

Note The definition of the peak value divided by $\sqrt{2}$ as the test voltage value is based on the fact that the breakdown of insulating material usually depends on the maximum voltage—apart from the thermal breakdown at constant load.

Occasionally, e.g. when investigating thermal effects, the *root-mean-square (RMS) value* is to be determined as the test voltage value:

$$\boxed{U_{\text{rms}} = \sqrt{\frac{1}{T}\int_0^T u^2 \mathrm{d}t},} \tag{2.3}$$

where T is the time of an integer number of cycles of the AC voltage. For an ideal sinusoidal voltage, both test voltage values according to Eqs. (2.2) and (2.3) are identical. The AC test voltages that are generated by transformers are generally not purely sinusoidal but superimposed by harmonics of the mains frequency. The voltage shape and thus the result of the voltage test is considered acceptable if the quotient of the peak value and RMS value corresponds to $\sqrt{2}$ within ±5%.

The test voltage value according to Eq. (2.2) or Eq. (2.3) shall be measured with a suitable measuring system with an uncertainty of not more than 3%. Further requirements concern the frequency response. If the measuring system is used for voltage measurement at a single frequency f_{nom}, the frequency response may only change by ±1% within f_{nom} to $7f_{\text{nom}}$. For a wider frequency range of the AC voltage to be measured, e.g. $f_{\text{nom},1}$ = 45 Hz to $f_{\text{nom},2}$ = 65 Hz, the frequency response must be constant within ±1% from 45 Hz to at least 7 × 65 Hz = 455 Hz. The frequency response above $7f_{\text{nom}}$ is subject to further specifications. The requirements on the frequency response of the measuring system are considered sufficient to determine the *Total Harmonic Distortion (THD)* value (see Sect. 2.2.1). However, there is no requirement for the THD value of a test voltage.

On-site testing with AC voltage are used to verify the correct installation of a complete operating system, the individual components of which have already been thoroughly tested in the high-voltage laboratory [3, 7]. For on-site tests, the permissible tolerances and measurement uncertainties are partly greater. The values are summarized in Table 2.1 compared to those when testing in high-voltage laboratories. The frequency range of test voltages for on-site tests is extended from 10 to 500 Hz, with low frequencies being preferred for cable testing.

Table 2.1 Requirements on the measuring system during tests in the laboratory and on site

AC voltage	Test in laboratory	On-site test
Tolerance		
$\hat{u}/\sqrt{2}$ (Test duration $\leq$ 1 min)	$\pm 1\%$	$\pm 3\%$
$\hat{u}/\sqrt{2}$ (Test duration >1 min)	$\pm 3\%$	$\pm 5\%$
Quotient $\hat{u}/u_{\text{eff}}$	$\sqrt{2} \pm 5\%$	$\sqrt{2} \pm 15\%$
Uncertainty		
Test voltage value $\hat{u}/\sqrt{2}$	$\leq 3\%$	$\leq 5\%$
Scale factor F	$\leq 1\%$	$\leq 2\%$

2.2 Alternating Currents

In connection with high AC voltages, high *alternating (AC) currents* also occur in the electrical power supply. The test specifications were published in 2010 in the new Publication IEC 62475, together with the specifications for DC currents and impulse currents [4]. The requirements apply to test currents in high-voltage and high-power test laboratories of more than 100 A, with a distinction between steady-state AC currents and short-time AC currents. The basis for the test specifications are the test and measurement methods in the major European high-power test fields. For this purpose, intercomparison measurements were performed with a coaxial shunt and a Rogowski coil as transfer standards for current measurements [8].

2.2.1 *Steady-State Alternating Current*

The standard test current is an alternating current with an approximately sinusoidal shape and a frequency that is generally between 45 and 65 Hz. For certain test objects, the frequency may have a different value. The value of the test current is the *true RMS value* according to:

$$\boxed{I_{\text{rms}} = \sqrt{\frac{1}{T}\int_0^T i^2(t)\,\mathrm{d}t}.} \tag{2.4}$$

where T is the time corresponding to an integer number of cycles. The tolerance for generating steady-state AC test currents is $\pm 3\%$. The difference between the positive and negative peak values shall be less than 2%.

For a more accurate characterization of the test current' shape, the *Total Harmonic Distortion* (*THD*) is used:

$$\boxed{\mathrm{THD} = \frac{\sqrt{\sum_{\mathrm{n}=2}^{\mathrm{N}} I_{\mathrm{rms,\,n}}^{2}}}{I_{\mathrm{rms,\,1}}}}. \tag{2.5}$$

In Eq. (2.5), $I_{\mathrm{rms},1}$ is the RMS value of the fundamental oscillation ($n = 1$) and $I_{\mathrm{rms,n}}$ is the RMS value of the nth harmonic with $n = 2$ to $n = 50$. The total harmonic distortion according to Eq. (2.5) shall not exceed 5% of the RMS value of the fundamental oscillation $I_{\mathrm{rms},1}$.

The requirements on the frequency response of AC current measuring systems are the same as for AC voltage measuring systems (see Sect. 2.2.1). Within the frequency range of $f_{\mathrm{nom},1}$ to $7f_{\mathrm{nom},2}$, the frequency response or the scale factor may only vary by ±1%. The frequency response outside the usable frequency range is subject to further specifications.

2.2.2 *Short-Time Alternating Current*

The test with *short-time AC currents* simulates the stress that the test object must withstand during a short circuit in the supply network. The switching angle ψ is the angle between the zero crossing of the voltage and the beginning of the short circuit. It has a decisive influence on the course of the short-time current, which lasts only for a certain number of periods. In general, the test current is asymmetric and characterized by an AC component with a superimposed transient DC component (Fig. 2.2a). In extreme case, the peak value $\hat{\imath}$ of the short-time current reaches almost double the amplitude of the steady-state AC current due to the superimposed DC component. The maximum current can therefore reach several 100 kA. After the exponential decay of the DC current component, the short-time AC current lags behind the voltage by the phase angle φ determined by the resistance and the inductance of the shorted circuit. Under certain switching and

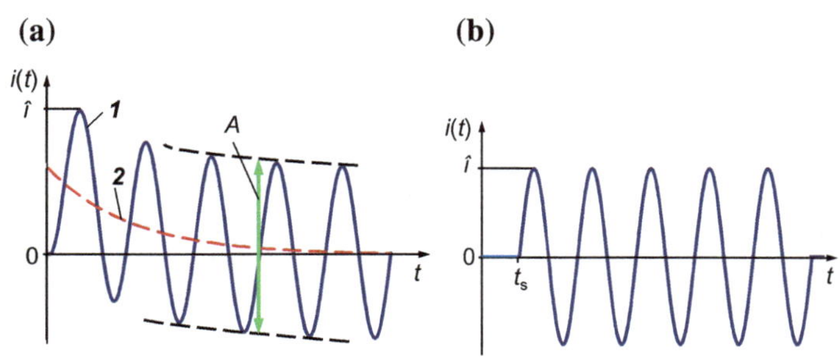

Fig. 2.2 Examples of short-time AC currents. **a** Asymmetrical short-time AC current *1* superimposed by a transient DC component **2**, **b** symmetrical short-time AC current without DC component

phase conditions, a symmetrical short-time AC current results without a DC component (Fig. 2.2b). The analytical representation of short-time AC currents is given in Sect. 8.3.

According to the main part of IEC 62475 [4], the short-time AC current is characterized by its *peak value* $\hat{i}$ and the *RMS value of the symmetrical AC component 1*. The latter results as the difference A between the upper and lower envelope of the short-time current, divided by $2\sqrt{2}$ (see Fig. 2.2a). In addition, the *true RMS value* I_{rms} is defined by:

$$\boxed{I_{rms} = \sqrt{\frac{1}{T}\int_0^T i^2(t)\,\mathrm{d}t}}, \tag{2.6}$$

which includes the contribution from the DC current ***2***. The integral in Eq. (2.6) is calculated from $t = 0$, i.e. when the current first deviates from zero, until $t = T$ when the current last deviates from zero. Other parameters of the test current are the frequency, the duration and the impedance angle $\varphi = \arctan(\omega L/R)$, where L is the inductance and R is the resistance of the short circuit. The informative Annex G of IEC 62475 [4] defines additional RMS values, e.g. the *conventional RMS value* of the AC current component, determined with the *three-crest method*. It is calculated from the difference between the peak value of a half-oscillation and the mean of the two adjacent peak values with opposite polarity divided by $2\sqrt{2}$.

In the generation of short-time AC currents, the tolerance is ±5% for the peak value $\hat{i}$ as well as for the RMS value of the symmetrical AC current component. The expanded measurement uncertainty for both measurement quantities must not exceed 5%. Depending on the test object, the required bandwidth of the measuring system ranges from 0 or 0.2 Hz up to $7f_{nom}$, where f_{nom} is the mains frequency.

2.3 Generation of High Alternating Voltages

Some generator systems for generating high AC voltages are briefly dealt with in this section, as their typical properties, such as the harmonic content of the voltage or the occurrence of partial discharges can have an influence on the test and measurement. High AC voltages are predominantly generated by inductive transformers. Depending on their intended use, they have cast resin, oil or SF_6 insulation. They are either single-step transformers or cascaded to generate voltages of more than 600 kV. In addition, there are resonance systems, also in cascade, which are often used for on-site testing due to their smaller dimensions and if the required excitation power is not too high. The AC voltages generated by transformers are used to test the equipment and components for the electrical power supply, but of course also for the calibration of the measuring systems used. Further applications include measurements of dielectric properties (see Chap. 11) and partial discharges (see Chap. 12). The highest AC voltages of up to 1000 kV, which are mainly found in Asia, require correspondingly high voltages for testing and calibration. These ultra-high test voltages (UHV) can only be generated in a few test laboratories in some industrialized countries. In addition, AC voltages are needed to generate high DC and impulse voltages for testing but also in many other fields of physics and engineering.

2.3.1 Types of Test Transformers

Test transformers for the generation of high AC voltages exist in different versions (Refs. [1–5] of Chap. 1). Figure 2.3 shows two conventional types of oil-insulated *test transformers* designed as single units. In the example in Fig. 2.3a, the magnetic core ***2*** with the low- and high-voltage windings ***3*** and ***4*** is located in an oil-filled metal tank K. The high-voltage winding is provided with control electrodes ***5*** for reducing the electric field strength. This type of transformer with oil-impregnated paper insulation requires a fairly complex, field-controlled oil-to-air bushing D to conduct the high voltage along the metal tube ***6*** to the high-voltage electrode ***7*** outside the tank. Large test transformers have high-voltage bushings that are arranged horizontally or diagonally to reduce the overall height.

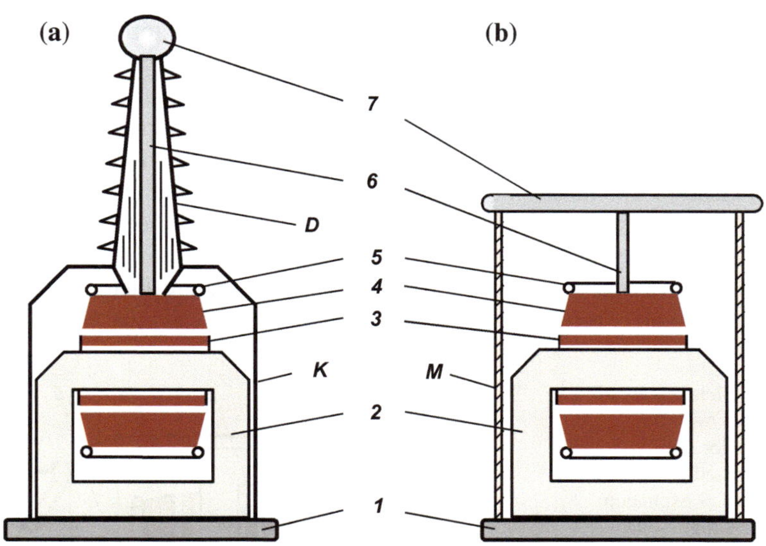

Fig. 2.3 Construction of oil-insulated test transformers (schematic). **a** Metal tank type, **b** insulated housing type. ***1*** Base plate, ***2*** iron core, ***3*** exciting winding, ***4*** high-voltage winding, ***5*** control electrode, ***6*** high-voltage metal tube, ***7*** high-voltage electrode, **D** high-voltage bushing, **K** metal tank, **M** insulating housing

The oil-insulated test transformer in Fig. 2.3b is housed in an insulating cylinder M between the grounded plate electrode ***1*** and the high-voltage plate electrode *7*. Therefore, this type of transformer does not require a high-voltage bushing. However, in comparison with transformers in a metal tank, such transformers have the disadvantage that heat dissipation to the environment is less efficient. They therefore have a reduced test power. Because of the possible risk of bursting or leakage of the insulating cylinder, it must be ensured that in the event of such an emergency, the considerable amount of leaking oil is collected in a sufficiently large safety tank.

In smaller transformers for indoor applications of up to 100 kV, the windings and even the iron core are often cast in epoxy resin. Since these *dry-type transformers* cannot be repaired after manufacture, the cast-resin insulation must be free from defects which can lead to partial discharges and thus impair the service life (see Chap. 12). In order to reduce the electric field strength in the insulation, the windings can be divided into two halves, which are connected in series and arranged on the right and left side of the core. The high voltage is then tapped between two bushings, which are either floating or grounded on one side. The core is connected with the mid-point of the two windings, i.e. at half potential, and must therefore be isolated from ground. This circuit is found, for example, in oil testers where grounding of the high-voltage winding is not necessary.

The primary voltage of the test transformers is often supplied by manually adjustable or motorized regulating transformers from the supply network. The primary voltage is slowly increased from zero to the desired test voltage. Alternatively, the primary voltage can be supplied by a machine set, then also with other frequencies, e.g. 16.7 Hz for the equipment of the railway system in Germany. The reduced permissible test voltage of the transformer at lower frequencies must be observed! Smaller test transformers can be excited with static (electronic) voltage generators, with the frequency being adjustable over a wide range. However, even with a purely sinusoidal primary voltage, the secondary voltage of the transformer has higher harmonics due to the non-linear magnetization characteristic of the transformer core, under both load and idle conditions.

Test transformers in compressed-gas vessels with SF_6-impregnated foil insulation have relatively small dimensions. However, when used in conjunction with air-insulated systems, they require large bushings. They are therefore preferable integrated in metal-enclosed gas-insulated systems. As an example, Fig. 2.4 shows

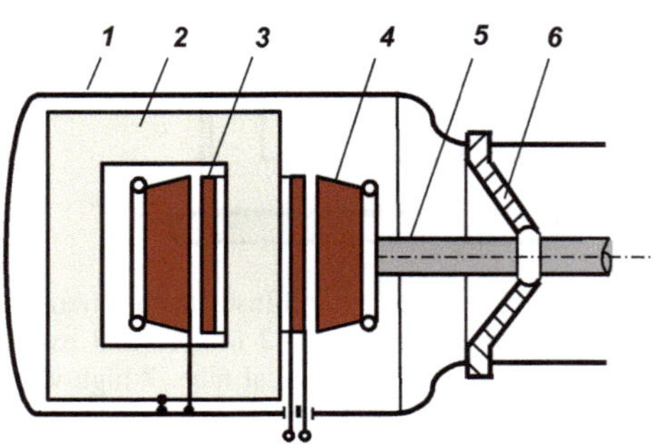

Fig. 2.4 Metal-enclosed transformer with SF_6 insulation for GIS. ***1*** Compressed-gas vessel, ***2*** iron core, ***3*** excitation winding, ***4*** high-voltage winding, ***5*** conductor, ***6*** support insulator

a compact version that is flanged directly to the *gas-insulated switchgear (GIS)* to be tested. The magnetic core and the low end of the high-voltage winding are connected to the vessel at ground potential. This transformer type is particularly suitable for on-site testing of GIS and GIL.

2.3.2 Cascaded Test Transformers

AC test voltages of more than 600 kV are usually generated with *cascaded transformers*. The principle of a three-stage cascade is shown in Fig. 2.5. The three high-voltage windings H are connected in series, which requires an isolated arrangement of the second and third transformer stages. The first and second stages have coupling windings K for the primary (excitation) windings E of the two upper stages. The connecting lines from the coupling windings K to the excitation windings E are each housed in the relevant high-voltage tubes of the bushings. At the output of the third stage, the maximum voltage $3U$ with the power P is available. The power of the excitation and coupling windings to be transmitted is $2P$ for the second stage and $3P$ for the first stage. The windings are dimensioned according to the higher load. Similarly, the dimensions of shields and toroidal electrodes are adapted to the higher voltages in the second and third stages. A three-stage cascade system with oil-insulated transformers in metal tanks probably has the world's highest test voltage of 3 MV [9].

A two-stage cascaded test transformer with cast resin insulation has already been described in Sect. 2.3.1. The cascade is fabricated as a single unit in such a way that a second excitation and high-voltage winding with a coupling winding is wound on the core in Fig. 2.3a. This particular design of a two-stage cascade is also found in

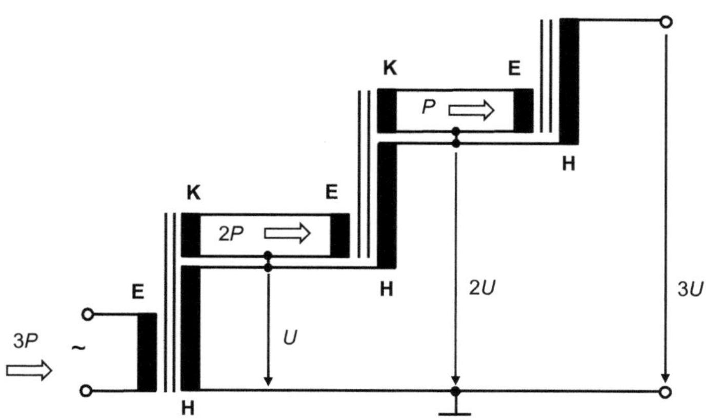

Fig. 2.5 Basic circuit of a three-stage cascade for AC voltages. **E** Excitation (primary) winding, **K** coupling winding, **H** high-voltage winding

the version in which the oil-insulated transformer is housed in a metal tank. The transformer has two high-voltage bushings, one of which is generally grounded. In this arrangement, the metal tank connected to the magnetic core is at half potential and therefore needs to be insulated. In the case that both bushings are not grounded and the tank is grounded, a symmetrical voltage is obtained across the two bushings.

Cascaded transformers can also be realized with oil-insulated transformers that are housed in insulating cylinders. The individual stages are arranged one above the other so that the cascade requires a small base area. A two-stage cascade of this type with a rated voltage of 800 kV is shown in the background of Fig. 2.6. The compressed-gas capacitor in the foreground serves as the measurement capacitor of a peak voltmeter (see Sect. 2.5.2.2). At the base of the cascade, an annular drip tray is installed to collect leaking oil in the case of leakage or explosion of the insulating housing. The collected oil is then conducted to a large container located below the hall floor.

Fig. 2.6 Two-stage 800 kV transformer cascade housed in insulating cylinders (in the background) with compressed-gas capacitor (right) and instrument transformer (left) in the foreground (PTB)

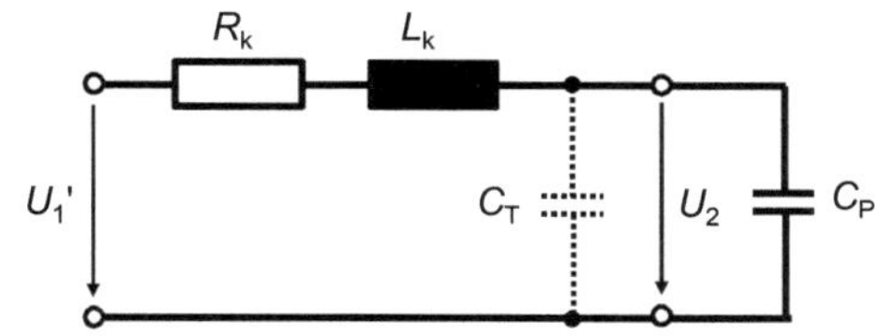

Fig. 2.7 Simple equivalent circuit diagram of a transformer with the short-circuit impedance R_k and L_k, the stray capacitance C_T and the capacitive load C_p of the test object

2.3.3 Simple Equivalent Circuit Diagram

In the equivalent circuit diagram, the test transformer appears as capacitance C_T at the high-voltage output in connection with the short-circuit impedance R_k and L_k (Fig. 2.7). Therefore, a series resonant circuit is formed with the mostly capacitive test object C_P. U_1' is the primary voltage relative to the secondary side of the transformer, in accordance with the transformer ratio. The capacitance C_T, which adds to C_P, takes into account the capacitances of the transformer windings, shielding electrodes and high-voltage leads. The series resonance leads to an increase in the secondary voltage U_2 of the transformer. This means that the test object is exposed to a higher voltage than that resulting from the product of the applied primary voltage and the transformer ratio. The voltage actually applied to the test object must therefore always be measured on the high-voltage side using a separate measuring system. The reactive power caused by the capacitive load is compensated on the primary side by reactors.

2.3.4 Resonant Circuits

Test voltages with nearly sinusoidal shape and high stability can be generated in *resonant circuits* with a high-voltage reactor. The equivalent circuit diagram of a *series resonant circuit* is similar to that in Fig. 2.7 and consists mainly of the inductance and resistance of the low-loss reactor in series with the capacitance of the test object. Other resonant circuits, e.g. the *parallel resonant circuit*, are discussed in (Refs. [4, 5] of Chap. 1; [10]). The resonant circuit is supplied from an exciter transformer with low voltage and power. In case of resonance in the tuned circuit, the voltage increases considerably at the test object according to the quality factor. The resonance condition at the *natural frequency* is achieved either by varying the inductance of the reactor at a constant frequency of the supply voltage or by a variable frequency at a constant inductance, the latter operating mode having a number of advantages. By series connection of several reactors, test voltages of 2 MV and more can be generated.

Due to their relatively small size and low weight, series resonant systems are suitable for mobile use, especially for on-site testing of power apparatus with a large capacitance such as cables and GIS [11]. The frequency of the test voltage can be selected in a wide range. Because of the large reactive power available in the case of resonance, cable tests can advantageously also be carried out at frequencies around 50 Hz. Cable testing with other high-voltage sources can only be performed at much lower frequencies because of the strong capacitive load.

In connection with resonant circuits, the *Tesla transformer* should also be briefly mentioned. Due to its very low output power, however, the Tesla transformer plays no special role in the field of electrical power engineering. The principle of the Tesla transformer is based on the resonance of two magnetically loosely coupled coils without a magnetic core, which are wound on top of each other. The oscillation is periodically excited by charging a capacitor in the primary circuit and then discharging it again by igniting a spark gap. With Tesla transformers, voltages of up to several megavolts with frequencies from 10 to 500 kHz can be generated [12–14].

2.4 Generation of High Alternating Currents

Steady-state AC test currents are generated with high-current transformers in which the primary current magnitude can be set in fine steps by a variable transformer. The variable transformer can either be fed from the mains power supply, from a generator set or from a static (electronic) generator. With the last two possibilities, the frequency can be varied. High-current generators for more than 50 kA in continuous operation are often equipped with an external connection for cooling. High currents are associated with strong magnetic fields that can be dangerous for people and equipment. The test set-up including the high-current supply and return conductors should be positioned as symmetrically as possible in order to avoid electromagnetic interference in measuring instruments.

Short-time AC test currents simulate the stress caused by a short circuit of power apparatus, e.g. circuit breakers. They can be generated in the test fields with powerful machine sets up to the highest currents of several 100 kA. The short-time AC current is limited to a few periods or half-periods during the test of circuit breakers, so that the maximum test duration is in the range of 1 s. The processes can be described with the simple equivalent circuit diagram in Fig. 2.8, where the resistance R and the inductance L simulate the components of the test object and the connecting leads [4]. When the switch S is closed at $t = t_0$, the AC voltage of generator G with the instantaneous value $u(t_0) = \hat{u}\sin\psi$ is applied to the test object,

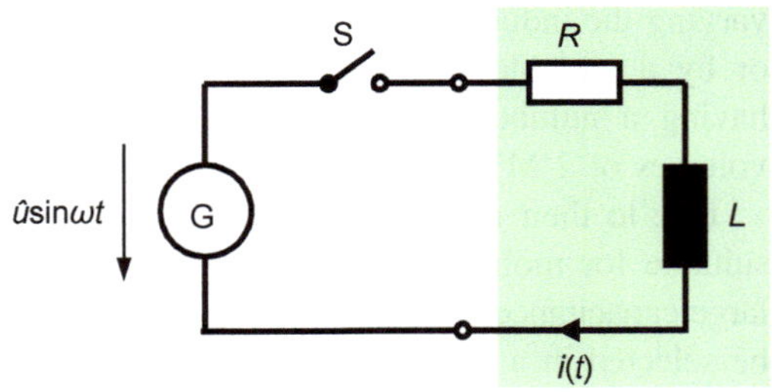

Fig. 2.8 Simple equivalent circuit diagram of the test arrangement with generator G for generating short-time AC currents

where ψ is the switching angle. For a predetermined duration or period number, the voltage $\hat{u}\sin(\omega t + \psi)$ drives the short-time AC current $i(t)$ through the test object. The analytical treatment of the short circuit is given in Sect. 8.3.

In stationary operation, the short-time current lags behind the AC voltage by the phase angle φ due to the inductive load. Depending on the switching angle ψ, the stationary short-time AC current is superimposed by a more or less large DC component, which decreases exponentially with time (see Fig. 2.2a). The peak value of the resulting short-time AC current can thus reach almost twice the steady-state value, which means a particularly high stress for the test object. Short-time AC currents with lower magnitudes can also be generated with a static generator that is fed by a digital-to-analog converter with the desired waveform.

2.5 Measurement of High Alternating Voltages

For measuring high AC voltages with the mains frequency of 50 Hz or 60 Hz—in special cases with different frequencies, e.g. 16.7 Hz for the equipment of the German railway system—there are several possibilities. The majority of the measuring systems consist of a capacitive high-voltage divider whose output voltage is recorded by a measuring instrument. Instead of a voltage divider, a single high-voltage capacitor can also be used to measure the capacitive AC current. The measuring instruments used today are predominantly digital ones on the basis of A/D converters. They allow a comprehensive, computer-assisted evaluation of all parameters of the AC voltage (see Sect. 2.2.1). This enables on-site testing and online monitoring to be carried out in connection with the use of special transportable voltage generators and measuring systems installed at the site of the test object. Due to the existing or planned energy transmission with ultra-high AC voltages of up to 1 MV (UHV range), especially in Asia, the requirements in the field of measurement technique are also increasing.

The direct measurement of high AC voltages is possible with instrument transformers, electrostatic voltmeters or sphere gaps, the latter two options being used more frequently in the past. In combination with current transformers, inductive and capacitive instrument transformers are preferably used for energy measurements in the network because of their small phase difference between input and output voltages. With field sensors, potential-free voltage measurements are possible, in which the measured values are sent to the measuring device on ground potential by means of fiber optic data transmission or wireless technology. Recent advances in the application of the Pockels effect and the Kerr effect have led to an increasing use of optoelectronic sensors for AC voltage measurements (see Sect. 6.1).

2.5.1 Capacitive High-Voltage Dividers

Capacitive high-voltage dividers generally consist of a number of series-connected capacitors arranged one above the other. The high AC voltage is applied to the top electrode of the divider, and a true-to-scale reduced voltage is available for measurement at the output terminal, i.e. at the lowest capacitor. The output voltage, which is usually limited to not more than 2 kV, is then evaluated by an analog or digital measuring device. The insulation of high-voltage capacitors usually consists of windings made of oil-impregnated paper or gas-impregnated plastic film (see Sect. 4.3.3.1). Certain ceramic plate capacitors have an excellent frequency response and are therefore preferably used in voltage dividers for the measurement of high-frequency voltages or impulse voltages (see Sect. 4.3.4.1). Compressed-gas capacitors of the Schering and Vieweg type have excellent characteristics with regard to the accuracy of the capacitance and tan δ up to the megavolt range, making them ideally suited for use in a reference system (see Sect. 11.5).

The series connection of several capacitors in the voltage divider reduces the total capacitance. For example, for n equal capacitors C_1, the total capacitance—without taking into account the stray capacitances to ground—is only C_1/n. Voltage dividers for measurement purposes usually have only a relatively small total capacitance of a few 100 pF. This is because their properties in terms of long-term stability, frequency behavior, temperature dependence and voltage dependence are better than those of a voltage divider with larger capacitance. In addition, a large capacitance puts higher load on the generator.

2.5.1.1 Stray Capacitances and Simple Equivalent Circuit Diagrams

The transfer behavior of unshielded high-voltage dividers is dealt with in detail in Sect. 4.3.1.4. The capacitive voltage divider has inductances of the components and high-voltage leads as well as *stray capacitances* to ground and to electrodes, e.g. to the torus electrode on the divider top. In the case of capacitive AC voltage dividers, the distributed stray capacitances C'_e must be taken into account because they affect the division ratio and the frequency behavior. The current i'_e flowing through C'_e, in particular the higher frequency components, does not reach the capacitor C_2 and is thus lost in the measurement result (Fig. 2.9). Therefore, the stray current leads to a division ratio different from the theoretical value $(C_1 + C_2)/C_1$ of the simple series connection with C_1 and C_2. All partial capacitances C'_e are assumed to be approximately equal. In the usual vertical arrangement of the voltage divider, the calculable value for the stray capacitance to ground is in the range 15–20 pF/m, depending on the diameter of the voltage divider [15].

In the frequency range of the mains frequency and its harmonics, the influence of inductive and resistive components of the capacitors is usually negligible. Based on this assumption, two *equivalent circuit diagrams* can be derived for the capacitive high-voltage divider. Both diagrams show that the effective high-voltage

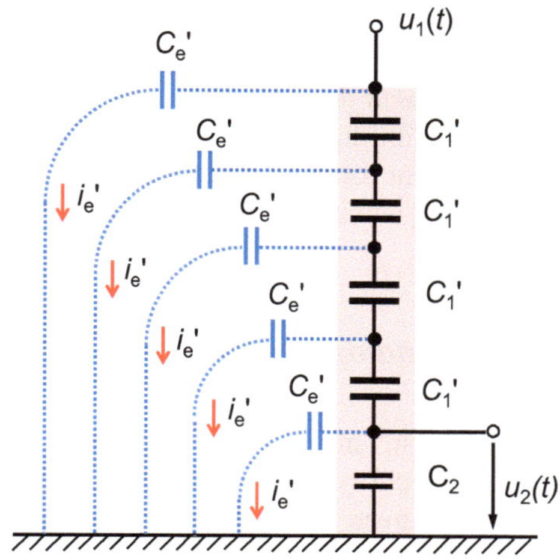

Fig. 2.9 Capacitive voltage divider with distributed stray capacitances C_e' to ground

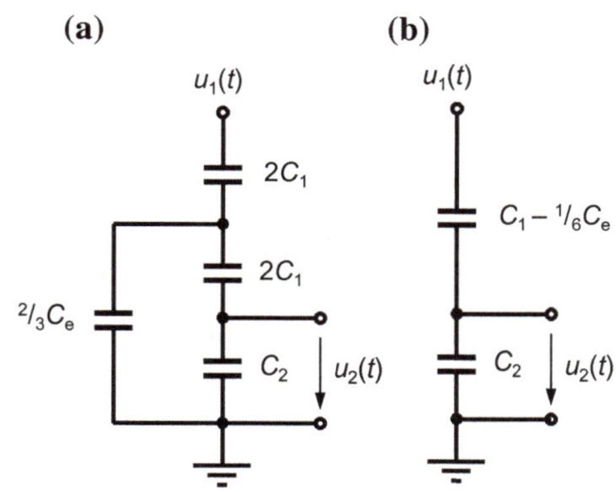

Fig. 2.10 Simple equivalent circuit diagrams of a capacitive voltage divider taking into account the stray capacitance C_e. **a** Parallel capacitance $^2/_3C_e$, **b** reduced high-voltage capacitance $C_1 - {}^1/_6C_e$

capacitance C_1 is reduced by a part of the ground capacitance C_e. In other words, the division ratio becomes larger and the output voltage u_2 is reduced. In the equivalent circuit diagram in Fig. 2.10a, a capacitance $^2/_3C_e$ is connected in parallel to half of the high-voltage capacitance. In Fig. 2.10b, C_1 is reduced by $^1/_6C_e$ so that the effective capacitance C_{eff} in the high-voltage branch is (Refs. [1, 4] of Chap. 1):

$$\boxed{C_{\text{eff}} = C_1 - \frac{C_e}{6}}. \tag{2.7}$$

Both equivalent circuit diagrams show that the transfer behavior of a capacitive voltage divider for frequencies up to the kHz range can be assumed to be approximately frequency-independent. Due to the stray capacitances, the exact division ratio u_1/u_2 cannot be calculated from C_1 and C_2, but must be determined from measurements. The division ratio of voltage dividers is predominantly dimensioned such that the maximum output voltage $u_2(t)$ at the rated input voltage is usually 1 kV or 2 kV. To measure the output voltage $u_2(t)$, analog and digital measuring devices are basically suitable. However, the trend towards digital data acquisition with software-based device control and data evaluation has already been largely implemented. With digital instruments and data processing, a complete analysis of the measurement data required for AC voltage tests including the comprehensive documentation within the framework of quality management, can be carried out.

2.5.2 *Analog Measuring Instruments*

In the past, measuring instruments for AC voltages in combination with high-voltage dividers were exclusively analog. They have been continuously improved over the years and extended in their measurement possibilities. Nowadays, new measuring instruments are built exclusively digitally, making analog meters and the corresponding measurement methods less and less common. Therefore, only a few basic principles of analog measurement techniques are dealt with in this section.

2.5.2.1 Simple Analog Peak Voltmeter

Figure 2.11a shows the principle of a simple *analog AC peak voltmeter* connected to the low-voltage capacitor C_2 of a capacitive high-voltage divider (Ref. [4] of Chap. 1; [16]). Via the rectifier G, the measuring capacitor C_m is charged to the positive peak value of the AC voltage $u_2(t)$, the voltage drop across the rectifier G1 being neglected. When the AC voltage $u_2(t)$ decreases again, C_m is slightly discharged and its voltage $u_m(t)$ decreases according to the time constant R_mC_m. In the next positive half-cycle, when $u_2(t) > u_m(t)$, C_m is recharged and $u_m(t)$ increases again to the peak value (Fig. 2.11b). The resistances R_2 and R_m as well as the capacitance C_m are chosen such that, on the one hand, the division ratio is as little affected as possible and, on the other hand, small changes in $u_2(t)$ and thus in the high voltage $u_1(t)$ can be detected. To display the DC voltage u_m with ripple, moving-coil instruments, electrostatic voltmeters or electronic analog circuits, optionally with digital display, can be used. The difference in the display of meters indicating the mean or RMS value is acceptable if the test voltage complies with the test standards.

The simple basic circuit in Fig. 2.11a can cause several measurement errors. Due to the ripple of u_m, the mean or RMS value displayed by the measuring instrument M is always slightly lower than the peak value and thus also frequency-dependent (see Sect. 2.2.1). During the charging phase of the rectifier, C_m is parallel to C_2, which increases the division ratio. Since the AC voltage can be asymmetrical, the measuring instrument must be able to measure both the positive and the negative peak values. Several circuit variants have been developed in the past, which led to improvements in the measurement behavior of the circuit and to the reduction of interference (Refs. [4, 6] of Chap. 1). As a result, the measurement uncertainty required by the test

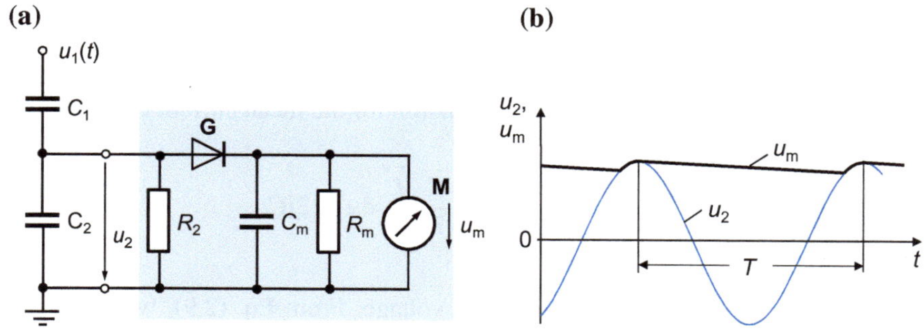

Fig. 2.11 Principle of an analog AC peak voltmeter with capacitive voltage divider. **a** Basic circuit of the measuring circuit at the divider output, **b** input voltage u_2–u_1 of the measuring instrument M and displayed voltage u_m

standards for the complete measuring system can be maintained. In summary, it can be said that analog peak voltmeters for the frequency range 16.7–300 Hz have a long tradition, but are increasingly being replaced by digital instruments.

2.5.2.2 Peak Voltmeter According to Chubb and Fortescue

The principle of the analog peak voltmeter developed by *Chubb and Fortescue* is remarkably simple (Ref. [4] of Chap. 1; [17]). The high AC voltage $u(t)$ applied to the capacitor C generates the current $i_c(t)$ which is proportional to the derivative of $u(t)$ (Fig. 2.12):

$$i_c(t) = C\frac{du(t)}{dt}. \tag{2.8}$$

By the rectifiers G1 and G2, the current $i_c(t)$ is divided into positive and negative components. In the conduction period of rectifier G1, the current measured by the moving-coil instrument M is $i_m(t) = i_c(t)$, neglecting the voltage drop across G1. In

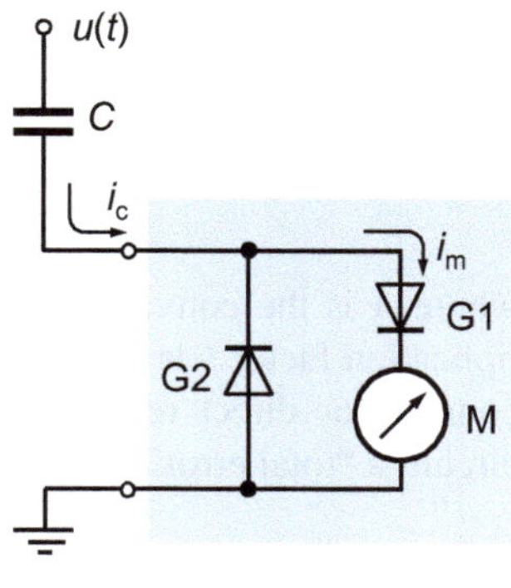

Fig. 2.12 AC peak value measuring circuit according to Chubb and Fortescue (basic principle). C High-voltage capacitor, G1, G2 rectifier, M moving-coil instrument

the negative half-cycle of the AC voltage, $i_c(t)$ flows through the second rectifier G2 in the parallel branch and $i_m = 0$. Due to its working principle, the moving-coil instrument M displays the arithmetic mean I_m of the AC current $i_m(t)$. Under simplifying assumptions, the following equation for the mean current can be given:

$$I_\mathrm{m} = \frac{1}{T}\int\limits_0^{T/2} i_\mathrm{c}(t)\mathrm{d}t = \frac{C}{T}\int\limits_{-\hat{u}}^{+\hat{u}} \mathrm{d}u = 2fC\hat{u}, \tag{2.9}$$

where T is the period duration of the AC voltage. From Eq. (2.9), we obtain the peak value of the AC voltage:

$$\boxed{\hat{u} = \frac{I_\mathrm{m}}{2fC}}. \tag{2.10}$$

Equation (2.10) applies to AC voltages without a saddle point where $\mathrm{d}u/\mathrm{d}t = 0$. Saddle points can occur by superimposing strong oscillations. According to Eq. (2.8), a saddle point would lead to an additional zero crossing of the current $i_m(t)$. In this case, the moving-coil instrument would display an incorrect mean or peak value. In test practice, however, the standardized test voltage usually has no saddle point.

The circuit according to Chubb and Fortescue basically allows an accurate measurement of the peak value of AC voltages. A *compressed gas capacitor* according to Schering and Vieweg, which is known to have good measurement properties, is very well suited as the high-voltage capacitor C (see Sect. 11.5). Compared with the circuit in Fig. 2.11a, the Chubb and Fortescue circuit has the advantage that no low-voltage capacitor is needed, which would also contribute to measurement uncertainty.

In an improved circuit, the moving-coil instrument M is replaced by a measuring resistor R_m [18]. The rectified current $i_m(t)$ flowing through R_m causes a voltage $u_m(t) = R_m\, i_m(t)$, which is converted by a voltage-to-frequency converter into an average pulse rate f_m. The subsequent gate circuit is controlled by an auxiliary voltage derived from the primary voltage of the high-voltage transformer. It opens for a predetermined number of p periods of the AC voltage, so that N pulses can pass through the gate to a frequency counter. The number N of counted pulses is proportional to the peak value $\hat{u}$ of the AC voltage according to the equation:

$$\boxed{\hat{u} = \frac{N}{2pA\,R_\mathrm{m}\,C} = k \cdot N}, \tag{2.11}$$

where A is the conversion factor of the voltage-to-frequency converter. The multiplication factor k is determined by the circuitry and may be input to the frequency counter for direct reading of the peak value $\hat{u}$. For the improved measurement circuit, a "total error" of 6×10^{-4} is given (compared to 3.4×10^{-3} for the simple

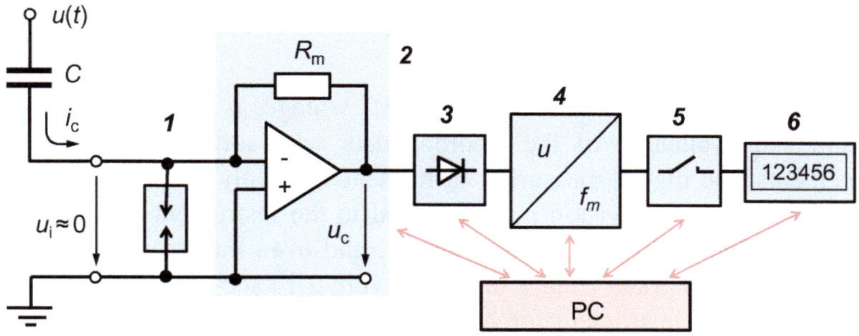

Fig. 2.13 Improved measuring circuit according to Chubb and Fortescue with PC control. ***1*** Overvoltage protection, ***2*** current-to-voltage converter with measuring resistor R_m, ***3*** full-way rectifier, ***4*** voltage-to-frequency converter, ***5*** gate circuit, ***6*** pulse counter

circuit in Fig. 2.12). In addition, it is particularly advantageous that measurements and calibrations can be performed by one person easily and efficiently.

Note The "total error" specified in [18] was determined by linear addition of the error contributions of the individual components of the measuring device.

Figure 2.13 shows a further development of the peak voltmeter according to Chubb and Fortescue, in which a PC takes over the control of the individual circuit components [19]. The capacitor current $i_c(t)$ according to Eq. (2.8) is fed via the overvoltage protection circuit ***1*** into the operational amplifier ***2*** with the feedback resistor R_m. At the output of ***2***, an AC voltage u_c proportional to i_c is generated and converted by the full-wave rectifier ***3*** into a pulsating DC voltage. Subsequently, this voltage is converted by the voltage-to-frequency converter ***4*** into a pulse train with the average pulse repetition rate f_m. An auxiliary voltage derived from u_c opens the gate circuit ***5*** for p periods so that N pulses reach the frequency counter ***6***. The display of the counter is again proportional to the peak value $\hat{u}$ according to Eq. (2.11). The peak voltmeter is programmed to measure both positive and negative peak values and their mean. Erroneous measurements due to saddle or inflection points of the AC voltage are displayed.

A particular advantage of the circuit in Fig. 2.13 is that the input of the operational amplifier ***2*** is at *virtual zero potential*. The overvoltage protection ***1*** and the usually long connecting cable to the high voltage capacitor C are thus also at zero potential. This means that no leakage currents flow from the coaxial cable to ground and falsify the measurement. Moreover, different cable lengths have no influence on the measurement result and even small AC voltages $u(t)$ down to 1 kV can be measured with the same low uncertainty as high voltages. Furthermore, DC voltages can also be measured by replacing the high-voltage capacitor C with a resistor. Careful estimates of all error influences result in a measurement uncertainty of 6×10^{-5} for the peak value of AC voltages and of 2×10^{-5} for the arithmetic mean of DC voltages. The measuring instrument is therefore ideal for automated and accurate calibration of other AC and DC voltage measuring devices.

2.5.3 Digital Measuring Instruments

A comprehensive, rational measurement of AC voltages is only possible by digital recording and evaluation of the recorded data with software. For AC voltage measurements, the digital measuring instrument is mainly used with a capacitive voltage divider whose division ratio is stored in the instrument. Often the digital instrument is designed to be suitable for DC and even impulse voltage measurements with the appropriate voltage divider. Figure 2.14 shows the basic structure of a digital measuring instrument for AC voltage measurements. The output voltage $u_2(t)$ of the capacitive high-voltage divider is first reduced in amplitude by an internal attenuator and then applied to the input of an A/D converter via an impedance converter and a programmable amplifier. The A/D converter is the main component of the digital measuring instrument with an amplitude resolution of at least 8 bit (see Chap. 7). The digital data record is stored and evaluated by an internal microprocessor or an external PC. The result values and the curve of the AC voltage are usually displayed on a screen. Extensive measures are taken to protect the entire electronics against overvoltages and interference, e.g. complex circuit layout, overvoltage protection and shielded housing.

The digital measuring instrument in Fig. 2.14 is more than a simple peak voltmeter, because in addition to the positive and negative peak values and the value of the test voltage, i.e. the RMS value, the ripple and the THD value are also indicated. In addition, the AC voltage curve is displayed on a monitor. The digital instrument also has the advantage that in the event of a breakdown or flashover in the test circuit, the voltage applied to the test object at this time is indicated.

Instead of using a capacitive voltage divider, the digital instrument can also be operated with a high-voltage capacitor for AC voltage measurements. A sophisticated variant of the measuring instrument for AC voltage is shown in Fig. 2.15. The modules of the low-voltage side are accommodated as plug-in cards in a well-shielded industrial PC [20]. The input circuit of the measuring instrument is in principle comparable to that shown in Fig. 2.13. The charging current $i_c(t)$ of

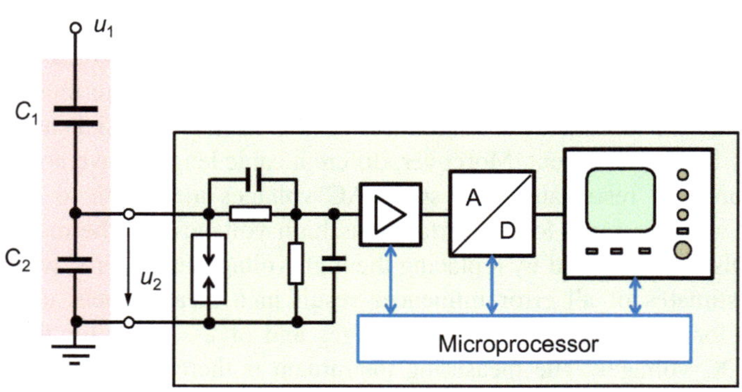

Fig. 2.14 Capacitive high-voltage divider with digital measuring circuit

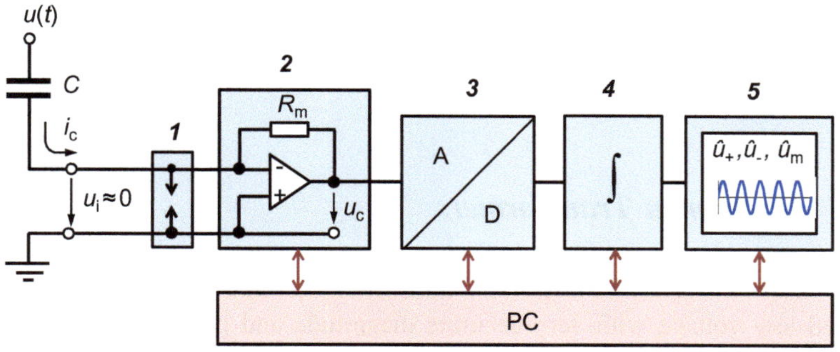

Fig. 2.15 Principle of a digital measuring instrument with compressed gas capacitor *C*, A/D converter and numerical integration unit. ***1*** Overvoltage protection, ***2*** current-to-voltage converter with measuring resistor R_m, ***3*** A/D converter, ***4*** software for numerical integration, ***5*** evaluation and display

the capacitor *C* first passes through the overvoltage protection circuit and is then converted into a proportional AC voltage $u_c(t)$ by the operational amplifier ***2*** with the feedback resistor R_m. In the next step, the analog voltage $u_c(t)$–$i_c(t)$ is digitized by a high-resolution A/D converter ***3*** (24 bit, 1 MS/s). According to Eq. (2.8), the desired high voltage $u(t)$ is given by:

$$\boxed{u(t) = \frac{1}{R_m C} \int u_c(t)\,dt} \tag{2.12}$$

and results as digital data set by numerical integration of $u_c(t)$ in ***4***. From the data set, the peak values $\hat{u}_+$, $\hat{u}_-$ and the mean RMS value $\hat{u}_m = (\hat{u}_+ + \hat{u}_-)/2$ as well as other measurement quantities, such as the THD value, are calculated. In addition, calculations using the FFT can be performed. The values and the curve of $u(t)$ are displayed on the monitor ***5*** of the PC.

2.5.4 Electrostatic Voltmeters

Electrostatic voltmeters utilize the force of the electric field between two electrodes. In the most widely used version of Starke and Schröder, a small metal plate is placed in the opening of one of the parallel plate electrodes so that it can be deflected by the applied voltage (Refs. [4, 6] of Chap. 1). The plate is provided with a mirror, on which a beam of light is directed and then reflected to a scale. The deflection of the reflected light beam on the scale increases with the square of the voltage. Electrostatic voltmeters according to Starke and Schröder are suitable for RMS measurements of both AC and DC voltages, the maximum measurement voltage being about 500 kV. With electrostatic voltmeters, voltages with frequencies in the MHz range can be measured. The voltage source is practically not loaded, which is advantageous for special measurement tasks. The main disadvantage is the quadratic scaling of the display so that only the upper part of the scale has satisfactory sensitivity.

2.5.5 Instrument Transformers

The *inductive instrument transformer* transforms the high AC voltage into a standardized low voltage with very accurate magnitude and phase values. Instrument transformers for voltage measurements—and also for current measurements—are preferably used in the supply network, more rarely in the high voltage test field. On the one hand, instrument transformers are suitable for accurate measurements and, on the other hand, are used for protection purposes in order to initiate the reliable grid disconnection in the event of a fault in the grid. Instrument transformers for voltages are used with current transformers for the metering of electrical energy in the power supply [21]. Instrument transformers for indoor use and voltages of not more than 100 kV are often designed with cast resin insulation. For higher voltages, instrument transformers have oil-impregnated paper insulation and are housed in oil-filled metal tanks equipped with porcelain or composite bushings. Instrument transformers for direct use in GIS are metal-enclosed and have SF_6-impregnated foil insulation.

Instrument transformers have galvanically isolated primary and secondary windings (high- and low-voltage windings), which are wound around a common iron core and thus magnetically coupled. The basic working principle is comparable to that of a test transformer housed in a metal tank. However, the primary winding is excited by the high AC voltage to be measured with low power consumption (Fig. 2.16a). Instrument transformers in the supply network are operated in the

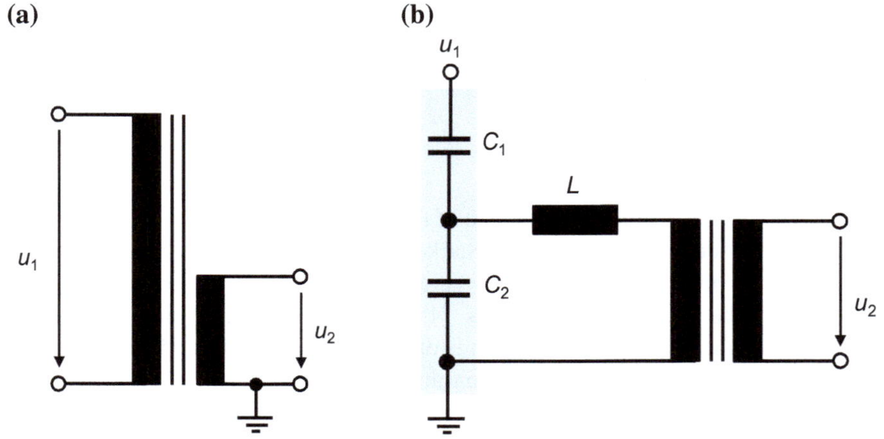

Fig. 2.16 Types of conventional transformers for high-voltage measurement. **a** Inductive voltage transformer, **b** capacitive voltage transformer

linear range of the magnetization characteristic with voltages between 0.8 and 1.2 times the rated voltage. On the secondary side, the measuring instrument, e.g. an analog voltmeter indicating the RMS value or an electronic device, is connected. As the highest operating voltage of an instrument transformer, the RMS value of the phase-to-phase voltage U is defined. It should be noted that when measuring the phase-to-ground voltage, the permissible voltage is only $U/\sqrt{3}$.

Oil-immersed instrument transformers with porcelain insulators are mainly used in substations for the medium voltage range. In the upper high-voltage range, instrument transformers are usually filled with SF_6 and have composite insulators. The secondary rated voltages of instrument transformers are standardized, e.g. 100 V or $100/\sqrt{3}$ V. These output voltages are needed to connect the old analog devices, while the newer digital instruments require voltages of only a few volts. The open circuit voltage ratio u_1/u_2 is approximately equal to the winding ratio $K = N_1/N_2$ of the primary and secondary windings, resulting in the following equation:

$$\boxed{\frac{u_1}{u_2} \approx \frac{N_1}{N_2}}. \tag{2.13}$$

Deviations from the simple form of Eq. (2.13) are caused by the resistances and inductances of both windings. In their measurement behavior, instrument transformers are characterized by the *ratio error* $\varepsilon = (Ku_2 - u_1)/u_1$ and the *phase displacement* δ, which is positive when the secondary voltage u_2 leads the primary voltage u_1. Both quantities are defined for the fundamental oscillations of u_2 and u_1 with mains frequency.

The advantages of inductive instrument transformers are the potential-free secondary voltage, high accuracy, insensitivity to electromagnetic interference, long-term stability and low phase displacement between the primary and secondary voltages. The low phase displacement is the prerequisite for accurate energy measurements with combined voltage and current transformers. The amplitude and phase errors remain low up to frequencies in the kHz range [22]. Traceable calibrations of the transformers are performed by comparison measurement with a very accurately designed instrument transformer and a transformer test set.

2.5.6 Capacitive Voltage Transformers

With increasing voltage, the effort involved in the construction of instrument transformers increases disproportionately. For voltages above 220 kV, *capacitive voltage transformers* are therefore preferably used for voltages up to more than 1 MV [23, 24]. According to Fig. 2.16b, the AC voltage to be measured is first divided by a capacitive voltage divider to an output value between 10 and 30 kV and further with an inductive transformer to the standardized secondary voltage. The high-voltage capacitance C_1 is in the range of 1000 pF to more than 10,000 pF. The inductance L and the resulting capacitance form a resonant circuit for the mains frequency, whereby the capacitive voltage divider is only minimally loaded. The secondary voltage of the transformer is approximately given by:

$$\boxed{u_2 \approx \frac{C_1}{C_1 + C_2} u_1}. \tag{2.14}$$

Amplitude and phase errors are low during resonance. The resonance circuit thus enables the non-reactive connection of an analog RMS voltmeter directly to the transformer output.

The capacitive voltage transformer is subject to a number of disturbing influences that do not occur with the purely inductive transformer. These include, for example, leakage currents, temperature influence and stray capacitances, so that the capacitive voltage transformer does not achieve the highest accuracy and long-term stability of an inductive transformer. In [25], the influence of stray capacitances for different arrangements of the capacitive voltage divider is investigated theoretically and experimentally. Calculations for a 765 kV capacitive voltage transformer with a height of 6 m clearly show that the larger the high voltage capacitance C_1, the lower the influence of the stray capacitances on the division ratio. Depending on the position of the voltage divider, the measurement deviation is ε = 2–3% for C_1 = 1500 pF and $\varepsilon \leq 0.5\%$ for $C_1 \geq 6000$ pF.

Before delivery, the capacitive voltage divider is calibrated in the manufacturer's laboratory, with the divider usually placed directly on the floor. In the substation, however, the capacitive voltage divider is installed in a higher position, resulting in a change in stray capacitance. With an installation height H = 6 m, the calculation yields ε values which are only slightly smaller than for H = 0. For example, the difference $\Delta\varepsilon$ is less than 0.18% for $C_1 \geq 2500$ pF, i.e. the different installation heights are often negligible for large capacitances.

2.5.7 *Electronic Voltage Transformers*

Conventional instrument transformers have secondary voltages of the order of 100 V and are loaded by analog measuring instruments. For some time now, there are strong efforts to introduce electronic transformers with secondary voltages of a few volts and process the analog or digital measurement data by software. Various methods are known, have been investigated and some have already been standardized. Figure 2.17 shows the principle of an *electronic voltage transformer* with an *electronic current transformer*, which together represent a *combined electronic instrument transformer* [26]. In this example, the high AC voltage is measured with a parallel-mixed RC voltage divider whose output voltage (or secondary voltage)—unlike a conventional voltage divider—is tapped from the RC element at the divider top and then fed into subassembly S1. For current measurement in the high-voltage line, a toroidal coil with magnetic core is used. Its secondary current is converted by a resistor (*burden*) into a small voltage, which is also fed into S1. Both voltages are

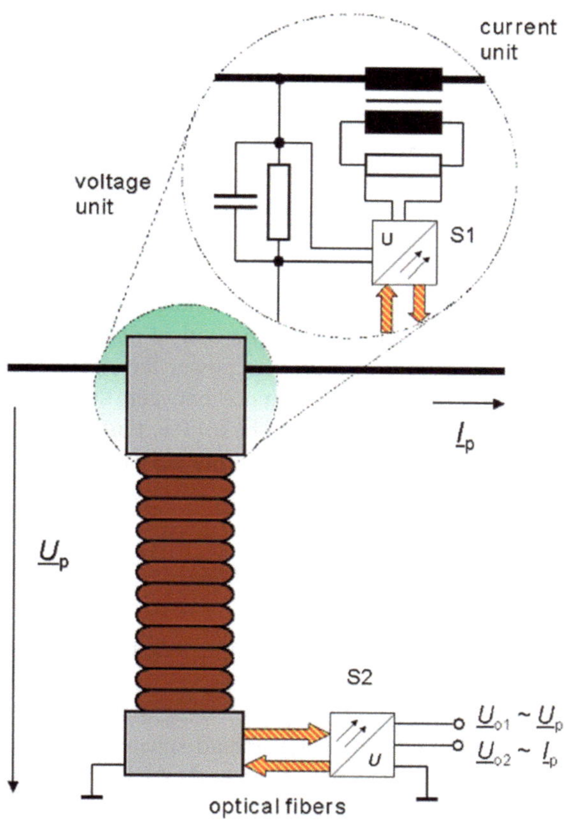

Fig. 2.17 Example of a combined electronic instrument transformer with data transmission via optical fibers to ground potential ([26]; Fig. 1)

digitized in assembly S1, converted into light pulses and sent via optical fibers to subassembly S2 on ground potential. In S2, the optical pulses are reconverted into analog voltages of a few volts, which are proportional to the secondary voltage or secondary current.

A particular problem of the electronic instrument transformer is the verifiability of its measurement uncertainty, which is a prerequisite for approval in the field of legal metrology. The decades-proven calibration methods are based on transformer test sets developed to measure the secondary voltages in the range of 100 V. They are now not suitable for evaluating the significantly lower output voltages of electronic transformers. Therefore, a new programmable two-channel AC voltage source with high-resolution D/A converters has been developed to calibrate electronic transformers (see Fig. 11.8). In the first step of the calibration process, the two subassemblies S1 and S2 and a commercially available transformer test set are each individually calibrated in the required measurement ranges. The programmable two-channel AC voltage source is also suitable for determining the different durations of the current and voltage signals through the long optical fibers, resulting in a large phase displacement. Finally, the complete electronic transformer is then calibrated by comparison with standard current and voltage transformers and the transformer test set [26].

Detailed experience on conventional and electronic energy measuring systems installed in a three-phase 135 kV network with a rated current of 1500 A is reported in [27]. The conventional measuring system consists of capacitive voltage transformers of accuracy class 0.2, inductive current transformers of accuracy class 0.2S and a three-phase energy meter of class 2S. In the electronic measuring system, capacitive voltage dividers and current transformers of the ring-core type with fixed burdens are used. The output voltages of the voltage dividers are digitized with A/D converters as well as the burden voltages at high potential. The data are sent to the interface at ground potential via optical fibers and evaluated with an industrial PC. Optical fibers are also used to power the A/D converters. Over a one-month trial period, the maximum difference between the conventional and electronic energy metering systems is found to be 0.11%. Further studies are needed to determine the long-term stability and temperature behavior of the system with digital output.

2.5.8 Standard Sphere Gap

The *standard sphere gap* according to IEC 60052 [5] is an approved IEC measuring device for the peak value of high AC voltages. In principle, sphere gaps are also suitable for measuring impulse voltages and—with increased measurement uncertainty—DC voltages. The standard sphere gap consists of two equally sized metal spheres, arranged vertically or horizontally, with one of the spheres connected to the point where the voltage is to be measured while the other is grounded (Fig. 2.18). Special requirements apply to the surface quality of the spheres,

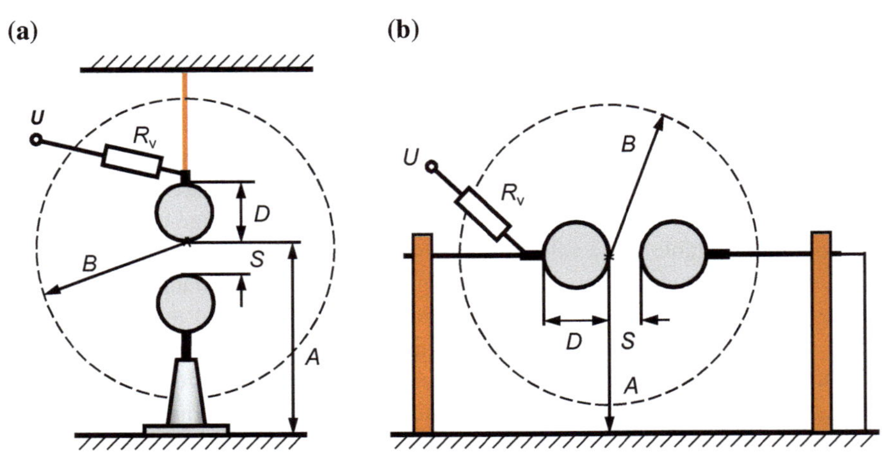

Fig. 2.18 Standard sphere gap according to IEC 60052 [5]. **a** Vertical arrangement of the spheres, **b** horizontal arrangement of the spheres

particularly in the region of the sparking points. The sphere diameter D ranges from 2 to 200 cm, the spacing S from 0.05 to 150 cm for voltages up to the 2 MV range. There are various requirements regarding the arrangement of a standard sphere gap. For example, the minimum distances A and B to the earth plane and adjacent objects depend on the sphere diameter, and the geometry of the sphere holder is fixed for each sphere diameter D and for the shanks. The series resistor R_v should be between 0.1 and 1 MΩ for AC voltage measurements in order to protect the spherical surfaces from damage caused by excessive currents and to dampen oscillations of higher frequencies.

A sufficiently high voltage and the availability of free electrons in the gap between the spheres are prerequisites for the reproducible ignition of the measuring spark gap. An additional ionization source is often required, especially for a small spacing S of the spheres. For example, this can be a mercury vapor quartz glass lamp with a spectrum in the far ultraviolet range (UVC). The softer irradiation in the UVA or UVB range has proven to be insufficient. Ionization can also be achieved by corona at AC voltage or pre-discharges from a DC corona source at negative voltage. The previously used ionizing preparations, e.g. α-emitters, are not recommended for the ionization of spark gaps or only under special safety precautions because of the potential radiation exposure of the employees.

The sphere gap is mainly used to establish the scale factor of a measuring system or to check its linearity. The AC voltage is initially set to a value immediately below the expected disruptive discharge voltage according to the selected sphere diameter D and spacing S. The voltage is then slowly increased and at the instant of the disruptive discharge, the value displayed by the measuring system used is noted. This value may be, for example, the primary voltage of the transformer used or the display of a suitable high-voltage measuring instrument. The procedure is repeated at intervals of at least 30 s until $n \geq 10$ consecutive values are obtained. If the standard deviation s is not greater than 1%, the mean of the n measured values is accepted as the disruptive discharge voltage specified for D ad S in IEC 60052 [5], possibly after correction according to the actual atmospheric conditions (see below). In Fig. 2.19, the disruptive discharge voltage U_{do} is plotted versus the gap spacing S for AC voltages and standard atmospheric conditions. For each spherical diameter D, there is initially an approximately linear relationship between U_{do} and S, which is lost, however, with increasing voltage.

The standard deviation s of the n disruptive discharge voltages is an important parameter for the acceptance of the measured values. If the requirement $s \leq 1\%$ is not satisfied, the cause may be an insufficient number of free electrons or inaccurate readings of the disruptive discharge voltages. In addition, flying insects, dust particles and other particles in the air, e.g. small floating plant particles, can cause early disruptive discharges and thus incorrect measurement values [28]. Recent investigations largely confirm the values of the disruptive discharge voltages and measurement methods that were defined long ago as the result of international comparison measurements [29, 30]. Under standard atmospheric conditions, the uncertainty of the disruptive discharge voltages for a gap spacing $S \leq 0.5D$ is estimated to be 3% (confidence level $p \geq 95\%$). The standard sphere gap thus

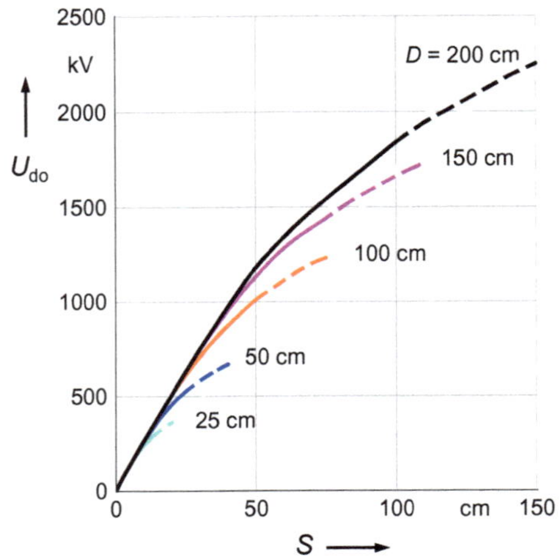

Fig. 2.19 Disruptive discharge voltage U_{do} of standard sphere gaps according to IEC 60052, valid for AC voltage and standard atmospheric conditions (temperature T_0 = 20 °C, air pressure p_0 = 101.3 kPa, absolute humidity h_0 = 8.5 g/m^3)

fulfills the requirement for an approved measuring system according to IEC 60060. For a spacing S > 0.5D, greater uncertainties are to be expected, as the dashed line of U_{do} in Fig. 2.19 indicates.

The disruptive discharge voltages U_{do} shown in the tables in IEC 60052 apply to *standard atmospheric conditions* (temperature T_0 = 20 °C, atmospheric pressure p_0 = 101.3 kPa = 1013 mbar). In the case of deviating atmospheric conditions, the corresponding disruptive discharge voltage U_d is obtained by multiplying the table values U_{do} by a correction factor equal to the relative air density δ:

$$\boxed{U_\mathrm{d} = \delta \cdot U_{\mathrm{d}_0}}. \tag{2.15}$$

The relative air density δ takes into account the deviations of the atmospheric pressure p and temperature T from the corresponding standard values p_0 and T_0:

$$\boxed{\delta = \frac{p}{p_0}\frac{273\ \mathrm{K} + T_0}{273\ \mathrm{K} + T}}. \tag{2.16}$$

Furthermore, the table values for the disruptive discharge voltage U_{do} apply to the reference value h_0 = 8.5 g/m^3 of the absolute humidity. For any other humidity h in the range of 5–12 g/m^3, U_{do} is multiplied by the correction factor k:

$$\boxed{k = 1 + 0.002\left(\frac{h}{\delta} - 8.5\right)}. \tag{2.17}$$

In industrialized countries, standard sphere gaps are now rarely used for AC voltage measurements. The reasons for this are the possible effects of electromagnetic interference on the electronic equipment used in the test laboratory, the high effort for the proper operation of the standard sphere gap and the impossibility of measuring the harmonic content of the AC voltage. Sphere gaps are, however, still used to check the scale factor and to prove the linearity of voltage dividers (see Sect. 10.3.5). Since the linearity test with the sphere gap is performed as a relative measurement within a short time, the atmospheric conditions remain approximately constant during the test time and a correction of the disruptive discharge voltage is unnecessary. With appropriate effort and careful experimental procedure, the linearity of a voltage divider can thus be confirmed within $\pm 1\%$.

2.6 Measurement of High Alternating Currents

AC currents in the conventional electrical power supply are nearly sinusoidal with a frequency of 50 Hz in Europe and 60 Hz in the US and other countries. Other frequencies are also possible, e.g. 16.7 Hz for the railway system in Germany. When measuring currents with higher harmonics, short-circuit currents and transient currents on transmission lines, high-frequency components must be considered as well as a transient DC component in short-circuit currents. Various measuring systems are used for different measurement tasks. For the measurement of AC currents with magnitudes of up to 10 kA, the devices for DC current measurement such as Hall sensors (see Sect. 3.5.2), zero flux DC current transformers (see Sect. 3.5.3) and magneto-optical sensors (see Sect. 6.2) can also be used. Basically, it must be distinguished whether the measurement point is at high-voltage or ground potential.

2.6.1 Measuring Resistors

The simple AC current measuring circuit with a resistor R_m is shown in Fig. 2.20. The output voltage u_m of the measuring resistor is usually recorded with a digital recorder for further data processing. From the simple equation $u_m = iR_m$, the current i can in principle be easily calculated. The use of low-ohmic resistors to measure AC currents has advantages and disadvantages. It is advantageous that resistors can measure both the DC component, e.g. in the short-time AC current, as well as higher harmonics of the AC current with mains frequency. The fundamental disadvantage is their self-heating at high current magnitudes, which leads to a change in resistance according to the temperature dependence. The properties of low-ohmic measuring resistors and the consideration of electromagnetically generated interference in the measurement of impulse currents are described in detail in Sect. 5.3.1. Low-inductive coaxial resistors, which are mainly designed for impulse current measurement, are also used for AC current measurements. In most measuring circuits, the resistor is grounded, but can also be used with an optical data transmission system for current measurements at high-voltage potential.

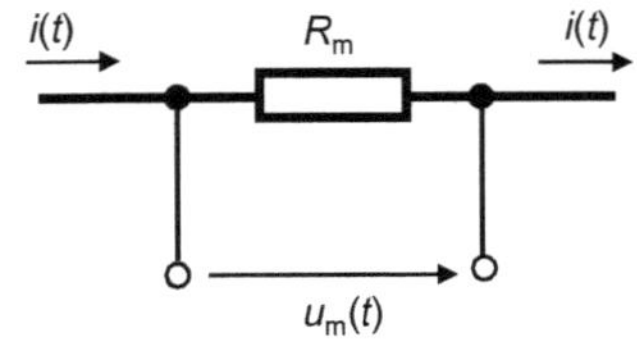

Fig. 2.20 Simple circuit with measuring resistor for current measurement

2.6.2 Inductive Current Transformers

The *inductive current transformer* with magnetic core basically resembles a voltage transformer, which is short-circuited on the secondary side. It transforms large primary currents into small measurable secondary currents corresponding to the quotient of the number of windings N_1 and N_2. Inductive current transformers are designed for AC currents with mains frequency and their harmonics. The basic equation results from the condition for the same flux of the magnetic core:

$$i_1 N_1 = i_2 N_2. \tag{2.18}$$

The magnitude of the current is usually displayed as RMS value. Standard rated currents on the secondary side are 1 and 5 A. The current transformer can be provided with two or more cores, each with its own secondary winding (Fig. 2.21). To avoid dangerously high output voltages, the secondary side must be practically short-circuited. High-current inductive transformers are also built as *ring-core transformers* in which the primary current conductor is inserted through the magnetic ring core, with the primary winding number N_1 = 1 (see Sect. 2.6.3).

Inductive current transformers with porcelain insulators have been used for decades in the supply network at high voltage for current and—in combination with voltage transformers—power measurements and for protective purposes. With a correspondingly qualitative design, there are only small current errors and phase displacements between the primary and secondary currents. These errors are determined by comparison with a high-precision *standard current transformer* and a transformer test set at ground potential [21].

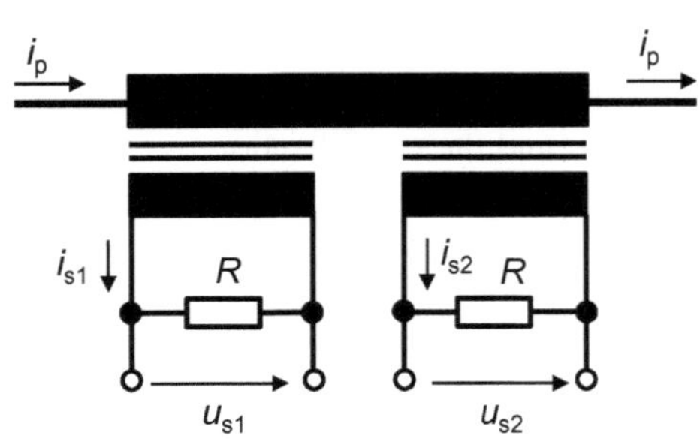

Fig. 2.21 Example of an inductive current transformer with two secondary windings

2.6.3 Measuring Coils with Electronic Data Transmission

The basic mode of operation of measuring coils with magnetic core, referred to as *ring core current transformer*, and of *Rogowski coils* with non-magnetic core is described in detail in Sect. 5.3.2, which deals with the impulse current measurement technique. There is a basic difference between the two measuring coils. The ring-core transformer is generally loaded on the secondary side with a low-ohmic resistor (*burden*) and provides a current-proportional output voltage. The Rogowski coil induces an open-circuit voltage proportional to the derivative of the current, which is to be integrated before evaluation. In some cases, integration takes place internally in the Rogowski coil through an internal passive RC element. The output voltage is then directly proportional to the current. Ring core current transformers and Rogowski coils are available as *clamp-on ammeter* or *split-core version* , which can be easily installed around the current conductor. Both embodiments of measuring coils, in particular the ironless Rogowski coil, are increasingly being used for potential-free current measurements in the high-voltage network in conjunction with an optical data transmission system.

2.6.3.1 Rogowski Coils for Alternating Current Measurements

To measure a current, the primary conductor is placed centrally through the Rogowski coil so that the number of primary turns is $N_1 = 1$ (Fig. 2.22). For a sinusoidal current $i_1 = \hat{\imath}_1 \sin\omega t$, the voltage $u_\text{i} = M\text{d}i_1/\text{d}t = \omega M\hat{\imath}_1\sin(\omega t + \pi/2)$ is induced at the output of the secondary winding (see Eq. 5.23), from which the RMS value $\hat{\imath}_1/\sqrt{2}$ of the current is calculated. The *mutual inductance* of the measuring coil with N_2 turns is $M = \mu_0 N_2 A/l_\text{m}$ (see Eq. 5.21), where A denotes the cross section of the ironless coil carrier ($\mu_\text{r} = 1$) and l_m is the mean field line length in the toroidal coil. The induced output voltage is smaller compared to the transformer with magnetic core and can only be loaded with high resistance. In general, the current in transmission lines is not purely sinusoidal, and the above equation for the induced sine voltage u_i is no longer valid. Therefore, the induced output voltage u_i must be

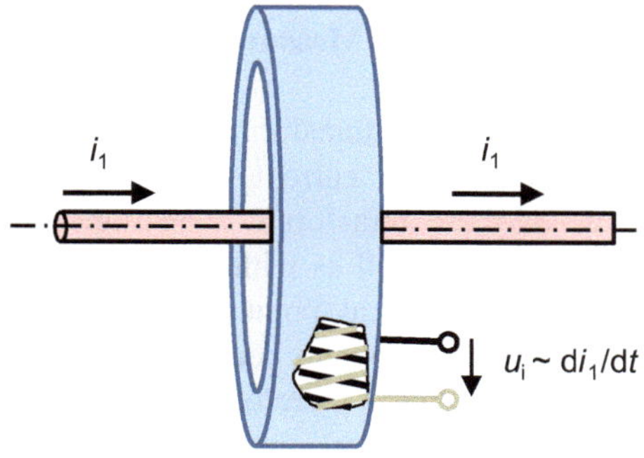

Fig. 2.22 Principle of a Rogowski coil with the conductor and secondary cross winding

determined by integrating the primary current i_1. Integration is done with passive or electronic circuits or by applying numerical algorithms (see Sect. 5.3.2.5.2). A further possibility for determining the primary current i_1 is given by Fourier analysis of the output voltage. In addition to the fundamental oscillation, the harmonics can then also be determined in terms of their magnitude and phase.

The basic behavior of Rogowski coils for measuring AC currents is treated in a number of publications. The results of the studies on the linearity of ten different Rogowski coils at power frequencies are presented in [31]. Because of the different types and rated currents, no comprehensive statement of linearity can be given here. A winding that is accurately wound on a rigid core is, in principle, the best prerequisite for low non-linearity. For example, a large exactly wound Rogowski coil showed a relative linearity deviation of only $\pm 25 \times 10^{-6}$ up to the maximum test current of 22.7 kA. Such a low nonlinearity is considered sufficient for most applications. In addition, this Rogowski coil consists of two core halves with two windings connected in series. It can therefore be easily opened and placed around the current conductor (see Fig. 5.30a).

The frequency behavior of Rogowski coils is discussed in [32]. In general, Rogowski coils have two types of windings. In one type, the turns are uniformly arranged on the rigid core and the end is returned to the beginning of the winding within the coil. The other type has a cross winding on the core, which proves advantageous because the influence of external magnetic fields is reduced. The paper deals with theoretical and experimental investigations on both Rogowski coils to determine the mutual inductance M, its frequency dependence and the phase displacement between the primary current and the induced output voltage.

A report on the suitability of an AC standard measuring system with Rogowski coil and digital voltmeter for on-site calibration of inductive current transformers can be found in [33]. During the measurements, the current transformer was separated from the 400 kV grid and the current was generated by a separate current generator at ground potential. The output voltage of the Rogowski coil was compared with that of a measuring resistor as a reference. Long-term studies with the Rogowski coil with AC currents up to 6 kA confirmed a very small relative measurement uncertainty of 0.02% for the current error and 0.2′–0.7′ for the phase displacement [34].

2.6.3.2 Current Measurement on High-Voltage Potential

Inductive current transformers with oil insulation have been used for decades for safe and accurate current measurement in high-voltage networks. However, the costs of these transformers are high and increase disproportionately with their dimensions as well as with the required amount of oil and the insulation voltage. From an ecological point of view, the oil in the transformer is an increasing environmental problem because of potential leaks and faulty disposal. The aim of several activities is therefore to replace the conventional current transformer in the high-voltage network by an oil-free converting device, if necessary with a potential-free data transmission. The converting device can be a ring-core transformer with a magnetic core or an ironless Rogowski coil. The A/D conversion of the signal measured at high-voltage potential, the data transmission to ground potential and the conversion back into a current proportional analog or digital signal are investigated in detail with very different techniques.

Two variants with ring-core transformers on high-voltage potential are already briefly described in Sect. 2.5.7, together with electronic converters [26, 27]. The voltage at the burden of the ring-core transformer, which is proportional to the AC line current, is digitized by an A/D converter, converted into light pulses and sent via optical fiber to the measuring device on ground potential (see Fig. 2.17). There, the light pulses are converted back into an analog voltage, which is proportional to the current for further evaluation.

The measuring system presented in [35] uses *clamp-on current transformers* for current measurements in a three-phase 13.8 kV network. They can be installed around the current conductors in a short time without interrupting the power supply. The output voltages of the current transformers are converted by a pulse frequency modulation into light pulses, which are transmitted to the receiver on ground potential via optical fibers. There, they are converted back into current-proportional voltages, digitally recorded and evaluated. The electronics on high-voltage potential is powered by a battery with an operating time of ten days.

Another version consists of a split-core current transformer with two magnetic cores and two secondary windings [36]. The current transformer employs the *zero-flux principle* described in Sect. 3.5.3 and is thus very accurate. It is constructed to be installed and removed from high-voltage transmission lines up to 765 kV using *live-line techniques*. The output voltage proportional to the current of the main coil winding is digitized with an A/D converter and the data is transmitted via Bluetooth wireless technology to ground potential for evaluation with a PC. This electronic reference system is intended to calibrate conventional inductive current transformers under operation. Data transmission takes place in conjunction with a *GPS-based temporal assignment* of the signals coming from the reference system and the conventional current transformer to be calibrated. This makes it possible to detect the phase displacement between the measured signals appearing at different points in the power line network. The electronics on the high-voltage side are powered via the second core winding. The three-phase reference system is suitable for a quick check of conventional current transformers at high voltage and for monitoring the currents on transmission lines in medium- and high-voltage networks. The results of extensive testing in the laboratory and in a 138 kV substation are described as good.

The test results of an *optoelectronic signal transmission* system for use with a Rogowski coil for current measurement in the high-voltage network are described in [37]. The output voltage of the Rogowski coil is first integrated and then converted by pulse frequency modulation into square-wave pulses, which are converted into optical pulses and transmitted via an optical fiber to the receiver at ground potential. After optoelectronic conversion, the frequency-modulated pulses are demodulated and amplified to the required voltage level for further evaluation. In the range between 3 and 2500 A, the ratio error of the optoelectronic signal transmission system is within ±0.3% and the phase displacement is less than 20′. The bandwidth of the measurement system ranges from DC to 1.2 MHz, so that even higher harmonics of the AC current as well as impulse currents and DC currents can be measured. The power supply of the electronics at high voltage is provided from the transmission line via a second Rogowski coil.

Another current measurement system with a Rogowski coil is presented in [38]. The precision Rogowski coil has a split-core design and can be easily installed around the current conductor. The output voltage is electronically integrated, digitized with a 16 bit A/D converter and converted into optical pulses that are transmitted to the receiver at ground potential via an *optical data link*. In the receiver, the optical pulses are converted back into an analog voltage proportional to the measured current. In addition, there is an optical power link with laser diodes to supply the electronics on high-voltage potential of up to several 100 mW. This hybrid current measurement system is intended for on-line monitoring of overhead transmission lines and as a reference system for the verification of ANSI/IEEE class 0.3 current transformers.

References

1. IEC 60060-1: High-voltage test techniques—Part 1: General definitions and test requirements (2010) [German edition: DIN EN 60060-1 (VDE 0432 Teil 1): Hochspannungs-Prüftechnik—Teil 1: Allgemeine Begriffe und Prüfbedingungen (2010)]
2. IEC 60060-2: High-voltage test techniques—Part 2: Measuring systems (2010) [German edition: DIN EN 60060-2 (VDE 0432 Teil 2): Hochspannungs-Prüftechnik—Teil 2: Messsysteme (2010)]
3. IEC 60060-3: High-voltage test techniques—Part 3: Definitions and requirements for on-site tests (2006) [German edition: DIN EN 60060-3 (VDE 0432 Teil 3): Hochspannungs-Prüftechnik—Teil 3: Begriffe und Anforderungen für Vor-Ort-Prüfungen (2006)]
4. IEC 62475: High current test techniques—General definitions, test requirements and measuring systems (2010) [German edition: DIN EN 62475 (VDE 0432-20): Hochstrom-Prüftechnik (2011)]
5. IEC 60052: Voltage measurement by means of standard air gaps (2002) [German edition: DIN EN 60052 (VDE 0432 Teil 9): Spannungsmessungen mit Standard-Luftfunkenstrecken (2003)]
6. IEC 61180: High-voltage test techniques for low-voltage equipment—Definitions, test and procedure requirements, test equipment (2016) [German edition: DIN EN 61180 (VDE 0432-10): Hochspannungs-Prüftechnik für Niederspannungsgeräte—Begriffe, Prüfung und Prüfbedingungen, Prüfgeräte (2017)]
7. Hauschild, W.: Der künftige IEC-Standard IEC 60060-3 "Hochspannungsprüfungen vor Ort" und seine Bedeutung für die off-line Diagnostik. ETG-Fachtagung "Diagnostik elektrischer Betriebsmittel", Köln, ETG-FB 97, VDE Verlag Berlin Offenbach, 35 (2004)
8. Weck, K.-H., et al.: European intercomparison of high-current measuring systems of high-power laboratories to establish traceability. In: ERA Conference Proceedings on "Measurement and Calibration in High Voltage Testing", London, 1.1.1–1.1.13 (1998)
9. Dietrich, M., Frank, H., Schrader, W., Spiegelberg, J.: A 3 MV A.C. voltage testing system for research work on UHV transmission. In: Proceedings of 5th ISH Braunschweig, Paper 62.11 (1987)

10. Ward, B.H., Reynolds, P.H., Holst, D.B.: A comparative analysis of resonant power supply design techniques for high voltage testing. In: Proceedings of 5th ISH Braunschweig, Paper 62.03 (1987)
11. Hauschild, W., Schierig, S., Coors, P.: Resonant test systems for HV testing of super-long cables and gas-insulated transmission lines. In: Proceedings of 14th ISH Beijing, Paper J-02 (2005)
12. Heise, W.: Tesla-Transformatoren. ETZ-A **85**, 1–8 (1964)
13. Damstra, G.C., Pettinga, J.A.J.: A six pulse kV Tesla transformer. In: Proceedings of 5th ISH Braunschweig, Paper 62.13 (1987)
14. Samagpong, T., Parinyakupt, U.: The design and construction of the Tesla transformer using a DC high voltage switching power supply as the primary source. In: Proceedings of 15th ISH Ljubljana, Paper T10-525 (2007)
15. Lührmann, H.: Fremdfeldbeeinflussung kapazitiver Spannungsteiler. ETZ-A **91**, 332–335 (1970)
16. Davis, R., Bowdler, G.W., Standring, W.G.: The measurement of high voltages with special reference to the measurement of peak voltages. IEE **68**, 1222 (1930)
17. Chubb, L.W., Fortescue, C.: Calibration of the sphere gap voltmeter. Trans. AIEE **32**, 739–748 (1913)
18. Boeck, W.: Eine Scheitelspannungs-Messeinrichtung erhöhter Messgenauigkeit mit digitaler Anzeige. ETZ A **84**, 883–885 (1963)
19. Marx, R., Zirpel, R.: Präzisions-Messeinrichtung zur Messung hoher Wechsel- und Gleichspannungen. PTB-Mitt. **100**, 119–123 (1990)
20. Schmidt, M., Meisner, J., Lucas, W.: Improvement and upgrading of the PTB standard measurement system for high alternating voltages. In: Proceedings of 17th ISH Hannover, Paper C-038 (2011)
21. Zinn, E.: Instrument Transformers. PTB Prüfregeln, vol. 12. Limbach Verlag, Braunschweig (1984)
22. Kunde, K., Däumling, H., Huth, R., Schlierf, H.-W., Schmid, J.: Frequency response of instrument transformers in the kHz range. etz Heft 6/2012, 122-125 (2012)
23. Kahnt, H.: Kapazitive Spannungswandler. ETZ-B **11**, 476–479 (1959)
24. Gertsch, G.A., Schlicht, D.: Zur Genauigkeit der Messung von Höchstspannungen mittels kapazitiver Wandler. In: Proceedings of 2nd ISH Zürich, pp. 205–210 (1975)
25. Mercure, H.P., et al.: Effect of stray capacitance on the accuracy of capacitive voltage transformers. In: Proceedings of 10th ISH Québec, Paper 3531 (1997)
26. Latzel, H.-G., Roeissle, G., Moser, H., Ramm, G.: Calibration scheme for electronic voltage and current transformers. In: Proceedings of 13th ISH Delft, Paper 067 (2003)
27. Juvic, J.I., Nilsson, M., Maalo, A.: On site, long term comparison between instrument transformers with digital output and a conventional measuring system. In: 19th ISH Pilsen, Paper 7-6 (2015)
28. Peier, D.-W., Groschopp, H.: Wechselspannungsmessungen mit Kugelfunkenstrecken auf Freiluftversuchsfeldern. PTB-Mitt. **87**, 396–398 (1977)
29. Gockenbach, E.: Measurement of standard switching impulse voltages by means of sphere-gaps (one gap earthed). ELECTRA **136**, 91–95 (1991)
30. Marinescu, A., Dumbrava, I.: A high voltage calibration method using the measuring spark gap. In: Proceedings of 11th ISH London, Paper 1.222.P4 (1999)
31. Djokíc, B.V., Ramboz, J.D., Destefan, D.E.: To what extent can the current amplitude linearity of Rogowski coils be verified. IEEE Trans. IM **60**, 2409–2414 (2011)
32. Prochaska, R., et al.: Rogowski coil parameters verification in wide frequency range. In: Proceedings of 16th ISH Johannesburg, Paper B-29 (2009)
33. Suomalainen, E.-P., Hällström, J.: Onsite calibration of a current transformer using a Rogowski coil. IEEE Trans. IM **58**, 1054–1058 (2009)
34. Havunen, J., Suomalainen, E.-P., Hällström, J.: High AC current calibration system based on Rogowski coil. In: Proceedings of 19th ISH Pilsen, Paper 144 (2015)

35. Werneck, M.M., Abrantes, A.C.S.: Fiber-optic-based current and voltage measuring system for high-voltage distribution lines. IEEE Trans. PWDR **19**, 947–951 (2004)
36. Doig, P., Gunn, C., Durante, L., Burns, C., Cochrane, M.: Reclassification of relay-class current transformers for revenue metering applications. In: Proceedings of IEEE T&D PES Conference (2005)
37. Fang, Z., et al.: Development of an opto-electrical system for application to high voltage measurement. In: Proceedings of 13th ISH Delft, Paper 021 (2003)
38. Andersson, A., Destefan, D., Ramboz, J. D., Weiss, S.: Precision EHV current probe and comparator for field CT/OCT verification and line monitoring. In: Proceedings of 3rd EPRI Optical Sensor Systems Workshop, Pittsburgh, pp. 1–7 (2001)

Chapter 3
High Direct Voltages and Currents

Abstract High direct (DC) voltages and currents are becoming increasingly important because of the increasing number of high voltage direct current (HVDC) transmission systems worldwide. Numerous other application fields exist in physics and technology, and DC voltages are also needed to generate impulse voltages and currents. The basic requirements for test and measurement methods for electrical power equipment are laid down in national and international test specifications. The introduction of digital measurement techniques with computer aided data processing has greatly improved the scope, quality and accuracy of the measurements. This marks the imminent end of the analogue circuits and instruments used so far, though analog measurement methods and devices, including the standard rod-rod gap, are still used. This chapter first defines the standardized measurement quantities of DC voltages and currents, briefly describes a few selected generator systems and then discusses in detail the predominantly used digital measurement circuits and methods with computer-aided data processing, allowing online monitoring and on-site tests.

High direct (DC) voltages and currents are becoming increasingly important because of the increasing number of high voltage direct current (HVDC) transmission systems worldwide. Numerous other application areas are found in physics and technology, and DC voltages are also required to generate impulse voltages and currents. The basic requirements for testing and measuring methods for electrical power equipment are laid down in national and international test specifications. The introduction of digital measurement technique with computer aided data processing has greatly improved the scope, quality and accuracy of the measurements. This marks the imminent end of the analogue circuits and instruments used so far. In the chapter, the standardized measurement quantities of DC voltages and currents are presented, some selected generator systems are briefly described and the predominantly used measuring circuits and methods are dealt with.

K. Schon, *High Voltage Measurement Techniques*, Power Systems,
https://doi.org/10.1007/978-3-030-21770-9_3

3.1 Direct Voltages

The test and measurement methods for equipment used for the transmission of electrical energy with high *direct (DC) voltages* are described in the test specifications already cited in Refs. [1–5] of Chap. 2. The increasing trend towards HVDC transmission systems with ever higher voltages of up to ±800 kV represents a special technical challenge and is increasingly discussed in the relevant international standards bodies. For test and development tasks, cascade circuits are in operation, which generate DC voltages of up to 3 MV. High DC voltages are also required for further tasks in physics and technology, e.g. particle accelerators, X-ray machines, electrostatic precipitators, coating systems and aluminum production. For devices with a rated DC voltage of not more than 1.5 kV, there are test specifications, which result in the appropriate manner from the high voltage test specifications (Ref. [6] of Chap. 2).

High DC voltages are usually generated by rectifying a mains-frequency AC voltage. A simple example shows the voltage curve $u(t)$ for the one-way rectification (Fig. 3.1). The *value of the test voltage* is the *arithmetic mean value*:

$$U = \frac{1}{T}\int_0^T u(t)\mathrm{d}t, \tag{3.1}$$

where T is the duration of the measurement.

The requirements for the test voltage are given in IEC 60060-1 (Ref. [1] of Chap. 2) and that of the measuring system in IEC 60060-2 (Ref. [2] of Chap. 2). Depending on the type of the rectifier circuit and the load of the test object, the DC voltage generated has a more or less large *ripple* which denotes the periodic deviation from the arithmetic mean value U. The *ripple amplitude* δu is defined as half the difference between the maximum and minimum voltage values (Fig. 3.1):

$$\delta u = \frac{u_{\max} - u_{\min}}{2}. \tag{3.2}$$

The quotient $\delta u/U$ is called *ripple factor*. It shall not exceed 3% for voltage tests, unless otherwise specified by the Technical Committee responsible for the apparatus.

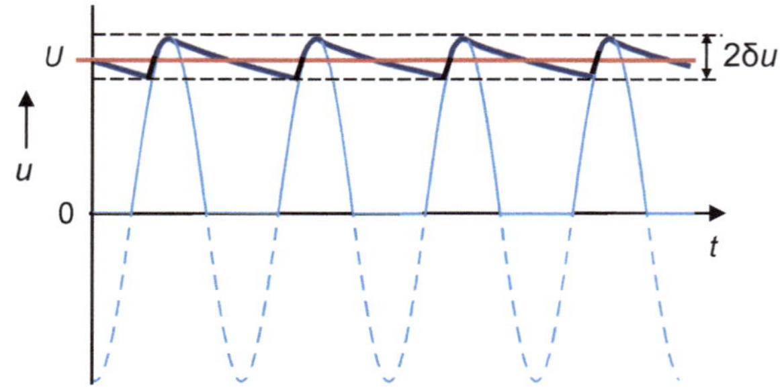

Fig. 3.1 DC voltage with the arithmetic mean value U and ripple amplitude δu

Table 3.1 Requirements for the DC test voltage U and the measuring system for tests in the high-voltage test field and for on-site tests

DC voltage	Test in the test field	On-site test
Tolerance		
U (duration $\leq$ 1 min)	$\pm 1\%$	$\pm 3\%$
U (duration > 1 min)	$\pm 3\%$	$\pm 5\%$
Ripple factor	$\leq 3\%$	$\leq 3\%$
Meas. uncertainty		
U	$\leq 3\%$	$\leq 5\%$
Ripple	$\leq 10\%$ of the ripple or $\leq 1\%$ of U	$\leq 10\%$ of the ripple

The tolerance for the value of the generated test voltage is $\pm 3\%$. The DC test voltage must not be applied immediately to the test object, but shall be slowly increased from an initial value to the desired maximum value. Starting from an initial voltage of $0.75U$, a voltage increase of 2% of the test voltage value U per second is generally considered appropriate. Inadmissibly high capacitive overvoltages as well as too long loading of the test object are thus avoided. If the duration of the applied test voltage is up to 1 min, the test voltage value may only change within $\pm 1\%$ ($\pm 3\%$ for durations longer than 1 min). After passing the withstand voltage test, the remaining DC voltage on the test object is to be slowly discharged to zero through a resistor. This controlled discharge of the capacitances of the test circuit is also necessary to avoid endangering the employees due to residual voltages.

The *expanded uncertainty of measurement* for the test voltage value according to Eq. (3.1) should not be more than 3% ($k = 2$, see Sect. 13.1.6). For measuring voltage changes in the range of 1% per second, the time constant of the measuring system must not be greater than 0.25 s. The amplitude of the ripple is to be determined with an uncertainty of no more than 1% of the test voltage value or of a maximum of 10% of the ripple amplitude, whichever is greater. The ripple measuring system with the fundamental frequency f_0 must have a frequency response ranging from $0.5f_0$ to $7f_0$ within 3 dB.

In the case of on-site tests with DC voltage according to IEC 60060-3 (Ref. [3] of Chap. 2), larger tolerances and uncertainties apply in some cases. They are summarized in Table 3.1 in comparison with the values specified for stationary tests in the high-voltage test field.

3.2 Direct Currents

For DC currents, a distinction is made between the *steady-state DC current* and the *short-time DC current*. The test specifications for DC currents are formally aligned with those for high DC voltages and published as IEC 62475 in 2010 (Ref. [4] of Chap. 2). The basis for standardization is the experience with the test methods used in the major international high-power laboratories. The requirements apply to equipment in the high and low voltage range with currents of generally more than 100 A.

3.2.1 Steady-State Direct Current

The *steady-state DC current* is not fixed or limited in time by definition. In analogy to the DC voltage, the value of the DC test current is defined as the *arithmetic mean value*:

$$I = \frac{1}{T}\int_0^T i(t)\mathrm{d}t, \tag{3.3}$$

where T is the duration of the measurement. Correspondingly, the following applies to the *ripple amplitude* of the current:

$$\delta i = \frac{i_{\max} - i_{\min}}{2}, \tag{3.4}$$

which is generally greater than the ripple of a DC voltage (see Sect. 3.1). The *ripple factor* of the current is $\delta i/I$. The following requirements and limits apply to the generation and measurement of DC currents:

- ±3% tolerance for the value of the test current generated
- 7% limit for the ripple factor $\delta i/I$
- 3% expanded uncertainty for the test current value I
- 10% expanded uncertainty for the ripple amplitude δi (or 1%, based on the value of the test current, if this value is greater).

The requirements for the measurement uncertainty of the ripple amplitude with the fundamental frequency f_0 are generally met if the measuring system used for this purpose has an upper −3 dB limit frequency of more than $10f_0$ and a lower −3 dB limit frequency of less than $0.1f_0$.

3.2.2 Short-Time Direct Current

The *short-time DC current* has a limited duration, e.g. less than 1 s. It is used for *short-circuit tests* on equipment in high-power laboratories. An example of the curve of a short-time DC current is shown in Fig. 3.2. The current rises exponentially with the time constant $\tau = L/R$ given by the inductance L and the resistance R of the test circuit. However, due to saturation effects in ferromagnetic

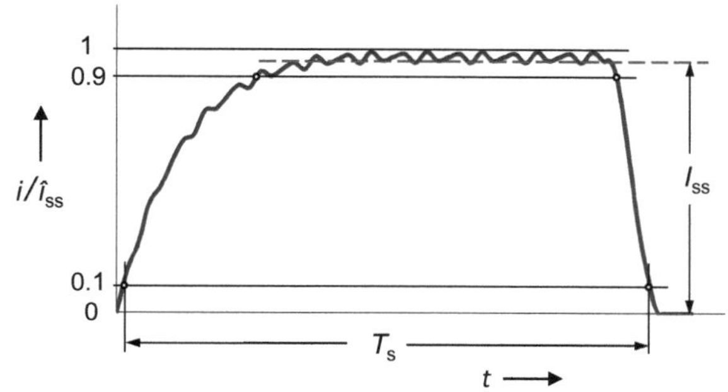

Fig. 3.2 Example of a short-time DC current. $\hat{\imath}_{ss}$: peak value, I_{ss}: steady-state current, T_s: time duration

materials, the inductance can change with the current magnitude. The current rise can be very steep so that an initial overshoot occurs due to the circuit inductance. After a certain time, the current reaches its steady-state value, before it drops more or less steeply. The drop of the test current takes place in a relatively short time so that the electronic measuring instruments in the high-power laboratory are exposed to strong electromagnetic interference.

Short-time DC currents are characterized by two current quantities and the current duration. The maximum magnitude of the current is the peak value $\hat{\imath}_{ss}$. The steady-state current I_{ss} is defined as the arithmetic mean value of the current above $0.9\hat{\imath}_{ss}$. The duration T_s is the time within which the current is greater than $0.1\hat{\imath}_{ss}$.

Note Figure 16 in IEC 62475 shows an example of a short-time DC current (Ref. [4] of Chap. 2). However, the quantities $\hat{\imath}_{ss}$ and I_{ss} are not drawn correctly, i.e. the positions indicated in the drawing do not correspond to the written definitions. The corresponding German edition contains a corrected drawing.

The following tolerances are specified for the generation of the test current:

- $\pm 5\%$ for the peak value $\hat{\imath}_{ss}$
- $\pm 5\%$ for the steady-state current I_{ss}
- 0–25 s for the time constant τ of the current rise.

The duration T_s is given by the Technical Committee responsible for the test in the relevant test specification. T_s may have, for example, a value of up to 1 s.

The expanded measurement uncertainty of the peak value $\hat{\imath}_{ss}$ and the steady-state current I_{ss} shall not be more than 5%. The measuring system must have a frequency response from 0 to at least 1 kHz (−3 dB upper limit frequency).

3.3 Generation of High DC Test Voltages and Currents

Some generator systems commonly used to generate high DC test voltages and currents are briefly discussed in this chapter. Their typical properties such as ripple or the occurrence of partial discharges can have an influence on tests and measurements. High DC test voltages are generated predominantly by rectifying an alternating voltage, either in a single-stage or to achieve higher voltages in a multi-stage circuit. Ultra-high DC voltages of up to several Megavolts can be generated by charge separation with band generators. High DC voltages generated by rectifier circuits are primarily used for the testing of equipment and components for the electrical power supply, but also, of course, for calibrating the measuring systems used. Further fields of applications of high DC voltages are partial discharge measurements and determination of the properties of dielectric materials. The highest DC transmission voltages of up to 800 kV, which are mainly found in Asia, require correspondingly high test voltages. They can only be generated in large test laboratories in a few industrial countries. DC voltage generators are also used in other areas of physics and technology.

3.3.1 Rectifier Circuits

High DC voltages for the testing of apparatus in power supplies are generally generated by rectifying AC voltages. In the simple *one-way rectifier circuit* according to Fig. 3.3, the rectifier G charges the smoothing capacitor *C* to the peak value of the AC voltage in the positive half-period of the transformer output voltage (Refs. [2, 4] of Chap. 1). Subsequently, the DC voltage drops by $2\delta u$ due to the load of the resistor R_P of the test object until the rectifier G again passes current for recharging *C*. The result is a pulsating DC voltage $u(t)$ with the mean value *U* and the ripple δu (see Fig. 3.1). The maximum value of the DC voltage in the one-way rectification is equal to the single peak value of the alternating voltage.

An almost twice as high DC voltage is achieved with two rectifiers in the *Greinacher voltage doubler circuit* (Fig. 3.4). The rectifier G1 causes the capacitor

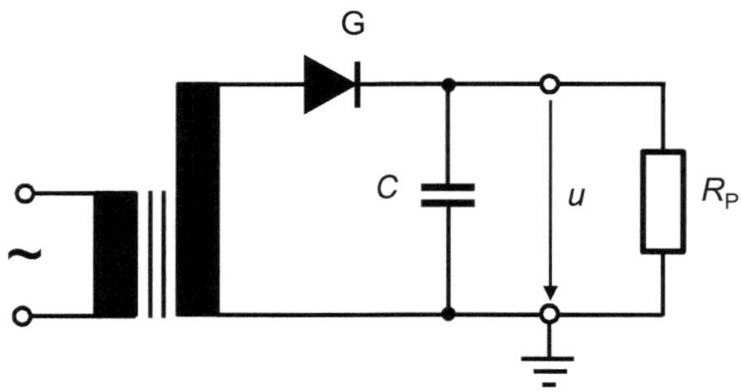

Fig. 3.3 One-way rectifier circuit. G: rectifier, *C*: smoothing capacitor, R_P: resistance of test object

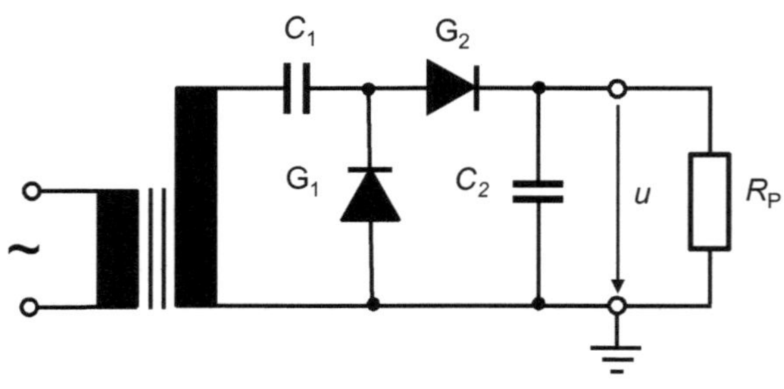

Fig. 3.4 Greinacher voltage doubler circuit

C_1 to be charged to a DC voltage corresponding to the peak value of the AC voltage, around which the AC voltage oscillates. At the input of G2, therefore, there is a mixed voltage with a maximum of twice the AC peak value, which charges the capacitor C_2 to this value. Depending on the load R_P of the test object, the capacitor voltage subsequently decreases again more or less until the next charging cycle begins. At the output of the Greinacher circuit, the mean value of the pulsating DC voltage is approximately $2U$ with the ripple δu, whose fundamental frequency is twice as high as that of the one-way rectification.

As rectifiers, semiconductors are used, which consist mainly of silicon, in the past also selenium, and which are connected in a correspondingly large number in series. They have almost completely replaced the high-vacuum valves used previously with glow cathode. The series circuit of a plurality of semiconductor rectifiers requires a potential control by parallel-connected capacitors.

As with AC voltage, a multiple higher DC voltage is achieved by cascading the basic circuit (Refs. [2, 4] of Chap. 1). Figure 3.5 shows the principle of the (unsymmetrical) *Greinacher cascade circuit*, later improved by Cockcroft and Walton, with $n = 3$ stages. For ideal circuit elements and no-load condition, the DC voltage at the output of the cascade is n times the peak value of the AC voltage applied to the input. The two upper capacitors C_1 are each charged to the double, and the lowermost capacitor to the simple peak value of the AC voltage. The lowermost capacitor with the capacitance $2C_1$ receives the same charge as the two overlying capacitors. The capacitors C_2 serve to smooth the DC voltage. Given a correspondingly large number of stages n, DC voltages of several megavolts can be generated. The output current is in the range of some 10 mA.

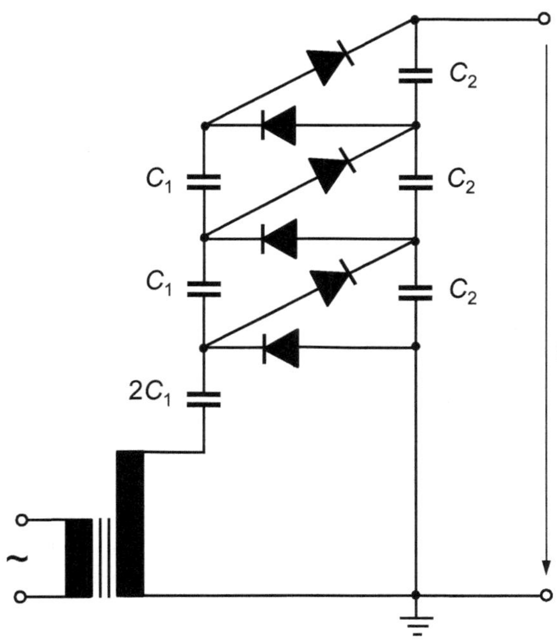

Fig. 3.5 Greinacher cascade circuit for the generation of high DC voltages (principle with $n = 3$ stages)

Under load, the DC voltage obtained by rectifying an AC voltage of power supply frequency is superimposed by a pronounced ripple. Smaller ripple can be achieved when the AC voltage to be rectified has a higher frequency. In this case, the voltage is generated electronically, amplified by a HF transformer with ferrite core, and then rectified. Rectifier systems are usually built in an open design and are therefore exposed to ambient conditions. Figure 3.6 shows a metal-enclosed SF_6-insulated 300 kV DC voltage supply accommodated in the metal tank rear left. The DC output voltage is very precisely controlled by means of a highly stable SF_6-insulated voltage divider arranged in the large metal tank in the middle of Fig. 3.6. The electronic control unit operated by a PC is located in the movable chassis. In the smaller SF_6-insulated metal tank on the right-hand side of Fig. 3.6, a high-precision 100 kV DC voltage divider is accommodated (see Sect. 3.4.5). On the left in Fig. 3.6 is an older 300 kV DC voltage divider in oil-insulated design [1].

With electrostatic generators, ultra-high DC voltages of several megavolts can be generated. They have virtually no ripple, but can deliver only low current magnitudes of usually less than 1 mA. The best known example of an electrostatic generator is the *van de Graaff generator*, which is mainly used in nuclear physics research laboratories, e.g. for accelerating particles. The functional principle is

Fig. 3.6 SF_6-insulated 300 kV DC voltage supply (left-hand metal tank, background) connected with a 300 kV voltage divider (large metal tank in the middle), a separate SF_6-insulated 100 kV voltage divider of the highest accuracy (right) and an older oil-insulated 300 kV voltage divider (left) (PTB)

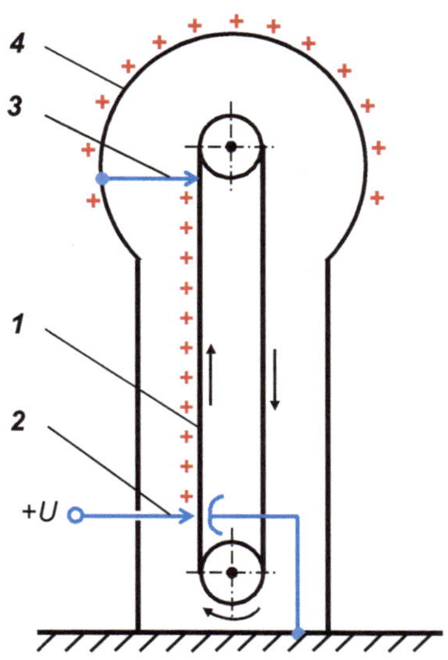

Fig. 3.7 Principle of the electrostatic generator according to van de Graaff for the generation of ultra-high DC voltages. ***1*** moving insulating belt, ***2*** spray electrode for charging, ***3*** discharging electrode, ***4*** high-voltage electrode

briefly explained with the help of Fig. 3.7. Electric charge from a corona source is sprayed on the lower part of an insulating belt ***1***, which runs over two motor-driven pulleys of different material. The spray electrode system ***2*** has a comb-like shape. The charge on the belt is transported to the upper insulating pulley where the belt is discharged through the electrode system ***3*** connected to the spherical high-voltage electrode ***4***. The discharged part of the belt then returns to the lower metal pulley of the generator and is recharged there. By further revolutions of the belt, the spherical high-voltage electrode is charged further and further until there is balance with the charge lost to the environment. An electrostatic generator built in a pressure tank and insulated with SF_6 generates DC voltages of up to 25 MV (Ref. [4] of Chap. 1).

3.4 Measurement of High DC Voltages

Most high DC voltage measurement systems consist of a high-ohmic divider that provides a scaled-down voltage for the measuring instrument at the divider output. Alternatively, a single high-voltage resistor can also be used to measure the DC current flowing through the resistor. The currently used measuring instruments are predominantly digitally structured and, with appropriate software, enable a comprehensive, computer-assisted determination of all parameters. A direct measurement of high voltages is possible with electrostatic voltmeters or *rod-rod gaps*, which measure the rms value or peak value. The difference to the arithmetic mean value is usually negligible, since the ripple factor is limited to 3%. Another measurement principle is found in the so-called *field mill* and the *rotary voltmeter*, in which the electric field generated by the DC voltage is converted in an AC voltage by rotation of a correspondingly shaped electrode. The use of *electro-optical sensors*, which exploit the *Pockels* or *Kerr effect*, is described in Sect. 6.1.

DC voltages and currents are particular important from a metrological point of view, since the units of measure "Ampere" and "Volt" belong to the base or derived units in the *International System of Units* (*SI units*). They are realized with highest precision in the *National Metrology Institutes* (*NMI*), usually at low voltages and currents, but also at high voltage.

Special precautions must be taken when working with high DC voltages and currents. In order to avoid dangerously high electrostatic charging of persons and objects by influence, the test area must be separated from the control room by a grounded protective grating. After switching off the DC test voltage, the capacitors in the test circuit must be discharged via ground rods with protective resistor.

3.4.1 Measuring Circuit with Resistive Divider

The resistive voltage divider in combination with a digital measuring instrument DM at its output is the most frequently used circuit for measuring high DC voltages (Fig. 3.8). The top of the voltage divider, possibly also the bottom, is equipped with a torus electrode for field control. The upper resistor R_1 has a high ohmic value, but should allow a current magnitude of at least 0.5 mA at rated voltage. R_1 is usually implemented as a series connection of a plurality of individual resistors, which are often arranged in the form of a helix. The resistors are mechanically held by a highly insulating construction in order to keep leakage currents as low as possible. The resistor chain R_1 is housed in a highly insulating tube filled with dry air or insulating oil, the insulating medium being circulated for better heat dissipation. The lower resistor R_2 is generally a parallel connection of a large number of individual resistors, often housed separately from R_1 in a metal box at the bottom of the voltage divider. This allows an easy change of the lower resistor R_2 to obtain a different *division ratio*.

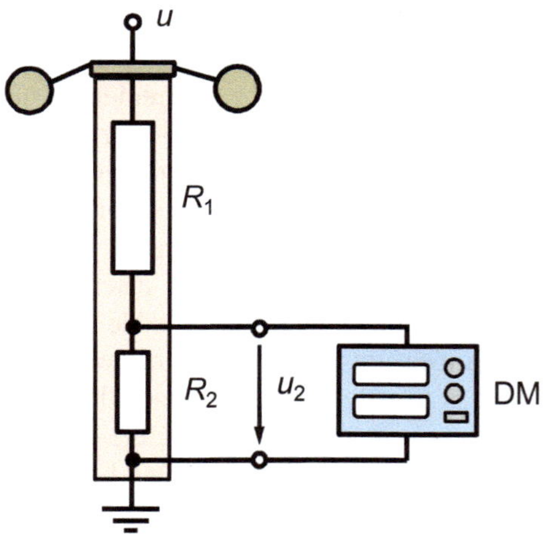

Fig. 3.8 Resistive voltage divider with digital meter for high DC voltage measurement (principle)

The individual resistors R_1 and R_2 may be wire-wound resistors, metal oxide film resistors, carbon film resistors or composition resistors. High-ohmic metal film resistors and carbon film resistors are available with a helical groove cut in the resistance layer. This increases or determines on the one hand, as desired, the value of R_2, but on the other hand there is the risk of flashover across the groove in the event of a transient overvoltage. As a consequence, the resistance can be increased or even damaged. Before installation, the individual resistors of any type are generally aged for several hours at a temperature of more than 100 °C. Usually, the temperature coefficients of R_1 and R_2 are chosen as equal as possible. Precise standard voltage dividers are occasionally composed of a special type of resistor, the *Park resistor* (Ref. [4] of Chap. 1; [2–4]). It consists of a single or a few wire-wound resistors, which are placed in a cylindrical metal housing for protection against external influences. This construction also leads to a reduction of the local field strength.

In [1], an oil-insulated 300 kV precision voltage divider is presented which consists of 300 wire-wound resistors each of 2 MΩ. The resistors are connected in series in the form of a *helix* with 50 windings. The pitch of the helix, however, is not uniform, i.e. the spacing between the upper turns of the helix is smaller than that between the lower turns (see Fig. 3.6, left side). This means that the voltage drop along the helix is greater in the upper part than in the lower one. By this design, the potential distribution along the axis of the resistor helix is matched to the electrostatic field between the high-voltage and ground electrodes. In the ideal case of this *field-conformal arrangement*, there exists no horizontal component of the resultant electric field so that leakage currents to the surroundings are negligible. Furthermore, capacitive stray currents normal to the helix are also reduced, resulting in an improvement of the bandwidth of the DC voltage divider. It is thus possible to measure low-frequency AC components superimposed on the DC voltage, e.g. the ripple. The principle of the field-conformal arrangement of resistors is also found in impulse voltage dividers with the aim of improving the transfer behavior (see Sect. 4.3.2.4).

An alternative to single resistors connected in series is the so-called *Schniewind resistor*. It consists of a resistance wire that is interwoven meander-shaped with insulation yarn to form a resistor web of almost any length and width. Due to its low inductance, this type of resistor is also used in impulse voltage dividers and damping resistors (Refs. [71, 72] of Chap. 4). For the production of high-ohmic resistors with sufficient mechanical strength, the resistor web is wrapped around a support tube and cast with epoxy resin to form a stable and stackable unit. However, the cast resistor web has poorer heat dissipation.

The digital measuring instrument DM at the output of the voltage divider replaces the previously used measuring devices, such as the moving-coil instrument or electrostatic measuring device, as well as the subsequent analog electronic circuits. The term "digital" refers here to the entire measurement principle and not only to the display, i.e. the divider output voltage $u_2(t)$ is digitized by an A/D converter and temporarily stored. The data set is then evaluated by means of software with regard to the arithmetic mean value, ripple and, if requested, other quantities. One of the first digital measuring devices with A/D converter for high-voltage application is described in [5]. Today's measuring instruments with high-resolution A/D converter enable fully automated measurements and fast computer-aided evaluation of the measurement data with high accuracy.

The arithmetic mean value U of the high voltage in the circuit of Fig. 3.8 results formally:

$$U = U_2\left(1 + \frac{R_1}{R_2}\right), \tag{3.5}$$

wherein the term in the bracket represents the division ratio or scale factor of the measuring system. Various methods can be used to determine the ratio. The development of high-resolution digital voltmeters and precise calibration generators now allows the direct and accurate measurement of the ratio with calibration voltages of up to 1000 V. Previously, each high-ohmic series resistor of a DC voltage divider was individually measured at low voltage and added to a total value R_1. The division ratio was obtained by calculation, but with limited accuracy. In the case of higher accuracy requirements and a large number of approximately equal individual resistances, the *Hamon principle* offers another possibility for determining R_1 at low voltage. For this purpose, the divider resistors are grouped into several parallel and series connections whose resistances are measured. This allows the high-voltage resistance R_1 to be calculated, and hence the division ration. The field-conformal voltage divider mentioned above has special devices to achieve these series-parallel connections of the helix resistors by pushing metal bars into the oil-filled vessel from above [1]. The uncertainty of the low-voltage ratio obtained according to the Hamon principle was estimated to be 12×10^{-6}.

In accordance with IEC 60060-2, the division ratio determined at low voltage is to be checked by a linearity test with at least five voltage values up to the maximum operating voltage. If the five individual ratios deviate, a mean value is set as the *assigned scale factor*, and the greatest deviation of the individual ratios from the mean is taken into account in the uncertainty budget (see Sect. 10.3.1.4). During the linearity test for obtaining the voltage dependence of the scale factor, the high voltage is applied only for a short time. With longer duration of the applied voltage, the voltage-dependent *self-heating effect* is noticeable. This leads to an increase in temperature of the resistors and thus also to a change of the division ratio (see Sect. 3.4.3).

3.4.2 Measuring System with Series Resistors

Figure 3.9 shows three principle circuits with series resistors for measuring high DC voltages. In the historical circuit of Fig. 3.9a, a moving-coil instrument M is connected in series with the known high-voltage resistor R_1, and because of its operating principle, the arithmetic mean value I of the current $i(t)$ flowing through

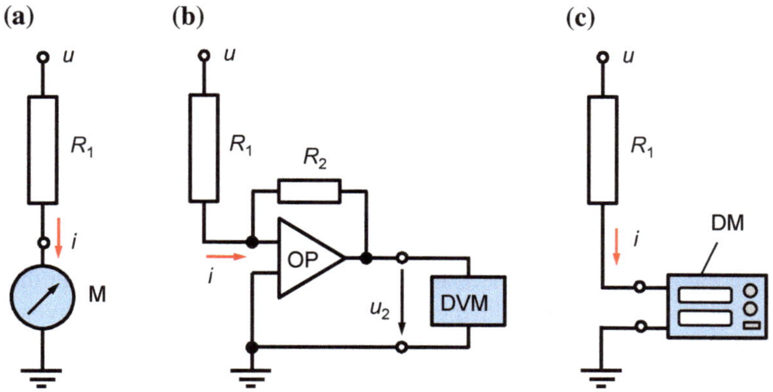

Fig. 3.9 Circuits for measuring the arithmetic mean value of high DC voltages using a resistor. **a** Simple circuit with moving-coil instrument M. **b** Circuit with operational amplifier OP and digital voltmeter DVM. **c** Circuit using a digital measuring device DM with A/D converter

R_1 is indicated. Neglecting the voltage drop on the coil windings of M, the arithmetic mean value of the DC voltage is given by:

$$U = I R_1. \tag{3.6}$$

Instead of the moving-coil instrument, an operational amplifier OP and digital voltmeter DVM is used in the circuit of Fig. 3.9b [6]. The arithmetic mean value of the high voltage is:

$$U = \frac{R_1}{R_2} U_2. \tag{3.7}$$

In addition to the arithmetic mean value, further quantities can be measured in this circuit, e.g. the ripple, if the high-voltage resistor R_1 has a sufficiently high bandwidth. The circuit in Fig. 3.9b also has the advantage that the input of the operational amplifier OP is practically on *virtual zero potential* due to the high amplification, i.e. there is no open-circuit voltage between the input terminals of the OP that could falsify the measurement. Nowadays, digital circuits with A/D converters (Fig. 3.9c) are used instead of the moving-coil instrument or analogous circuit, just as is the case with AC voltage circuits.

The series resistor R_1 must be high-ohmic in order to limit its self-heating and not to load the high-voltage source too much. However, a low measurement current increases the influence of leakage currents, which flow to ground through the support of R_1 and do not contribute to the display of the measuring instrument M. The measurement current through R_1 should therefore not fall below 0.5 mA. The same types of resistor are used for R_1 as for the high-voltage resistors in the voltage divider (see Sect. 3.4.1), including the Schniewind resistor. To protect the operator and instrument, a surge arrester (not shown in Fig. 3.9) parallel to the instrument is essential.

3.4.3 Temperature Behavior

The individual resistors of a voltage divider or series resistor have a *temperature coefficient* (*TK*) depending on the type. With increasing temperature, the total resistance R_1 and correspondingly the division ratio change, so that the voltage reading is incorrect. The same applies to the Schniewind resistor (see Sect. 3.4.1). By selecting a suitable low-voltage resistor R_2, at least a partial compensation of the temperature effect can be achieved. The temperature increase of a resistor divider can have two causes. In the simple case, the ambient temperature changes evenly over the entire length of the voltage divider. With the same TC values of R_1 and R_2, the division ratio and thus the indicated voltage value are temperature-independent. The other cause of the temperature rise is self-heating of R_1, expressed by the *Joule heat Q*:

$$Q = P\Delta t = \frac{U^2}{R_1}\Delta t, \tag{3.8}$$

where P is the electrical power converted into heat in the time unit Δt.

The self-heating effect is as follows. The high-voltage resistors are usually housed in an insulating cylinder, which is filled with dry air or transformer oil and bounded above and below by the electrodes. During operation of a DC voltage divider, Joule heat rises upwards and causes a noticeable increase in temperature at the top of the divider. At the same time, heat is dissipated to the surroundings through heat conduction and convection. After some time, a temperature gradient results in the voltage divider. The upper part of R_1 and the lower resistor R_2 therefore have different temperatures, which cause a change in the division ratio. An approximately equal temperature distribution in the voltage divider can be achieved when the air or the insulating oil is circulated.

How the temperature distribution affects the division ratio also depends on the type of resistors used. In the case of sheet and film resistors, the resistance layer is separated from direct contact with the air or the insulating oil only by a thin protective coating. The Joule heat of the resistors can therefore be transferred well to the insulating medium. In contrast, high-ohmic wire-wound resistors have a relatively thick insulating casing which impedes heat transfer. The internal temperature of the resistance wire is therefore not necessarily equal to the temperature on the outside of the resistor. The internal temperature is decisive for the resistance, but in general only the temperature on the outside is measurable.

The temperature distribution in a voltage divider can be approximately calculated using the classical laws of heat generation and heat transfer by conduction and convection [7]. An example is given for a 300 kV standard voltage divider consisting of a helix with 300 wire-wound resistors à 2 MΩ as described in [1]. The calculation was carried out for three different operation conditions, i.e. the resistors were assumed to be embedded in stationary air or oil and in circulated oil. Ambient temperature outside the divider was always constant 20 °C. Figure 3.10 shows the simplified construction of the voltage divider and the temperature increase $\Delta\vartheta(U)$ in the uppermost and lowest resistor [8]. As expected, the internal resistor temperature increases most with the voltage when air is the insulating medium. At 300 kV, the temperature increase is $\Delta\vartheta \approx 48$ K in the top resistor (Fig. 3.10b, curve ***1***) and $\Delta\vartheta \approx 41$ K in the bottom resistor (Fig. 3.10c, curve ***1***). When the resistors are embedded in insulating oil, the heat dissipation to the environment improves and the temperature increase $\Delta\vartheta$ is only about half that in air (curves ***2***). The circulation of the insulating oil obviously causes only a small further reduction in the resistor temperatures (curves ***3***).

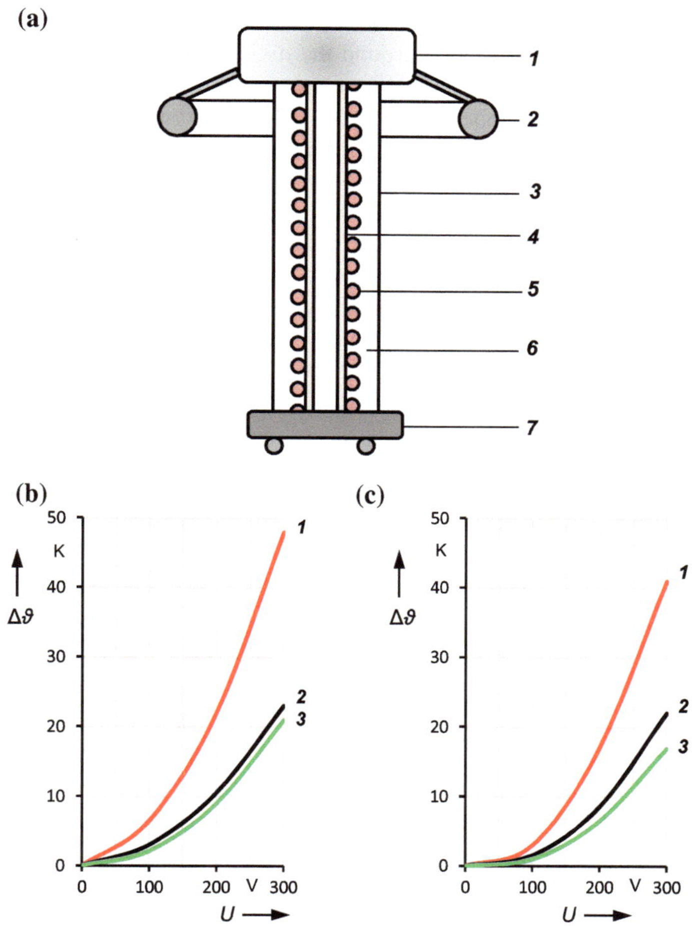

Fig. 3.10 Calculated self-heating of a 300 kV DC voltage divider. **a** Construction of the voltage divider with wire-wound resistors (principle), ***1*** high-voltage electrode, ***2*** torus electrode, ***3*** insulating tube, ***4*** insulating support, ***5*** wire-wound resistors, ***6*** air or transformer oil, ***7*** chassis. **b** Temperature increase $\Delta\vartheta$ (U) in the top resistor (calculation), ***1*** resistors in air, ***2*** resistors in transformer oil, ***3*** resistors in transformer oil with circulation. **c** Temperature increase $\Delta\vartheta$ (U) in the lowest resistance (calculation). ***1, 2, 3*** as in case **b**

The calculations for the divider were supplemented by extensive experimental investigations [8]. In the first step, the temperatures of the oil or air near the top and bottom resistors were measured at voltages up to 200 kV. During extensive revision work on the divider, the resistor helix was pulled out of the divider housing. Then, a wire-wound resistor of 2 MΩ was inserted as dummy on the divider top as well as on the divider bottom. The dummy resistor was provided with a bore and a temperature sensor was inserted as close as possible to the resistor wire. The test conditions at low voltage were largely adapted to the use of the complete voltage divider. That is, the DC current flowing through the dummy resistor had the same magnitude and the insulating medium the same temperature as in the corresponding use of the divider at high voltage. The temperature of the insulating media was achieved by heating bands wound around the divider housing.

For stationary oil or air, the results of the experimental investigation confirm very well the computed temperatures of the top and bottom resistors according to Fig. 3.10b, c. In the case of circulated oil, the calculated temperatures are slightly but significantly higher than the measured ones. This may be due to the fact that the equations derived for heat transfer apply to natural, but not to forced convection. It has already been pointed out that the internal temperature of a wire-wound resistor can deviate from the temperature of the surrounding insulating medium. Thus, the temperature of the wire in the top resistor was about 10 K higher than the temperature of the surrounding air at 200 kV.

Note The measurements were performed during a revision of the voltage divider because the division ratio had changed after a long period of use. The insulating oil was drained and the resistor helix was taken out from the divider housing. The change of the division ratio was caused by semiconducting deposits on the surface of the resistors due to decomposition products of the transformer oil. After careful cleaning of all resistors, the original division ratio could be exactly restored. The voltage divider with air insulation can only be used up to approx. 200 kV. The values calculated for higher voltages serve for comparison with the values for oil insulation.

The self-heating of the resistors, of course, affects the division ratio X of the divider according to the applied voltage. The ratio change with voltage was calculated for rather small temperature coefficients α_1 for R_1 and α_2 for R_2, which are typical of selected wire-wound resistors in precise voltage dividers. For the three operating conditions, the results for four quotients α_1/α_2 are shown in Fig. 3.11.

The conclusions are that the influence of the self-heating effect on the relative change $\Delta X/X$ of the division ratio:

- is particularly large in stationary air, both for $\alpha_1 = \alpha_2$ and $\alpha_1 \neq \alpha_2$,
- can be reduced by selecting a suitable TK value α_2 for R_2, and
- is almost negligible for $\alpha_1 \approx \alpha_2$ and oil circulation.

Conventional voltage dividers for tests usually have resistors with higher TC values so that larger changes in the division ratio are to be expected.

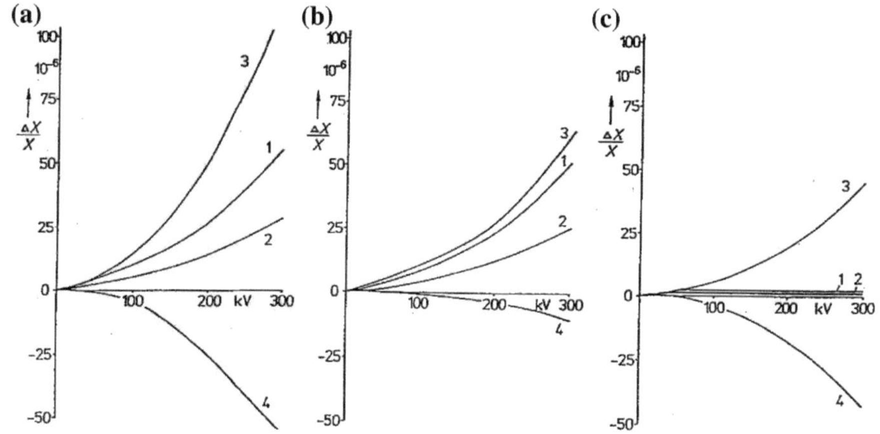

Fig. 3.11 Relative change $\Delta X/X$ of the division ratio due to self-heating, calculated for different temperature coefficients α_1 und α_2 of the wire-wound resistors embedded in: **a** Stationary air, **b** stationary oil, **c** circulated oil, *1* $\alpha_1/\alpha_2 = 10/10 \times 10^{-6}\ \mathrm{K}^{-1}$, *2* $\alpha_1/\alpha_2 = 5/5 \times 10^{-6}\ \mathrm{K}^{-1}$, *3* $\alpha_1/\alpha_2 = 6/4 \times 10^{-6}\ \mathrm{K}^{-1}$, *4* $\alpha_1/\alpha_2 = 4/6 \times 10^{-6}\ \mathrm{K}^{-1}$

3.4.4 Transfer Behavior of DC Voltage Dividers

The high-ohmic design and large dimensions of DC voltage dividers limit their bandwidth, so that the ripple of the DC voltage can often not be measured correctly. This is mainly due to the distributed *stray capacitances* C_e' of the voltage divider to ground, through which the AC component of the DC voltage flows (Fig. 3.12a). Depending on the dimensions of the voltage divider, the stray capacitances are in the order of 15–20 pF/m. The parallel capacitances C_{p1}' in the equivalent circuit diagram are determined by the type of resistor, the pitch of the resistor helix and the shape of the intermediate electrodes at the individual divider stages. Inductances of the high-ohmic resistors and leads do not need to be considered because of the small currents. If the parallel capacitances C_{p1}' are negligible, a simplified *equivalent circuit diagram* of the high-ohmic DC voltage divider can be used in which a capacitance $^2/_3 C_e$ is connected in parallel to $(R_1/2 + R_2)$ as shown in Fig. 3.12b. The theoretical basics for the *transfer behavior* of a voltage divider, which is considered as a *ladder network*, are discussed in detail in Sect. 4.3.2.2. Due to the charged stray capacitances, high transient voltages can occur in the event of a sudden switch-off or failure of the DC voltage supply, resulting in damage or destruction of the resistors, in particular in the upper region of the voltage divider.

In order to improve the transfer behavior, the capacitors C_1' and C_2 are occasionally added in parallel to the individual resistor stages R_1' und R_2 of the voltage divider (Fig. 3.13a). The general requirement for the time constants of the upper and lower divider parts is $R_1C_1 = R_2C_2$. The parallel capacitor chain of this *parallel-mixed RC voltage divider* reduces the effect of stray capacitances and thus the danger that transient overvoltages destroy the resistors. An even better transfer

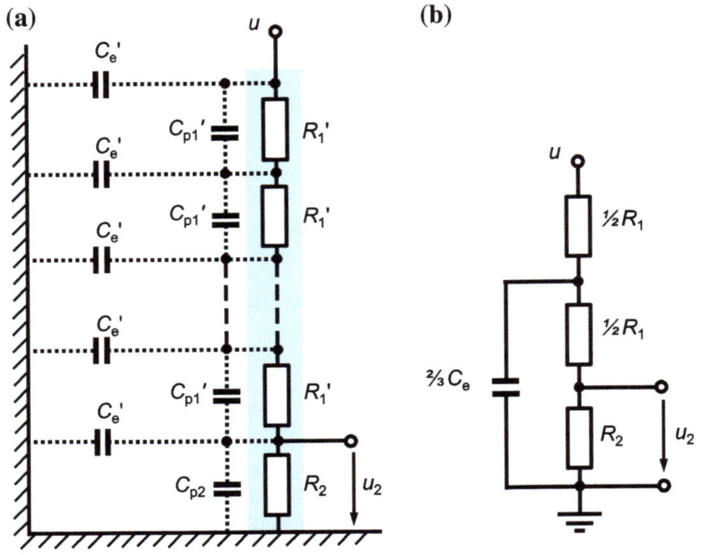

Fig. 3.12 Equivalent circuit diagrams of a high-ohmic DC voltage divider. **a** Circuit with distributed capacitances C_e' to ground and parallel capacitances C_{p1}' and C_{p2}. **b** Simplified equivalent circuit for negligible parallel capacitances C_{p1}' and C_{p2}

behavior is achieved when the capacitors C_1' and C_2 are connected in series with damping resistors. The result is a *damped capacitive voltage divider* which can also be used for AC and impulse voltage measurements (see Sect. 4.3.4).

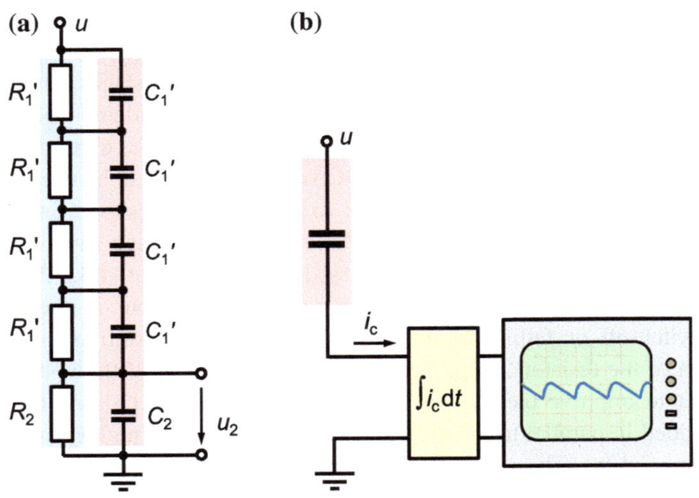

Fig. 3.13 Principle circuits for measuring the ripple of a high DC voltage. **a** Parallel-mixed RC voltage divider, **b** circuit with high-voltage capacitor, integrator and digital recorder

3.4.4.1 Measurement of Ripple

The parallel-mixed RC voltage divider shown in Fig. 3.13a also provides the possibility of measuring the ripple of DC voltages. The ripple itself can, of course, also be measured with a separate capacitive voltage divider. Another option is provided by the measurement circuit with a high-voltage capacitor, which decouples the AC component of the pulsating DC voltage (Fig. 3.13b). The capacitor current i_c must first be integrated in order to obtain a voltage corresponding to the ripple, which is then recorded by the digital recorder. The ripple is evaluated with regard to its amplitude and frequency f_0. The 3 dB bandwidth of the ripple measuring system must be at least $0.1f_0$ to $10f_0$ (see Sect. 3.1).

3.4.5 High-Precision DC Voltage Divider

For special measurement tasks, e.g. for checking particle accelerators and X-ray equipment or for the detection of certain elementary particles as in the *tritium neutrino experiment*, highly stable DC voltage dividers with relative uncertainties of a few 1×10^{-6} are required. These high requirements can only be met by a sophisticated design and special measures in the selection of the resistors, temperature stabilization and shielding [1, 9–11]. As an example, the construction and characteristics of a very accurate and stable 100 kV voltage divider are described in [10]. The entire divider developed at the PTB in Germany is housed in a metal pressure vessel filled with compressed SF_6 (Fig. 3.14a). The main part of the voltage divider consists of a resistor helix surrounded by five copper ring electrodes. They are connected to a parallel-mixed RC auxiliary divider and serve for potential control (Fig. 3.14b). The resistor helix has a constant pitch and is arranged in five sections (Fig. 3.14c).

The total high-voltage resistance is 1 GΩ, resulting from the series connection of 100 wire-wound resistors of 10 MΩ. The nominal division ratios are 100:1 for calibrations at low voltage and 10,000:1 for high voltage measurements. The 100 resistors of 10 MΩ each were carefully selected from a batch of 500 resistors with regard to voltage dependence, temperature behavior and self-heating. They are selected in pairs and connected in series, so that the run-in behavior of each resistor pair is practically zero. The resulting run-in behavior of the complete divider with the matched resistors is thus very short. The extraordinary constancy of the division ratio M at 20, 60 and 100 kV is shown in Fig. 3.15, with the first 2 min being necessary for the preparation of the measurement and the start-up of the high DC voltage.

The internal temperature of the pressure vessel is kept constant at 26 ± 0.15 °C by means of a sophisticated control unit, the heart of which is a *Peltier element* for heating and cooling. The divider output voltage is measured using a high-ohmic digital voltmeter or a null detector in combination with a reference voltage source. The settings of the high-voltage source, the measurement and evaluation of the

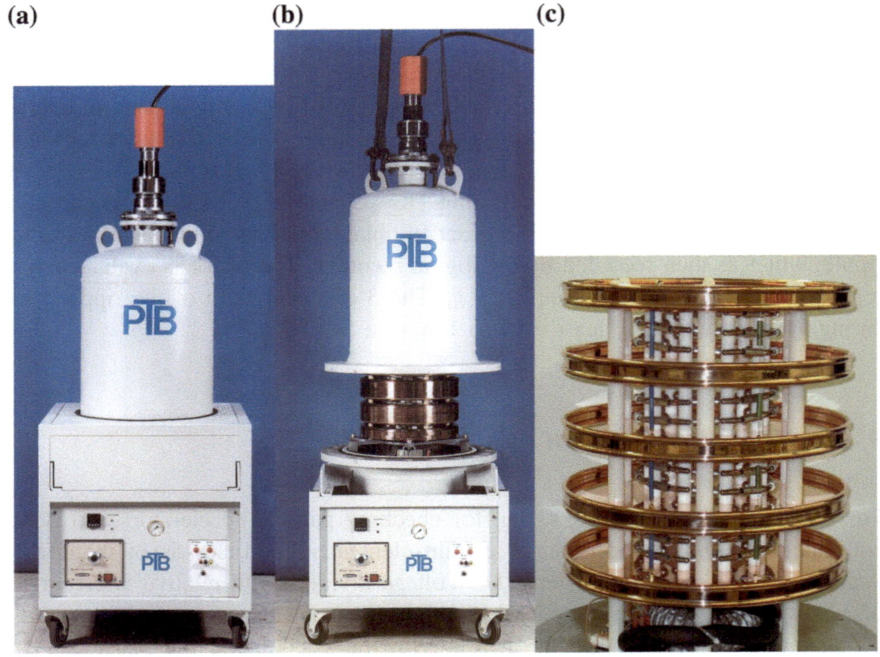

Fig. 3.14 100 kV voltage divider of highest accuracy (PTB). **a** View of the SF_6-insulated voltage divider in the pressure vessel with underframe. **b** View of the lower Cu electrodes for potential control of the resistors. **c** Arrangement of the resistors in a 5-stage helix (control electrodes dismantled)

Fig. 3.15 Relative temporal change of the division ratio *M* at 20, 60 and 100 kV (the first 2 min are for preparing the measurement and starting-up the high DC voltage)

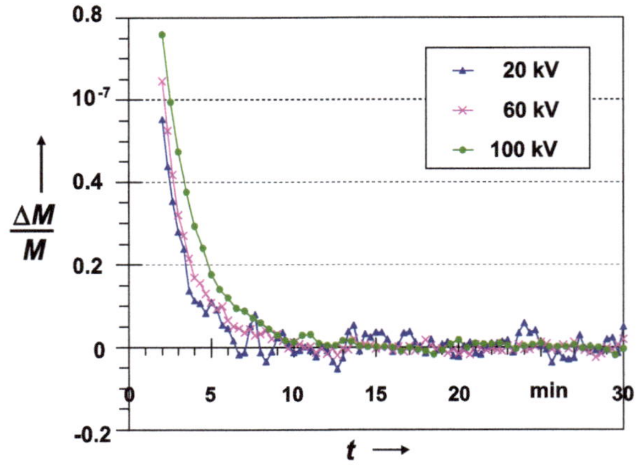

recorded data are program-controlled. In-depth investigations on the complete voltage divider, including a linearity test compared to a 300 kV voltage divider (see Fig. 3.6, center) and taking into account the possible influence quantities, resulted in an expanded relative uncertainty of 2×10^{-6} for the division ratio.

Comparison measurements between PTB's standard divider and a 150 kV standard divider developed at the NML Australia showed excellent agreement within $\pm 2 \times 10^{-6}$ for DC voltages up to 100 kV [12]. During the comparison measurements, however, there was a significant effect. As the high voltage was increased step by step, the division ratio of the NML divider initially decreased by up to 10×10^{-6}, but then gradually returned to its initial value within the following 30 min. The metal oxide film resistors used in the NML voltage divider can be regarded as the main cause of this behavior. Their voltage and temperature coefficients cause about the same but opposite effect. In the case of a voltage change, the division ratio is immediately affected according to the negative voltage coefficient, while the subsequent compensation by the self-heating of the resistors takes place only slowly.

At the time of preparing the manuscript, a divider for DC voltages of up to 1000 kV was developed as part of a European joint project for HVDC technology. It serves as a transportable reference divider for calibrating the voltage dividers in converter stations [13]. The high-voltage unit consists of five stages of 200 kV each, each stage consisting of four modules of 50 kV. The reference divider is surrounded by three resistive-capacitive voltage dividers for shielding and potential control. The grading of the capacitances in the shielding divider was determined by field calculations so that the field distribution in the immediate vicinity of the 8 m high reference divider is as linear as possible with the divider height and thus equal to the electric flow field. The relative measurement uncertainty of the 1 MV reference divider is stated to be less than 1×10^{-4}. The reference divider has a bandwidth of a few 10 kHz and is therefore also suitable for measuring the ripple of DC voltages. In the context of the HVDC project, methods and devices for generating and measuring of the ripple have also been developed with high accuracy [14].

Due to their size and their high-ohmic design, DC voltage dividers with rated voltages in the 1 MV range have the basic problem that leakage currents can flow to ground through the numerous supports of the individual resistors. They are therefore lost in the voltage measurement at the bottom resistor. Although the total leakage current is taken into account during the calibration of the voltage divider, one must assume that it is non-linear with voltage and temperature changes. This behavior affects the scale factor and causes a measurement error in the output voltage.

The total leakage current can be determined as the difference between the current entering the top of the voltage divider and the current leaving the divider on its ground side. For this purpose, two measuring resistors of 15 kΩ each were connected to the high- and low-voltage side of a 1.4 GΩ voltage divider and their voltages were measured with precision digital voltmeters (DVM) [15]. The DVM at high-voltage potential was powered by batteries. Both DVM were connected to a PC via *wireless local area networks* (*WLAN*). This enabled simultaneous measurements at both measurement points and data transfer to the PC. Due to the simultaneous recording, the influence of instability and ripple of the DC voltage (1.2% at 800 kV) could be largely excluded. The investigation showed that with increasing voltage the current through the bottom resistor became disproportionately smaller than that through the top resistor. At 1 MV, the relative current deviation and thus the relative error of the measured high voltage amounted to -65×10^{-6}.

3.4.6 Addition of Partial Voltages

The measurement of very high DC voltages is impaired by leakage currents in the resistive voltage divider and discharge currents due to corona. The large voltage dividers used are generally unshielded and of modular design. It is therefore possible to measure the partial voltages of the individual modules and obtain the total voltage by an exactly timed addition [16]. The measurement principle is described using the example of a three-stage system shown in Fig. 3.16. In addition to the main resistor, each stage ***1*** is provided with a small series resistor whose voltage is converted by a voltage-frequency converter ***2*** into a pulse sequence. The pulses are sent via an optical transmission path with transmitter ***3***, optical waveguide and receiver ***4*** to a parallel-to-serial converter ***5*** on low-voltage potential. The sum of the pulses coming from the three stages is detected by a pulse counter ***6*** and displayed as the total voltage.

A test model of this measurement principle was thoroughly tested at DC voltage up to 250 kV under various conditions. The sum of the partial voltages representing the total voltage showed excellent linearity. The voltage indication remained completely unaffected, even when corona was artificially generated by means of small metal tips on the intermediate electrodes, causing leakage currents to ground

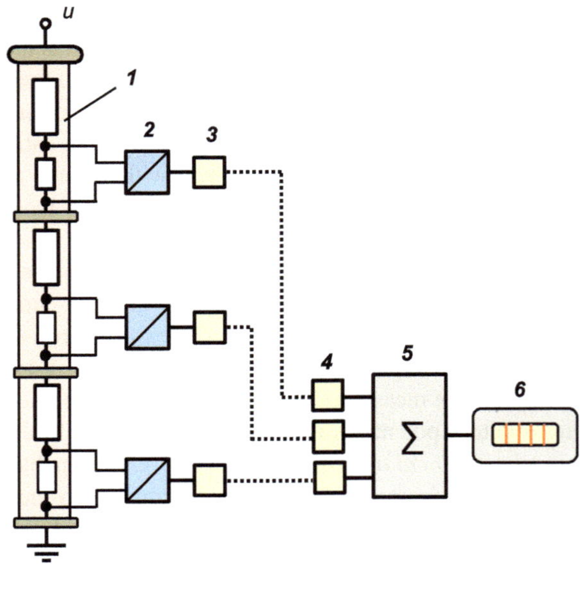

Fig. 3.16 Addition of partial voltages (principle arrangement of a three-stage measuring system). ***1*** divider stage with lower and upper resistor, ***2*** voltage-to-frequency converter, ***3–4*** optoelectronic transmission path, ***5*** summation of pulses, ***6*** pulse counter and voltage display

and the high-voltage electrode. On the other hand, the indication of a measuring system with a conventional DC voltage divider was significantly affected by corona. Further advantages of the measurement principle are lower costs due to the smaller high-voltage electrode and the possibility of increasing the number of stages as needed.

3.4.7 Fixed Points of the High Voltage Scale

The general requirement of each measurement is that the measured quantities are traceable to the units of measurement in the SI system (see Sect. 10.1). The unit of voltage is available with the highest possible accuracy by means of the *Josephson effect*, but only at low voltage of 1 V or 10 V [17]. Traceability of voltages of several 100 kV or even 1 MV to this small voltage unit can, in principle, be achieved by using resistor dividers. However, this implies a loss of accuracy that may be tolerable for standardized testing on power apparatus, but not for high-precision voltage measurements. In this context it is worth mentioning that calibration laboratories and national metrology institutes regularly perform multi-lateral or bilateral comparison measurements even in the field of high DC voltages. The aim is to detect possible measurement deviations of the high-voltage dividers used in comparison with an accurate transfer standard. Although the quality of transfer dividers has in fact increased over decades, the achievable accuracy of the comparative measurements is limited for a number of reasons.

As an example of a so-called *absolute measurement standard* in the high-voltage range, the sphere gap can also be considered, with which high voltage measurements are traced back to the SI base unit of length (spark gap, sphere diameter). However, the influence of several disturbance quantities only allows measurement uncertainties in the percentage range. Another, more precise absolute standard is the *electrostatic voltage balance* that has been studied in some national metrology institutes in the past. With the voltage balance, the force between two voltage electrodes is compensated by weights. Voltages of up to several 10 kV can thus be traced back to the SI unit of mass with a relative uncertainty of about 10×10^{-6}.

Another method deals with the realization of *calibration values* or *fixed points* of the voltage scale in the high-voltage range, comparable to fixed points in the temperature scale for melting ice or evaporating water. Using the Maxwell equations, the interaction of accelerated electrons with the microwave field inside a rectangular cavity resonator is calculable. According to the theory, the electrons generally undergo a deflection from their original path, but for certain calculable values U_i of the acceleration voltage, they can leave the cavity resonator in the same direction as at the entrance [18, 19]. The equation for determining U_i contains two resonant frequencies of the high-frequency field and the quotient e/m_0 (elementary charge e per electron rest mass m_0). Since frequencies can in principle be measured very accurately, at the beginning of the experimental work, a relative measurement uncertainty of 5×10^{-7} was considered quite realistic for the determination of these fixed points U_i.

Figure 3.17 shows the setup for the experimental confirmation of the novel theoretical approach in the voltage range up to 100 kV [19, 20]. The electrons emerging from the electron source ***1*** are accelerated by a controllable highly stable DC voltage U in the z direction and pass through a small aperture into the superconducting rectangular *cavity resonator* 2 made of niobium. The microwave field of a $H_{1,0,p}$ mode in the resonator, generated by the microwave generator G, causes a deflection of the electrons by Coulomb and Lorenz forces. In general, the deflected electrons strike the inner walls of the resonator. However, if the electrons ***3*** are accelerated by one of the calculable voltages U_i, they can leave the resonator again in the z direction through a second aperture at the resonator exit. These electrons are collected by the Faraday Cup ***4*** and displayed as current I_e. Figure 3.17b schematically shows a typical measurement curve of I_e versus the acceleration voltage U. The location of the maximum of I_e at the calculable voltage U_i is given by the intersection of the extrapolated flanks of I_e. Theoretically, the shape of I_e should be a single vertical line at the calculable voltage U_i, but this is not experimentally achievable due to the not infinitely small diameter of the aperture and further inadequacies.

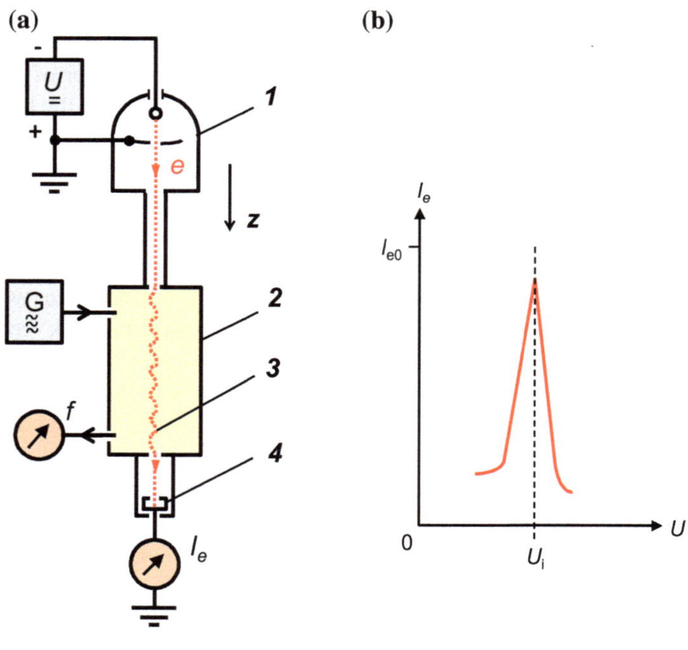

Fig. 3.17 Realization of fixed points of the high voltage scale (schematic). **a** Experimental setup with simulated electron path at one of the calculable acceleration voltages, ***1*** electron source, ***2*** superconducting cavity resonator, ***3*** electron path, ***4*** Faraday cup. **b** Measured detector current I_e whose maximum marks one of the fixed points of the voltage U_i

The cavity resonator thus acts as an *electron speed filter*, i.e. the electrons can pass through the resonator via small apertures only for well-defined velocities that correspond to the calculable voltages U_i. Moreover, the filter characteristic is independent of the phase of the electrons, i.e. the time of their entry into the resonator. As a result of the experimental investigations, five maxima of the current I_e, i.e. five fixed points U_i were detected in the voltage range between 40 and 100 kV. The reproducibility of the fixed points was 1×10^{-4}, which was considered as still improvable by various measures. However, the experimentally determined fixed points differed by about 10% from the theoretical fixed points U_i. The cause of this large deviation could not be determined within the given timeframe [21, 22].

3.4.8 Rotary Voltmeter

The *rotary voltmeter* belongs to the group of field sensors and voltmeters called *generating voltmeter*. Originally, these devices were used as *field mills* in a handy design for the measurement of electrostatic fields, even in the atmosphere (Ref. [4] of Chap. 1). The measurement principle is that the capacitance of an electrode arrangement located in an electric field is periodically changed so that the influenced charge periodically changes between a minimum and a maximum value. The oscillating charge causes a dielectric displacement current that generates a measurable AC voltage across a resistor. The electrodes of generating voltmeters are often plate-shaped and provided with sectoral recesses, but other electrode arrangements as described below for high DC voltage measurements are also possible.

The generating voltmeter shown in Fig. 3.18 has cylindrical electrodes, the high DC voltage being applied directly to the outer cylindrical electrode ***1*** [23]. The eccentrically arranged low-voltage electrode consists of two half cylinders ***2*** and ***3*** as well as the grounded end sections. The two half cylinders are connected to each other via an AC current meter with the internal resistance R_i and an equal resistor $R = R_i$, the junction between the two resistors being grounded. Since R_i is very small, the two half cylinders ***2*** and ***3*** are practically at ground potential. The complete low-voltage electrode is motor driven and rotates about its own axis. As a result, the partial capacitances of the two half cylinders to the surrounding high-voltage electrode change periodically. In the drawing in Fig. 3.18, the capacitance C_{12} of the half cylinder ***2*** has its maximum value and the capacitance C_{13} of the half cylinder ***3*** has its minimum value. After half a turn in the time $T/2$, the two half cylinders have exchanged their positions and capacitances. The charge ΔQ transported thereby from one half cylinder to the other is:

$$\Delta Q = UC_{\max} - UC_{\min} = (C_{12} - C_{13})U = \Delta CU. \tag{3.9}$$

The rotation of the half cylinders with the frequency $f_r = 1/T$ causes a periodic change of charge and thus a measurable AC current i:

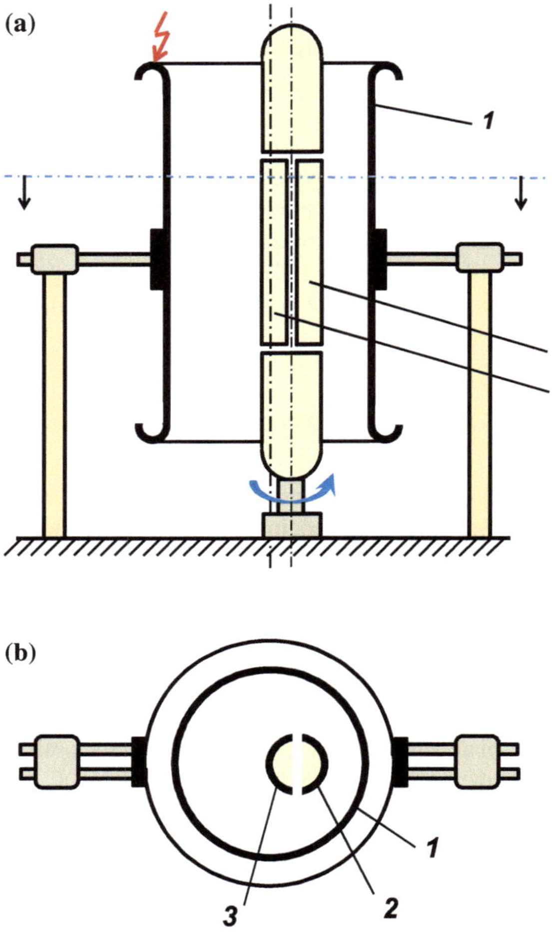

Fig. 3.18 Construction of the rotary voltmeter with cylindrical electrodes (principle). **a** High-voltage electrode ***1*** and low-voltage electrodes ***2*** and ***3***. **b** Cross-sectional drawing through the high- and low-voltage electrodes

$$i = \frac{\Delta Q}{T/2} = 2\Delta C U f_{\mathrm{r}}, \tag{3.10}$$

from which the DC voltage U can be determined.

An advantage of the rotary voltmeter is that the two half cylinders of the low-voltage electrode move in an almost homogeneous field and are largely shielded from external disturbances. The capacitances C_{12} and C_{13} of the half cylinders can be determined by field calculation and are thus traceable to the SI unit "length". Alternatively, the capacitances can be measured, as well as the rotation frequency f_{r}. Another advantage is the powerless measurement without loading of the DC voltage source. Two prototypes of the rotary voltmeter for 30 kV and 200 kV have been constructed and tested. The achievable measurement uncertainty was considered to be better than 1%. Several improvements in the design and smaller dimensions, which can be achieved with SF6 insulation, have been proposed to further reduce measurement uncertainty.

3.4.9 Standard Rod-Rod Gap

The determination of the peak value of high voltages up to about 2 MV using *standard sphere-gaps* has a long tradition. According to recent studies, the *standard rod-rod gap* is the better, more precise alternative for DC voltage measurements of more than 135 kV (Ref. [5] of Chap. 2; [24]), while the *sphere-gap* is intended for AC and impulse voltage measurements (see Sect. 2.5.8). The two rod-shaped electrodes made of steel or brass can be arranged vertically or horizontally (Fig. 3.19). A minimum distance of 5 m must be maintained between the tip of the high voltage electrode and grounded objects or walls, except the ground plane. For the rods, a solid cross-section with side lengths between 15 and 25 mm is required. The rod ends shall be cut at right angles to the axis with sharp edges, which is the prerequisite for a reproducible discharge mechanism. However, reproducibility of the *streamer pre-discharges* is not given for small gap spacing, so that the standard rod-rod gap can only be used as an approved measuring system for $d \geq 250$ mm corresponding to a voltage of at least 135 kV.

When carrying out the measurements, the positive or negative DC voltage is increased from 75 to 100% of the expected discharge voltage in about 1 min. The voltage at the instant of sparkover is taken with the indicating device used, e.g. for the control voltage of the DC voltage generator. The procedure is repeated until a

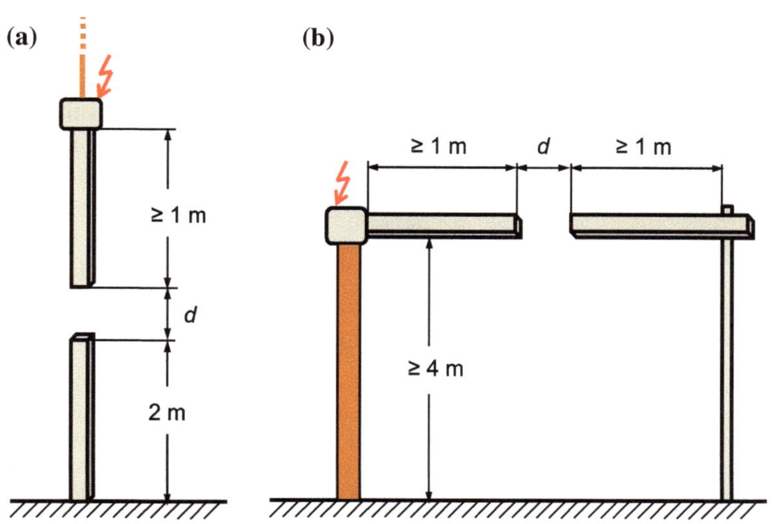

Fig. 3.19 Standard rod-rod gap for DC voltage measurement (250 mm $\leq d \leq$ 2500 mm) **a** Vertical arrangement **b** Horizontal arrangement

total of $n = 10$ successive readings. The mean of the ten individual values gives the disruptive discharge voltage. For positive and negative DC voltages, the disruptive discharge voltage can be approximated by the numerical equation [24]:

$$U_0 = 2 + 0.534d, \tag{3.11}$$

where U_0 is in kilovolts and the gap spacing d in millimeters. Equation (3.11) applies to vertical and horizontal gap arrangements with d between 250 and 2500 mm. Furthermore, the U_0 values in Eq. (3.11) are valid for *standard atmospheric conditions* (temperature $T_0 = 20$ °C, air pressure $p_0 = 1013$ mbar = 101.3 kPa) and the quotient h/δ in the range 1–13 g/m^3, where h is the absolute humidity in g/m^3 and δ the air density according to Eq. (2.16):

$$\delta = \frac{p}{p_0}\frac{273\ \mathrm{K} + T_0}{273\ \mathrm{K} + T}.$$

Under these conditions, the reference values of the disruptive discharge voltage U_0 have an estimated uncertainty of 3% for a confidence level of at least 95%.

In the case of deviating standard atmospheric conditions, the measured breakdown voltage U is converted by division with the factors δ and k into the value U_0 for standard conditions:

$$U_0 = \frac{U}{\delta \cdot k}. \tag{3.12}$$

The humidity correction factor k is given by the numerical equation:

$$k = 1 + 0.014\left(\frac{h}{\delta} - 11\right), \tag{3.13}$$

where h/δ is in g/m^3. Humidity therefore affects the breakdown voltage of the rod-rod gap more than that of the sphere gap (see Sect. 2.5.8). The equations given for U_0 and k apply to the investigated values h/δ in the range 1–13 g/m^3.

Several studies have also been conducted with rods having circular cross-section and rounded ends [24–26]. However, the breakdown voltages for positive and negative DC voltages are different, so that separate equations for the respective polarity should have been taken into account. In addition, the spread of the breakdown voltage values for round rods is greater than for rectangular rods.

Note For measuring positive and negative DC voltages, the use of sphere gaps is not recommended by IEC 60052 because of erratic disruptive discharges. Nevertheless, with some precautions, sphere gaps can also be used for DC voltage measurements, with the breakdown values being the same as for AC voltages and negative impulse voltages. The IEC edition of 2002 (Ref. [5] of Chap. 2), valid at the time of printing, gives no indication of the uncertainty achievable for DC voltages with sphere gaps, whereas an uncertainty of 5% was stated in previous IEC editions.

3.5 Measurement of High DC Currents

With the exception of the DC currents supplied by batteries and accumulators, the technically generated DC currents are associated with a more or less large ripple. A DC measurement system should therefore also be designed to measure the components with ripple frequency, or the ripple factor is determined by another measurement system. In general, the measurement system consists of a current sensor and a digital device, often a digital recorder. DC current measurements on high voltage potential, e.g. on overhead lines, require a potential-free measurement system. This is no longer a fundamental problem due to the data transmission facilities with optical fibers. Conventionally, low-ohmic shunts are used for current measurement, but also Hall sensors and DC transformers based on the zero-flux principle. Advances in the technical application of the Faraday effect have led to increased use of magneto-optical current sensors, e.g. for current measurement in aluminum production (see Sect. 6.2).

3.5.1 *Low-Ohmic Measuring Resistors*

The conventional measurement principle for high DC currents is shown in Fig. 3.20. The current i flows through the low-ohmic resistor R_m and generates the voltage u_m, which is indicated by the measuring device M connected via a coaxial cable. In general, digital voltmeters are used to measure steady-state DC currents and digital recorders to measure short-time DC currents. For very large DC currents, R_m is less than 50 μΩ to limit the measurement voltage u_m and the thermal load. For the measurement of short-time DC currents, *coaxial shunts* are preferably used as they are generally present in testing laboratories for AC current and impulse current measurements. Further details on coaxial shunts can be found in Sect. 5.3.1.4. The voltage is simply $u_m = iR_m$ if the input resistance of the measuring instrument is $R_e \gg R_m$. In contrast to rapidly changing currents, the coaxial cable does not have to be terminated by its wave impedance when measuring steady-state DC currents.

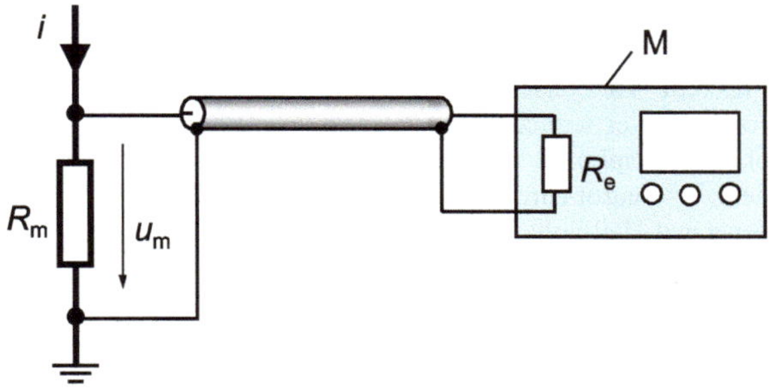

Fig. 3.20 Measurement of a DC current i with a low-ohmic measuring resistor R_m. M: digital voltmeter or digital recorder with input resistance R_e

According to IEC 62475 (Ref. [4] of Chap. 2), the measuring instruments used for testing must indicate the arithmetic mean value, the ripple and, in addition, for short-time DC currents, their duration and rise time. If the short-time DC current drops rapidly, the test object is heavily loaded, possibly in conjunction with electromagnetic influence on the measuring system. A double shielding of the measuring system is preferred, comparable to impulse current measurements (see Sect. 5.3.1.1).

A potential-free hybrid-optical measuring transformer (HOCT) for current measurement up to 300 A in a 220 kV DC transmission line is described in [27]. The current sensor on high-voltage potential is a Manganin shunt for measuring the DC component. In addition, the AC component of the current is measured with a Rogowski coil. The output voltages of both sensors are digitized and sent via two optical fiber links to the measuring instruments and the PC on ground potential. The temperature dependence of the shunt is corrected by calculation. The power supply of the optocouplers is around 1 W, which is sent to high-voltage potential using a power laser diode and a third optical fiber link. The preliminary results of the investigation are satisfactory and the measured values are in accuracy class 0.2.

3.5.2 Hall Current Sensors

Current sensors of this type exploit the *Hall effect*, which is caused by moving charge carriers in a magnetic field due to the Lorentz force. The principle of the Hall effect is shown in Fig. 3.21. When a semiconducting plate of thickness d is positioned in a vertical magnetic field with the inductance B and the control current I_s flows through it, the electrons are deflected by the Lorentz force perpendicular to the original current direction and perpendicular to the magnetic field. Due to the charge displacement, the *Hall voltage* u_H is generated at the outer edge of the plate:

$$u_H = R_H \frac{I_s B}{d} \sim B. \tag{3.14}$$

The *Hall coefficient* R_H is inversely proportional to the charge carrier concentration n_e in the plate material and the elementary charge e_0. For metallic conductors with a high charge carrier concentration, the Hall voltage u_H is therefore relatively

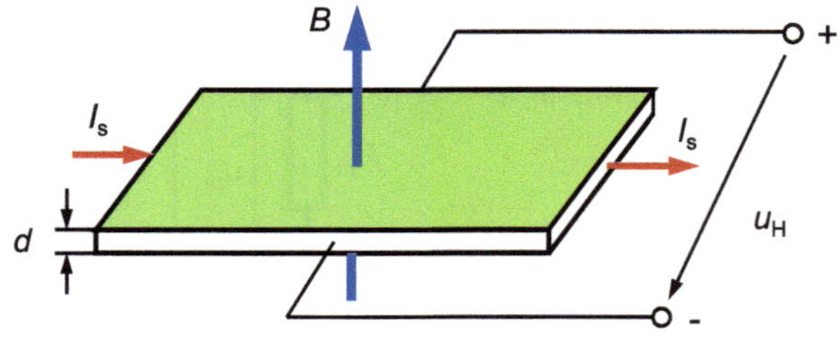

Fig. 3.21 Hall effect in a conductive or semiconducting plate (principle). B: magnetic field, I_s: control current, u_H: generated Hall voltage

small. Only the use of semiconductors with orders of magnitude smaller charge carrier concentrations and thus higher values of R_H and u_H has led to a wider application of the Hall effect. By vapor deposition of a thin semiconductor layer on a carrier material, the effective thickness d can be kept very small, whereby u_H becomes larger. The Hall effect exists for DC fields as well as AC fields and impulse fields. The same applies to the control current I_s.

The basic principle of a Hall current sensor is explained with reference to Fig. 3.22. The current-carrying conductor is passed through the opening of the magnetic core and generates a magnetic field with the induction B_{Fe} in the core. In the air gap of the core is the Hall element, where it is exposed to the magnetic field H_{air}. Because of the continuity of induction in the magnetic core and in the air gap of the width δ, the following applies:

$$B_{Fe} = \mu_r \mu_0 H_{Fe} = B_{air} = \mu_0 H_{air}. \tag{3.15}$$

The relationship between the current i and the magnetic field strength H is determined by the law of magnetic flux (Ampere's law) according to Eq. (5.16). The line integral of the magnetic field strength extends over the magnetic core of the length l_{Fe} and the air gap with $l_{Luft} = \delta$ and is expressed in a simplified form:

$$i = \oint H \mathrm{d}s = H_{Fe} l_{Fe} + H_{air} \delta. \tag{3.16}$$

For a highly permeable magnetic core, a simple expression for the induction B_δ in the air gap results from Eqs. (3.15) and (3.16):

$$B_\delta = B_{air} = \frac{\mu_0 i}{\frac{l_{Fe}}{\mu_r} + \delta} \approx \frac{\mu_0}{\delta} i. \tag{3.17}$$

Inserting Eq. (3.17) in Eq. (3.14) gives the equation for the Hall voltage:

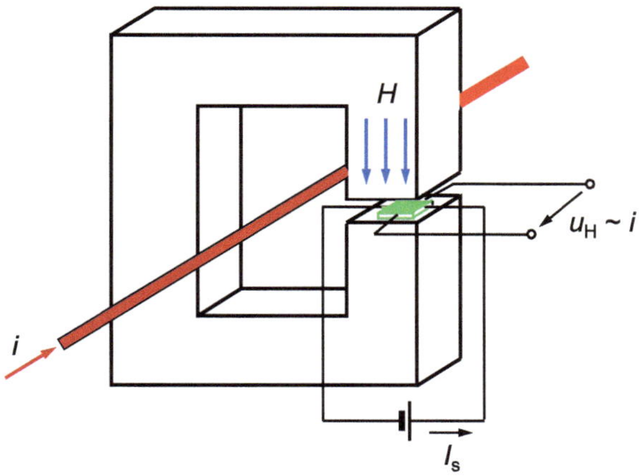

Fig. 3.22 Principle of a simple Hall sensor with semiconductor plate in the air gap of a magnetic core

$$\mu_{\mathrm{H}} = R_{\mathrm{H}} \frac{I_s}{d} \frac{\mu_0}{\delta} i = K_{\mathrm{H}} i, \tag{3.18}$$

where K_{H} is the Hall coefficient. Under the conditions mentioned above, the Hall voltage u_{H} is proportional to the induction B and thus to the current i to be measured.

The simple basic circuit in Fig. 3.22 has the disadvantage that the measurement accuracy is not satisfactory because of the non-linearity of the magnetic core and the Hall sensor. Improvement is achieved by a compensation winding N_{k} on the magnetic core (Fig. 3.23). Through the winding, the compensation current i_{k} flows so that the induction in the core becomes zero (*zero-flux principle*). The compensation current i_{k} is derived from the Hall voltage and sent via an operational amplifier OP through the compensation winding N_{k}. The voltage u_{m} across the resistor R_{m} is proportional to the compensation current and thus also to the current i to be measured [28–30].

Depending on the design, Hall current sensors can be used to measure DC, AC and impulse currents with peak values of up to 30 kA. In the version without compensation winding (see Fig. 3.22), the achievable bandwidth is between 1 and 10 kHz. Higher bandwidths of up to some 100 kHz are achieved by Hall sensors with compensation winding, with the sensor in this version operating above 1 kHz as a current transformer. The zero-flux principle ensures relative measurement uncertainties of less than 1%. Hall sensors are also installed in handy clip-on current transformers, which can be conveniently opened and placed around the current carrying conductor. It should be added that the Hall effect is also used in sensors for measuring magnetic fields.

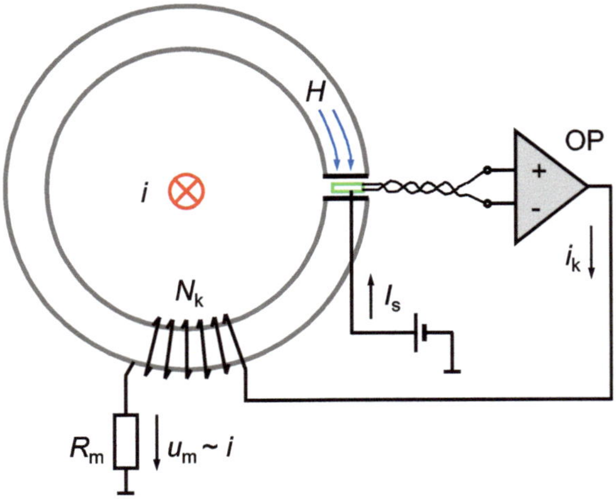

Fig. 3.23 Hall current sensor in the air gap of a magnetic core with compensation winding N_{k} (zero-flux principle)

3.5.3 Zero Flux DC Current Transformer

DC currents as well as AC and impulse currents can also be measured with *DC current transformers* operating on the *zero-flux principle* [29, 30]. The DC current transformer consists of three magnetic cores, a common secondary winding, several auxiliary windings and an electronic module with the zero-flux detector (Fig. 3.24). The transformer is usually constructed with toroidal cores, through whose openings the primary conductor (shown as a winding in Fig. 3.24) is guided. The primary current i_p produces in the three magnetic cores a magnetic field which is compensated by the compensation current i_k in the winding W2. For this purpose, the AC and DC components of the measured current are supplied separately to the two inputs of the operational amplifier OP. The AC current component i_{AC} is generated in the auxiliary winding W1. The DC current component i_{DC} including the low-frequency component is provided by means of an electronic module in the zero-flux detector and the symmetrical auxiliary windings W3 and W4.

Via an oscillator in the zero-flux detector and the symmetrical auxiliary windings W3 and W4, the two magnetic cores are driven into saturation in opposite direction, so that the resulting flux in the main core is zero. In the case where the primary current has a DC component, a corresponding magnetic flux is generated in the two cores. This means that the cores are no longer in the same state of saturation and the currents through W3 and W4 are then different. The current difference is proportional to the primary DC current and fed into the positive input of the operational amplifier OP. The resulting compensation current i_k with the DC and AC components is therefore a scaled image of the primary current. The corresponding voltage at the measuring resistor R_m can be evaluated with a digital measuring device.

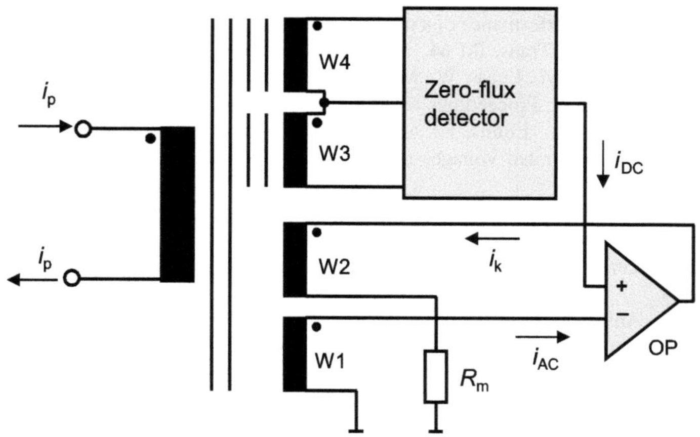

Fig. 3.24 Principle diagram of a DC current transformer with zero-flux detector

Due to the zero-flux principle, the resolution of the DC current transformer with electronic module according to Fig. 3.24 is very high, and relative measurement uncertainties in the range of some 10^{-6} can be achieved. Depending on the design of the DC current transformer, the maximum bandwidths are between 10 and 500 kHz and the maximum current amplitudes between 5 and 25 kA. Due to the low measurement uncertainty, the DC current transformer is particularly well suited for the precise calibration of other current measuring systems.

References

1. Peier, D., Graetsch, V.: A 300 kV DC measuring device with high accuracy. In: Proceedings of 3rd ISH Mailand, Paper 43.08 (1979)
2. Park, J. H.: Special shielded resistor for high-voltage DC measurements. J. Res. NBS **66C**, 19–24 (1962)
3. Rungis, J., Brown, D.E.: Modular DC resistance divider for 500 kV. AJEEE Australia (1985)
4. Jiang, G., Li, Y., Deng, H.: A 600 kV DC resistive divider with relative high accuracy. In: Proceedings of 5th ISH Braunschweig, Paper 73.04 (1987)
5. Strauss, W.: Ein Gleichspannungsmessgerät speziell für die Hochspannungstechnik. etz Archiv **1**, 255–257 (1979)
6. Berril, J., et al.: A high precision 300 kV D.C. measuring system. In: Proceedings of 3rd ISH Mailand, Paper 43.03 (1979)
7. Holman, J.P.: Heat Transfer, 5th edn. McGraw-Hill Book Company, New York (1981)
8. Wu, S.L., Schon, K.: Investigation into the self-heating effect of HV DC dividers. In: Proceedings of 5th ISH Braunschweig, Paper 73.05 (1987)
9. D'Emilio, S., et al.: Calibration of DC voltage dividers up to 100 kV. IEEE Trans. IM **34**, 224–227 (1985)
10. Marx, R.: New concept of PTB's standard divider for direct voltages of up to 100 kV. IEEE Trans. IM **50**, 426–429 (2001)
11. Thümmler, Th., Marx, R., Weinheimer, Ch.: Precision high voltage divider for the KATRIN experiment. New J. Phys. **11**, 1–20 (2009)
12. Marx, R., Li, Y., Rungis, J.: Comparison of two ultra-precision DC high voltage dividers developed at PTB and NML. In: Proceedings of 13th ISH Delft, Paper 362 (2003)
13. Hällström, J., et al.: Performance of a modular wideband HVDC reference divider for voltages up to 1000 kV. IEEE Trans. IM **64**, 1390–1397 (2015)
14. Meisner, J., Schmidt, M., Lucas, W., Mohns, E.: Generation and measurement of AC ripple at high direct voltage. In: Proceedings of 17th ISH Hannover, Paper D-010 (2011)
15. Li, Y., Ediriweera, M., Emms, F.: A digital system for measuring leakage current in DC voltage dividers with rated voltages up to 1000 kV. CIGRE SC D1 Colloquium Budapest (2009)
16. Kind, D., Lührmann, H., Weniger, M.: Messung hoher Spannungen durch Überlagerung von Teilspannungen. Bull. ASE/UCS **65**, 224–228 (1974)
17. Kind, D.: Accurate measurements over wide parameter ranges. IEEE Trans. DEI **3**, 473–481 (1996)
18. Schulz, B.: Absolute Messung hoher Spannungen durch Laufzeitmessungen an beschleunigten Elektronen in einem Hohlraumresonator. Thesis TU Braunschweig (1982)
19. Peier, D., Schulz, B.: Absolute determination of high DC voltages by means of frequency measurement. Metrologia **19**, 9–13 (1983)
20. Lucas, W., Schon, K., Hinken, J.H.: A superconducting electron speed filter for establishing high DC voltages. IEEE Trans. IM **34**, 227–231 (1985)

21. Lucas, W.: Untersuchungen zur Eignung supraleitender Hohlraumresonatoren als Geschwindigkeitsfilter für durch Hochspannung beschleunigte Elektronen. Thesis TU Braunschweig (1990)
22. Lucas, W., Kind, D.: Investigation into the deflection of relativistic electrons inside a superconducting cavity resonator for measuring high voltages. IEEE Trans. IM **40**, 288–290 (1991)
23. Claußnitzer, W.: A generating voltmeter with calculable voltage-current ratio for precise high-voltage measurements. IEEE Trans. IM **17**, 252–1968 (1968)
24. Feser, K., Hughes, R.C.: Measurement of direct voltage by rod-rod gap. ELECTRA **117**, 23–34 (1988)
25. Peschke, E.: Der Durch- und Überschlag bei hoher Gleichspannung in Luft. Thesis TH München (1968)
26. Gobbo, R., Pesavento, G.: Rod-rod gaps under DC voltage. In: Proceedings of 12th ISH Bangalore, Paper 7–25 (2001)
27. Zhao, Q., Zhang, G, Luo, C.: A hybrid-optical direct current transformer. In: Proceedings of 14th ISH Beijing, Paper J-47 (2005)
28. Kuhrt, F., Maaz, K.: Messung hoher Gleichströme mit Hallgeneratoren. ETZ-A **77**, 487–490 (1956)
29. Engelade, Th, Neergard, A., Bezolt, H.: Breitbandige Präzisionswandler für Messungen an Umrichtersystemen. Antriebstechnik **42**, 64–71 (2003)
30. LEM-Fachinformation: Galvanisch getrennte Strom- und Spannungswandler. Eigenschaften – Anwendungen – Dimensionierung. Druckschrift CH 24101, 3. Edition (2006)

Chapter 4
High Impulse Voltages

Abstract In power supply networks for the transmission and distribution of electrical energy at high voltage, transient voltages of more than 1 MV can occur due to lightning or switching operations. Because of their high magnitudes and short rise times, ranging from a fraction of a microsecond to milliseconds, the overvoltages cause enhanced stress on the insulation of the affected high-voltage apparatus, such as power transformers, switchgears, arrestors, insulators, power cables, etc. Therefore, before commissioning, each high-voltage apparatus must undergo acceptance tests with standardized impulse test voltages. The chapter introduces the standardized quantities and measurement methods and briefly describes the basic principles of impulse voltage generators. The various systems for measuring impulse voltages with resistive, capacitive or mixed voltage dividers and digital recorders are discussed in detail. The main focus is on digital measuring systems with computer-aided data processing that enable automated data acquisition, data filtering of impulses with overshoot, online monitoring and on-site tests. A well-founded measurement technique is essential and is achieved mainly by comparison measurement with a reference system. One important property of an impulse measuring system is its step response, which characterizes the transfer behavior. Short descriptions of analog measurement methods and devices that are still in use, including the standard sphere gap, are included. High impulse voltages are also used for applications in other fields of physics and engineering, such as plasma physics, electric spot-welding, electro-shock weapons, etc.

High impulse voltages are required for testing electrical apparatus used for the generation and transmission of electrical energy at high voltage, for example power transformers, switchgears, arrestors, insulators, power cables, etc. The test simulates the transient voltage stress with peak values of even more than 1 MV, which can be caused by lightning strokes, short circuits or switching operations in the supply network. The chapter presents the internationally standardized high-voltage impulses and measurement quantities, describes the measurement methods and briefly discusses some generator circuits for generating impulse voltages. The various systems for measuring impulse voltages with voltage dividers and digital

K. Schon, *High Voltage Measurement Techniques*, Power Systems,
https://doi.org/10.1007/978-3-030-21770-9_4

recorders are discussed in detail in theory and practice. Further measurement options such as the conventional standard sphere gap and newer capacitive field probes are described. For information on electro-optical sensors that exploit the Pockels or Kerr effect, see Sect. 6.1.

4.1 Definitions and Parameters of Standard Impulse Voltages

Standard impulse voltages used in high-voltage tests are characterized according to IEC 60060 by three or more parameters (Refs. [1–3] of Chap. 2). Their definitions differ in part from those common to low-voltage impulses (Refs. [1–3] of Chap. 2). Reasons for this are the inadequacies in the generation and measurement of high-voltage impulses, but also in the special stress of high-voltage insulation by impulse voltages with superimposed peak oscillation. As a rule, high-voltage impulses are recorded by means of a wide-band voltage divider and a digital recorder, stored and analyzed with regard to the parameters by software. Data processing with the right software is not only very helpful for the correct evaluation of the relevant impulse parameters, but also appears indispensable in many cases. For example, lightning impulse voltages with superimposed peak oscillation are evaluated by using the standardized filtering method. Another example is the numerical convolution calculation to prove the dynamic behavior of a measuring system. The analytical representation of some impulse voltages by their shape and spectrum is given in Chap. 8. For devices used in the low-voltage range with a rated voltage of 1 kV maximum, special test standards derived from high-voltage test standards apply (Ref. [6] of Chap. 2).

4.1.1 Lightning Impulse Voltages

Lightning impulse voltages are used to test the dielectric strength of high-voltage apparatus against external transient voltages, which can be caused by lightning strokes in the supply network. A distinction is made here between *full lightning impulse voltages* and *chopped lightning impulse voltages*. A standard *aperiodic lightning impulse* rises to its *peak value* $\hat{u}$ within a few microseconds and then falls more slowly to zero (Fig. 4.1a). The rising part of the impulse voltage is referred to as the *front*, the maximum as the *peak*, and the falling part as the *tail*. The impulse shape can be approximated by superposing two exponential functions with different time constants (see Sect. 8.1).

Chopping of lightning impulse voltages in the test field is achieved by the disruptive discharge of a chopping gap. A distinction is made between chopping on the tail (Fig. 4.1b), chopping at the peak and chopping on the front (Fig. 4.1c). The

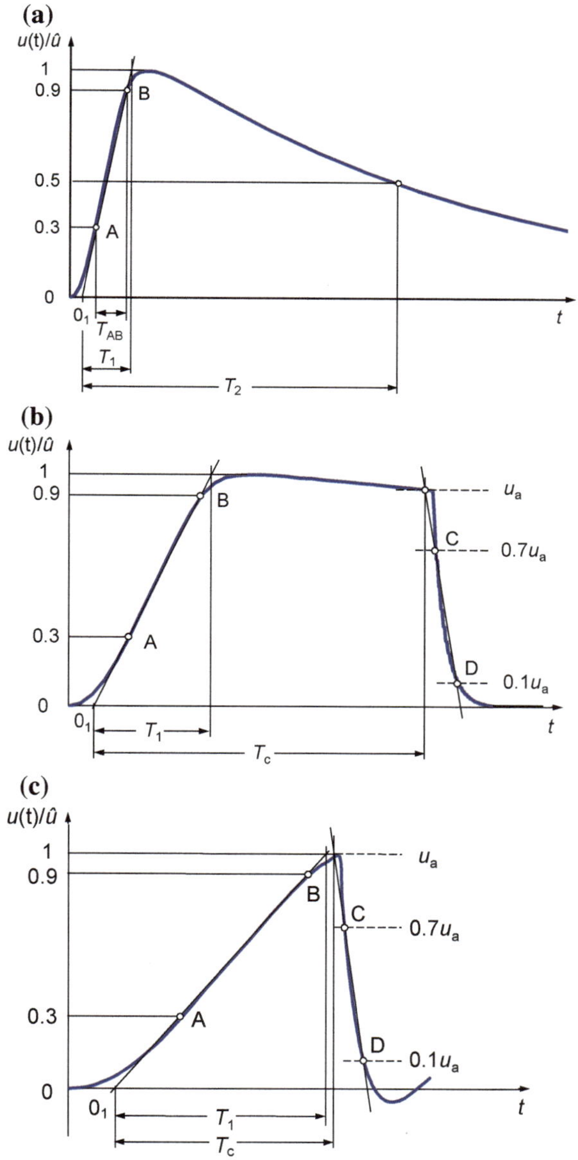

Fig. 4.1 Aperiodic lightning impulse voltages according to IEC 60060 (Ref. [1] of Chap. 2). **a** Full lightning impulse voltage. **b** Lightning impulse voltage chopped on the tail. **c** Lightning impulse voltage chopped on the front or wedge-shaped impulse voltage

standard chopped lightning impulse voltage has a *time to chopping* between 2 μs (chopping at the peak) and 5 μs (chopping on the tail) (Fig. 4.1b). The voltage collapse on the tail shall be significantly faster than the initial voltage rise. Due to the rapid voltage collapse, the test object is exposed to a very high stress.

Lightning impulse voltages chopped on the front have times to chopping between 2 and 0.5 μs (Fig. 4.1c). For short times to chopping, the time course of the front between $0.3\hat{u}$ and the voltage at the chopping instant is approximately linear. If the deviations from linearity lie within ±5% of the front time, a *wedge-shaped impulse voltage* is obtained with the *virtual steepness*:

$$\boxed{S = \frac{\hat{u}}{T_c}}. \tag{4.1}$$

In the test specifications, the different lightning impulse voltages are characterized by the following *time parameters*:

- *front time* T_1 and *time to half-value* T_2 for full lightning impulse voltages,
- *front time* T_1 and *time to chopping* T_c for standard chopped impulse voltages ($2\ \mu s \leq T_c \leq 5\ \mu s$),
- *time to chopping* T_c for lightning impulse voltages chopped on the front ($T_c < 2\ \mu s$),
- *front time* T_1 and *virtual steepness* S for wedge-shaped impulse voltages.

The starting point for determining the time parameters is the *virtual origin* O_1, which precedes the instant of point A of the impulse voltage by the time $0.3T_1$ (Fig. 4.1a–c). Graphically, O_1 is obtained as the intersection of the straight line through the points A and B with the zero line. The introduction and definition of the virtual origin O_1 is necessary because the origin O of the recorded impulse voltage is often not recognizable due to superimposed interference voltages and limited bandwidth of the measuring system.

The impulse parameters are defined for idealized smooth time courses, as shown in Fig. 4.1. For a smooth impulse, the peak value $\hat{u}$ is the *value of the test voltage* to which all requirements apply. In the test practice, however, the peak region of a lightning impulse voltage may be superimposed by an oscillation or overshoot. Depending on the frequency of the oscillation or the duration of the overshoot, the insulation of the test object is stressed differently. The value of the test voltage is then determined for a fictitious curve of the test voltage, which results from the application of a special evaluation method to the recorded data of the oscillating lightning impulse voltage (see Sect. 4.1.1.2).

The *front time* T_1 is the time between the virtual origin O_1 and the intersection of the front line through the points A and B with the peak line (Fig. 4.1):

$$\boxed{T_1 = \frac{1}{0.6} T_{AB}}, \tag{4.2}$$

where T_{AB} is the time between the instants when the impulse is $0.3\hat{u}$ (point A) and $0.9\hat{u}$ (point B) on the test voltage curve. Lightning impulse voltages are defined by the front time $T_1 < 20\ \mu s$, otherwise it is a *switching impulse voltage* (see Sect. 4.1.2). The *time to half-value* T_2 is the time between the virtual origin O_1 and the instant when the test voltage curve has decreased to half the test voltage value (Fig. 4.1a).

The *time to chopping* T_c is defined as the time interval between the virtual origin O_1 and the *instant of chopping*. For an impulse voltage chopped on the tail, the instant of chopping is given by the intersection of the straight line through the points C at $0.7u_a$ and D at $0.1u_a$ with the impulse voltage immediately before the collapse (see Fig. 4.1b). This intersection also defines the voltage u_a. In the case of an impulse voltage chopped on the front, the instant of chopping is defined by the intersection of the straight line through C and D with the horizontal peak line (see Fig. 4.1c). The determination of the virtual instant of chopping takes into account that the beginning of the chopping is not always clearly evident from the recorded impulse. This is due to the finite duration of the chopping and a limited bandwidth of the measuring system, resulting in a rounded shape of the recorded impulse voltage in the chopping region [1]. Furthermore, during the firing of the chopping spark gap, electromagnetic disturbances occur, which may be superimposed on the impulse at the time of chopping. The duration of the voltage collapse is defined as $T_{CD}/0.6$, where T_{CD} is the time interval between the points C and D.

A full lightning impulse voltage is characterized by its front time and time to half-value given in microseconds as abbreviation. The *standard 1.2/50 impulse voltage* accordingly has a front time $T_1 = 1.2$ µs and a time to half-value $T_2 = 50$ µs. In the generation of the impulse voltages, certain tolerances apply for the peak value and the time parameters.

4.1.1.1 Tolerances and Measurement Uncertainties in Tests

Deviations from the parameter values of the generated impulse voltages are permissible within the tolerances specified for each impulse type. For the parameters of the lightning impulse, the tolerances are (Ref. [1] of Chap. 2):

- $\pm 3\%$ for the value of the test voltage,
- $\pm 30\%$ for the front time T_1, and
- $\pm 20\%$ for the time to half-value T_2.

The reason for the large tolerances of the time parameters lies in the varying degrees of interaction of the test objects with the generator circuit, as a result of which the impulse shape and thus the parameters of the generated lightning impulse voltage are more or less strongly influenced. Therefore, the elements of the impulse generator with which the impulse shape is set need not be adapted each time the load of the test object is slightly changed. There are no tolerances defined for the time to chopping T_c.

The value of the test voltage and the time parameters shall be determined within the specified limit values of the *expanded measurement uncertainty* ($k = 2$). The measurement uncertainty consists of the uncertainty of the *approved measuring system* and, in most cases, of further uncertainty contributions. These contributions take into account the actual conditions during the voltage test, which often differ from those during the calibration of the measuring system, such as ambient temperature, deviations in the impulse shape or long-term drift (see Chap. 13). The permissible measurement uncertainties are (Ref. [2] of Chap. 2):

- 3% for the value of the test voltage of full and chopped lightning impulse voltages with times to chopping $T_c \geq 2$ µs,
- 5% for the value of the test voltage of lightning impulse voltages chopped on the front with times to chopping 0.5 µs $\leq T_c < 2$ µs, and
- 10% for the time parameters.

Note Uncertainties are given without polarity sign but are to be understood as positive and negative limits.

4.1.1.2 Lightning Impulse Voltage with Overshoot

Lightning impulse voltages actually generated in the test circuit may be superimposed by an *overshoot* or *oscillation* at the peak and on the front. Such overshoot or oscillation is caused by the inductances and capacitances of the impulse voltage generator as well as of the test and measurement circuit including the high-voltage leads. Further reasons are a non-optimal sequence in the firing of the generator spark gaps or reflection phenomena in the test circuit. In order to detect the oscillations correctly, the measuring system must have a sufficiently large bandwidth, at least 10 MHz for front oscillations and 5 MHz for peak oscillations. Oscillations in the test circuit must be clearly distinguished from natural oscillations of the voltage divider itself, which can be excited due to unfavorable construction. If in the test circuit oscillations occur with the natural frequency of such a voltage divider, these are reproduced at the divider output with amplified amplitude. The voltage divider is therefore not suitable for measuring oscillating test voltages. According to IEC 60060-1 (Ref. [1] of Chap. 2), lightning impulse voltages with oscillations in the peak region require a special evaluation method for determining the test voltage value, which is decisive for the stressing of the insulation of the test object (see Sect. 4.1.1.2.2).

Former Evaluation Procedure

It has long been known that the stress of the insulation of high-voltage apparatus depends on the frequency of superimposed peak oscillations or the duration of an overshoot. Accordingly, a lightning impulse voltage with a high-frequency oscillation in the peak region does not stress the insulation as much as an impulse voltage with a low-frequency peak oscillation, with both impulse voltages having the same extreme value. In previous test standards, the extreme value of a lightning impulse voltage with superimposed peak oscillation of the frequency $f < 500$ kHz was defined as the value of the test voltage. On the other hand, for $f \geq 500$ kHz, the peak value $\hat{u}$ of the mean curve ***2*** through the oscillating impulse ***1*** was the

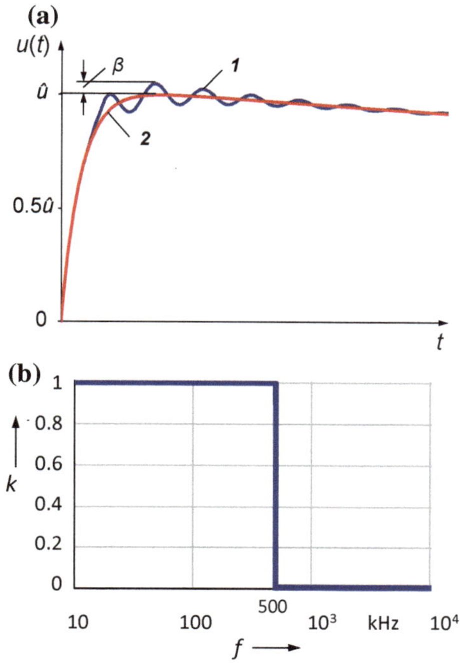

Fig. 4.2 Former evaluation of a lightning impulse voltage with superimposed peak oscillation of the frequency $f \geq 500$ kHz. **a** Oscillating lightning impulse ***1*** and mean curve ***2*** with the value $\hat{u}$ of the test voltage. **b** Definition of the k factor in dependence of the frequency of the peak oscillation

value of the test voltage (Fig. 4.2a). In other words, for determining the test voltage value, the amplitude of the oscillation was multiplied by a factor $k = 1$ for $f < 500$ kHz or $k = 0$ for $f \geq 500$ kHz (Fig. 4.2b). This evaluation procedure was valid for an overshoot $\beta \leq 5\%$.

This evaluation was also unsatisfactory from a metrological point of view, since the frequency of the peak oscillation in the critical range about 500 kHz cannot be determined exactly. It was therefore not possible to make a clear decision as to which evaluation method, i.e. $k = 1$ or $k = 0$, should be used. In addition, the course of the mean curve to be drawn through the peak oscillation was not precisely defined, but was dependent on the visual impression of the operator.

Revised Evaluation Procedure

Detailed investigations have been carried out in several high-voltage test laboratories with the aim of determining the dielectric strength of gaseous, liquid and solid insulations against lightning impulse voltages with superimposed peak oscillations [2]. In carrying out the extensive series of tests, the breakdown values of similar test samples for impulse voltages with and without peak oscillation were measured. Here, the amplitude, frequency and phase of the superimposed oscillation were varied within wide limits. Summarizing the test results, it can be stated that the

Fig. 4.3 Oscillating lightning impulse ***1*** and equivalent smooth impulse voltage ***2***, which both lead to the breakdown of comparative samples according to [2]

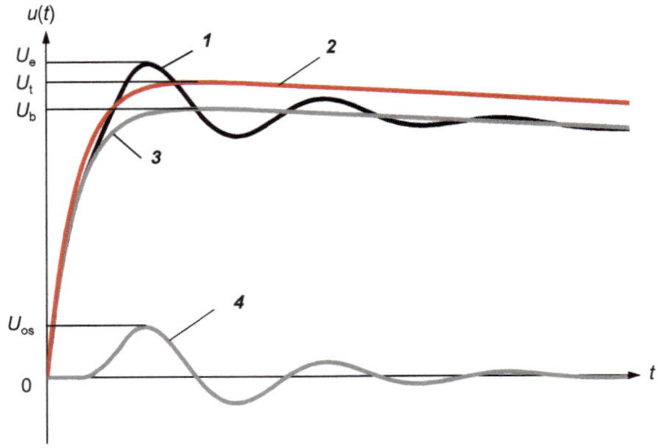

frequency-dependent stress of the insulation was confirmed in principle, but in a modified form.

The example in Fig. 4.3 shows schematically the voltage curves shortly before the breakdown of a sample. Curve ***1*** shows the damped oscillating impulse voltage obtained experimentally by superposition of the smooth impulse voltage ***3*** (*base curve*) and the oscillation ***4***. Curve ***2*** with the peak value U_t is the *equivalent smooth impulse voltage*, which also leads to the breakdown of a comparative sample. The equivalent smooth impulse voltage ***2*** thus represents the test voltage curve of the oscillating impulse voltage ***1*** and its peak value U_t is the test voltage value of both impulse voltages.

By introducing the frequency-dependent factor *k*(*f*), the results of the investigation can be expressed by the following relationship for the test voltage value U_t:

$$\boxed{U_t = U_b + k(f) \cdot U_{os} = U_b + k(f) \cdot (U_e - U_b)}, \tag{4.3}$$

where U_b is the peak value of the base curve ***3***, U_{os} the amplitude of the superimposed peak oscillation ***4***, and U_e the extreme value of the oscillating impulse voltage ***1***. The value of the frequency-dependent factor *k*(*f*) is therefore decisive for the extent to which the oscillation amplitude U_{os} enters into the determination of the test voltage value U_t.

For all investigated insulations, test samples and frequencies of the peak oscillation, the experimentally determined values of the *k*-factor are summarized in Fig. 4.4. Despite the scattering of the values for the different insulating materials, it can be clearly seen that the *k*-factor—and thus the influence of the peak oscillation on the breakdown—steadily decreases with the frequency above 100 kHz and becomes approximately zero for $f \geq 5$ MHz. In the semi-logarithmic representation in Fig. 4.4, a straight line can be drawn through the empirically values that characterizes the basic frequency dependence of the *k*-factor. Instead of the previously accepted abrupt change in the evaluation of peak oscillations at 500 kHz (see Fig. 4.2b), a gradual transition in the frequency range from 100 kHz to 5 MHz has proven to be correct.

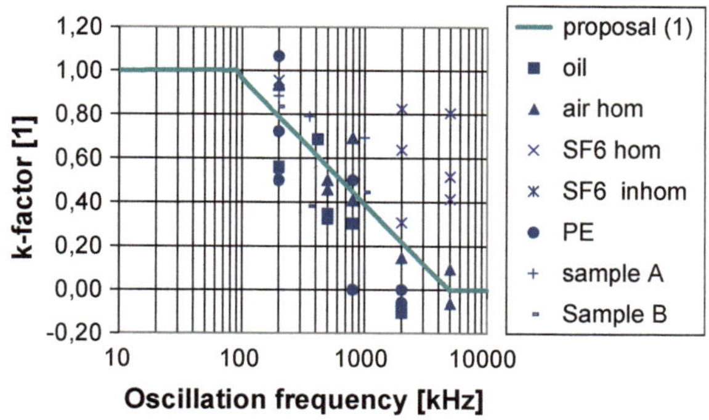

Fig. 4.4 *K*-factor values experimentally determined for solid, liquid and gaseous insulations versus the frequency of the peak oscillation superimposed on the lightning impulse voltage [2, Fig. 127]

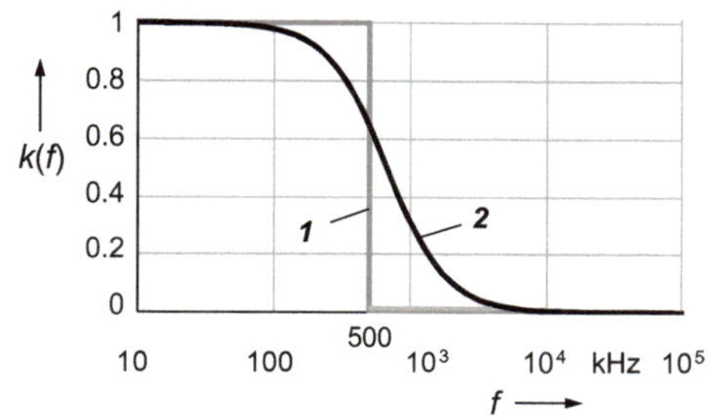

Fig. 4.5 Test voltage function *k(f)* versus the frequency *f* of peak oscillations superimposed on lightning impulse voltages. ***1*** *K*-factor of the old definition, ***2*** test voltage function *k* (*f*) according to Eq. 4.4

Further studies dealt with the adoption of the experimental results into the test standards [3–8]. A good approximation of the experimentally obtained *k*-factor values is—apart from the simple straight line in Fig. 4.4—given by the equation:

$$\boxed{k(f) = \frac{1}{1 + 2.2f^2}} \tag{4.4}$$

with the oscillation frequency *f* in megahertz (curve 2 in Fig. 4.5). This function *k* (*f*), called *test voltage function* with the advantage of continuity, replaces the previous, several decades valid *k*-factor according to curve 1 in Fig. 4.5.

The test voltage function *k(f)* is the basis for the standardized *filtering method* in IEC 60060 (Ref. [1] of Chap. 2) for calculating the fictitious test voltage, with which the high-voltage apparatus is actually stressed by a lightning impulse voltage with a superimposed peak oscillation. The results of the breakdown tests in [2] are transferred to the stress of a high-voltage apparatus during voltage tests, i.e. with voltages below the breakdown value. The method is briefly described with reference to the curves in Fig. 4.3. The starting point of the evaluation is the data record of the oscillating impulse voltage ***1*** measured during the high-voltage test of the apparatus. The corresponding base curve ***3*** is calculated as a double exponential curve using Eq. (8.8) and adapted to the curve ***1***. The difference between the two curves ***1*** and ***3*** results in the superimposed oscillation ***4*** with the frequency *f*. The data of the oscillation ***4*** are filtered with the test voltage function *k(f)* according to

Eq. (4.4). The result of the filtering is an oscillation whose amplitude is determined by the oscillation frequency. The filtered oscillation is then superimposed on the base curve ***3***, which finally yields the fictitious test voltage curve. Its maximum U_t is the sought value of the test voltage. The test voltage curve computed with the filtering method thus has a superimposed oscillation, which stands in contrast to the experimentally determined equivalent smooth impulse voltage ***2*** without oscillation (see Fig. 4.3).

The time parameters are also determined from the fictitious oscillating test voltage curve obtained by the filtering method. In the case of an oscillating impulse voltage chopped on the tail, the filtering is performed for the corresponding full oscillating impulse voltage, but with a reduced voltage level. The result is then transferred to the chopped impulse curve in the corresponding voltage and time formats. A lightning impulse voltage chopped on the front is basically defined as the test voltage.

Note The responsible IEC working group has discussed quite intensively whether the smooth or the oscillating impulse shape of the fictitious test voltage should be preferred. Finally, the decision was made in favor of the described filtering method. i.e. in favor of the oscillating test voltage curve. In any case, the value U_t of the test voltage remains the same, since it is identical for the smooth and the oscillating test voltage. With regard to the time parameters, however, different values are to be expected for the smooth and the oscillating curve.

An alternative to the comprehensive filtering method is the so-called *manual evaluation method*, which can also be applied to impulse shapes in graphic formats (Ref. [1] of Chap. 2). Starting from the recorded oscillating impulse voltage ***1***, the mean curve is drawn through this curve, which is the base curve ***3*** with the peak value U_b (see Fig. 4.3). The difference between the two curves ***1*** and ***3*** gives the superimposed oscillation ***4*** with the amplitude U_{os}. The frequency f of this oscillation can be evaluated from the duration of the first half-oscillation of curve ***4***. With the knowledge of f, the corresponding k-factor can be calculated according to Eq. (4.4) and, finally, the test voltage value U_t according to Eq. (4.3). The complete test voltage curve results from the base curve ***3*** multiplied by the k-factor. It is thus a smooth test voltage curve comparable to the curve ***2*** in Fig. 4.3. The time parameters are also determined from the smooth test voltage curve. Since the manual evaluation procedure depends on the subjective perception of the responsible person, an additional uncertainty contribution could occur (Ref. [1] of Chap. 2). On the other hand, one can image that the manual evaluation can easily be replaced by computer-aided data evaluation with appropriate software.

Both evaluation methods also eliminate the digitalization noise of the recorder (see Sect. 7.2.7) and the front oscillation of lightning impulse voltages, but with the filtering method only for oscillation frequencies above 10 MHz, because then $k(f) = 0$. The experimental determination of the frequency-dependent k-factors in [2] as well as the approximation by the test voltage function $k(f)$ according to Eq. (4.4) are associated with uncertainties. In order to limit the resulting uncertainty contribution (see Appendix B.2.2) in the determination of the test voltage value and the time parameters, the application of both evaluation methods is limited to a relative overshoot $\beta = (U_e - U_b)/U_e$ of 10% of the extreme value.

Oscillation on the Front

Oscillations on the front of a lightning impulse voltage can influence the determination of the virtual origin O_1 and thus also the time parameters. With the two above-mentioned evaluation methods for peak oscillations, it is also possible to eliminate front oscillations in whole or in part [6]. There are further methods to eliminate front oscillations, for example digital filtering of the recorded data, limitation of the frequency spectrum, or step-wise substitution of the impulse by exponentials, parables or straight lines [9–11]. As a result, a mean curve through the front oscillation is obtained, comparable to the earlier graphical evaluation, and the points at $0.3\hat{u}$ and $0.9\hat{u}$ are used for the determination of the virtual origin O1 and the front time T_1 (Fig. 4.6). Front oscillations occur mainly in the initial part of a lightning impulse voltage and then affect only the determination of the point A at $0.3\hat{u}$. If the evaluation of the front at $0.3\hat{u}$ is ambiguous as shown in the example in Fig. 4.6, it is suggested to take the middle of the three intersection points as a simple approximation solution. The calculation of the complete mean curve is then superfluous [12].

Investigations on calculated impulses with and without superimposed front oscillation show that the impulse shape is more or less strongly distorted by the application of smoothing methods. The front time of the smoothed impulse is therefore not identical to that of the original impulse without front oscillation. Decisive for the quality of the smoothing is the frequency difference in the spectra of the front oscillation and the impulse voltage. High-frequency oscillations can be easily eliminated by filtering without a significant distortion of the impulse. On the other hand, the filtering of a low-frequency oscillation whose frequency is in the characteristic frequency range of the impulse voltage causes a strong distortion of the impulse. In the case of an impulse voltage chopped on the front, the superimposed front oscillation may extend to the peak. In the peak region of front-chopped impulse voltages, smoothing methods should be used with caution to avoid peak distortion.

4.1.2 Switching Impulse Voltages

Tests with *switching impulse voltages* simulate the stress of high-voltage apparatus due to internal overvoltages as a result of switching operations in the supply

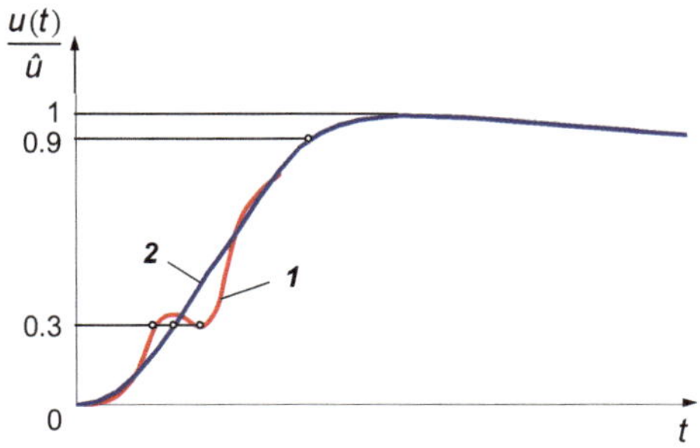

Fig. 4.6 Lightning impulse voltage with front oscillation. *1* Original record with three intersection points at $0.3\hat{u}$, **2** mean curve through the front oscillation

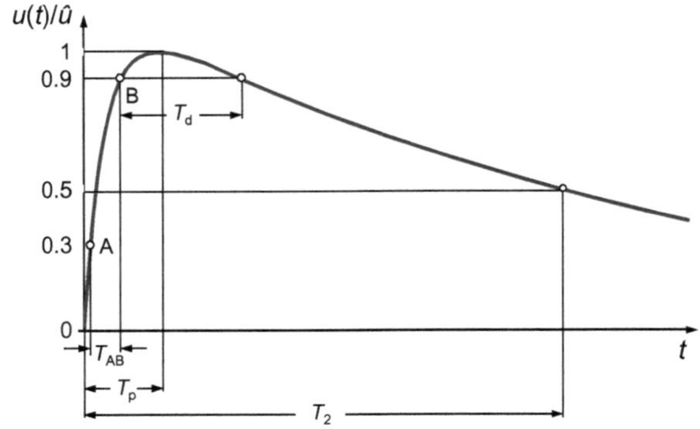

Fig. 4.7 Switching impulse voltage and its impulse parameters (aperiodic form)

network. The idealized form of an *aperiodic switching impulse voltage* as that of an aperiodic lightning impulse voltage is represented by the superposition of two exponential functions, but their time constants are much greater (see Sect. 8.1.3). The switching impulse voltage is characterized by the peak (or maximum) value $\hat{u}$ as the *test voltage value* and two time parameters. These are related to the *true origin* O, which is defined as the time at which the recorded curve increases monotonically (Fig. 4.7). Similar to lightning impulse voltages, the beginning of the switching impulse voltage is distorted, but due to the larger time scale, the distortion is barely noticeable and therefor generally negligible. The *time to peak* T_p is defined as the time interval from the true origin O to the time of maximum value $\hat{u}$, the *time to half-value* T_2 as the time interval between O and the instant when the impulse has first decreased to $0.5\hat{u}$.

In addition to T_p and T_2 or as a substitute, further time parameters are defined. The *time above 90%* is the time T_d, during which the impulse voltage is greater than $0.9\hat{u}$. In special circuits, switching impulse voltages can swing below zero in the tail region. Therefore, it may be necessary to specify the *time to zero* T_z between the virtual origin O and the instant of the first zero crossing in the tail. In addition to the time to peak T_p, the front time T_1 of switching impulse voltages is defined according to Eq. (4.2). It serves as a criterion for distinguishing between lightning and switching impulse voltages. Switching impulse voltages are defined to have a front time of 20 μs or longer.

Switching impulse voltages are characterized by the numerical values of the time parameters T_p and T_2 in microseconds. The standard 250/2500 impulse has a time to peak T_p = 250 μs (tolerance: ±20%) and a time to half-value T_2 = 2500 μs (tolerance: ±60%). The large tolerances make it possible to test different apparatus without having to adapt the elements of the impulse voltage generator to the changed load. Corresponding to the lightning impulse voltages, the permissible uncertainties of measurement (k = 2) are:

- 3% for the value of the test voltage, and
- 10% for the time parameters.

Each measurement uncertainty consists of the uncertainty of the approved measuring system determined by calibration and possibly further uncertainty contributions during the impulse test (see Sect. 13.1.4).

4.1.2.1 Evaluation of the Time to Peak

According to its definition, the *time to peak* T_p of switching impulse voltages appears to be an easily measurable quantity. In automated data evaluation, however, the long duration of the peak region is superimposed by small digitization errors of the recorder or by disturbances, which can lead to incorrect values of the time to peak. The uncertainty of measurement given in the test specifications for T_p is then often not complied with. Since the time to peak is to be retained as an important time parameter in the test practice, it is not determined directly, but from the time interval T_{AB} between $0.3\hat{u}$ and $0.9\hat{u}$, multiplied by the factor K:

$$\boxed{T_p = K \cdot T_{AB}}. \tag{4.5}$$

For the 250/2500 impulse with double exponential impulse shape according to Eq. (8.8), the calculation yields $T_{AB} = 99.1\ \mu s$ and thus $K = 2.523$. For other values of T_p and T_2 within the permissible tolerances of the 250/2500 impulse, K can be calculated approximately according to the following equation (Ref. [1] of Chap. 2):

$$K = 2.42 - 3.08 \times 10^{-3} T_{AB} + 1.51 \times 10^{-4} T_2, \tag{4.6}$$

wherein the measured values of T_{AB} and T_2 are to be inserted in microseconds. The error in the calculation of T_p with K according to Eq. (4.6) is within $\pm 1.5\%$, which is usually negligible in tests. For other switching impulses, Eq. (4.6) is not valid. The factor $K = T_p/T_{AB}$ is then obtained from the impulse shape calculated with Eq. (8.8) and having the same value T_{AB} as the measured impulse. For on-site testing with oscillating switching impulse voltages, a uniform value $K = 2.4$ applies (see Sect. 4.1.3).

4.1.3 On-site Testing and Oscillating Impulse Voltages

Impulse voltage tests on electrical power apparatus are not only carried out in the high-voltage laboratory, but more and more on site during installation (Ref. [3, 7] of Chap. 2). *On-site testing* makes it possible to check the proper assembly, the error-free initial commissioning and recommissioning after successful repair or the long-term behavior. For these on-site tests, difficult test conditions are often prevalent. In general, other generation and measuring systems than those used in a test laboratory are more suitable. In addition to the aperiodic lightning and

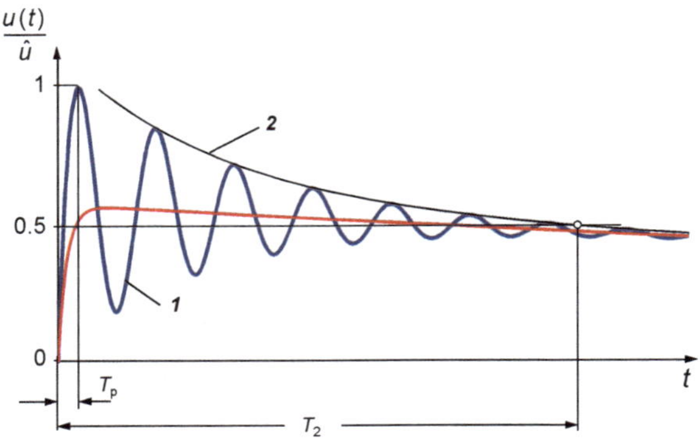

Fig. 4.8 Example of an oscillating switching impulse voltage ***1*** for on-site testing with its upper envelope ***2***, which is decisive for the determination of the time to half-value T_2

switching impulse voltages shown in Figs. 4.1a and 4.7, oscillating impulse voltages can be used. As an example, Fig. 4.8 shows an *oscillating switching impulse voltage* (curve ***1***) and its upper envelope (curve ***2***). As a result of the oscillation, almost a doubling of the peak value of the aperiodic impulse voltage is achieved. This is advantageous for on-site testing since the required transportable test generator is correspondingly smaller.

The definitions for the origin and the front time of oscillating lightning or switching impulse voltages are the same as for the corresponding aperiodic impulse voltages. For example, O_1 is the virtual origin of oscillating lightning impulse voltages, and O is the true origin of oscillating switching impulse voltages. The time to half-value T_2 is defined as the time interval between O_1 (or O) and the time when the upper envelope of the oscillating impulse voltage has fallen to 50% of the peak value (Fig. 8.8). For on-site testing, the time to peak is $T_p = 2.4T_{AB}$ for all standard switching impulses (see Fig. 4.7).

Owing to the more difficult test conditions during on-site testing, larger tolerances in the generation and partly greater uncertainties in the measurement of aperiodic or oscillating impulse voltages are permitted than in stationary tests in the laboratory. The tolerance for the value of the test voltage is ±5%. For lightning impulse voltages, the permissible front times are between 0.8 and 20 μs, the times to half-value between 40 and 100 μs and the oscillation frequencies between 15 and 400 kHz. Switching impulse voltages are defined by times to peak between 20 and 400 μs, times to half-value between 1000 and 4000 μs and oscillation frequencies between 1 and 15 kHz. The maximum permissible expanded measurement uncertainties are 5% for the value of the test voltage, 10% for the time parameters and 10% for the oscillation frequency.

4.1.4 Steep-Front Impulse Voltages

Steep-front impulse voltages occur, for example, during disconnecting switch operations in gas insulated equipment, which are also referred to as *very fast*

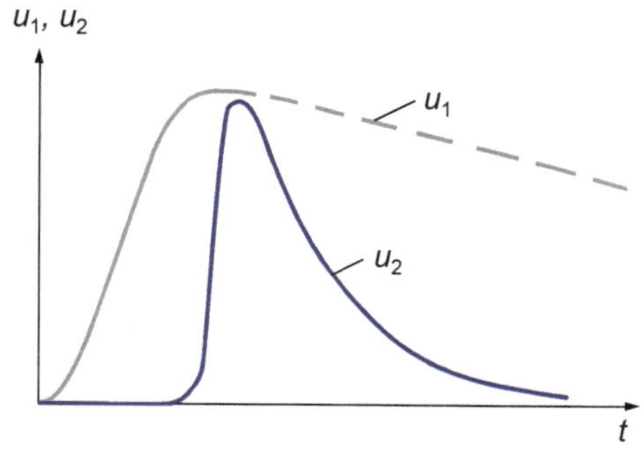

Fig. 4.9 Output voltage u_1 of a lightning impulse voltage generator and steep-front impulse voltage u_2 of the secondary peaking circuit connected to the generator output

transient overvoltages. These transient voltages with rise times down to 5 ns are dangerous to the equipment connected to the substation. The standardization of steep-front impulse voltages used in tests is not uniform, but is left to the Technical Committees responsible for the individual power apparatus. Steep-front impulse voltages can be generated with a conventional impulse voltage generator in conjunction with a spark gap or an exploding wire and a *secondary peaking circuit* (see Sect. 4.2.4). With an appropriate design of the circuit, steep-front impulses with slopes of up to 100 kV/ns are achieved, which corresponds, for example, to a rise time of 5 ns at 500 kV.

Figure 4.9 shows schematically the output voltage u_1 of a lightning impulse voltage generator and the steep-front impulse voltage u_2 at the output of the secondary peaking circuit connected to the impulse generator (Ref. [2] of Chap. 1). With optimal configuration of the elements of the impulse voltage generator, the switching element and the secondary peaking circuit, the steep front of u_2 begins at the instant of the peak u_1 when the switching element is closed. The time course in the tail depends on the configuration of the secondary peaking circuit and the test object, including the voltage divider. An approximately rectangular impulse voltage with a steep front is generated by means of a spark gap, which is located at the output of the secondary circuit and fired shortly after the peak of u_2. Due to the inductances of the elements in the test circuit and by reflection phenomena, high-frequency oscillations can superimpose the steep-front impulse voltage.

4.2 Generation of High Impulse Voltages

The basic principle of the predominantly used generator circuits for generating high impulse voltages is that a storage capacitor is charged relatively slowly and then quickly discharged to an RC network and the test object. The shape of the generated lightning or switching impulse voltage is determined by both the RC network and the connected test object. In conjunction with a chopping spark gap, further impulse shapes, such as chopped lightning impulse voltages or steep-front impulse voltages, can be generated. For on-site testing, voltage generators are preferably used which can generate oscillating impulse voltages. The generators are designed with low-inductance elements and for high impulse stress. In addition to the generator circuits with capacitive storage, there are other ways to generate switching impulse voltages, for example by using a test transformer to increase the impulse amplitude (Refs. [2, 4] of Chap. 1).

4.2.1 Generators for Lightning and Switching Impulse Voltages

Two basic circuits are mainly used to generate lightning and switching impulse voltages (Ref. [2] of Chap. 1). In both circuits, the impulse capacitor C_s is charged relatively slowly by a rectified AC current via the resistor R_L (Fig. 4.10). When the DC voltage U_0 at C_s reaches the firing voltage of the sphere gap FS, C_s discharges in a short time via the discharge circuit formed by the load capacitor C_b, the damping resistor R_d and the discharge resistor R_e. As a result, the impulse voltage $u(t)$ is generated at C_b with the peak value $\hat{u}$ depending on the spacing between the two spheres. When the capacitors C_s and C_b are discharged, the spark extinguishes and C_s can be recharged via R_L. The impulse repetition rate depends on the charging voltage or the charging current amplitude. In small generators with impulse voltages of up to a few 10 kV, mechanical or electronic switches are preferably used instead of the switching spark gap.

The two circuits in Fig. 4.10 are distinguished by the position of the discharge resistor R_e. In the circuit A, R_e is arranged behind R_d, and in the circuit B, R_e is before R_d. While R_d is mainly decisive for the charging of C_b and thus for the front time T_1 of the impulse voltage, R_e influences discharge of C_b and thus the time to half-value T_2. From the basic circuits, the equations for the double exponential impulse can be derived, with which the impulse shape of the generated lightning or switching impulse voltage is calculated (see Sect. 8.1).

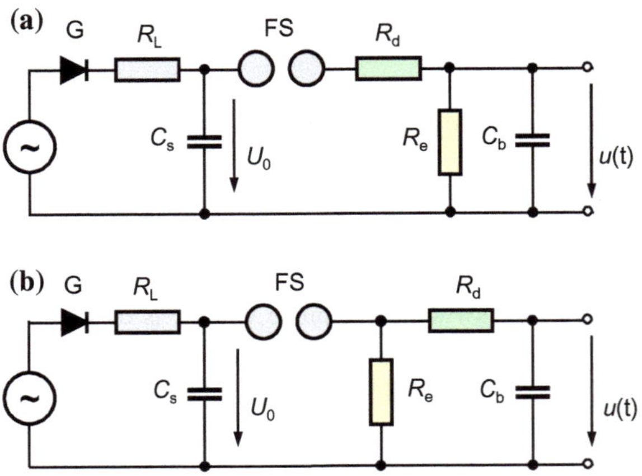

Fig. 4.10 Single-stage basic circuits for the generation of impulse voltages. **a** Basic circuit A. **b** Basic circuit B

The unavoidable stray capacitances and inductances of the circuit elements and high-voltage leads are not shown in Fig. 4.10. They can be taken into account in the extended equivalent circuit diagram (see Sect. 4.2.1.3). The impedance of the test object also affects the circuit and influences more or less the shape of the generated impulse voltage.

The maximum energy stored in the impulse capacitor C_s is:

$$\boxed{W = \frac{1}{2} C_s U_0^2}, \tag{4.7}$$

which is the most significant parameter of impulse voltage generators. The *efficiency* (or: *utilization) factor* η is defined as the quotient of the peak value $\hat{u}$ of the generated impulse voltage and the charging voltage U_0:

$$\boxed{\eta = \frac{\hat{u}}{U_0} = \mathrm{f}\left(\frac{C_s}{C_b}\right)}. \tag{4.8}$$

In order to achieve a high efficiency factor and hence a large peak value, $C_s \gg C_b$. For example, in the circuit B according to Fig. 4.10b with $C_s = 5C_b$, $\eta \approx 0.8$ for a 1.2/50 lightning impulse voltage. The efficiency factor of circuit B is generally greater than that of circuit A. Furthermore, η is greater for lightning impulse voltages than for switching impulse voltages. Information on the efficiency factor of an impulse voltage generator is supplied by the manufacturer in the form of diagrams.

4.2.1.1 Multiplier Circuit

The single-stage basic circuits according to Fig. 4.10 are realized for impulse voltages of up to about 300 kV. The multiplier circuit patented by E. Marx makes it possible to build up relatively compact generators for lightning and switching impulse voltages—also known as *Marx generators*—with peak values of up to several megavolts (Ref. [1] in Chap. 1). Figure 4.11 shows schematically a multi-stage impulse voltage generator on the basis of circuit B, which consists of $n = 4$ equal stages. The principle of the multiplier circuit is as follows. The parallel-connected impulse capacitors C_s' are first slowly charged to the DC voltage U_0' and then suddenly connected in series by the firing of the sphere gaps, so that the individual stage voltages add up to the total charging voltage $n\,U_0'$. As a result, the external load capacitor C_b is charged via the series connection of all damping resistors R_d' and then discharged via all R_e' and R_d'. Compared to the basic circuit in Fig. 4.10b, $R_e = n\,R_e'$, $R_d = n\,R_d'$, $C_s = C_s'/n$ and $U_0 = n\,U_0'$.

During operation, the test object is connected in parallel with the load capacitor C_b. Due to the modular design of the impulse generator, individual stages can be connected in parallel for adaptation to increased test loads. The influence of

Fig. 4.11 Multiplier circuit of basic circuit B for generating impulses in the megavolt range

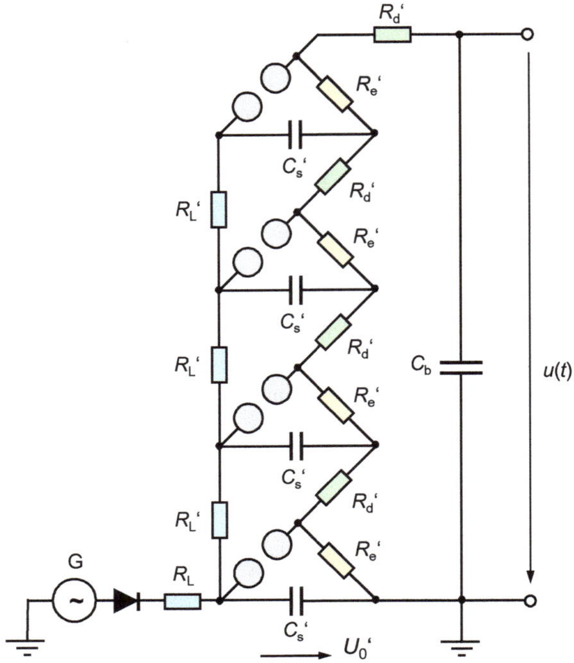

different test object capacitances on the shape of the impulse voltage is kept small when the generator is operated with the largest possible C_b.

Other multiplier circuits with a modification or combination of the two basic circuits A and B in Fig. 4.10 are also used. Usually, impulse voltage generators are equipped with interchangeable sets of resistors and capacitors for generating both lightning and switching impulse voltages. During the charging process, external discharges may occur which can be suppressed by various measures if necessary. Figure 4.12 shows two different versions of impulse voltage generators with a total charging voltage of about 3 MV. The individual stages of the generators are clearly recognizable. While impulse voltage generators usually consist of rectangular stages with a metallic frame (Fig. 4.12a), the generator in Fig. 4.12b has triangular stages with insulating side plates [13]. The efforts to achieve ever higher transmission voltages, especially in the Asian region, have led to the development of impulse generators with much higher rated voltages. In [14], a 24-stage impulse voltage generator with a total charging voltage of 7.2 MV is presented for use in the open air test field.

4.2.1.2 Operation of the Impulse Voltage Generator

Important precondition for the proper functioning of the multiplier circuit is the reliable and time-delayed firing of the sphere gaps arranged one above the other.

Fig. 4.12 Two types of impulse voltage generators. **a** Total charging voltage 3.2 MV, 320 kJ (HIGHVOLT Prüftechnik Dresden GmbH). **b** Total charging voltage 3 MV, 300 kJ (Haefely Test AG)

For this purpose, the lowest sphere gap has a slightly reduced spacing so that it fires slightly earlier than the other sphere gaps. This can also be achieved by a triggered auxiliary discharge. When the lowest sphere gap is fired, almost twice the voltage is momentarily applied to the sphere gap above it, resulting in a rapid firing. The other sphere gaps are fired accordingly. For firing the next sphere gap, it is also important that a sufficiently large number of initial electrons be generated by photoemission, resulting from the preceding firing of a sphere gap. A non-optimal firing of the individual generator stages causes a damped oscillation in the front of the impulse voltage with a frequency that may be above 1 MHz.

With increasing number of stages and with small charging voltages, i.e. less than 20% of the total charging voltage, reliable firing of one or more sphere gaps in conventional impulse voltage generators is not always ensured. In specially designed impulse generators, the controlled firing of all sphere gaps is achieved by electrical or optical trigger signals, the latter by means of potential-free laser sources. Impulse voltage generators with triggered spark gaps are required for composite AC and impulse voltage tests, whereby the impulse voltage is triggered at a defined phase of the AC voltage. The reproducibility of the impulse voltage also depends quite considerably on the stability of the charging voltage [15]. When the spark gaps fire, electromagnetic fields are generated which can affect the measuring instruments and falsify the measurement result. Interference can be prevented only to a limited extent by shielding the measuring instruments (see Sects. 4.3.1.7 and 5.3.1.1).

The polarity of the generated impulse voltage is changed by simply reversing the rectifier G in Fig. 4.11. After a voltage impulse has been generated or when the charging process has been interrupted, dangerously high residual charges can remain on the capacitors. It is then not sufficient to ground the capacitors of the lowermost stages for a short time, as they subsequently recharge. In modern impulse voltage generators, the residual charges of all capacitors are automatically short-circuited and grounded via a continuously rotating metal strip after switching off. The repetition rate of a generator at maximum charging voltage is limited to one or two impulses per minute in order not to overload the generator elements thermally.

The load capacitor C_b in the generator circuit for lightning impulse voltages according to Fig. 4.11—as well as the discharge resistor R_e in circuit A—is occasionally provided with a low-voltage unit and then used as a capacitive or resistive voltage divider. This allows the generator output voltage to be measured, but not the actual impulse voltage at the test object. For the correct measurement of lightning impulse voltages, the prescribed sequence is: generator—test object—measuring divider (see Sect. 4.3.1.1). Also, the dynamic behavior of such a voltage divider formed with C_b would generally not be suitable for measuring lightning impulse voltages, since the capacitors required in the microfarad range have large inductances.

Using a simple equivalent circuit diagram for the impulse generator, the equation for the double exponential impulse is developed in Sect. 8.1. However, the impulse voltages actually generated differ more or less from the calculated impulse. This is because the test object and the generator elements, including stray capacitances and inductances, affect the shape of the impulse voltage. The generated impulse voltage therefore often has an undesirable overshoot in the region of the peak (see Sect. 4.2.1.3). The influence can theoretically be studied with various methods and software for the calculation of linear circuits. The aim is to optimize the generator circuit and to keep the parameters of the impulse voltage within the permissible tolerances [16–20]. A very detailed equivalent circuit diagram is available on the internet, which allows the calculation of the impulse shape after entering the values for the generator elements, test objects, stray capacitances, inductances, etc. [21]. The other way, namely the calculation of the corresponding circuit elements of the impulse voltage generator for given values of the time parameters T_1 and T_2, is described in [22].

High-voltage impulses comparable to switching impulse voltages can also be generated with testing transformers, which are excited with a rapid voltage change [23, 24]. In one method, an initially charged capacitor is suddenly discharged into an RC circuit and the low-voltage winding of a transformer. In the other method, the rectified mains AC voltage is switched to the low voltage winding at the time of its maximum. The shape of the impulse voltage occurring on the high-voltage side of the transformer depends on the low-voltage circuit and the load. It is different from the standardized switching impulse voltage; in particular the time to peak and the time to half-value are longer.

4.2.1.3 Lightning Impulse Voltage with Overshoot

The large dimensions of an impulse voltage generator include unavoidable inductances L_S and stray capacitances C_e. Together with the capacitance C_P of the connected test object, they influence the shape of the generated impulse voltages, in particular lightning impulse voltages. Figure 4.13 shows a simple equivalent circuit diagram of the generator circuit B with the stray inductance L_S and the resulting capacitance $C_b^* = C_b + C_e + C_P$. With increasing size of the high-voltage circuit, consisting of the generator, test object, high-voltage leads and voltage divider, L_S is also larger. Typical values for L_S are between 20 and 150 μH. Because of this inductance, the generated lightning impulse voltage can be superimposed by a damped oscillation (see Sect. 4.1.1.2). In particular in the generation of impulses with short front times, a pronounced overshoot in the peak is to be expected, since in this case the damping resistance R_d is reduced and the inductances in the test circuit are more effective. A large overshoot in the order of $\beta = 20\%$ or more occurs in the test of transformers.

In model calculations, the influence of R_d, L_S and C_b^* on the overshoot β and the front time T_1 is studied in detail and graphically illustrated [25]. In principle, β increases with increasing inductance L_S, which can be at least partially compensated by larger values of R_d and C_b^*. On the other hand, the front time T_1 increases with increasing R_d and decreases with increasing C_b^*. There are therefore critical values R_d, L_S and C_b^*, for which the overshoot meets the former IEC limit $\beta \leq 5\%$ (see Sect. 4.1.1.2.1), but T_1 is above the IEC tolerance limit of 1.56 μs. A similar study for the generator circuit A with the inductance L_S can be found in [26].

According to Sect. 4.1.1.2, an overshoot β or oscillation superimposed on the peak of the lightning impulse voltage can lead to an increased stress on the test object. Although this is taken into account in the data evaluation of the recorded impulse voltage with the frequency-dependent test voltage function $k(f)$ (see Sect. 4.1.1.2.3), it is generally better to prevent or reduce the overshoot from the outset by suitable circuit measures. A simple measure for reducing the oscillation is to increase the attenuation resistance R_d. However, the front time T_1 of the generated impulse voltage is thereby longer and may then lie outside the tolerance. In addition, the measurement of the overshoot or oscillation places higher demands on

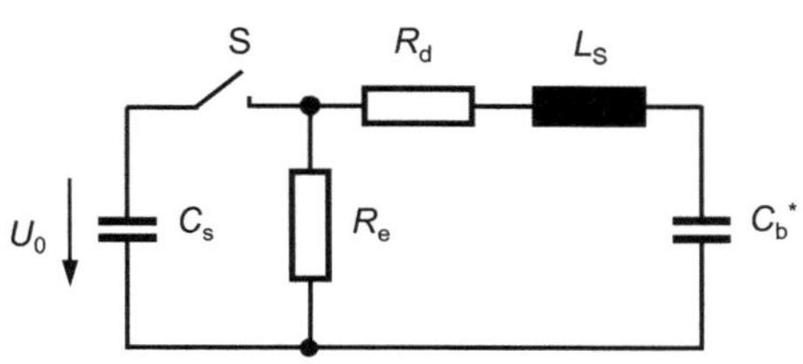

Fig. 4.13 Simple equivalent circuit diagram of an impulse voltage generator with stray inductance L_S and capacitance $C_b^* = C_b + C_e + C_P$

Fig. 4.14 Equivalent circuit diagram of the impulse voltage generator in circuit A with serial compensation circuit R_C, L_C and C_C to reduce the overshoot of lightning impulse voltages

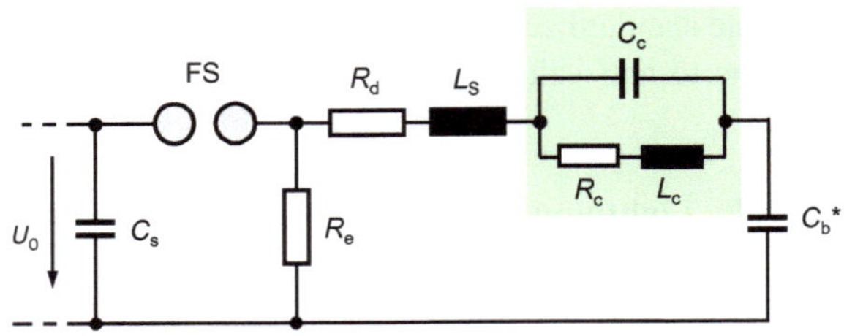

the dynamic behavior of the measuring system. If the bandwidth of the measuring system is insufficient, the overshoot is not recorded correctly, so that the maximum value of the test voltage is measured too low.

The *serial compensation circuit* with R_C, L_C and C_C according to Fig. 4.14, which was first proposed in [27], can be used successfully to reduce a strong overshoot. A reduction of the overshoot is also achieved with the parallel compensation circuit, in which R_C, L_C and C_C are connected in parallel to the load capacitor C_b [28, 29]. Model calculations and measurements show that with both compensation circuits even large overshoots can be reduced to the previously admissible values $\beta \leq 5\%$. In this case, however, an influence on the front time T_1 is to be expected so that it can then lie outside the IEC tolerance. In the interpretation of the results, a number of advantages are shown for the series circuit in [28], whereas according to [29] both compensation circuits are equivalent. In general, however, the application of compensation circuits for various test objects in the UHV range is rather time-consuming and costly, so that other ways for reducing overshoots are sought.

The trend to significantly higher transmission voltages, especially in the Asian region, is not yet taken into account in IEC 60060 (2010). For this reason, the experts have begun to discuss the possibility of introducing extended tolerance limits for the overshoot β and the front time T_1 of lightning impulse voltages in future regulations. According to extensive experimental investigations in [30], the 50% breakdown voltage of all tested high-voltage insulations is only slightly influenced by the front time in the range 1.2 μs $\leq T_1 \leq$ 4.8 μs. For the UHV range, therefore, a limit of more than the currently valid value of 1.56 μs could be allowed for the front time T_1. The advantage is that a large overshoot β on the peak of a lightning impulse voltage can then be reduced to the permissible value simply by increasing the damping resistor R_d, i.e. an expensive compensation circuit would be superfluous in the case of a permissible longer front time.

4.2.1.4 Low-Inductance Load

When testing low-inductance loads with L_b < 40 mH, for example inductors and low-voltage windings of power transformers, the time to half-value of lightning impulse voltages may be significantly shorter than the permissible lower tolerance limit of 40 μs. In addition, the impulse shows an oscillation in the tail, so that the voltage can swing below zero. In the simple equivalent circuit diagram of the impulse generator, the inductance L_b of the test object can be placed parallel to C_b* (see Fig. 4.13). For different variants of this circuit with and without the additional inductance L_S, the solutions of the differential equations of higher order are given in [31]. Diagrams are provided for the dimensioning of generator circuits to generate lightning and switching impulse voltages within the allowable tolerances.

The shortening of the time to half-value of lightning impulse voltages can be largely compensated by connecting an inductance L_p parallel to the damping resistor R_d in Fig. 4.13 [32, 33]. In order to compensate for the influence of a very small load inductance $L_b < 4$ mH, a resistor $R_p = R_d L_b / L_p$ parallel to the load capacitor C_b* is required in addition to the inductance L_p [33]. A detailed theoretical and experimental treatment of both circuit variants when testing the low-voltage winding of transformers can be found in [34]. A further contribution in this field deals with the reduction of oscillations on the peak of lightning impulse voltages by means of the compensation circuit shown in Fig. 4.14 [35].

4.2.2 Generation of Oscillating Impulse Voltages

Oscillating lightning and switching impulse voltages according to Fig. 4.8 are usually generated with impulse voltage generators, in which the damping resistor R_d in the basic circuit A of Fig. 4.10b is replaced by an inductance. The equivalent circuit diagram is then similar to that shown in Fig. 4.14, where $R_d \approx 0$ and L_S represents this inductance in addition to the other inductances. The oscillating impulse voltage is triggered by the firing of the spark gap FS. The natural frequency f_0 of the generated oscillation is calculated as (Ref. [5] of Chap. 1):

$$f_0 = \frac{1}{2\pi\sqrt{L_S \dfrac{C_s C_b^*}{C_s + C_b^*}}}. \tag{4.9}$$

The attenuation of the oscillating impulse voltage is caused by the ohmic losses in the generator circuit and possibly by an additional resistor in series with L_S. Due to the superimposed oscillation, the maximum value of an oscillating impulse voltage is considerably greater than the peak value of the corresponding aperiodic impulse voltage at the same charging voltage. For an oscillating lightning impulse voltage, the amplification factor is 1.7–1.8 and that for an oscillating switching impulse voltage 1.3–1.4. Oscillating impulses are preferably used for on-site testing, as mobile generators are smaller than the conventional generators for aperiodic impulse voltages. Requirements for oscillating impulse voltages are specified in IEC 60060-3 (Ref. [3] of Chap. 2). Oscillating switching impulse voltages can also be generated with test transformers [36].

4.2.3 Generation of Chopped Impulse Voltages

Chopped impulse voltages are generated by means of a sphere gap, which is in parallel to the load capacitor C_b of the lightning impulse voltage generator. A triggered sphere gap is necessary for reproducible chopping on the tail of impulse voltages. Impulse voltages chopped on the front can be generated without triggering, if the sphere gap is irradiated with UVC light. The UVC irradiation of the gap produces a sufficiently large number of initial electrons to fire the sphere gap. This improves the reproducibility of chopping, which should be sufficient for most applications, including the calibration of measuring systems within the required uncertainty. In order to achieve different slopes of the impulse voltage with the same peak value, the spacing of the sphere gap must be adjusted. The atmospheric ambient conditions also influence the firing behavior of the sphere gap and thus the peak value (see Sect. 4.3.6)

For the generation of chopped impulse voltages of more than 600 kV, the use of a *multiple spark gap* is recommended in [37]. The device consists of n sphere gaps arranged one above the other. They obtain the same potential difference by means of a parallel-connected n-stage resistive or capacitive voltage divider. Firing of the spark gaps is initiated by triggering the lowest two or three sphere gaps. During firing, transient overvoltages occur in the voltage divider, through which the upper sphere gaps are also fired. Triggering can also be effected electronically or with laser pulses. In order to achieve a very rapid voltage collapse, gas-filled sphere gaps or multiple-plate gaps are used.

4.2.4 Generation of Steep-Front Impulse Voltages

With conventional low-inductive generators, impulse voltages can only be generated with a slope of not more than 2.5 kV/ns. Due to the unavoidable inductances of the generator elements of more than 1 μH per stage and of the lead inductances, greater slopes cannot be achieved directly. To generate *steep-front impulse voltages* with significantly greater slopes, the conventional impulse voltage generator is operated in conjunction with a *secondary circuit*, the *peaking circuit* [38]. Figure 4.15 shows the basic circuit consisting of the 1.2/50 impulse voltage generator ***1*** and the peaking circuit ***2*** with the fast multi-plate spark gap FS. The impulse capacitor C_1 of the 1 MV generator with a capacitance of 1–2 nF is connected in series with a part of the damping resistor R_1. To limit the inductance L_S of the secondary circuit to 2–3 μH, HF ceramic capacitors are used for C_1 and carbon composition resistors for R_1' and R_1 in a series-parallel circuit.

The secondary circuit is dimensioned such that the fast spark gap FS fires at the peak of the lightning impulse voltage u_1 and rapidly charges the capacitive test object C_2. For $C_2 \approx (0.1–0.2)C_1$, the output voltage u_2 thus increases within 20 ns to a maximum of 700 kV and then decreases more or less quickly depending on the

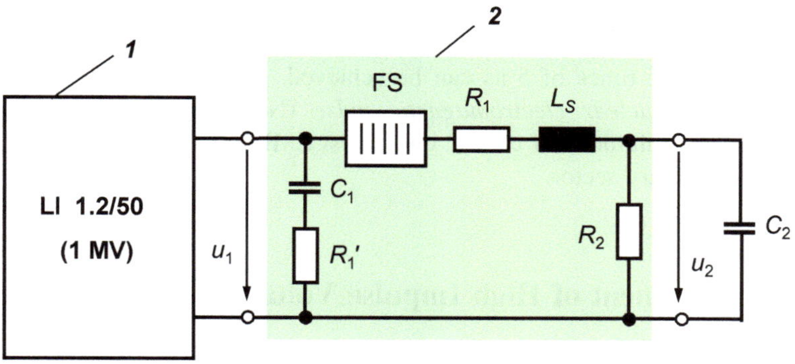

Fig. 4.15 Generation of steep-front impulse voltages with impulse voltage generator *1* and peaking circuit *2* with multi-plate spark gap FS

damping resistor R_2 (see Fig. 4.9). With a chopping spark gap, which is additionally connected to the output of the secondary circuit, the steep-front impulse voltage can be chopped in the tail, so that an approximately rectangular time course results.

Not only in high-voltage test practice, but also in many areas of physics and in the military field, various variants of the basic circuit according to Fig. 4.15 have been developed. An overview of the test and measurement technology for steep impulse voltages can be found in [39]. With careful, mostly coaxial design, peak values of up to 1 MV, rise times down to a few nanoseconds and steepness in the order of 100 kV/ns can be achieved [40–45]. A metal-enclosed, compressed-gas insulated steeping circuit for the generation of *very fast front (VFF) impulse voltages* is described in (Ref. [5] of Chap. 1).

Steep-front impulse voltages can also be generated with exploding wires as switches [46, 47]. A copper wire connected to the output of an impulse voltage generator evaporates explosively when a lightning impulse voltage is generated. In conjunction with the circuit inductances and capacitances, a steep impulse voltage is generated whose peak value and time parameters are determined by the length and the diameter of the wire. The peak value of the steep impulse voltage generated by the exploding wire can be a multiple of the total charging voltage of the impulse generator. The maximum attainable steepness of the voltage is in the range of 10 kV/ns. The test circuit with the exploding wire also serves to commutate an impulse current with a steep front to a test object that is parallel to the wire and the impulse voltage generator.

Often there is the obligation to prove the *electromagnetic compatibility* (*EMC*) of electronic equipment or to demonstrate the screening effect of electronic control cabinets. For this purpose, the test object is exposed to the electromagnetic radiation of steep impulse voltages (*electromagnetic pulse, EMP*), which are applied between ground and a horizontal plate or strip electrode connected to the secondary circuit in Fig. 4.15. Depending on the application, the electrode arrangement can have large dimensions so that entire assemblies from the distribution box of the power supply to motor vehicles or even airplanes can be tested. With the EMP test facility described in [48], electric and magnetic field strengths of up to 200 kV/m or 500 A/m with rise times of 5 ns can be achieved. Comparable values are to be expected for a *nuclear electromagnetic pulse* (NEMP) caused by a nuclear explosion at high altitude. Of course, the largest EMP test systems of this type are found in the military sector.

4.3 Measurement of High Impulse Voltages

For conventional measurement of impulse voltages during high-voltage tests, measuring systems with an impulse voltage divider are predominantly used. At its output terminals, the high-voltage divider is intended to provide a true-to-scale replica of the impulse voltage, which can be recorded and evaluated by the measuring instrument. As measuring instruments on the low-voltage side, mainly digital recorders with computer-aided data evaluation of the recorded impulse voltage are used (see Chap. 7). Occasionally, standard sphere gaps are also used for measuring high voltages, mainly for the linearity check of impulse voltage dividers. The disruptive discharge voltage of sphere gaps is standardized up to 2 MV depending on the sphere diameter and spacing.

The trend towards ever higher transmission voltages outside Europe leads to considerable difficulties in the development of accurate impulse voltage dividers with correspondingly high rated voltages. Capacitive field probes provide another, potential-free measurement option for determining high impulse voltages through field measurements. The Pockels and Kerr cells based on electro-optic effects that exploit the effect of the electric field on the optical properties of crystals and other materials are described in Sect. 6.1.

4.3.1 Measuring Systems with High-Voltage Impulse Dividers

Measuring systems with resistive, capacitive or resistive-capacitive voltage dividers are used for impulse voltage measurements of up to several megavolts. In special cases, voltage dividers with aqueous solutions are also used. This section deals with the different types of high-voltage dividers, their individual components and general characteristics. Important characteristics are the linearity up to the maximum applied voltage and the transfer behavior for measuring fast rising voltages. Attention is drawn to the importance of the step response for characterizing the dynamic behavior of impulse voltage dividers. With some limitations, the numerical convolution algorithm can be well used to study the dynamic behavior of a measurement system. A general equivalent circuit diagram of impulse voltage dividers, including the stray capacitances to ground and inductances of the components, allows the calculation of

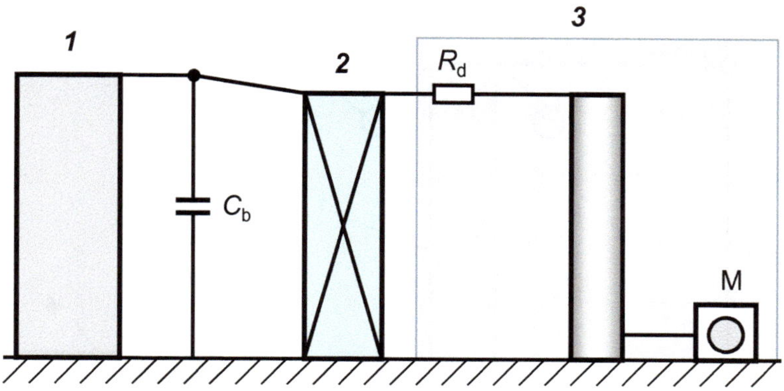

Fig. 4.16 Basic set-up of a high-voltage impulse test circuit (schematic representation). ***1*** Impulse voltage generator with load capacitor C_b and the unavoidable high-voltage leads, ***2*** test object, ***3*** measuring system with damping resistor R_d and measuring instrument M

the basic transfer characteristics. Examples of the design and performance of highly accurate reference dividers are presented, which will be taken to calibrate other reference or approved voltage dividers by comparison measurement.

4.3.1.1 Basic Arrangement of the Test and Measurement Circuit

The basic set-up of a high-voltage impulse test circuit consists of the impulse voltage generator ***1*** with the load capacitor C_b, the test object ***2***, the measuring system ***3*** with damping resistor R_d, the measuring instrument M and the high-voltage leads (Fig. 4.16). In this circuit, the measuring system ***3*** with its high-voltage lead is arranged so that it measures the impulse voltage actually applied to the test object ***2***. Occasionally, the load capacitor C_b is supplemented with a low-voltage capacitor and used as a measuring divider with a measuring instrument. In this arrangement, the measuring divider is positioned between generator ***1*** and test object ***2***. The separate measuring system ***3*** would thus obviously be dispensable. For the measurement of lightning impulse voltages, however, the measuring circuit with C_b as the voltage divider does not comply with IEC 60060-2. In addition, the transfer behavior of load capacitors is usually unsatisfactory, since the necessarily large capacitances of about 1 μF can only be produced with high inductance. There is no objection to using C_b with low-voltage capacitor as a voltage divider for switching impulse voltage and AC voltage measurements.

4.3.1.2 Components of an Impulse Voltage Measuring System

The complete high-voltage impulse measuring system consists of several components (Fig. 4.17). The conventional impulse voltage divider ***1*** has a high-voltage

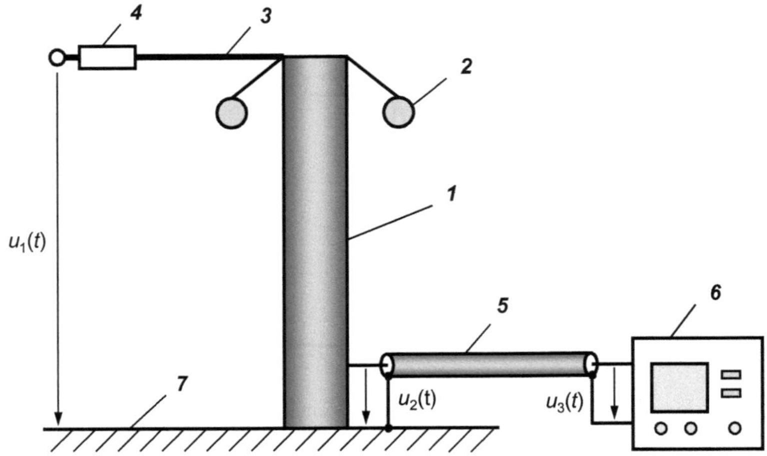

Fig. 4.17 Principle arrangement of a high-voltage impulse measuring system (without cable terminating resistor). *1* Impulse voltage divider, *2* toroidal electrode, *3* high-voltage lead, *4* damping resistor, *5* coaxial cable, *6* digital recorder, *7* low-inductance grounded area

part, generally not shielded, and a shielded low-voltage part. As components, resistors or capacitors or a combination of both connected in series or in parallel are used. One or more toroidal electrodes *2* are attached to the divider top for field control. Calculations have shown that the electric field in the vicinity of the top of the voltage divider has a maximum value and a high field gradient to ground. The capacitive coupling of the toroidal electrodes makes the electric field around the divider top more uniform and reduces the influence of the high-voltage lead *3*. In the simplest case, the lead *3* is a metal wire or, to reduce the lead inductance, a conductive strip, pipe or tube. The length of the lead should correspond approximately to the divider height.

The external *damping resistor* *4* attenuates oscillations in the circuit, which would otherwise be superimposed on the impulse voltage. These are caused, on the one hand, by reflections of traveling waves on the high-voltage lead and, on the other, by oscillatory *LC* components of the measurement circuit [49–52]. In order to avoid flashovers, the voltage divider including the toroidal electrode, connecting lead and damping resistor should have sufficient clearance from walls and adjacent objects, at least according to its height. Typical values for this are 3 m for 1 MV lightning impulse voltages and 5 m for 1 MV switching impulse voltages.

An impulse voltage $u_1(t)$ applied to the top of a voltage divider produces at the divider output the scaled-down voltage $u_2(t)$, which is fed via the measuring cable *5* to the measuring instrument *6* as input voltage $u_3(t)$ for further evaluation. For impedance matching, a resistance equal to the cable impedance is placed either at the beginning or the end of the usually long coaxial measurement cable. Reflection phenomena that occur with rapidly changing voltages are thus suppressed. The voltage divider, the test object and the impulse generator are generally positioned in the test hall on the low-inductance ground conductor ***7***. The measuring instrument ***6*** is usually located in a separate measurement and control room, which allows the operator to look at the high-voltage test circuit. The voltages $u_2(t)$ and $u_3(t)$ should, as far as possible, be a true-to-scale copy of the impulse voltage $u_1(t)$. Deviations from the exact record result from the quality of the measurement arrangement, and for which limits are specified in the test standards.

The elements (resistors, capacitors) of the high-voltage part of an impulse voltage divider are housed in an insulating cylinder made of laminated paper, Plexiglas or glass fiber reinforced plastic. The insulating cylinder is thus decisive for both the mechanical stability and the electrical strength. In order to prevent flashover inside and improve heat dissipation to the outside during long-term loading, impulse voltage dividers are sometimes insulated by pressurized gas or by transformer oil. As a rule, impulse voltage dividers for higher voltages have a modular structure with several identical units of 1 m to 2 m height mounted one above the other. The minimum overall height is determined by the flashover voltage of the insulating cylinder.

Some voltage dividers have *toroidal electrodes* on the top as well as on the low-voltage side. These toroidal electrodes can be dimensioned to act as an external voltage arrester, and if the applied voltage exceeds the rated voltage, a flashover is triggered. The voltage divider itself is thereby protected, but not the external damping resistor ***4*** arranged at the beginning of the high-voltage lead and through which the entire short-circuit current would flow.

For special measurement tasks, smaller impulse voltage dividers up to several 100 kV are sometimes needed, which are built into a metal housing and are therefore completely shielded. Larger impulse voltage dividers are generally not completely shielded. If one still speaks of a "shielded" impulse voltage divider, it is usually a voltage divider with large toroidal electrodes at the top and bottom. However, their shielding effect is limited, particularly against high-frequency interferences, for example when spark gaps fire.

High-Voltage Lead

The approximately horizontal high-voltage lead connects the voltage divider to the test object. The lead and the usually existing top electrode of the voltage divider contribute to the stray capacitance to ground. The stray capacitance $C_{e,h}$ of a horizontal thin wire with the diameter d and the length l at a height h above ground is given under the condition $(4\,h)^2 \gg l_2$ by the equation (Ref. [2] of Chap. 9):

$$\boxed{C_{e,h} = \frac{2\pi\varepsilon_0 l}{\ln\left(\frac{2l}{d}\right)}}. \tag{4.10}$$

The stray capacitance of the high-voltage lead is therefore not dependent on the height h above ground and only slightly dependent on the diameter d of the wire or tube with $d < l$. For a tubular lead of length $l = 1$ m and diameter $d = 2$ cm, the stray capacitance is 12 pF according to Eq. (4.10). Since Eq. (4.10) also applies to curved wires and tubes with not too small bending radius, the stray capacitance of toroidal electrodes can also be estimated. The high-voltage lead and the toroidal electrode also have stray capacitances to the voltage divider. This influence of the toroidal electrode is used directly to control the electric field of the voltage divider, mainly for reducing the high field in its top region.

When measuring the step response of voltage dividers, the measurement circuit may also have a vertical lead (see Fig. 9.16b). The stray capacitance of the vertical lead, whose lower end is slightly above ground, is approximately given by Eq. (4.16). The length-related capacitances of the horizontal and vertical leads are therefore not very different.

In addition, the horizontal high-voltage lead of length l and diameter d has an inductance (Ref. [2] of Chap. 9):

$$\boxed{L_\mathrm{h} = \frac{\mu_0 l}{2\pi} \ln \frac{4h}{d}}, \tag{4.11}$$

where h is the height above ground. A lead with $d = 2$ cm and $h = 1.5$ m, therefore, has an inductance $L_\mathrm{h} = 1.14$ μH/m.

For an infinitely long horizontal high-voltage lead with the diameter d in cm and the height h in m above ground, the *wave impedance* is determined by:

$$\boxed{Z = \sqrt{\frac{L'}{C'}} = \frac{1}{2\pi}\sqrt{\frac{\mu_0}{\varepsilon_0}} \ln \frac{4h}{d} \approx 60 \ln \frac{4h}{d}\ \Omega}, \tag{4.12}$$

where L' and C' are the inductance or stray capacitance per unit length of the lead. Accordingly, the wave impedance increases only slightly with the height h of the horizontal lead, i.e. with the size of the voltage divider.

Damping Resistor

As already mentioned, the external *damping resistor* R_d at the beginning of the high-voltage lead (see Fig. 4.17) has two tasks: avoidance of high-frequency traveling wave phenomena on the high-voltage lead due to reflections, as well as attenuation of the oscillations caused by inductances and capacitances in the high-voltage circuit. The influence of the damping resistor on the step response of different configurations of impulse voltage dividers is shown in [53]. From the wave impedance of the idealized high-voltage lead according to Eq. (4.12), an approximate value for the reflection-free termination of the lead can be calculated. For an infinitely long lead with $d = 2$ cm and $h = 1.5$ m, the wave impedance is $Z = 342\ \Omega$. In terms of its dimensions, this would correspond to the horizontal high-voltage lead of a 500 kV impulse voltage divider with the height h.

Measurement Cable and Its Termination

The measurement cable **5** from the divider output to the measuring instrument is a single or double shielded coaxial cable (see Fig. 4.17). When measuring high-frequency oscillations or fast transient voltages, in particular step responses, chopped impulse voltages or steep-front impulse voltages, the *wave impedance* of the cable:

$$\boxed{Z = \sqrt{\frac{L}{C}}} \tag{4.13}$$

must be taken into account, where L is the inductance and C is the capacitance of the coaxial cable. Typical values for Z are 50, 60 and 75 Ω. In order to avoid reflection phenomena when measuring high-frequency signals, the coaxial cable must be terminated at least at one end by a resistance $R = Z$. If the terminating resistor is missing, the voltage to be measured is completely or partially reflected at the high-impedance input of the measuring instrument and travels back to the beginning of the cable. Here again, there is a reflection in the case of mismatch. As a result, along the lossy cable runs a traveling wave, which is superimposed on the voltage to be measured as a damped oscillation.

The propagation velocity of a traveling wave in the coaxial cable is:

$$\boxed{c = \frac{c_0}{\sqrt{\varepsilon_r}}}, \tag{4.14}$$

where c_0 is the speed of light in vacuum and ε_r is the permittivity of the cable dielectric. For insulation made of polyethylene ($\varepsilon_r = 2.25$) or Teflon ($\varepsilon_r = 2$), the signal propagation time τ in the cable is approximately 5 ns/m. For a 10 m long coaxial cable, the period of oscillation is 100 ns, which results from the double travel time 2τ of the traveling wave moving back and forth. This time is short compared to the front time of a full lightning impulse voltage. Therefore, even when the cable termination is not matched, the traveling wave oscillation, being damped by the cable losses, will hardly affect the evaluation of a full lightning impulse voltage. As a rule, however, the terminating resistor will not be dispensed with in order to correctly record high-frequency oscillations or fast transients. Low-loss coaxial cables whose inner conductor is held by an insulating spiral have air or insulating gas as the dielectric. Due to $\varepsilon_r \approx 1$, the cable propagation time is about 3.3 ns/m.

In large test fields, the measurement cable can reach a length of 50 m and more. As the length increases, the quality of the cable must be considered. A long coaxial cable of poor quality has an ohmic resistance which, in conjunction with a low-ohmic input resistance of the measuring instrument, causes a non-negligible voltage drop along the cable. The input voltage u_3 at the measuring instrument is then about 1–2% smaller than the divider output voltage u_2 (see Fig. 4.17). The capacitance of coaxial cables with rated voltages of a few kilovolts ranges from 60 to 100 pF/m. The cable capacitance is parallel to the low-voltage part of the divider and, in the case of a long cable, represents a significant load on the divider output. Depending on the design of the voltage divider, the scale factor of the measuring system and the shape of the recorded voltage are affected. For this reason, the measuring system must always be used in tests and calibrations with the same coaxial cable or with another cable of similar length and type.

Measuring Instrument

As measuring instrument ***6***, preferably a *digital recorder* with an amplitude resolution of 8 bit to 16 bit is used (see Chap. 7). Digital recorders allow a largely automated data acquisition, digitization and computer-aided evaluation of the voltage to be measured. To avoid interference caused by ground loops, the measuring instrument is grounded not directly but via the shield of the measuring cable, which is connected to the voltage divider (see Fig. 4.17). Occasionally, analogue impulse oscilloscopes are still in use, however, without additional equipment, they allow only a manual evaluation of the recorded impulses with insufficient accuracy. If only the test voltage value is to be determined, analogue or digital impulse voltmeters can be used. In this case, compliance with the standard impulse shape with regard to the time parameters and oscillations is to be checked with an oscilloscope.

4.3.1.3 Scale Factor

The displayed or recorded output voltage $u_2(t)$ of the measuring system shall be a true-to-scale replica of the impulse voltage $u_1(t)$ within the permitted tolerances (Ref. [2] of Chap. 2). In measurement practice, the quotient of the two peak values gives the *impulse scale factor F*:

$$\boxed{F = \frac{\hat{u}_1}{\hat{u}_3} \approx \frac{\hat{u}_1}{\hat{u}_2}}. \tag{4.15}$$

In this case, it is assumed that the measuring system is linear and the impulse voltage has no superimposed oscillation on the peak (see Fig. 4.17). The impulse scale factor is usually a number without unit. When evaluating a measurement, the instrument reading is multiplied by the scale factor in order to obtain the test voltage.

The preferred method for determining the scale factor is the *comparison measurement* at impulse voltage with an accurate *reference measuring system* (see Sect. 10.3.1). The measurement practice shows that the scale factor is not a constant, but depends on a number of influencing quantities. These include the peak value and the time parameters of the impulse voltage themselves, furthermore the ambient temperature, the clearance to adjacent objects, etc. The scale factor may only vary within the limits defined for the intended measurement task. A measuring system with a universal voltage divider for all types of voltages can have different scale factors for DC, AC, lightning and switching impulse voltages. It is also possible to limit the validity of the scale factor to a certain range of front times or times to chopping.

As an alternative to the comparison with a reference system, the individual scale factor of each component of a measuring system such as voltage divider, digital recorder and external attenuator can be determined by calibration, and their product gives the scale factor of the complete measuring system. Sometimes, other terms such as divider ratio, transformation ratio or gain factor are common. The calibration of the individual components shall be carried out under conditions comparable to those for the complete measuring system. For determining the scale factors of the components, the measurement methods known in the low-voltage range can be applied. The scale factor calculated from the individual components must be confirmed by a linearity test of the complete measuring system up to the rated voltage or maximum operating voltage (see Sect. 10.3.5). This also proves that the measuring system has no external partial discharges, which could affect the measurement accuracy.

4.3.1.4 Stray Capacitance to Ground

First, the influence of the *stray capacitance* of an unshielded impulse voltage divider on the transfer behavior is investigated in more detail. In the general equivalent circuit diagram, the high-voltage divider is represented by a series connection of N equal impedances Z_1' (Fig. 4.18). From the divider column, distributed stray capacitances C_e' lead to ground and grounded walls. In a first approximation, all values of C_e' are assumed to be equal. Looking at the voltage

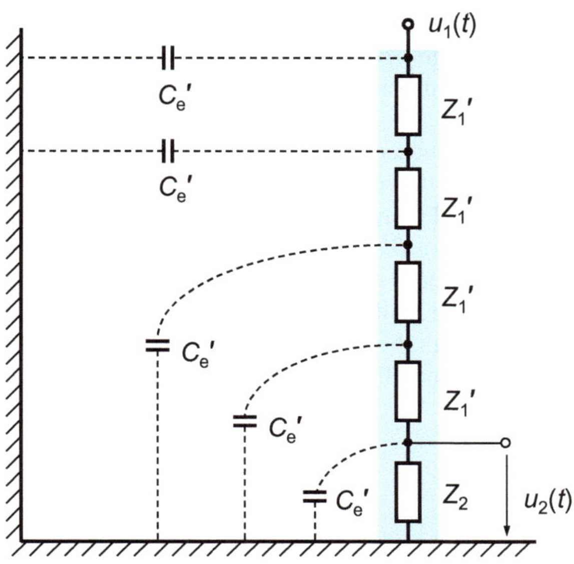

Fig. 4.18 Equivalent circuit diagram of a high-voltage divider with series-connected impedances Z_1' and distributed stray capacitances C_e' to ground and grounded walls

divider as a slim, vertical cylinder above ground, its total stray capacitance to ground is given by (Ref. [2] of Chap. 9):

$$\boxed{C_{\mathrm{e}} = \frac{2\pi\,\varepsilon_0\,l}{\ln\left(\frac{2}{\sqrt{3}}\frac{l}{d}\right)}}, \tag{4.16}$$

where l is the length and d is the diameter of the cylinder. According to Eq. (4.16), the stray capacitance per unit length is in the range 10–15 pF/m for tall, slim voltage dividers and approx. 20 pF/m for short, thick voltage dividers.

Furthermore, the capacitance per unit length varies only slightly with the divider height (Ref. [6] of Chap. 1), which confirms the assumption that the stray capacitances C'_{e} in Fig. 4.18 have approximately the same value. Via these stray capacitances, a frequency-dependent leakage current flows to ground. The consequence is that the high-frequency signal components do not reach the low-voltage part of the voltage divider and are therefore absent in the spectrum of the output voltage $u_2(t)$. How strongly this affects the transfer behavior at high frequencies depends on the type of the voltage divider and the values of the impedances Z' in relation to C'_{e}.

4.3.1.5 General Consideration of the Transfer Behavior

The voltage divider is usually the most important—and often most critical—component of a high-voltage impulse measuring system. Its transfer behavior can be characterized in the frequency domain by the *transfer function* $H(\mathrm{j}\omega)$ or in the time domain by the *step response* $g(t)$ (see Chap. 9). In the measurement practice with impulse voltages, the step response is generally preferred, while for theoretical investigations also the transfer function is of interest. Both forms of representation can be transformed into each other. They offer the possibility of calculating the output voltage of a voltage divider with known transfer behavior for any input voltages by means of *convolution*. Convolution may be a complement or even alternative to the comparative measurement with an accurate *reference divider* to study the influence of different front times or times to chopping on the scale factor. In this case, the peak value errors and the time parameter errors of the examined voltage divider are calculated for the respective impulse voltage form and represented in *error diagrams* (see Sects. 9.7.4 and 10.3.9).

For a perfectly constructed homogeneous voltage divider, a simple rule of thumb for the upper limit frequency can be given. Assuming that the voltage impulse passes through the voltage divider at the speed of light, the maximum value for the upper frequency limit is [53]:

$$\boxed{f_2 = \frac{150}{h}\ \mathrm{MHz}}, \tag{4.17}$$

where h is the divider height in meters. According to this rule of thumb, the upper limit frequency or the bandwidth of an ideally constructed voltage divider decreases with increasing divider height and thus with higher rated voltage. For example, a 1.5 m high impulse voltage divider for 500 kV achieves according to Eq. (4.17) a bandwidth of 100 MHz. However, the actual values of real voltage dividers are considerably lower. For very large voltage dividers, the bandwidth is still considered sufficient to be able to measure full impulse voltages within the given uncertainty limits, but not lightning impulse voltages chopped on the front (Ref. [34] of Chap. 9). If the voltage divider is not optimally dimensioned, larger measurement deviations in the peak value and in the time parameters result.

Prerequisite for the good transfer behavior of a real impulse voltage divider is the optimal frequency balance of the high-voltage and low-voltage parts, taking into account all components. This includes not only the visible components, but also the unavoidable stray capacitances and self-inductances. The term *compensated voltage divider* is well known from the low-voltage technique, i.e. the upper and lower parts of a shielded voltage divider are built up identically in their components, taking into account the parallel capacitances and self-inductances. Such a compensated structure can hardly be achieved with large unshielded high-voltage dividers. This is primarily due to the stray capacitances of the high-voltage components to ground and to the walls, through which an increasing proportion of the measurement signal is derived with increasing signal frequency. There is no equivalent for such compensation in the small low-voltage part. Compensation of the large high-voltage divider is therefore possible only in a limited frequency range.

For the theoretical analysis of the transfer behavior, the spatially extended impulse voltage divider can be represented as a *ladder network* with homogeneously distributed elements, which are successively passed through by a fast transient signal. The advantage of the ladder network as an equivalent circuit lies in the uniform representation for the different types of voltage dividers, which allows general statements about the transfer behavior. However, the effect of the high-voltage lead, the damping resistor and the coaxial cable, which is connected to the divider output with terminating resistor, must be considered separately. The same applies to any deviation from the homogeneity of the voltage divider, which inevitably results from the low-voltage part or is intended to improve the transfer behavior.

For the standardized impulse voltages with time parameters in the microsecond range, it is not absolutely necessary to consider the voltage divider as a spatially extended ladder network. The signal propagation time in the impulse voltage divider is short even with a large overall height in comparison to the time parameters of a full lightning impulse voltage. A simplified equivalent circuit diagram with concentrated elements of the impulse voltage divider, including the unavoidable stray capacitances and inductances, should therefore be sufficient in many cases. This simplification applies to voltages whose time parameters are large compared to the signal propagation time through the voltage divider. Only the voltage collapse of a chopped lightning impulse voltage occurs in a time comparable to the propagation time of the signal in the voltage divider and should therefore be investigated using the *ladder network*. However, the exact evaluation of the voltage collapse is not the subject of a standard impulse voltage test.

4.3.1.6 General Ladder Network Diagram and Step Response

To investigate the propagation of fast transient signals, the spatially extended high-voltage divider is approximately represented as a ladder network with homogeneously distributed longitudinal and transverse impedances (Fig. 4.19). This approach is common in many other areas, for example in the transmission line theory at higher frequencies. The signal to be measured enters the voltage divider at its top and passes through the individual ladder elements one after the other down to the last element lying at ground potential. The total transit time through the ladder network is no longer negligible compared to the duration of a fast signal.

The ladder network as an equivalent circuit makes it possible to establish a uniform theory that provides general information about the transfer behavior of impulse voltage dividers. The general equivalent circuit diagram of Fig. 4.19a shows a large number n of equal longitudinal impedances Z_1' and transverse impedances Z_q', each related to the length. The longitudinal impedances Z_1' represent the real elements of the voltage divider (resistors, capacitors) as well as their inevitable parasitic elements such as inductances and parallel capacitances (Fig. 4.19b). The transverse impedances Z_q' represent the distributed stray capacitances of the voltage divider to ground. The low-voltage part of the voltage divider with the output voltage $u_2(t)$ is also shown as a similar element of the ladder network.

The voltage impulse $u_1(t)$ applied to the upper input of the ladder network passes through the network elements to the lowest grounded network element at which the output voltage $u_2(t)$ can be tapped. From the quotient of the output and input voltages, the *complex transfer function* F(s) of the ladder network is calculated and expressed by the Laplace transform:

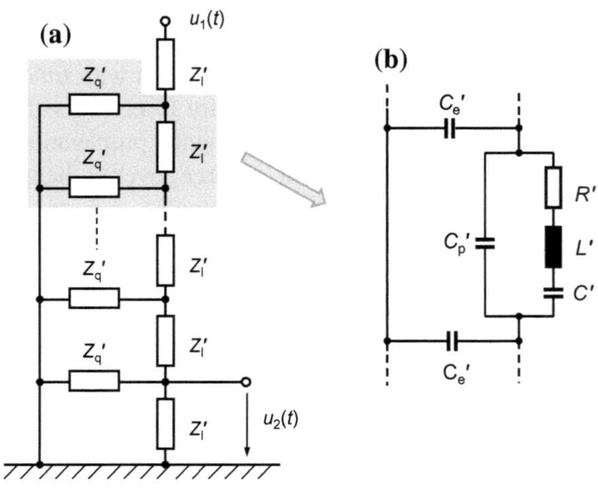

Fig. 4.19 Equivalent circuit diagram of a high-voltage divider. **a** General ladder network with longitudinal impedances Z_1' and transverse impedances Z_q'. **b** Individual element of the ladder network

$$\boxed{F(s) = n\frac{u_2(s)}{u_1(s)} = n\frac{\sinh\frac{1}{n}\sqrt{\frac{Z_l(s)}{Z_q(s)}}}{\sinh\sqrt{\frac{Z_l(s)}{Z_q(s)}}}}, \tag{4.18}$$

where s is a complex number [49, 53–55]. In Eq. (4.18), Z_l and Z_q are the accumulated complex longitudinal and transverse impedances of the n-stage ladder network (see Fig. 4.19):

$$Z_l(s) = n\,Z_l'(s) \quad \text{and} \quad Z_q(s) = \frac{1}{n}Z_q'(s).$$

With F(s) according to Eq. (4.18), the normalized step response of the impulse voltage divider results formally by inverse Laplace transformation (see Table A.2):

$$\boxed{g(t) = L^{-1}\left\{\frac{1}{s}F(s)\right\}}. \tag{4.19}$$

The general solutions of Eqs. (4.18) and (4.19) with the longitudinal and transverse impedances according to Fig. 4.19 are not dealt with here. In the following sections, typical values of the impedances Z_l and Z_q for the resistive and damped capacitive impulse voltage dividers are introduced into the equations to calculate $F(s)$ and $g(t)$. However, because of the simplified representation of the impulse voltage divider by the ladder network with equal stray capacitances to ground and the same end element for the low-voltage part, the result of the calculations can only give a general overview of the dynamic behavior of impulse voltage dividers. In the practical construction of an impulse voltage divider, an optimized design of the low-voltage part can often improve the dynamic behavior [56–58].

The advantages and disadvantages of representing a voltage divider through the ladder network using an analytical approach have already been briefly discussed above. Nowadays, commercial or user-developed software for network analysis of electrical circuits and for the calculation of electric fields provides a practical and effective method for theoretical investigations on impulse voltage dividers. In this way, the transfer behavior of a voltage divider can be determined based on its individual equivalent circuit diagram or by field calculations. The presentation and calculation as a ladder network with individual elements that better corresponds to reality is, of course, also possible with such software. The influence of the high-voltage lead with the damping resistor, the grading electrodes of the voltage divider and the measuring cable with the terminating resistor can be taken into account. The correctness of the calculation model assumed for the impulse voltage divider is checked by comparing the calculated step response with the recorded experimental step response [59–64]. This comparison provides valuable information for improving the model approach and thus to improve the construction of the voltage divider itself. However, it should be noted that inadequacies in the experimental setup and interference effects affect the measurement of the step response (see Sect. 9.8).

4.3.1.7 Influence of Disturbances and Countermeasures

When firing the sphere gaps of an impulse voltage generator or a chopping spark gap, strong *electromagnetic fields* are generated, which interfere with the measuring system in many ways. Measuring instruments specially developed for impulse voltage measurements are largely protected against conducted and electromagnetically coupled disturbances by suitable electronic circuitry and shielding. To improve the signal-to-noise ratio, maximum input voltages of up to 1000 or 2000 V corresponding to the maximum divider output voltage are common. If the instrument is not designed for such high input voltages, it will be supplemented by an external attenuator.

Measuring instruments that are not adequately protected by their design are operated in a shielded cabin, the *Faraday cage*. Power is supplied to the instrument via an isolating transformer outside the shielded cabin and a mains filter mounted on the outer cabin wall. This enables potential-free operation of the measuring instrument and makes the penetration of transient interferences into the mains supply more difficult. The shielded cabin is grounded so that the source voltage induced by the electric field on the cabin surface is dissipated and does not reach the measuring instrument inside the cabin. Grounding the cabin is essential as some mains filters charge the (ungrounded) cabin to half the mains voltage. In order to dissipate the heat generated by the measuring instrument, the shielded cabin has ventilation openings covered with a honeycomb grid or a fine metal mesh to keep out interference fields.

When the measuring instrument is operating in a shielded cabin and there is a strong magnetic field outside the cabin, additional shielding of the measurement cable is recommended. Well suited for this purpose are metal tubes embedded in the floor or flexible corrugated metal pipes, through which the measuring cable is guided. Also suitable are double-shielded coaxial cables, although their shielding braid allows a certain punch-through of the external interference field on the inner conductor increasing with frequency. The outer shield is connected to the grounded cabin at the location of the cable bushing, while the inner shield is connected to the shield of the instrument inside the cabin. In general, it is advantageous to interconnect the outer and inner shields of the cable at the divider output (see Fig. 5.12). Transient potential rises in the ground circuit or source voltages induced by magnetic fields can then discharge via surface currents in the closed ground circuit and cannot reach the measuring instrument in the shielded cabin through the *coupling impedance* of the cable [65–69]. There must be no cable or other electrically conductive connection to devices such as PC, printer, etc. outside the screened cabin at the time of generating the impulse voltage. Otherwise, external interference is coupled into the measuring instrument [70]. Optoelectronic data transmission via optical fibers from the measuring instrument to the PC and peripheral devices outside the cabin are very practical and effective, while retaining the full shielding effect of the cabin.

Electromagnetic interference fields also affect the high-voltage divider that is usually unshielded and thus acts as an antenna. Only the low-voltage part of the divider is usually shielded. The interference signal picked up by the voltage divider superimposes itself on the measured signal at characteristic times. For example, the firing of the generator sphere gaps is noticeable at the beginning of the recorded impulse voltage, so that the determination of the 30% point and thus the front time of a lightning impulse voltage could be influenced. In the case of an impulse voltage chopped on the front, the firing of the chopping gap affects the peak region. Frequently, the disturbance already shows up before the peak of the chopped impulse voltage, since the interference coupled electromagnetically into the voltage divider has a shorter propagation path in air than the conducted measurement signal. Depending on the phase of the disturbance, the recorded peak value of the chopped impulse voltage may become larger or smaller.

The magnitude and temporal assignment of the total interference coupled into the measuring system can be detected by two measurements with a two-channel digital recorder. First, the impulse voltage is recorded as usual in the complete measurement circuit. The second record is taken at the same charging voltage of the generator, but the high-voltage lead connecting the voltage divider to the impulse voltage generator is removed. For synchronization of the two records taken with the first recorder channel, a trigger signal is supplied to the second channel, either directly from the trigger output of the impulse voltage generator or from a wire acting as an antenna. The comparison of the two records yields the magnitude and temporal assignment of the interference.

4.3.2 *Resistive Impulse Voltage Divider*

Resistive impulse voltage dividers with a total resistance of 1–20 kΩ are mainly used to measure lightning impulse voltages up to 2 MV and steep-front impulse voltages up to several 100 kV. These low-ohmic voltage dividers are less suitable for the measurement of switching impulse voltages. This is partly due to the higher power consumption because of the longer impulse duration, but also to the deformation of the impulse tail by the low-ohmic load. Occasionally, high-ohmic voltage dividers and those with low-ohmic aqueous solutions are also used for lightning impulse voltage measurements. The resistive voltage divider can be represented in the equivalent circuit diagram as a ladder network and calculated with regard to its transfer behavior. Special versions of the resistive voltage divider with non-linear distribution of the resistances and optimal voltage tap are intended to improve the transfer behavior.

4.3.2.1 Impulse Measuring System with Resistive Voltage Divider

Figure 4.20 shows the general circuit of the complete high-voltage measuring system with a resistive impulse voltage divider. Typical values for the high-voltage resistor R_1 are 1–20 kΩ depending on the rated voltage and impulse type. The lower

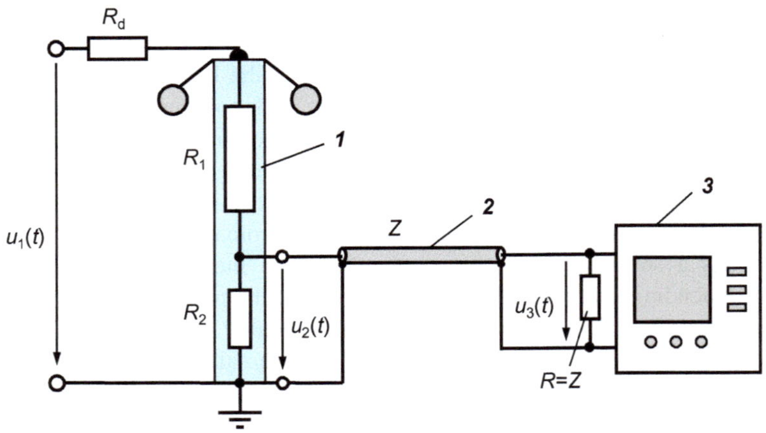

Fig. 4.20 Impulse voltage measuring system with resistive voltage divider and digital recorder. ***1*** Resistive impulse voltage divider, ***2*** coaxial cable with wave impedance *Z*, ***3*** digital recorder

resistances apply to voltage dividers suitable for measuring very short impulse voltages of a few 100 kV. At the divider top, a more or less large torus electrode is attached. The damping resistor R_d at the beginning of the high-voltage lead, whose task is to suppress or at least reduce traveling wave oscillations on the lead, is in the order of 100–400 Ω. Typical values for the resistor R_2 in the low-voltage part are 0.5–10 Ω. The transient output voltage $u_2(t)$ of the voltage divider passes via the coaxial cable ***2*** with the wave impedance $Z = 50\ \Omega$ or 75 Ω to the input of the measuring instrument ***3***, usually a digital recorder. To avoid reflection phenomena, the coaxial cable is terminated at the recorder input with the resistance $R = Z$. This resistance and the cable capacitance C_k are for low-frequency signals parallel to R_2, which may affect the division ratio for larger values of R_2.

The use of high-ohmic voltage dividers for measuring impulse voltages is relatively rare. A high-ohmic 400 kV voltage divider as a reference for the measurement of switching impulse voltages is described in [70]. The high-voltage resistor consists of a series and parallel connection of individual resistors with a total resistance of 150 kΩ. With a response time $T_N = 0.22\ \mu s$ and settling time $t_s = 3\ \mu s$, the reference divider has low measurement errors for the peak value and the time parameters. To evaluate the measured switching impulse voltage, the curve in the peak region is fitted to a double exponential function.

For a resistive impulse voltage divider, different types of resistors come into consideration. For larger voltage dividers, a meandering resistor wire of NiCr or CuNi is often used, which is woven with textile threads into a resistor band with a width of up to 1 m ("Schniewind resistor") [71, 72]. The resistor band is either flexible with several meters in length or rigid in the form of a rod or tube cast with epoxy resin. Small damping resistors as well as large voltage dividers with the highest rated voltages can thus be produced efficiently. The meandering arrangement of the resistance wire reduces the inductance, which typically ranges from less than 1–30 μH/m. With an optimal layout of the meander, the effect of the inductance is at least partially compensated by the capacitance of the parallel wire sections. In another embodiment, a resistor wire is wound around an insulating tube and returned in the opposite winding direction, thereby forming a *bifilar winding* to reduce the inductance [73].

For particularly "fast" voltage dividers with rated voltages of less than 500 kV, series-connected low-ohmic individual resistors are also used. For example, carbon film resistors have a good high-frequency behavior and high impulse withstand voltage. The carbon film deposited on the insulating body, however, must not have any groove in the form of a helix or partial helix, with which a defined resistance is usually to be achieved. With a groove incised, there is a risk that electrical discharges will form across the groove, damaging the carbon film and leading to a local flashover. Film resistors also have a good high-frequency behavior, but their current loading capability is significantly lower compared to carbon film resistors and wire-wound resistors. The required higher current carrying capacity is achieved by connecting several resistors in parallel, which also reduces the total inductance. Chip and ceramic resistors with a resistance layer deposited on a quartz substrate have an even better frequency behavior. Series-connected carbon composite resistors can also be used for impulse voltage dividers [74]. Liquid voltage dividers consisting of low- and high-ohmic aqueous solutions are briefly described in Sect. 4.3.2.7.

The low-voltage part of an impulse voltage divider is usually housed in a shielded box at the bottom of the voltage divider. It consists of a parallel connection of individual resistors to achieve a high current loading and low inductance. The resistors used in the high- and low-voltage parts are often similar in construction to obtain the same temperature behavior, but they are differently loaded. In the low-voltage measurement technique, the principle of the compensated voltage divider is well-known, according to which the ratio of the inductances L_1/L_2 of the high- and low-voltage parts should be equal to that of the resistances R_1/R_2. This requirement cannot be fulfilled with large voltage dividers. On the one hand, a correspondingly small inductance cannot be achieved in the low-voltage part and, on the other hand, such balancing of the voltage divider is not necessarily advantageous. The low-voltage inductance can partially compensate for the effect of stray capacitances on the high-voltage part and improve the transfer behavior of the voltage divider in the desired frequency range. As a result, the step response of the voltage divider has a slightly increased overshoot, but shorter rise and response times [75].

To compensate for parallel capacitances of the resistors in the high-voltage part, it may be advantageous to connect a capacitance corresponding to the division ratio in parallel to the low-voltage resistor. In this context, the capacitance of the coaxial cable connected to the divider output must be considered. The input resistance of the measuring instrument is also parallel to the low-voltage part. If the recorder input is terminated with the wave impedance of the coaxial cable, i.e. low-ohmic, the conductor resistance must be taken into account in the case of a longer, not so high-quality coaxial cable. Coaxial cables typically have a conductor resistance between 15 and 150 mΩ/m. Using a longer coaxial cable of poor quality may result in a voltage drop of a few percent of the impulse voltage to be measured. The measuring instrument will then indicate too low a voltage if the voltage drop is not taken into account by calibration.

Considering the resistive impulse voltage divider alone, i.e. without coaxial cable, measuring instrument and terminating resistor R, its division ratio is determined by the quotient of the applied high voltage u_1 and the divider output voltage u_2. For DC and low-frequency voltages, the division ratio is formally given by (see Fig. 4.20):

$$\boxed{\frac{u_1}{u_2} = \frac{R_1 + R_2 + R_d}{R_2}}. \tag{4.20}$$

The division ratio determined by the resistors according to Eq. (4.20) is, however, no longer valid at higher frequencies. The unavoidable stray capacitances to ground and inductances of the divider components affect the transfer behavior and must be taken into account in the equivalent circuit diagram.

4.3.2.2 Step Response of the Resistive Voltage Divider

The basic transfer behavior of a resistive voltage divider can theoretically be investigated by means of the ladder network with a total of N homogeneously distributed elements (see Fig. 4.19). Besides the resistance R', the single element of the ladder network in consists of the inductance L', the parallel capacitance C_p' and the stray capacitance C_e' which is divided into two halves (Fig. 4.21). The capacitance C' in series with R' and L', as shown in Fig. 4.19b, is of course omitted here. Assuming that the low-voltage part is comparable to the last element of the ladder network, the step response of the resistive voltage divider results from the general solution in Eq. (4.18) as [53–56]:

$$\boxed{g(t) = 1 + 2e^{-at} \sum_{k=1}^{\infty} (-1)^k \frac{\cosh(b_k t) + \frac{a}{b_k} \sinh(b_k t)}{1 + \frac{C_p}{C_e} k^2 \pi^2}} \tag{4.21}$$

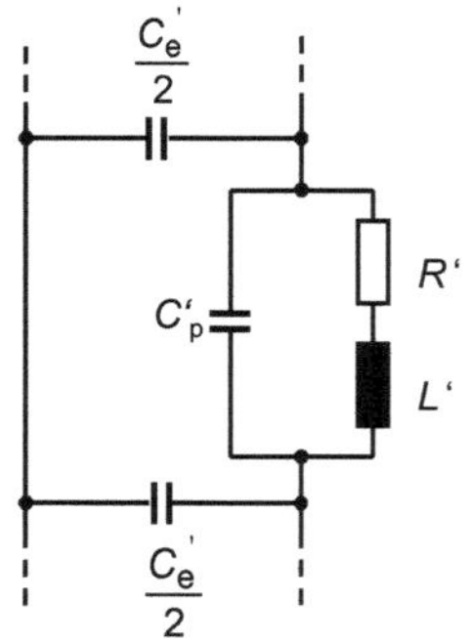

Fig. 4.21 Individual element in the ladder network of the resistive impulse voltage divider

with:

$$a = \frac{R}{2L}$$

$$b_\mathrm{k} = \sqrt{a^2 - \frac{k^2\pi^2}{LC_\mathrm{e}\left(1 + \frac{C_\mathrm{p}}{C_\mathrm{e}}k^2\pi^2\right)}}, \tag{4.22}$$

$R = nR', L = nL', C_e = nC_e', C_p = C_p'/n.$

Figure 4.22 shows the step responses $g(t)$ of a resistive voltage divider calculated according to Eq. (4.21) for the total resistances $R = 2$, 5 and 10 kΩ. The inductive time constant $L/R = 10$ ns is assumed to be constant, i.e. as the resistance increases, so does the inductance. The resulting stray capacitance to ground is $C_\mathrm{e} = 40$ pF, the parallel capacitance $C_\mathrm{p} = 1$ pF. These values are typical for an impulse voltage divider with a rated voltage of 1 MV. When R is low, the influence of C_e is small and the series inductance L causes a pronounced overshoot of the step response (curve ***1*** in Fig. 4.22). A small resistance R means that in Eq. (4.21) one or more b_k values become imaginary. The hyperbolic functions then turn into the corresponding trigonometric functions with the absolute values of b_k in the argument. As R becomes larger, the influence of the stray capacitance C_e increases. As a result, the step response rises more slowly and the overshoot is reduced (curve ***2***). Above a critical value of R, only hyperbolic functions remain in Eq. (4.21) and the step response asymptotically approaches the final value $g(\infty) = 1$ (curve ***3***).

At this point it should be pointed out again that in the measurement practice a limited overshoot of the step response is quite acceptable or even desirable. Due to the brief overshoot, the response time according to Eq. (9.31) is reduced, which is one of several criteria for the correct peak value measurement of impulse voltages. Decisive here is that the overshoot of the step response has subsided at a time that is clearly before the occurrence of the impulse peak, i.e. for the settling time $t_\mathrm{s} \ll T_\mathrm{p}$ must apply (see Sect. 9.8.2). A slight overshoot of the step response usually affects at first the front time of impulse voltages, while a large overshoot then also affects the peak value.

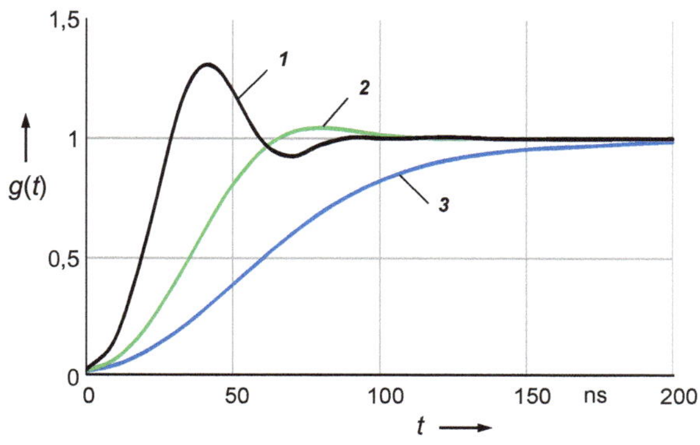

Fig. 4.22 Calculated step responses $g(t)$ of the resistive voltage divider represented as a ladder network according to Eq. 4.21 for different resistances R ($L/R = 10$ ns, $C_\mathrm{e} = 40$ pF, $C_\mathrm{p} = 1$ pF). ***1***: $R = 2$ kΩ, ***2***: $R = 5$ kΩ, ***3***: $R = 10$ kΩ

The calculated step responses in Fig. 4.22 show an initial jump at time $t = 0$. This is due to the parallel capacitances C'_p in the high and low voltage parts of the ladder network, which together with the stray capacitances C'_e form a capacitive voltage divider. At time $t = 0$, i.e. when the step is applied, the voltage at the lowest capacitance C'_p of the ladder network immediately jumps to the value determined by the capacitive division ratio. This initial voltage jump is the more pronounced the larger the ratio C_p/C_e. In measurement practice, however, such an initial jump has not been observed.

With the simple ladder network, not all details can be correctly replicated. For example, the individual ladder elements are identical both in the high- and in the low-voltage part, but this is not the case in the voltage divider. In addition, the skin effect as well as the high-voltage lead and the measuring cable are not considered. Also, the step voltage actually generated is not infinitely steep as is assumed in the calculation, which primarily affects the initial form of the step response. In order to improve the equivalent circuit diagram in Fig. 4.21, there are various suggestions, for example, to supplement each parallel capacitance C'_p with an inductance in series [53, 54].

To better account for the difference between the low-voltage and high-voltage parts in the ladder network, both parts are considered separately [56]. For this purpose, at first the impulse current flowing through the high-voltage part is calculated, whereby the low-voltage part is short-circuited. In the second step, this current is injected into the low-voltage part as an impressed current, thus generating the output voltage $u_2(t)$. The influence of the high-voltage lead with damping resistor has also been theoretically investigated in a similar way [57].

4.3.2.3 Simple Equivalent Circuit with Concentrated Elements

A resistive voltage divider with a resistance $R \geq 10\ \text{k}\Omega$ has a step response according to curve ***3*** in Fig. 4.22, which asymptotically approaches the final value without overshoot. For large values of R, the related inductance L/R and the parallel capacitance $C_p \ll C_e$ are negligible. Therefore, Eq. (4.21) turns into the simplified expression for the step response of a high-impedance ladder network [53–55]:

$$\boxed{g(t) = 1 + 2\sum_{k=1}^{\infty} (-1)^k \exp\left(-\frac{k^2\pi^2}{RC_e}t\right)}. \tag{4.23}$$

From Eq. (4.23), the response time T results:

$$\boxed{T = \frac{RC_e}{6}}. \tag{4.24}$$

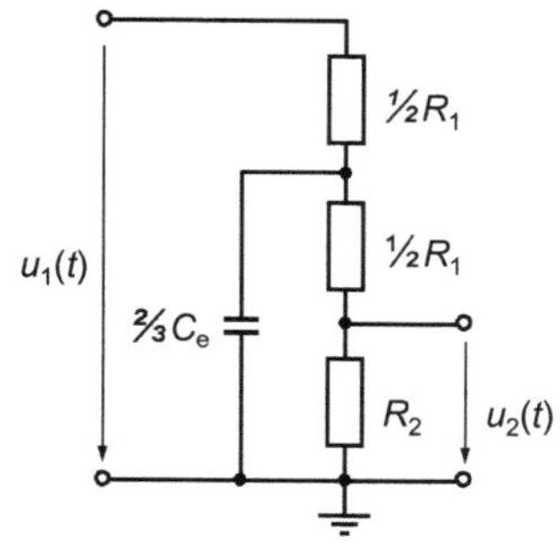

Fig. 4.23 Simplified equivalent circuit of a high-ohmic DC voltage divider with stray capacitance C_e

The response time according to Eq. (4.24) is equal to that of a simple circuit with lumped elements and RC behavior whose time constant is $\tau = T$ and whose step response is given by Eq. (9.20). Figure 4.23 shows the corresponding equivalent circuit diagram. It is frequently used not only for high-ohmic DC voltage dividers (see Sect. 3.4.4), but also in general, when the voltage to be measured is not too fast, instead of the ladder network. It can be shown that the step response of the simplified equivalent circuit according to Fig. 4.23 is approximately the same as that of the ladder network according to Eq. (4.23) with asymptotic approximation to the final value (Ref. [1] of Chap. 1).

A realistic estimate of the stray capacitance C_e to ground can be made with Eq. (4.16). For a typical voltage divider, a rough guideline value of 15 pF/m results. The negative effect of the stray capacitance on the transfer behavior can be reduced or partially compensated by constructive field controlling measures, for example, by attaching large torus electrodes to the voltage divider at the top and bottom, or by connecting additional capacitors with graded capacitances parallel to the divider resistors. Due to the influence of the stray capacitance, resistive voltage dividers with more than 20 kΩ can hardly be found for the measurement of lightning impulse voltages. Too small resistances are also considered unsuitable, since the overshoot of the step response and the load on the impulse voltage generator may become too large. The use of particularly low-ohmic, low-inductive resistive voltage dividers is therefore reserved primarily for the measurement of steep-front impulse voltages with small impulse duration.

4.3.2.4 Field-Conformal Resistive Voltage Divider

So far, it has always been assumed that the resistance in the high-voltage part is arranged linearly with the divider height. If the voltage divider is considered as a homogeneous resistance column through which the current flows, the voltage and field distribution along that column would also be linear without taking into account the stray capacitance. On the other hand, the field distribution between the high-voltage electrode and ground by itself, i.e. without the resistance column, is highly nonlinear. For example, in the vicinity of a spherical electrode, the field strength decreases with the square of the distance from the sphere. This means that the field strength in the immediate vicinity of the divider top is particularly large compared to the field strength in the lower region of the voltage divider. In extreme cases, more than half of the field strength accounts for the upper 20% of the divider height. In the upper part of the voltage divider, therefore, there is a marked difference between the two fields, which leads to a strong normal component of the resulting field, i.e. perpendicular to the divider column. This normal component of the field is the driving force for a frequency-dependent leakage current that flows through the stray capacitances of the voltage divider to ground and worsens its transfer behavior at high frequencies.

The normal field component and thus the capacitive leakage current to ground becomes zero if the current distribution of the voltage divider is adapted to the undisturbed field pattern of the high-voltage electrode. The required resistance per unit length results from field calculations or field measurements. It is quite high in the region of the top electrode and decreases quickly towards the bottom of the voltage divider. The step response of such a *field-conformal voltage divider* actually shows a steeper rise and therefore a shorter rise time than a comparable voltage divider with linear resistance distribution [76, 77]. The principle of field-conformal resistance matching can be applied particularly effectively to high-ohmic voltage dividers with rated voltages of 1 MV and more. Such voltage dividers are therefore equally suitable for measuring lightning and switching impulse voltages [78]. The high-ohmic voltage divider with deionized water described in Sect. 4.3.2.7 is also of the field-conformal type, which is achieved by appropriate division into differently sized sections.

The principal disadvantage of the impulse voltage divider with field-conformal arrangement of the resistors, however, is that in the region of the top electrode the comparatively high-ohmic resistor is subjected to an even higher voltage stress than in the linear voltage divider. The risk of a flashover or breakdown in the upper region of the field-conformal voltage divider is thereby increased. The field-conformal voltage divider can therefore not be operated at the same maximum voltage as a linear voltage divider of the same height. If one tries to remedy this disadvantage by increasing the height of the field-conformal voltage divider, its transfer behavior deteriorates and the advantage over the smaller voltage divider with linear resistance distribution decreases or is completely lost.

Instead of adapting the ohmic to the capacitive potential profile along the voltage divider, the reverse approach is used in [79]. The high-voltage part of the described 1 MV voltage divider consists of a resistance wire with a total resistance of 6.56 kΩ, which is wound linearly around a 3.70 m high insulating tube with a diameter of 13 cm. The voltage divider is supplied with three torus electrodes which are optimized by a field computation program so that the resulting capacitive potential profile is also approximately linear with the divider height. The voltage divider is suitable for measuring full and front-chopped lightning impulse voltages within the permissible IEC error limits.

4.3.2.5 Optimized Measurement Tap

Another proposal for improving the transfer behavior relates to the positioning of the voltage tap in the resistive voltage divider with linear resistance distribution. The tapping is not carried out as usual at the low voltage part of the voltage divider, but at a point which lies between the top and bottom of the voltage divider [80–82]. The advantage of the *optimized measurement tap* lies in the fact that the high-frequency signal components which flow off via the distributed stray capacitance and do not reach the low-voltage part of the divider, can still largely be detected (Fig. 4.24). The optimized position of the measurement tap is defined here as the point at which the response time of the calculated step response is $T = 0$. According to theoretical investigations, the optimized measurement tap of a voltage divider of height H lies at $z_{opt} = 0.57H$, i.e. just over half of the total resistance.

The theoretical results have been largely confirmed by detailed measurements on an 800 kV impulse voltage divider. For this purpose, the measurement tap was inserted at different heights of the divider between $z = 0$ and $z = H$. Since in this method the measurement tap is at high-voltage potential, the measuring instrument must either also be operated at high potential or connected via an optoelectronic data transmission system.

The experimental step response shows a small overshoot and a short settling time for the optimized measurement tap at $z_{opt} = 0.57H$, while at the divider top ($z = H$) a large overshoot and at the low-voltage side ($z = 0$) an asymptotic approach to the final value without overshoot occur. The step response measured at the optimized measurement tap has a minimum rise time of 10 ns. However, oscillograms of full and chopped lightning impulses measured at the optimized tap and at the conventional low-voltage tap do not show any visible difference [81]. An advantage of the optimized measurement tap is that at the point z_{opt} the influence of external fields disappears. The impulse voltage divider can therefore be constructed as a very high-ohmic one.

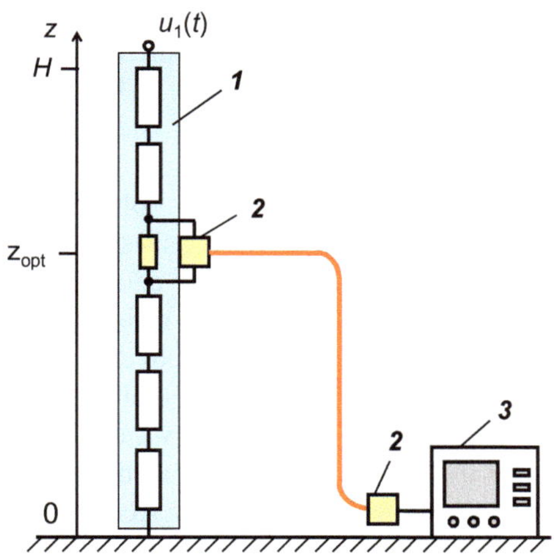

Fig. 4.24 Optimized measurement tap at $z_{opt} = 0.57H$. ***1*** Resistive voltage divider, ***2*** measurement tap and data transfer with optoelectronic data transmission system, ***3*** oscilloscope

4.3.2.6 Capacitive Signal Coupling into the Low-Voltage Part

The transfer behavior of a voltage divider can be improved by capacitive coupling of a portion of the impulse voltage from the top of the divider into the low-voltage output. The principle of *capacitive signal coupling* is described for a gas-insulated, metal-enclosed 1 MV impulse voltage divider in [83]. Thanks to the metal housing, the voltage divider with a total resistance of 10.5 kΩ is thus optimally shielded against electromagnetic interference from the outside. However, the large stray capacitance to the metal housing worsens the transfer behavior at higher frequencies. To improve the frequency response, the higher frequency components of the impulse voltage are capacitively decoupled at the divider top and fed with appropriate signal delay to the measurement tap at the low-voltage side, where they are superimposed on the output voltage. The coupling capacitance is simply realized by a circular plate electrode of 20 cm diameter, which is mounted opposite the central high-voltage conductor in the bushing (see Sect. 4.3.7). Via a coaxial cable and a delay network, the decoupled sensor signal is fed to the low-voltage divider output and superimposed on the output voltage. As a result, the rise time of the experimental step response of the voltage divider is reduced from the original 70 ns to less than 10 ns.

4.3.2.7 Voltage Divider with Aqueous Solution

Low- and high-ohmic impulse voltage dividers with aqueous solutions are intended for special measurement tasks. With a two-stage, quite elaborately constructed 80 kV impulse voltage divider, the discharges of a TEA CO_2 laser with a pulse width of about 100 ns are measured [84]. The first, potential-controlled divider stage with a total of 50 Ω consists of a coaxial insulating tube filled with a one-molar aqueous $CuSO_4$ solution. At its output, the second divider stage is connected, which is composed of a coaxial arrangement of carbon mass resistors of a total of 500 Ω. The 50 Ω output resistance of the second stage corresponds to the wave impedance of the measurement cable, which is also terminated with 50 Ω at the input of the oscilloscope. The step response has a rise time of 4 ns and is superimposed by a decaying oscillation with initial amplitude of 5% in the first 40 ns.

Another 600 kV impulse voltage divider with a height of 0.75 m, which is simple and inexpensive, consists in the high-voltage part of a one percent NaCl solution with a resistance of about 300 Ω [85]. The low-voltage part consists of parallel-connected metal film resistors with a total of 0.2 Ω. The step response has a rise time of 5 ns and a short overshoot within the first 25 ns. Due to the different structure of the high- and low-voltage part, the temperature dependence is relatively large. The temperature of the voltage divider must therefore be measured before and after its use and considered in the division ratio. The voltage divider is used to measure the voltage during dielectric strength tests on insulators with an uncertainty of less than 5%.

A high-ohmic 500 kV voltage divider filled with deionized tap water is used to measure the high-frequency output voltage and discharges of a Tesla transformer [86]. The 2.7 m voltage divider with a total of 1.6 MΩ is suspended from the ceiling and connected directly to the toroidal electrode of the Tesla transformer. By varying the dimensions of the water column in terms of length and diameter, a field-conformal resistance profile along the divider is achieved (see Sect. 4.3.2.4). The capacitance of the voltage divider is estimated at 0.76 pF. The transfer behavior of the voltage divider can be optimized by applying a convolution algorithm in the frequency domain (see Sects. 9.2 and 9.3), which is underlined by examples of the step response and a discharge pulse.

4.3.3 Capacitive Impulse Voltage Divider

Capacitive voltage dividers are preferably used to measure switching impulse voltages up to the highest test voltages. They are not or only slightly damped and therefore less suitable for measuring lightning impulse voltages or fast transient voltages. Compared to resistive voltage dividers, they have the advantage that the tail of a generated switching impulse voltage is not significantly affected. In addition, self-heating can be largely excluded except from the effect of the dissipation factor of the capacitors used. Theoretical investigations on the basis of the ladder network show that the step response of the undamped capacitive voltage divider tends to strong high-frequency oscillations. As a rule, it is therefore used with an external damping resistor. A special type of the capacitive voltage divider is the *damped capacitive voltage divider* with internally distributed damping resistors, which will be discussed in Sect. 4.3.4.

4.3.3.1 Impulse Measuring System with Capacitive Voltage Divider

The basic circuit of a complete impulse voltage measuring system with a capacitive voltage divider is shown in Fig. 4.25. The high-voltage capacitor C_1 usually consists of a series connection of oil-impregnated capacitors with a total capacitance of several 100 pF. On the low-voltage side, C_2 also consists of oil-impregnated capacitors or low-inductance plastic film capacitors connected in parallel. When selecting C_2, the temperature coefficient should be of the same order of magnitude as C_1. The large capacitance and current carrying capacity of C_2 can be achieved by connecting a large number of capacitors in parallel, which also reduces the inductance on the low-voltage side. A perfect compensation of the inductive component of the high-voltage capacitor C_1 is not always intended even for the capacitive voltage divider. A larger inductance on the low voltage side may be advantageous since it leads to a slight overshoot of the step response and thus to a shorter rise time.

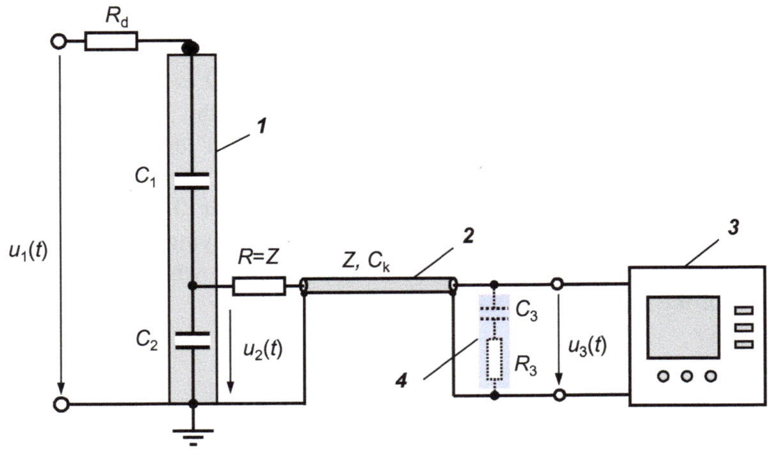

Fig. 4.25 Measuring system with capacitive impulse voltage divider and digital recorder. *1* Capacitive impulse voltage divider, *2* coaxial cable with impedance Z and capacitance C_K, *3* digital recorder. *4* Burch termination C_3 R_3 in case of a longer coaxial cable

For measuring impulse voltages, the capacitive voltage divider is usually operated with an external resistor R_d to attenuate the oscillations propagating in the divider and on the high-voltage lead. The origin of the oscillations is of a different nature. The high-voltage capacitors used and their inductances, on the one hand, represent an oscillating LC network in the MHz range. On the other hand, high-frequency oscillations are formed due to *reflection phenomena* of traveling waves in the capacitive voltage divider. They are caused when a steep voltage impulse enters the divider top and passes through the voltage divider, is reflected at the divider bottom, returns to the divider top almost undamped, is reflected there again, and so on. As a result, oscillations occur in the frequency range up to 100 MHz by traveling waves. The step response of the capacitive voltage divider therefore shows violent oscillations, which are only weakly damped by the ohmic losses of the capacitors and the connecting lead and by the skin effect [53–55]. With an external damping resistor R_d of 300–600 Ω at the beginning or end of the high-voltage lead, the oscillations can be reduced to such an extent that the capacitive voltage divider is occasionally also used to measure full lightning impulse voltages.

4.3.3.2 Low-Voltage Measuring Circuit

Although the pure capacitive voltage divider is not suitable for measuring fast transient voltages, the low-voltage circuit is usually designed according to high-frequency criteria. At the divider output, therefore, the coaxial cable is connected not directly but via a series resistor R equal to the cable impedance Z of

50 Ω or 75 Ω. For fast transient voltages, the coaxial cable acts at first like a resistor with the wave impedance Z. The steep voltage impulse entering the coaxial cable from the divider output and travelling to the recorder input is thus halved by the series connection of R and Z. After the cable delay τ, the voltage wave reaches the other end of the cable at the recorder input, where it is doubled due to the high-ohmic input resistance of 1 MΩ or more, so that the full voltage is applied to the recorder. The quasi-open cable end at the recorder input causes the voltage wave to be reflected to the voltage divider. After twice the double cable delay 2τ, the reflected wave arrives again at the voltage divider and finds there a reflection-free cable termination with the series resistance $R = Z$. The reflected wave is thus completely absorbed at R, since the large low-voltage capacitance C_2 is practically a high-frequency short circuit.

For low-frequency signals, the cable delay is negligible and the wave impedance Z in Fig. 4.25 is not effective. Since the series resistance R is negligible compared to the high-ohmic recorder input, almost the total impulse voltage lies on the recorder. Thus, the capacitive impulse voltage divider is equally balanced for low- and high-frequency voltages. By the way, the cable termination must not be implemented as a parallel resistor at the recorder input, as is the case with the ohmic voltage divider in Fig. 4.20. Such a termination would cause a rapid discharge of the low-voltage capacitor C_2 and thus a shortening of the time to half-value of switching impulses.

4.3.3.3 Burch Termination

In the case of a long coaxial cable, the *Burch termination* is advantageous for accurate measurement of fast transient signals. It consists of the series connection of the resistor R_3 and capacitor C_3, which are connected in parallel to the recorder input, as shown by the dashed lines in Fig. 4.25. Without the Burch termination, the coaxial cable loads the divider output with the cable capacitance C_k after double the cable delay time 2τ of the signal. For larger cable lengths, C_k is no longer negligible compared to C_2, so that the effective capacitance of the low-voltage part of the divider increases. The division ratio at the beginning of the signal transmission and after the double cable delay time is therefore different. This leads to an initial overshoot of the measured signal until time $t = 2\tau$ when the reflected signal arrives at the divider output. With the Burch termination with C_3 and R_3 parallel to the recorder input, this overshoot can be compensated. An optimized transfer behavior with reduced overshoot is obtained for $R_3 = Z$ and $C_k + C_3 = C_1 + C_2$ [58, 87]. The long coaxial cable is thus terminated at both ends with its high-frequency wave impedance.

4.3.3.4 Simple Equivalent Circuit Diagrams with Stray Capacitance

A rudimentary treatment of the transfer behavior of purely capacitive voltage dividers on the basis of the ladder network in Fig. 4.19 can be found in [53]. The transfer behavior is characterized by the already mentioned high-frequency oscillation within the voltage divider, which can be very large and long lasting in the absence of a damping resistor. The purely capacitive voltage divider without damping resistor is therefore less suitable for measuring fast transient voltages than other voltage dividers. Therefore, details of the ladder network will not be discussed in detail here.

The capacitive voltage divider has stray capacitances to ground of the same order of magnitude as the resistive voltage divider. In analogy to the resistive voltage divider, the resulting stray capacitance C_e can be taken into account in the simplified equivalent circuit diagram by a concentrated capacitance of $^2/_3C_e$, whereby inductances and resistances are neglected (Fig. 4.26a). The equivalent circuit diagram can be further simplified by reducing the high-voltage capacitance C_1 by an amount $C_e/6$ (Fig. 4.26b). Both equivalent circuit diagrams show that the stray capacitance reduces the current through the low-voltage capacitor C_2, so that the output voltage $u_2(t)$ at C_2 becomes smaller. The division ratio $\hat{u}_1/\hat{u}_2$ is increased accordingly.

In order to limit the influence of the stray capacitance and the cable capacitance, the high-voltage capacitance C_1 should be as large as possible. On the other hand, the frequency behavior of high-voltage capacitors with large capacitance is not very good due to the manufacturing process. In addition, the load of the impulse voltage generator is increased, which affects the impulse shape. Comparing the effect of the stray capacitance of capacitive and resistive voltage dividers, a decisive difference can be detected. The frequency behavior of the capacitive voltage divider is not affected by the stray capacitance according to the simplified equivalent circuit diagrams in Fig. 4.26. The division ratio thus appears to be the same for low and high frequencies. In a correspondingly extended equivalent circuit diagram, however, inductances and resistances are also to be considered. Then, the transfer behavior and thus the division ratio again become frequency-dependent.

The results for the optimized measurement tap in resistive voltage dividers can be transferred to capacitive voltage dividers (see Sect. 4.3.2.6). Theoretical investigations show that the optimized height of the measurement tap is again $z_{opt} = 0.57H$. At this point again the response time $T = 0$ and the influence of

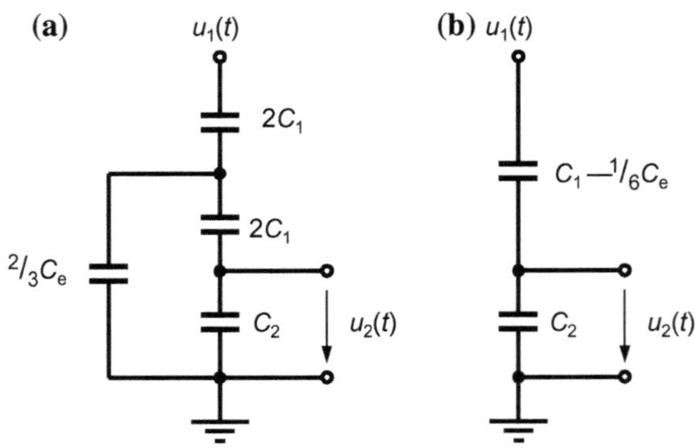

Fig. 4.26 Simplified equivalent circuits of the capacitive voltage divider in which the capacitance C_e to ground is taken into account by a: **a** parallel capacitance $^2/_3C_e$, **b** reduced capacitance $C_1 - {^1/_6}C_e$ in the high-voltage arm

external fields disappears. A capacitive voltage divider with a measurement tap at z_{opt} could therefore be built with a relatively small capacitance on the high-voltage side, resulting in lower production costs [82].

4.3.3.5 Variants of the Capacitive Impulse Voltage Divider

Compressed-gas capacitors after Schering and Vieweg are often used in the AC voltage measurement technique because of their excellent properties (see Sect. 11.5). They can also be used as impulse voltage dividers in combination with a low-inductance capacitor in the low-voltage part [88]. The tendency to oscillate can be reduced by a damping resistor inserted between the compressed-gas capacitor and the low-voltage capacitor. The concentric arrangement of the high- and low-voltage electrodes basically promises wide-band transfer behavior and effective shielding against external interferences. The investigated capacitive voltage divider with an 800 kV compressed-gas capacitor has a step response without overshoot and a rise time of 50 ns. Due to the low capacitance of compressed-gas capacitors, which is mainly in the range of 10–100 pF, the load of the impulse voltage generator is minimal.

In other measuring systems with compressed-gas capacitors, the impulse voltage is obtained by integrating the capacitive current $i = C \cdot du/dt$. The basic measurement principle is comparable to that in which an AC voltage is measured with a compressed-gas capacitor and a subsequent integrating circuit (see Sect. 2.5.2.2). The capacitor used in [89] is a coaxial electrode arrangement that is part of a metal-enclosed SF_6-insulated switchgear installation and used up to 200 kV. The measuring system is equipped with active or passive integration circuits. The step response shows a rise time of 120 ns and minimal overshoot of 5%, the step voltage being generated by firing a sphere gap. In [90], a circuit with a plate capacitor and a passive integrator is described to measure steep front impulse voltages up to 250 kV. With a 240 Ω damping resistor, the system's step response has a response time of 2.8 ns, a rise time of 8 ns and an overshoot of 4%.

The load capacitor C_b of the impulse voltage generator is occasionally equipped with an additional low-voltage capacitor and then used as a voltage divider for measurement purposes. The total capacitance of C_b is in the range of 1–10 nF, which is usually achieved by the series connection of several individual capacitors. The correspondingly required high capacitance is the cause that the frequency behavior is not very good. Special constructions of the load capacitor are internally equipped with low-ohmic damping resistors and are therefore better suited for measuring impulse voltages. However, as has already been pointed out in Sect. 4.3.1.1, such a voltage divider with C_b between the impulse generator and the test object shall not be used to measure lightning impulse voltages (Ref. [2] of Chap. 2).

4.3.4 Damped Capacitive Impulse Voltage Divider

The *damped capacitive impulse voltage divider*, also called *Zaengl divider*, is particularly well suited for measuring lightning impulse voltages and other fast transient voltages up to the UHV range, usually in conjunction with a digital recorder at the divider output. The Zaengl divider consists of a large number of series connected resistors and capacitors in the high- and low-voltage parts. Together with an external damping resistor, oscillations are successfully suppressed. The structure of the damped capacitive impulse voltage divider thus approximately resembles the homogeneous ladder network, which can well be used as equivalent circuit to calculate typical characteristics of the step response and derive basic properties. Special software for further optimization of the measurement behavior is available. Examples of the construction of accurate reference dividers and their behavior are given.

4.3.4.1 Measuring System with Damped Capacitive Voltage Divider

Figure 4.27 shows the basic arrangement of the complete measuring system with a damped capacitive impulse voltage divider. Characteristic are the internal damping resistors R_1' in series with the capacitors C_1' in the high-voltage part and R_2' in series with C_2' in the low-voltage part. Due to the distributed damping resistors, the oscillations which occur with rapidly changing voltages in the purely capacitive

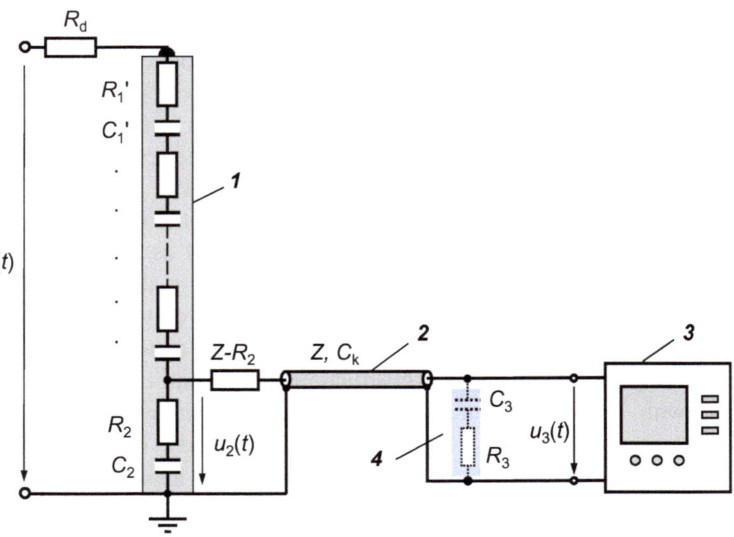

Fig. 4.27 Measuring system with damped capacitive voltage divider and digital recorder. ***1*** Damped capacitive impulse voltage divider (Zaengl divider), ***2*** coaxial cable with wave impedance Z and cable capacitance C_k, ***3*** digital recorder, ***4*** Burch termination C_3R_3 for longer coaxial cable

voltage divider by the traveling waves are successfully suppressed. A number of ten $R_1'C_1'$ elements in the high-voltage part are considered sufficient. The damped capacitive impulse voltage divider works as a capacitive divider for low signal frequencies and as a resistive divider for high frequencies [53–55, 91].

In the high-voltage part of damped capacitive impulse voltage dividers, usually oil-filled capacitors are used and in the-low voltage part also oil-filled capacitors or plastic film capacitors. Particularly suitable for use in "fast" damped capacitive dividers with rated voltages below 1 MV are the class 1 power *ceramic HF plate capacitors* with *Rogowski profile*, which have a permittivity $\varepsilon_r \leq 60$. They are characterized by a very low inductance and good long-term stability. Depending on its size, the capacitance of a single plate capacitor with this permittivity amounts to a maximum of 2000 pF and the permissible impulse load up to 50 kV. Due to the series connection of the plate capacitors, the total high-voltage capacitance C_1 of a voltage divider with a rated voltage of up to 1 MV reaches only a few 100 pF at. The capacitance should not be less than 100 pF, in order to minimize the influence of external fields. If the requirement for the measurement accuracy is higher, the influence of the stray capacitance to ground must be investigated separately (see Sect. 10.3.10). Class 2 power ceramic plate capacitors have a different dielectric with higher permittivity. However, they are not used in impulse voltage dividers because of their greater voltage and temperature dependence, and their long-term stability is not satisfactory.

As internal damping resistors R_1' between the individual capacitors C_1' in the high-voltage circuit, the types mentioned in Sect. 4.3.2.1 come into consideration. Carbon film resistors without any incised helical groove are well suited for smaller voltage dividers because of their low inductance and high impulse load capacity. The voltage stress of the resistors in the voltage divider is particularly high with chopped impulse voltages and comparable fast voltage changes, since then the capacitors act in the first approximation as a short circuit. The required current carrying capacity of R_1' is achieved by connecting several individual resistors in parallel.

When dimensioning the low-voltage part, the external damping resistor R_d must also be taken into account in order to obtain a voltage divider that is balanced for both fast and slow voltages. For the conventional balance of a compensated voltage divider with external damping resistor, the following relationship holds without considering the stray capacitance:

$$\boxed{\frac{R_1 + R_2 + R_d}{R_2} = \frac{C_1 + C_2}{C_1} = \frac{L_1 + L_2}{L_2}}, \tag{4.25}$$

wherein the index 1 denotes the elements on the high-voltage side and the index 2 the elements on the low-voltage side. If a long coaxial cable is used, its capacity may need to be considered in Eq. (4.25). For common division ratios between 500:1 and 2000:1, the resistance R_2 in the low-voltage part is only a few tenths of an ohm, while C_2 is in the range of 0.5 μF. By connecting many RC series elements in parallel in the low-voltage part, the required values of R_2 and C_2 are achieved, while the inductance L_2 is reduced. In order to compensate for the influence of the stray capacitance, an increased inductance on the low-voltage side can once again be advantageous.

4.3.4.2 Equivalent Circuit Diagram and Step Response

The equivalent circuit diagram of the damped capacitive impulse voltage divider can be represented as a ladder network (see Fig. 4.19a) with n individual sections as shown in Fig. 4.28. In the individual sections of the ladder network, the elements R' and C' represent the actual resistors and capacitors of the voltage divider. Further elements are the longitudinal inductances L' of the elements and their connecting lines, the stray capacitances C'_p parallel to the elements and the stray capacitances $C'_e/2$ to ground at the beginning and end of each ladder section.

From the general solution of Eq. (4.18) for the transfer function of a ladder network, the step response $g(t)$ of the damped capacitive voltage divider is obtained by Eq. (4.19) [53–55]:

$$\boxed{g(t) = 1 - \frac{C_e}{6(C + C_p)} + 2\,e^{-at} \sum_{k=1}^{\infty} (-1)^k \frac{\cosh(b_k t) + \frac{a}{b_k}\sinh(b_k t)}{\left(1 + \frac{C_p}{C} + \frac{C_e}{C\,k^2\pi^2}\right)\left(1 + \frac{C_p}{C_e}k^2\pi^2\right)}}. \tag{4.26}$$

where: $R = nR'$, $L = nL'$, $C = C'/n$, $C_e = n\,C'_e$, $C_p = C'_p/n$,

$$a = \frac{R}{2L},$$

$$b_k = \sqrt{a^2 - \frac{k^2\pi^2\left(1 + \frac{C_p}{C} + \frac{C_e}{C\,k^2\pi^2}\right)}{L\,C_e\left(1 + \frac{C_p}{C_e}k^2\pi^2\right)}}.$$

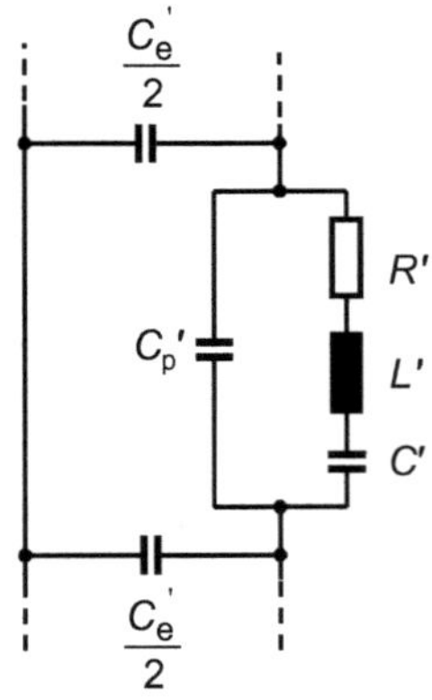

Fig. 4.28 Individual section of the equivalent circuit diagram of a damped capacitive impulse voltage divider

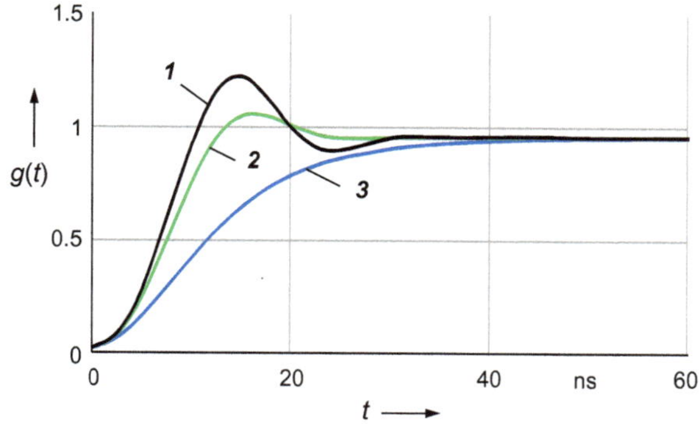

Fig. 4.29 Step response *g* (*t*) of a damped capacitive voltage divider, represented as a ladder network with sections according to Fig. 4.28 and calculated using Eq. 4.26 (C = 150 pF, C_e = 40 pF, C_p = 1 pF, L = 2.5 μH). ***1***: R = 0.75 kΩ, ***2***: R = 1 kΩ, ***3***: R = 2 kΩ

Figure 4.29 shows three examples of the step response of a damped capacitive voltage divider, calculated according to Eq. (4.26) with the capacitance C = 150 pF and the resistances R = 0.75, 1 and 2 kΩ. Further values in the calculation are: stray capacitance to ground C_e = 40 pF, parallel capacitance C_p = 1 pF and inductance L = 2.5 μH. The basic form of the step response of the damped capacitive voltage divider is similar to that of the resistive voltage divider in Fig. 4.22, but with a different time scale. Compared to the resistive voltage divider, the resistance in the damped capacitive voltage divider is much smaller, so that its stray capacitance to ground has an effect only at significantly higher frequencies. Therefore, the damped capacitive impulse voltage divider basically has a better transfer behavior than the resistive voltage divider and the load on the test circuit is lower.

For small resistances, the coefficients b_k in Eq. (4.26) become imaginary and the hyperbolic functions are converted into trigonometric functions. This means that the step response tends to oscillate. For example, for R = 0.75 kΩ, the step response $g(t)$ of the damped capacitive voltage divider shows a pronounced oscillation (curve ***1*** in Fig. 4.29). The oscillation or overshoot of $g(t)$ is significantly lower for R = 1 kΩ (curve **2**). For even larger resistances, the step response runs asymptotically to the final value without any oscillation (curve ***3***).

As with the resistive voltage divider, the calculated step response $g(t)$ in Fig. 4.29 shows an initial jump at the beginning. Again, this can be explained by the capacitive division ratio of the parallel capacitances C'_p to the stray capacitances C'_e (see Sect. 4.3.2.2). The final shape of the step response remains below one because a part of the step current flows via the stray capacitances to ground, thereby increasing the effective division ratio. The difference between the calculated final value $g(t = \infty)$ and the value one is time- or frequency-independent and is expressed on the right side of Eq. (4.26) by the second term $C_e/6(C + C_p)$. For the limiting case $C = \infty$, the resistive voltage divider is obtained with $g(t = \infty) = 1$.

The transfer behavior of real damped capacitive impulse voltage dividers cannot be fully described by the general ladder network. As with the resistive voltage divider, the influence of the low-voltage part and the high-voltage lead can be analytically taken into account by additional and improved terms [54, 55]. Procedures for experimental optimization of the low-voltage part are dealt with in Sect. 4.3.4.4.

4.3.4.3 Measurement Cable and Burch Termination

The output of the voltage divider is connected to the high-ohmic input of the recorder via the series resistance $R = Z - R_2$ and the coaxial measurement cable with its characteristic impedance Z (see Fig. 4.27). With R and the resistance R_2 in the low-voltage part of the divider, the coaxial cable at the divider output is terminated with its characteristic impedance, since C_2 represents a short circuit for fast signals. A voltage impulse which is reflected at the high-ohmic recorder input and travels back to the divider output is thus absorbed. Long coaxial cables have a capacitance C_k that is no longer negligible compared to C_2. As with the purely capacitive voltage divider, a *Burch termination* with R_3 and C_3 at the recorder input is then recommended [87]. Theoretical investigations show that for an assumed capacitance $C_k = 0.1(C_1 + C_2 + C_3)$ the transfer behavior becomes optimal for $R_3 = k \cdot Z$ and $k(C_k + C_3) = (C_1 + C_2)$ with $k = 1.25$. The initial overshoot of the step response during the cable delay time 2τ is thereby significantly reduced. Taking cable losses into account, a reduced value of $1 < k < 1.25$ is recommended [54].

4.3.4.4 Optimization of the Low-Voltage Part

The analytical calculation of the transfer behavior of a voltage divider with the ladder network is useful for general statements. However, optimization for a particular voltage divider in the UHV range can only be achieved by a meaningful combination of measuring the step response and applying special calculation methods. With a comparatively small change in the circuit or the structure of the low-voltage part, a significant improvement in the transfer behavior is often achieved. In general, the aim is to achieve a slight overshoot of the step response by adjusting the low-voltage circuit. The experimental step response is recorded either for the complete voltage divider or only for the low-voltage part into which a step current is injected [92]. However, detailed investigations on a 6 MV impulse voltage divider show that the measurement and evaluation of the step response of measuring systems in the UHV range is difficult (Ref. [34] of Chap. 9).

Damped capacitive impulse voltage dividers with rated voltages of more than 800 kV usually consist of oil-impregnated plastic film capacitors whose permittivity can be frequency-dependent. The step response of these voltage dividers often shows a slower increase after about 95% of the final value. This behavior is called *creeping*. Such a voltage divider is therefore not suitable as a reference divider. The step response can be optimized by connecting a resistor R in series with a capacitor C in parallel with the low voltage side. To additionally compensate for the influence of a large cable capacitance, the RC series circuit is implemented as a Burch termination on the high-ohmic recorder input (see Sect. 4.3.4.3).

If the output voltage of the divider is too large for the recorder, a secondary damped capacitive divider will be connected in parallel with the divider output. The values of the secondary low-voltage divider are optimal for a slight overshoot of the step response. The elements are calculated analytically using the ladder network as an equivalent circuit diagram [93].

The individual equivalent circuit diagram of a voltage divider and thus its step response can be calculated with special software. The calculated step response is then compared with the measured step response, which helps to further improve the equivalent circuit. Using the example of a 1 MV divider for switching impulse voltages, it is shown in [94] that the pronounced creeping of the step response can be completely eliminated by optimizing the circuit of the low-voltage part. This reduces the experimental response time of the impulse voltage divider from 1 μs to 32 ns, the settling time from 26 μs to 2.4 μs and the overshoot from 26 to 2.4%.

4.3.4.5 Optimally and Weakly Damped Voltage Dividers

For the damped capacitive voltage divider as well as for the resistive voltage divider, it is quite desirable that a small, time-limited overshoot of the step response be achieved to reduce the rise time and response time. From this point of view, the *optimally damped capacitive impulse voltage divider* should have a total internal damping resistance of [53]:

$$\boxed{R_{\mathrm{opt}} = (3\text{--}4)\sqrt{\frac{L}{C_{\mathrm{e}}}}}. \tag{4.27}$$

For a damped capacitive voltage divider with $L = 2.5$ μH and $C_{\mathrm{e}} = 40$ pF, the values for R_{opt} are between 750 and 1000 Ω. The step responses calculated for these limits are shown as curves ***1*** and ***2*** in Fig. 4.29. Since both L and C_{e} increase approximately linearly with the height of the voltage divider, R_{opt} remains constant. Equation (4.27) thus applies to both small and large voltage dividers, i.e. independent of the rated voltage. When dimensioning a damped capacitive impulse voltage divider, the lower limit of R_{opt} in Eq. (4.27) or an even smaller value is chosen. As a result, the overshoot of the step response is greater, which is acceptable in the case of a short duration, and the response time is shorter, which is often favorable.

A variant is the *low damped capacitive impulse voltage divider* with a rated voltage of usually more than 1 MV. The value of the internally distributed damping resistor is calculated as [95–97]:

$$\boxed{R \approx (0.25\ldots1.5)\sqrt{\frac{L}{C}}}, \tag{4.28}$$

where L is the inductance of the measuring circuit and C is the capacitance of the voltage divider. For low damping, Eq. (4.28) gives a typical resistance of 50–200 Ω, distributed in the high-voltage part. In the low-voltage part as well as in the high-voltage lead, no additional damping resistors are provided. The apparent advantage of the low damped capacitive voltage divider is that, due the small time constant RC, it can be used simultaneously as the measurement capacitor and as the load capacitor of the impulse voltage generator. However, the combination of load capacitor and voltage divider is not permitted for standard-compliant testing of power apparatus with lightning impulse voltages (Ref. [2] of Chap. 2).

The internal damping of a capacitive impulse voltage divider with distributed damping resistors has a further advantage. Calculations of the voltage distribution within a multistage high-voltage capacitor show that locally and temporally varying oscillations of different magnitudes are generated when a voltage step is applied. In this case, the capacitor represented as a ladder network according to Fig. 4.19a is stressed at the top by an overvoltage whose maximum value is a multiple of the step amplitude. With the internally distributed damping resistors, the initial magnitude of the overvoltage cannot be significantly reduced, but its duration is considerably shortened. Thus, with a distributed damping resistance of 1 kΩ in total, an approximately linear voltage distribution along the capacitor column is achieved after only 50 ns. The shorter voltage stressing of the capacitors is, of course, advantageous for their service life [98].

4.3.4.6 Examples of Reference Dividers

With appropriate dimensioning and use of low-inductance elements, damped capacitive impulse voltage dividers have excellent transfer behavior. Therefore, they are well suited as *reference dividers* for the calibration of other impulse voltage dividers with lightning and switching impulse voltages. For lightning impulse voltage measurements, low-ohmic resistive dividers are also suited as reference dividers. Figure 4.30a shows in the foreground a damped capacitive impulse voltage divider as a reference divider up to 500 kV. The high-voltage part consists of 20 stages with ceramic HF plate capacitors in series with carbon film resistors without any incised groove. The total capacitance is 150 pF, the internal damping resistance 400 Ω. Each damping resistor between two plate capacitors consists of a parallel connection of six carbon film resistors. The two-stage version in Fig. 4.30b also serves as a reference divider for impulse voltages up to 1 MV. The stages are similar in construction to the single-stage version in Fig. 4.30a.

Reference systems with reference dividers are characterized by reduced measurement uncertainties, which are as follows (Ref. [2] of Chap. 2):

- 1% for the test voltage value of full and chopped impulse voltages with times to chopping $T_c \geq 2$ μs,
- 3% for the test voltage value of front-chopped impulse voltages with times to chopping 0.5 μs $\leq T_c < 2$ μs, and
- 5% for the time parameters.

Reference systems are typically used only for the calibration of other voltage measuring systems in the high-voltage laboratory and are therefore not subject to

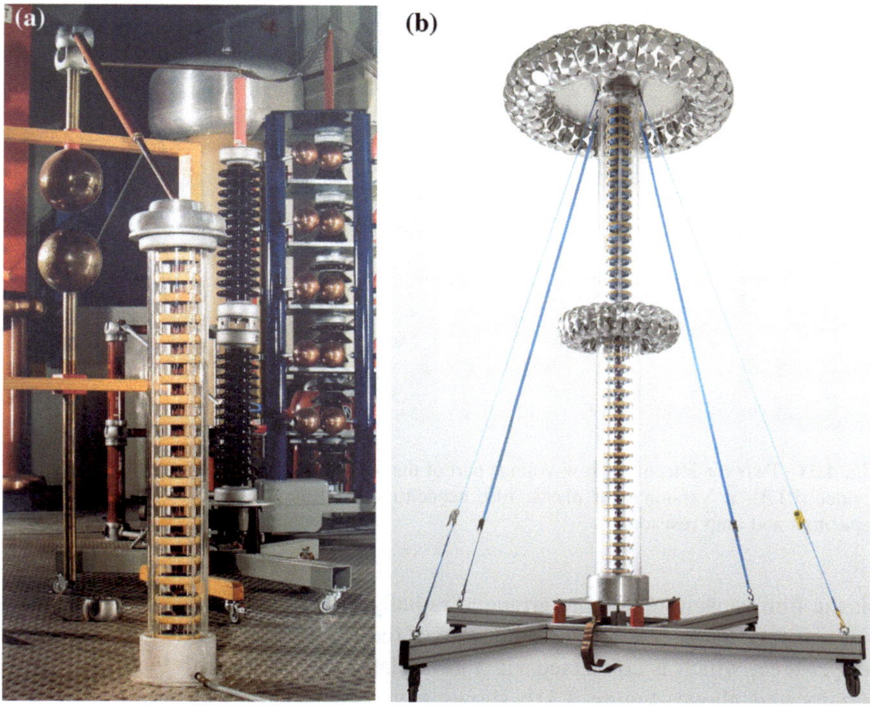

Fig. 4.30 Two examples of damped capacitive impulse voltage dividers (PTB). **a** Single-stage version for 500 kV (in the background: chopping spark gap on the left and impulse voltage generator on the right), **b** two-stage version for 1 MV

harsh test conditions, for example due to weather or mechanical stress. The long-term stability is verified by regular performance tests and checks.

Low-Voltage Part

The low-voltage part is housed in a metal box at the bottom of the damped capacitive voltage divider. For the single-stage reference divider in Fig. 4.30a, a division ratio of about 2000 is established according to Eq. (4.25). Carbon film resistors and plastic film capacitors are well suited because of their high-frequency characteristics and impulse voltage strength. Basically, even better characteristics show chip resistors and chip capacitors. By parallel connection of a larger number of resistors R_2 in series with the capacitors C_2, the current load is distributed and the total inductance of the low-voltage part is kept small.

Figure 4.31 shows two low-inductive versions of the low-voltage part, which are housed in metal boxes and arranged just below the high-voltage part. The first version in Fig. 4.31a consists of conventional carbon film resistors in series with

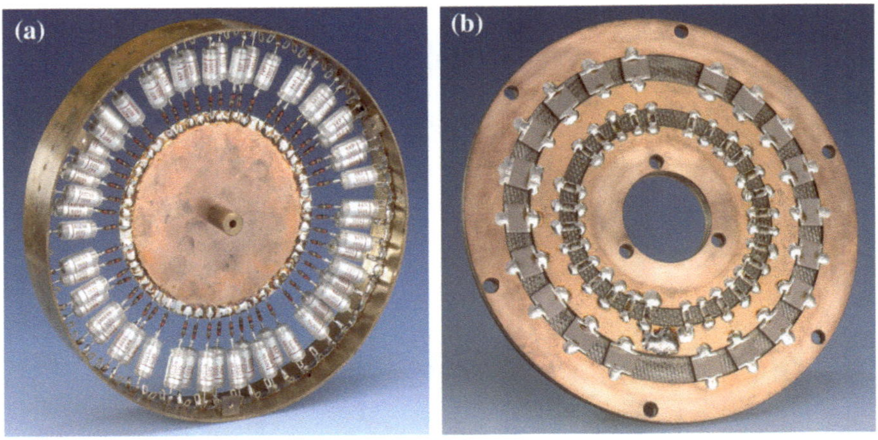

Fig. 4.31 Two versions of the low-voltage part of the 500 kV damped capacitive impulse voltage divider (PTB). **a** Version with plastic film capacitors and film resistors. **b** Version with chip capacitors and chip resistors

plastic film capacitors that are arranged radially in a circular plane. The capacitors and resistors are soldered to the outer ring of the grounded metal box or to the inner copper area, which is connected via a gold-plated plug contact to the high-voltage part located above. Figure 4.31b shows a variant of the low-voltage part in an extremely low inductance arrangement using chip capacitors and chip resistors. They are soldered between two concentric Cu rings on a printed circuit board.

External Damping Resistor

The external damping resistor R_d of the reference divider in Fig. 4.30a is a series and parallel connection of carbon film resistors or, as an alternative, a low-inductive resin-cast resistor web (see Sect. 4.3.2.1). The damping resistor is arranged at the beginning of the tubular high-voltage lead. The lead length is variable by using individual tubes that can be plugged into each other. The resistance $R_d = 330\ \Omega$ approximately corresponds to the wave impedance of the horizontal lead, so that almost no traveling wave phenomena occur along the lead. The internal and external damping resistors have a total of 730 Ω, which is close to the lower limit of optimal damping recommended in Eq. (4.27).

Step Response and Response Parameters

Figure 4.32 shows the step response of the single-stage reference divider in Fig. 4.30a with external damping resistor $R_d = 330\ \Omega$, recorded in the circuit with a horizontal lead of length $L = 1.5$ m according to Fig. 9.16a. The reference divider

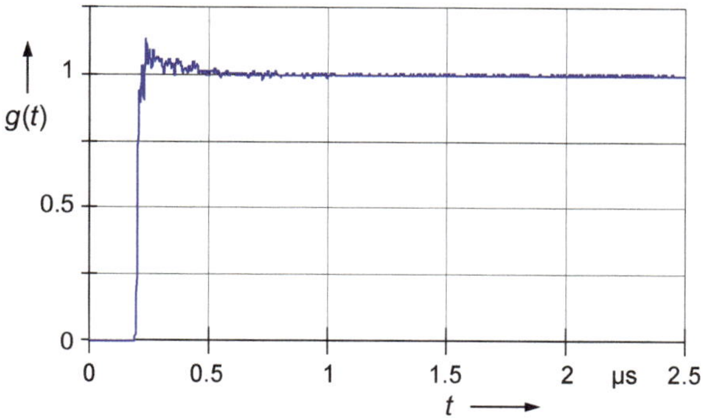

Fig. 4.32 Step response of the 500 kV damped capacitive voltage divider in Fig. 4.30a (damping resistor R_d = 330 Ω, lead length L = 1.5 m)

is matched such that the absolute values of the partial response times T_α and T_β of the step response are approximately equal. The result is a small overshoot and a negligible response time $T_N \approx 0$. Since the damping resistance corresponds approximately to the wave impedance of the high-voltage lead, its length has only a small influence on the step response and thus on the response parameters. A significant difference between the step responses of the reference divider with the two differently structured low-voltage parts in Fig. 4.31 is not recognizable.

The reference divider is used to calibrate other impulse voltage dividers, which usually have a larger overall height. For the comparison measurement, therefore, the reference divider must be connected via a correspondingly long high-voltage lead to the common measurement point of the *Y circuit* (see Fig. 10.2). However, the measurement behavior of the reference divider must not change significantly due to the longer connecting lead. This can be checked on the basis of the recorded step response and their response parameters. Figure 4.33 gives an impression of the dependence of the experimental response time T_N and the settling time t_s on several parameters. These are the external damping resistor R_d, height H of the step voltage generator, length L and arrangement of the high-voltage lead and bandwidth B of the digital recorder used. Table 4.1 provides further information on the experimental set-up for the step response measurement [99].

The results of the investigation show that the values of the experimental response time differ only within ±8 ns from the rated value $T_N = 0$. Another important parameter is the settling time, the mean value of which results in t_s = 180 ns with deviations within ±20 ns. Therefore, the small dispersion of the two response parameters will only affect the measurement of front-chopped lightning impulse voltages. For example, an increase of about 5 ns in the response time will result in a theoretical peak value error of about 1% for an impulse voltage chopped after 0.5 µs. The investigated variations in the measurement set-up also have only minimal influence on the other response parameters. Altogether, the reference divider is well suited for the calibration of other high-voltage dividers with standard full and front-chopped lightning impulse voltages within the permissible error limits, even with different designs of the high-voltage lead.

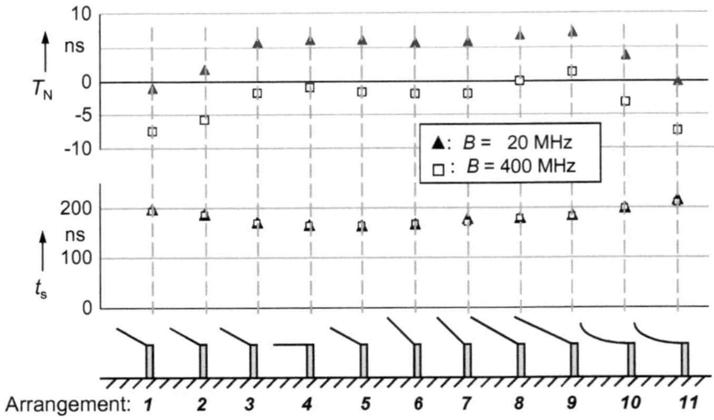

Fig. 4.33 Experimental response time T_N and settling time t_s of the 500 kV damped capacitive voltage divider with different measurement bandwidth B and set-ups (see Table 4.1). R_d external damping resistor, L length of the high-voltage lead, H height of the step generator B bandwidth of the recorder

Table 4.1 Further information on the experimental set-up for the step response measurement

Set-up (Fig. 4.33)	1	2	3	4	5	6	7	8	9	10	11
R_d in Ω	333	346	366	366	366	366	366	366	366	366	366
H in m	2.7	2.7	2.7	1.4	2.7	3.3	3.3	3.3	3.3	3.3	3.3
L in m	2.5	2.5	2.5	2.5	2.5	2.5	2.5	4.5	4.5	5.5	5.5
Lead	Cu pipe, 2 cm in diameter									Hose	Wire

4.3.5 Parallel-Mixed Resistive-Capacitive Voltage Divider

The *parallel-mixed resistive-capacitive voltage divider* is a high-ohmic voltage divider with parallel-connected capacitors to improve the transfer behavior at higher frequencies. The frequency-dependent influence of the stray capacitance is thus reduced by the parallel capacitance. The mixed voltage divider is resistive at low frequencies and capacitive at higher frequencies. The realization of the additional parallel capacitance can be done in different ways. In Fig. 4.34, each of the n series resistors on the high-voltage side is connected in parallel with a capacitor C_1', and a capacitor C_2 is connected in parallel with the low-voltage resistor. Taking into account the damping resistance R_d and the cable capacitance C_k and neglecting the stray capacitances and inductances, the division ratio of the balanced voltage divider at low frequency is:

$$\boxed{\frac{u_1}{u_2} = \frac{C_1 + C_2 + C_k}{C_1} = \frac{R_1 + R_2 + R_d}{R_2}} \tag{4.29}$$

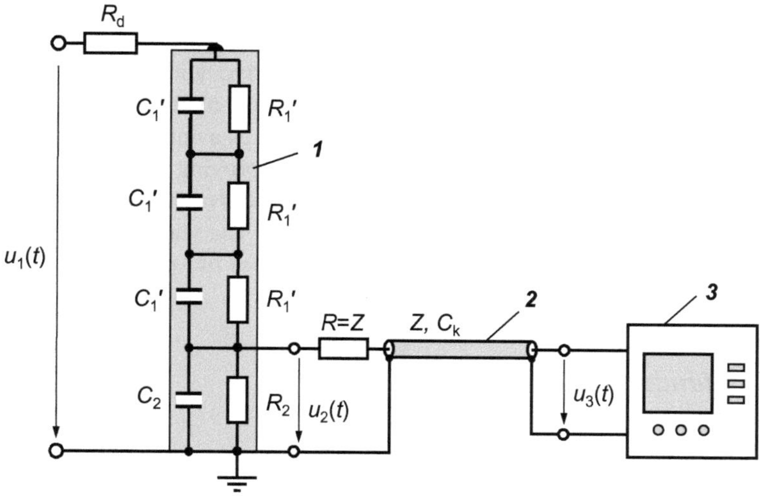

Fig. 4.34 Measuring system with parallel-mixed resistive-capacitive voltage divider. ***1*** Parallel-mixed resistive-capacitive voltage divider, ***2*** coaxial cable with wave impedance Z and cable capacitance C_k, ***3*** digital recorder

with $C_1 = C_1'/\mathrm{n}$ and $R_1 = \mathrm{n}R_1'$. Further, the individual parallel capacitances C_1' can be graded along the divider height. In order to reduce the influence of the stray capacitance C_e, a ratio of $C_1/C_e \geq 3$ is desirable (Ref. [6] of Chap. 1). The reflection-free termination of the coaxial cable at high frequencies is realized as in the purely capacitive voltage divider by a series resistor R equal to the cable impedance Z at the output of the voltage divider.

In a variant of the parallel-mixed resistive-capacitive voltage divider, there are no galvanic cross-connections between the resistors and capacitors on the high-voltage side, apart from the connections at the uppermost and lowest partial resistors R_1'. However, the particular, approximately concentric arrangement of the resistive and capacitive branches causes a capacitive coupling between them. Capacitive coupling is also achieved by several torus electrodes distributed along the divider height, which enclose the resistor column without galvanic connection to each other and to the resistors (Ref. [6] of Chap. 1).

The transfer behavior of the parallel-mixed resistive-capacitive voltage divider can be examined in more detail on the basis of the ladder network (Ref. [4] of Chap. 1). The frequency dependence of the scale factor caused, for example, by the change from capacitive to resistive behavior, is noticeable in the step response by an amplitude change (Ref. [4] of Chap. 1). The requirement that the parallel capacitance should be at least three times the stray capacitance in order to improve the transfer behavior, leads to large parallel capacitances C_1', in particular in the case of a multistage voltage divider. However, the available, mostly oil-impregnated high-voltage capacitors have a considerable inductance, whereby the desired improvement in the transfer behavior is more or less canceled.

The parallel-mixed voltage divider represents an additional capacitive load of the impulse voltage circuit. Not least because of the increased costs due to the additional capacitor, the parallel-mixed resistive-capacitive voltage divider has lost significance for impulse measurements in comparison to the damped capacitive voltage divider. The mixed voltage divider still serves as a universal voltage divider for measuring DC, AC and impulse voltages. For improving the transfer behavior, the capacitive arm can be designed as a damped capacitive voltage divider, i.e. the capacitors are connected in series with resistors [100]. The scale factor and the rated voltage are usually different for each of the three types of voltage.

4.3.6 Standard Sphere Gaps

Standard sphere gaps are used in a vertical or horizontal arrangement to determine the 50% *disruptive discharge voltage* U_{50} according to IEC 60052, which corresponds to the peak value of impulse voltages (Ref. [5] of Chap. 2). For large gap spacing, i.e. for large voltages, it should be noted that positive and negative impulse voltages have different U_{50} values. The arrangement of the sphere gaps for impulse voltages is identical to that for AC voltage measurements (see Sect. 2.5.8). However, the low-inductive series resistor R_v needed to eliminate oscillations in the test circuit during the voltage collapse should not exceed 500 Ω (see Fig. 2.18). Because of the short impulse duration, a sufficiently large number of free electrons in the spacing of the spheres are required for reproducible firing, in particular at a small spacing. The firing condition can usually be satisfied by impulse voltage generators used with open switching spark gaps. Generators with encapsulated switching spark gaps often require additional ionization, for example by corona discharges or UV light in the UVC wave range with a mercury vapor lamp. The thermal radiation generated by the UV lamp in continuous operation sometimes leads to an increasing air temperature in the vicinity of the sphere gap, whereby the disruptive discharge voltage is changed.

Note: The softer UVA or UVB irradiation is considered to be insufficient for ionization. The ionizing preparations (α emitters) used in the past are no longer permitted or only under special safety precautions because of the potential radiation hazard of the employees.

There are two methods to use a standard sphere gap with the sphere diameter D and the spacing S. In the first method, five series of at least ten equal impulse voltages are generated. In the first series, the charging voltage of the impulse voltage generator is set so that the peak value of the generated impulse voltage is slightly below the U_{50} value given in the IEC table for D and S (Ref. [5] of Chap. 2). At this voltage, the sphere gap must not fire yet. The time between the individual impulse voltages of the series shall be at least 30 s. In each of the following series, the peak value of the impulse voltage is increased by approximately 1%. The U_{50} value is obtained when half of the generated impulse voltages in a series lead to a disruptive discharge. The corresponding generator voltage or the peak value from the measuring

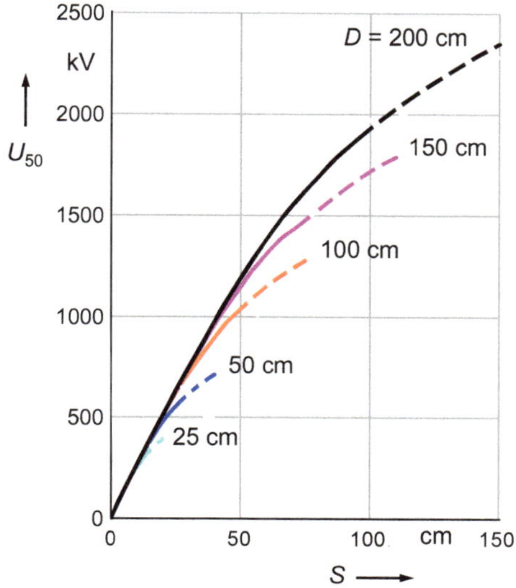

Fig. 4.35 Disruptive discharge voltage U_{50} of sphere gaps for positive lightning and switching impulse voltages versus spacing S at standard atmospheric conditions (temperature: 20 °C, atmospheric pressure: 101.3 kPa, absolute humidity: 8.5 g^{-3})

instrument used is recorded. This value, however, still has to be related to the standard atmospheric conditions. The alternative method of determining the 50% disruptive discharge voltage U_{50} is carried out as an up-and-down test with at least 20 V applications in steps of approximately 1% of the expected U_{50} value.

Whichever method is used, the reliability of the result obtained must be proven by an additional test. For this purpose, a series of 15 impulse voltages are generated with a peak value that is 1% below the U_{50} value obtained for lightning impulse voltages or 1.5% below the U_{50} value obtained for switching impulse voltages. Overall, there shall be no more than two disruptive discharges in this series.

Figure 4.35 shows the course of the disruptive discharge voltage U_{50} over the gap spacing S at standard atmospheric conditions for positive impulse voltages and different sphere diameters D. The corresponding U_{50} values for negative impulse voltage are up to 5% lower and identical to those for AC voltage (see Fig. 2.19). They also apply to DC voltage, but with greater uncertainty. The uncertainty of the U_{50} values for impulse voltages of both polarities is given as 3% for spacing $S \leq 0.5D$ (confidence interval $\geq 95\%$). For spacing $S > 0.5D$, larger uncertainties are to be expected, which is indicated in Fig. 4.35 by the dashed curve.

The U_{50} standard values apply to standard atmospheric conditions (temperature: 20 °C, air pressure: 101.3 kPa, absolute humidity: 8.5 g^{-3}). In the case of deviating atmospheric conditions, correction factors for the relative air density and humidity according to Eqs. 2.15–2.17 shall be applied to the measured disruptive discharge voltage U_{50} (Ref. [5] of Chap. 2).

The measurement methods for sphere gaps and the disruptive discharge voltages were established more than half a century ago as the result of international comparison measurements. They have been largely confirmed and also supplemented by recent investigations and comparison measurements with AC and impulse voltages (Ref. [28, 29] of Chap. 2). In industrialized countries, sphere gaps are nowadays rarely used for voltage measurements except for verification of the scale factor and to prove linearity of voltage dividers. On the one hand, the firing of the sphere gap causes serious EMC problems and, on the other hand, the time parameters cannot be determined. With appropriate effort and careful execution of the test, the linearity of an impulse voltage divider can be checked within ±1%.

4.3.7 Capacitive Field Sensors

The classic impulse voltage divider is not suitable for every test set-up and test object due to its large dimensions and limited bandwidth. Examples are measurements in gas-insulated switchgear, oil-insulated high-voltage equipment and water-cooled power pulse generators. In such arrangements, *capacitive field sensors* with small dimensions advantageously applicable in many cases. They detect the displacement current caused by the transient electric field and are suitable for voltage measurement if properly wired and calibrated. Various variants of the measuring arrangement with fixed or freely movable sensors are in use, including potential-free spherical sensors for three-dimensional measurements.

4.3.7.1 Measurement Principle and Equivalent Circuit Diagram

The principle of the capacitive field sensor in a coaxial gas-insulated electrode arrangement is shown in Fig. 4.36. The sensor ***1*** consists of a plastic foil metallized on both sides and is mounted on a metal support plate. The sensor is positioned either in the wall of the grounded outer conductor ***3*** or in a flange opening. Depending on the stray capacitance C_1, the upper electrode of the sensor captures a portion of the electric field between the inner conductor ***2*** at high voltage $u_1(t)$ and the grounded outer conductor ***3***. Together with the capacitance C_2 of the metallized foil, a capacitive voltage divider with the output voltage $u_2(t)$ is formed. The sensor is also used in a coaxial design in which a metallized plastic foil about 10 cm wide covers the inner wall of the outer conductor.

Figure 4.37 shows the equivalent circuit diagram of the field sensor as a capacitive voltage divider whose output voltage $u_2(t)$ is fed to the measuring

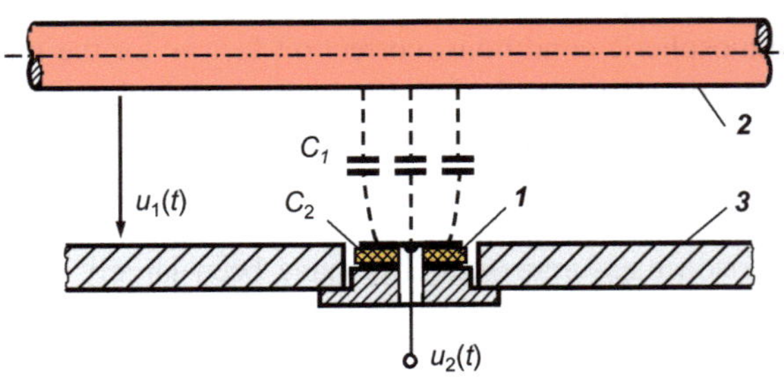

Fig. 4.36 Capacitive field sensor in a gas-insulated switchgear (schematic). ***1*** Field sensor, ***2*** inner conductor, ***3*** outer conductor

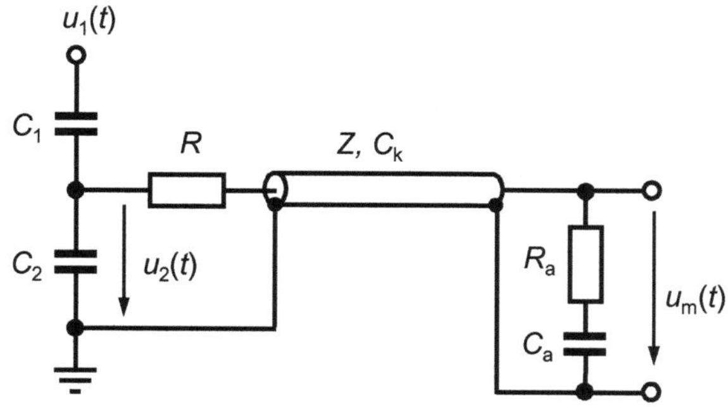

Fig. 4.37 Electrical equivalent circuit diagram of the sensor with the measurement set-up

instrument via the resistor $R = Z$ and the coaxial cable with the impedance Z. The terminating impedance R_a with C_a at the cable end is dimensioned as Burch termination (see Sect. 4.3.3.3), which is intended to compensate for the influence of the cable capacitance C_k. Since C_2 is not very large, C_k must not be neglected to accurately measure fast transient voltages. For dimensioning the terminating impedance, $R_a = kZ$ and $k(C_a + C_k) = (C_1 + C_2)$, where $k = 1$ according to [87] or $k = 1.25$ according to [57]. Depending on the size and bandwidth of the measurement set-up, such field probes allow transient voltages to be measured even in the nanosecond range [101–107].

In another version of the field sensor, a measuring resistor R_2 is used instead of C_2. Again, a small measuring electrode represents the stray capacitance C_1 to the inner conductor. The capacitive current flowing through C_1 and R_2 is:

$$\boxed{i = C_1 \frac{\mathrm{d}u_1}{\mathrm{d}t}}, \tag{4.30}$$

which generates the voltage $u_2(t) = iR_2$ at R_2. Thus, $u_2(t)$ is proportional to the derivative of the desired voltage $u_1(t)$. This sensor type is also referred to as a $\dot{E}$ (E-dot) sensor, where the superscript point represents the differentiated field strength (Ref. [40] of Chap. 5). Integrating $u_2(t)$ yields the desired measurement quantity:

$$\boxed{u_1(t) = \frac{1}{R_2 C_1} \int_0^t u_2(t)\mathrm{d}t}. \tag{4.31}$$

There are several ways to integrate the voltage $u_2(t)$. Integration can be carried out passively by a capacitor in parallel with R_2, actively with an integration amplifier or, if the impulse form is available as digital data set, numerically by software. For small dimensions of the *$\dot{E}$-sensor* and very high-frequency electric fields or voltages, integration is preferably carried out passively by connecting a capacitor C_2 in parallel with the high-ohmic measuring resistor R_2. Alternatively, C_2 can also be realized by a defined stray capacitance. This integration circuit can in turn be interpreted as a capacitive voltage divider according to Fig. 4.37, in which

C_2 is loaded by the high-ohmic input resistor $R_i = R_2$ of the measuring instrument. The voltage at C_2 is then directly proportional to the voltage $u_1(t)$ between the inner and outer conductors.

For extremely fast transient voltages in the range of nanoseconds and below, the classical formulae and equivalent circuit diagrams of electrical engineering are not sufficient to describe exhaustively the processes when using field sensors. Fast transient electrical fields are always associated with transient magnetic fields that can influence voltages in the measurement circuit and induce currents. The geometry of the field sensor determines whether the electric field acts as a useful signal and the magnetic field as an interference signal or vice versa. The measuring set-up must then be optimized so that the respective useful signal is as large as possible and the interference signal is negligibly small. The relationship between the electric and magnetic fields is determined by *Maxwell's equations*, which lead to simple equations and equivalent circuit diagrams for simple geometric arrangements [108]. If the wavelength of the fields is no longer short compared to the dimensions of the sensor, effects due to the transit time must not be ignored. Considering the field sensor in the UHF range of a few gigahertzes as an antenna, the transfer behavior and the optimal sensor form can be determined with the antenna theory. Field sensors for voltage measurements are calibrated directly on site and used without changing its position [109].

4.3.7.2 Electric Field Sensors for Linearity Proof of Voltage Dividers

Even in conventional test set-ups, simple planar field sensors with correspondingly larger dimensions can be used to measure impulse voltages. They can be easily made from a double-sided copper-coated circuit board, as it is used for electronic circuits. With a diameter of 0.5 m and plate thickness of 0.5–2 mm, a capacitance C_2 in the range of 1–10 nF is achieved. The plate capacitor C_2 and the stray capacitance C_1 to the high-voltage circuit form a capacitive voltage divider, whose output is connected via a shielded measurement cable to the high-ohmic input of a digital recorder. A damping resistor equal to the cable impedance is connected between the plate capacitor C_2 and the measurement cable to prevent traveling wave oscillations.

After on-site calibration, this measurement arrangement can completely replace the voltage divider during a test. However, it mainly serves to prove the linearity of impulse voltage dividers up to the highest operating voltage. For this purpose, the plate capacitor is placed on the hall floor or on a stool near the impulse voltage divider to be tested. Since the linearity test is only a relative measurement with different voltage levels, it is not absolutely necessary to calibrate the field probe on site. Prerequisite for successful use is the charge-free testing and measuring arrangement, which sometimes requires a thorough and quite time-consuming elimination of all possible sources of interference, for example partial discharges. Under this condition, the linearity proof is possible within ± 1% up to a few megavolts [110].

For the direct measurement of impulse voltages, a different arrangement with a field sensor is described in [111]. It consists of a large circular plate capacitor with a diameter of 1.15 m and a plate spacing of 1 m. Each of the plate electrodes is provided with a toroidal shield. The upper plate electrode is connected directly to the test object or impulse voltage generator. In the center of the lower grounded plate, a small plate electrode is installed as an isolated field sensor, thus located in a largely homogeneous field. The upper plate electrode and the sensor electrode form the high-voltage capacitance C_1. The field sensor has the low-voltage capacitance C_2 to ground. The signal measured at C_1 is amplified and recorded by a digital recorder installed in a shielded box under the lower plate electrode. The digital data are sent via optical fiber to a PC in the measurement room for further evaluation. The transfer behavior of the measuring arrangement is characterized by a response time $T_N < 10$ ns and settling time $t_s < 150$ ns, which fulfills the requirements for measuring lightning impulse voltages. Due to the large diameter and spacing of the plate capacitor, however, the influence of external fields, i.e. the proximity effect, cannot be completely avoided. Therefore, a distance of more than 5 m from adjacent test equipment is required. This means that the field sensor must be recalibrated every time the test set-up is changed within this distance or every time the location is changed.

4.3.7.3 Three-Dimensional Field Sensors

The electric field sensors described so far have in common that they only detect one field direction and that one electrode is at ground potential. Potential-free field measuring devices which can be positioned anywhere in the high-voltage field, are much more complex. They consist of a spherical sensor head, an analog (or digital) transmission link with optical fiber and a measuring instrument at ground potential. Two-dimensional and three-dimensional sensors with a diameter of 4 cm to 10 cm are in use. Depending on the design of the spherical sensor and the electronic circuit, bandwidths of up to 350 MHz can be achieved, with the lower limit frequency being 20 Hz [112–114]. In the wideband version, it is even possible to measure the EMP fields which are generated in a strip-line test arrangement by steep impulse voltages with rise times of a few nanoseconds (see Sect. 4.2.4).

Figure 4.38 shows the principle of a *three-dimensional field sensor* with two opposite measuring electrodes for each of the three field directions in the x-, y- and z-direction. The particular arrangement and shape of the measuring and reference electrodes are the result of field calculations in order to optimize the measurement accuracy of the spherical sensor in the inhomogeneous field. Inside the sphere, the two measuring electrodes for each field direction are connected by HF capacitors in series, whereby voltage dividers are formed whose output voltages are supplied to the evaluation device via optical fibers. The connection of the reference electrodes with each other supplies the reference potential zero of the electronic circuit. Power is supplied to the optical transmission system inside the sphere via batteries for several hours.

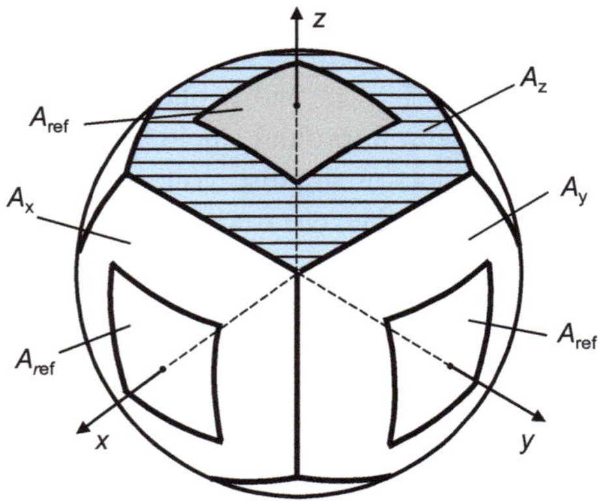

Fig. 4.38 Three-dimensional electric field sensor in spherical form (principle). A_x, A_y, A_z: measurement surfaces, A_{ref}: reference areas for zero potential

Potential-free field sensors can be used not only for the spatial measurement of transient electric fields, but also for the measurement of impulse voltages in the UHV range, in particular for the linearity test of impulse voltage dividers. The dimensions of the test set-up for voltages above 1 MV are much larger than those of the field sensor, so that field distortions caused by the introduction of the sensor head are negligible. With broadband field sensors, high-frequency oscillations in the impulse front and tail can be detected, in particular of chopped lightning impulse voltages [115]. Impulse voltage dividers for UHV measurements, on the other hand, have only limited transfer behavior due to their large dimensions. Prerequisite for the use of potential-free field sensors is—as in the case of the planar sensors described in Sect. 4.3.7.2—that there are no space charges in the environment, such as those generated by partial discharges or the firing of open spark gaps in the impulse voltage generator.

The calibration of a field measuring device itself essentially comprises the determination of the scale factor and its dependence on the strength and frequency of the electric field. For calibration, the potential-free field sensor is positioned in the nearly homogenous, calculable AC field between two large plate electrodes. The amplitude and frequency of the applied sine voltage is varied within the limits given by the electrode geometry and generator circuit [116]. The indication of the sensor is verified for all three field directions by aligning the sensor axes. The linearity test up to the highest measurable field strength can be performed in the high-voltage hall. For this purpose, the sensor is positioned in the AC field between the top electrode of a high-voltage transformer and the grounded wall or floor of the hall. For frequencies above 1 MHz, the sensor is calibrated in the transverse electromagnetic field of a *TEM* or *GTEM cell*. If the potential-free field sensor is to be used for impulse voltage measurements, the scale factor and the time parameters are determined by an on-site comparison measurement with a reference divider at a reduced impulse voltage. The position of the sensor during calibration and testing must be identical as far as possible.

References

1. Schon, K.: Korrektur des Scheitelwertes von Keilstoßspannungen unter Berücksichtigung des genauen Abschneidezeitpunktes. etz-Archiv **5**, 233–237 (1983)
2. Berlijn, S.: Influence of lightning impulses to insulating systems. Thesis, TU Graz (2000)
3. Li, Y., Rungis, J.: Analysis of lightning voltage with overshoot. In: Proceedings of 14th ISH Beijing, Paper B-08 (2005)
4. Hällström, J., et al.: Applicability of different implementations of K-factor filtering schemes for the revision of IEC 60060-1 and -2. In: Proceedings of 14th ISH Beijing, Paper B 32 (2005)
5. Simón, P., Garnacho, F., Berlijn, S.M., Gockenbach, E.: Determining the test voltage factor function for the evaluation of lightning impulses with oscillations and/or overshoot. IEEE Trans. PWRD **21**, 560–566 (2006)
6. Nilsson, A., Mannikoff, A., Hällström, J., Li, Y.: New procedures for determination of parameters of lightning impulse voltage waveforms. In: Proceedings of 15th ISH Ljubljana, Paper T10-399 (2007)
7. Lewin, P.L., Tran, T.N., Swaffield, D.J., Hällström, J.: Zero phase filtering for lightning impulse evaluation: a K-factor filter for the revision of IEC 60060-1 and -2. IEEE Trans. PWRD **23**, 3–12 (2008)
8. Pfeffer, A., Tenbohlen, S.: Analysis of full and chopped lightning impulse voltages from transformer tests using the new k-factor approach. In: Proceedings of 16th ISH Johannesburg, Paper A-10 (2009)
9. Schon, K.: Digital filtering of hv lightning impulses. IEEE Panel Session "Digital Techniques in HV Tests", Long Beach, California (1989)
10. Sato, S., Harada, T.: Lightning impulse parameter determination by means of moving average method. In: Proceedings of 13th ISH Delft, Paper 807 (2003)
11. Li, Y., Rungis, J.: Precision digital filters for high voltage impulse measuring systems. IEEE Trans. PWRD **14**, 1213–1220 (1999)
12. Gockenbach, E.: A simple and robust evaluation procedure for high-voltage impulses. In: IEEE International Symposium on Digital Techniques in High Voltage Measurement. Toronto, Session 3 (1991)
13. Wolf, J., Gamlin, M.: A new modular design for a new generation of impulse voltage generators. In: Proceedings of 13th ISH Delft, Paper 797 (2003)
14. Ai, X., Ding, J.: An outdoor 7.2 MV impulse generator. 16. ISH Johannesburg, paper G-16 (2009)
15. Stolle, D., Peier, D.: Reproducibility of Marx generators. In: Proceedings of 5th ISH Braunschweig, Paper 61.04 (1987)
16. Etzel, O., Helmchen, G.: Berechnung der Elemente des Stoßspannungskreises für die Stoßspannungen 1,2/50, 1,2/5 und 1,2/200. ETZ-A **85**, 578–582 (1964)
17. Heilbronner, F.: Firing and voltage shape of multistage impulse generators. IEEE Trans. PAS **90**, 2233–2238 (1971)
18. Del Vecchio, R.M., Ahuja, R., Frenette, R.: Determining ideal impulse generator settings from a generator-transformer circuit model. IEEE Trans. PWRD **17,** 1–4 (2002)
19. Goody, R.W.: OrCAD PSpice for WINDOWS, vol. I–III. Prentice Hall, New Jersey (2001)
20. Heinemann, R.: PSPICE – Einführung in die Elektroniksimulation. Carl Hanser Verlag München (2004)
21. Schufft, W., Hauschild, W., Pietsch, R.: Determining impulse generator settings for various test cases with the help of a www-based simulation program. In: Proceedings of 14th ISH Beijing, Paper J58 (2005)
22. Sato, S.: Automatic determination of circuit constants fulfilling the given impulse time parameters. In: Proceedings of 15th ISH Ljubljana, Paper T10-313 (2007)
23. Kind, D., Salge, J.: Über die Erzeugung von Schaltspannungen mit Hochspannungsprüftrans formatoren. ETZ-A **86**, 648–651 (1965)

24. Anis, H., Trinh, N.G., Train, D.: Generation of switching impulses using high voltage testing transformers. IEEE Trans. PAS **94**, 187–195 (1975)
25. Okabe, S., Tsuboi, T., Takami, J.: Influence of test equipment capacitance and residual inductance on waveform front generated in lightning impulse voltage test. In: Proceedings of 16th ISH Johannesburg, Paper A-55 (2009)
26. Matsumoto, S.: Analysis of waveform parameters for oscillating impulse voltage. In: Proceedings of 15th ISH Ljubljana, Paper T10-621 (2007)
27. Wolf, J., Voigt, G.: A new solution for the extension of the load range of impulse voltage generators. In: Proceedings of 10th ISH Montreal, Paper 3375 (1997)
28. Schwenk, K., Gamlin, M.: Load range extension methods for lightning impulse testing with high voltage impulse generators. In: Proceedings of 14th ISH Beijing, Paper B-78 (2005)
29. Hinow, M., Steiner, T.: Influence of the new k-factor method of the IEC draft 600060-1 on the evaluation of lightning impulse parameters in relation to ultrahigh-voltage testing. In: Proceedings of 16th ISH Johannesburg, Paper G-14 (2009)
30. Ishikura, T., et al.: Insulation characteristics with respect to the front time extension in lightning impulse voltage test for UHV equipment. In: Proceedings of 18th ISH Seoul, Paper OC2-02 (2013)
31. Papachristos, G., Woschnagg, E.: Norm- und Schaltstoßspannungsprüfung von Transformatoren und Kompensations-Drosselspulen. Digitale Berechnung der Spannungswellenform und Diagramme zur Dimensionierung des Stoßkreises. Bull. SEV **65**, 721–731 (1974)
32. Kannan, S.R., Narayana Rao, Y.: Prediction of the parameters of an impulse generator for transformer testing. Proc. IEE **120**, 1001–1005 (1973)
33. Glaninger, P.: Stoßspannungsprüfung an elektrischen Betriebsmitteln kleiner Induktivität. In: Proceedings of 2nd ISH, pp. 140–144 (1975)
34. Feser, K.: Auslegung von Stoßgeneratoren für die Blitzstoßspannungsprüfung von Transformatoren. Bull. SEV **69**, 973–979 (1978) (Translated version available as Haefely Scientific Document E1-41: Circuit design of impulse generators for the lightning impulse voltage testing of transformers)
35. Lakshmi, P.V., Sarma, S., Singh, B.P., Tiwari, R.K.: Determination of tuning parameters for reducing the overshoot during impulse test of power transformer. In: Proceedings of 13th ISH Delft Paper 87 (2003)
36. Schrader, W., et al.: The generation of switching impulse voltages up to 3.9 MV with a transformer cascade of 3 MV. In: Proceedings of 6th ISH New Orleans, Paper 47.39 (1989)
37. Feser, K., Rodewald, A.: Eine triggerbare Mehrfachabschneidefunkenstrecke für hohe Blitz- und Schaltstoßspannungen. In: Proceedings of 1st ISH München, pp. 124–131 (1972) (Translated version available as Haefely Scientific Document: A triggered multiple chopping gap for lightning and switching impulses)
38. Kärner, H.: Erzeugung steilster Stoßspannungen hoher Amplitude. Bull. SEV **58**, 1096–1110 (1967)
39. Feser, K.: Gedanken zur Prüf- und Messtechnik bei Steilstoßspannungen. HIGHVOLT Kolloquium Dresden, Paper 1.4 (1997)
40. McDonald, D.F., Benning, C.J., Brient, S.J.: Subnanosecond risetime multikilovolt pulse generator. Rev. Sci. Instr. **36**, 504–506 (1965)
41. Feser, K., Modrusan, M., Sutter, H.: Steep front impulse generators. In: Proceedings of 3rd ISH Mailand, Paper 41.06 (1979)
42. Dams, J., Dunz, T., Küchler, A., Schwab, A.: Design and operation of a Terawatt pulse-power generator. In: Proceedings of 5th ISH Braunschweig, Paper 61.02 (1987)
43. Cadilhon, B., et al.: Design, realisation and experimental study of a 200 kV, 1 ns rise-time Marx generator. In: Proceedings of 15th ISH Ljubljana, Paper T2–397 (2007)
44. Chang, Y.-M.: Steep front pulse generation using coaxial Marx circuit for EMC test. In: Proceedings of 18th ISH Seoul, Paper PB-29 (2013)
45. Yashima, M., et al.: Development of 1 MV steep-front rectangular impuls voltage generator. In: Proceedings of 11th ISH London, Paper 5.394.S26 (1999)

46. Salge, J., Peier, D., Brilka, R., Schneider, D.: Application of inductive energy storage for the production of intense magnetic fields. In: Proceedings of 6th Symposium on Fusion Technology, Aachen (1970)
47. Kind, D., Salge, J., Schiweck, L., Newi, G.: Explodierende Drähte zur Erzeugung von Megavolt-Impulsen in Hochspannungsprüfkreisen. ETZ-A **92**, 46–51 (1971)
48. Feser, K.: MIGUS – EMP-Simulator für die Überprüfung der EMV. etz **108**, 420–423 (1987)
49. Bellaschi, P.L.: The measurement of high-surge voltages. Trans. A.I.E.E. **52**, 544–567 (1933)
50. Feser, K.: Messung hoher Stoßspannungen. ETZ Bd. **104**, 881–886 (1983)
51. Modrusan, M.: Hochspannungsteiler: Typen, Messeigenschaften und Einsatz. Bull. ASE/UCS **74**, 1030–1037 (1983)
52. Yao, Z.G.: The standard impulse voltage waveforming in a test system including HV lead. In: Proceedings of 5th ISH Braunschweig, Paper 63.15 (1987)
53. Zaengl, W.: Das Messen hoher, rasch veränderlicher Stoßspannungen. Thesis, TH München (1964)
54. Zaengl, W., Feser, K.: Ein Beitrag zur Berechnung des Übertragungsverhaltens von Stoßspannungsteilern. Bull. SEV **55**, 1249–1256 (1964)
55. Zaengl, W.: Ein neuer Teiler für steile Stoßspannungen. Bull. SEV **56**, 232–240 (1965)
56. Feser, K.: Einfluss des Niederspannungsteiles auf das Übertragungsverhalten von Stoßspannungsteilern. Bull. SEV **57**, 695–701 (1966)
57. Zaengl, W.: Der Stoßspannungsteiler mit Zuleitung. Bull. SEV **61**, 1003–1017 (1970)
58. Zaengl, W.: Ein Beitrag zur Schrittantwort kapazitiver Spannungsteiler mit langen Messkabeln. etz-a Bd. **98**, 792–795 (1977)
59. Malewski, R., Maruvada, P.S.: Computer assisted design of impulse voltage dividers. IEEE Trans. PAS **95**, 1267–1274 (1976)
60. Di Napoli, A., Mazzetti, C.: Time-analysis of H.V. resistive divider from the electromagnetic field computation. In: Proceedings of 3rd ISH Mailand, Paper 42.10 (1979)
61. Kato, S.: Analysis of voltage divider response by finite element method. In: Proceedings of 5th ISH Braunschweig, Paper 71.07 (1987)
62. Kawaguchi, Y., Murase, H., Koyama, H.: Unit step response simulation of impulse voltage measuring system by EMTP. In: Proceedings of 8th ISH Yokohama, Paper 51.07 (1993)
63. Zucca, M., Sardi, A., Bottauscio, O., Saracco, O.: Modeling H.V. reference dividers for lightning impulses. In: Proceedings of 11th ISH London, Paper 1.70.S21 (1999)
64. Baba, Y., Ishii, M.: Numerical electromagnetic analysis of unit step responses of impulse voltage. In: Proceedings of 13rd ISH Delft, Paper 659 (2003)
65. Schwab, A.J., Herold, J.: Electromagnetic interference in impulse measuring systems. IEEE Trans. PAS **93**, 333–339 (1974)
66. Schwab, A.J.: Elektromagnetische Verträglichkeit. Springer, Berlin (1990)
67. Peier, D.: Elektromagnetische Verträglichkeit. Hüthig Buch Verlag, Heidelberg (1990)
68. van Waes, J.B.M., et al.: EMC analysis of voltage measuring systems in high power laboratory. In: Proceedings of 10th ISH Montréal, Paper 3331 (1997)
69. Smolke, M., Engelmann, E., Kindersberger, J.: Shielding effectiveness of boxes with apertures with respect to magnetic lightning impulse fields. In: Proceedings of 10th ISH Montreal, Paper 3250 (1997)
70. Wenger, P., Yan, W., Li, Y.: A reference resistive voltage divider for switching impulses. In: Proceedings of 19th ISH Pilsen, Paper 307 (2015)
71. Mahdjuri-Sabet, F.: Das Übertragungsverhalten von mäanderförmig gewebten Drahtwiderstandsbändern in Hochspannungsversuchsschaltungen. Arch. E-techn. **59**, 69–73 (1977)
72. Sundermann, U., Peier, D.: Transfer characteristics of meander wound resistors. In: Proceedings of 7th ISH Dresden, Paper 61.14 (1991)

73. Campisi, F., Rinaldi, E., Rizzi, G., Valagussa, C.: A new wire wound resistive divider for steep front impulse tests: design criteria and calibration procedure. In: Proceedings of 11th ISH London, Paper 1.164.P4 (1999)
74. Bossi, S., Rizzi, G., Valagussa, C., Garbagnati, E.: A special screened resistor-type divider for the measurement of the fast front-chopped impulses. In: Proceedings of 6th ISH New Orleans, Paper 47.42 (1989)
75. Harada, T., Wakimoto, T., Sato, S., Saeki, M.: Development of national standard class reference divider for impulse voltage measurements. In: Proceedings of 11th ISH London, Paper 1.13.S1 (1999)
76. Goosen, R.F., Provoost, P.G.: Fehlerquellen bei der Registrierung hoher Stoßspannungen mit dem Kathodenstrahl-Oszillographen. Bull. SEV **37**, 175–184 (1946)
77. Peier, D.: Ohmscher Stoßspannungsteiler mit kleiner Antwortzeit. PTB-Mitt. **88**, 315–318 (1978)
78. Peier, D., Stolle, D.: Resistive voltage divider for 1 MV switching and lightning voltages. In: Proceedings of 5th ISH Braunschweig, Paper 73.02 (1987)
79. Sfakianakis, Z.: 1000 kV resistive divider for measuring non-standard lightning impulses. In: Proceedings of 5th ISH Braunschweig, Paper 73.03 (1987)
80. Harada, T., Itami, T., Aoshima, Y.: Resistor divider with dividing element on high voltage side for impulse voltage measurements. IEEE Trans. PAS **90**, 1407–1414 (1971)
81. Harada, T., Aoshima, Y., Kawamura, T., Ohira, N., Kishi, K., Takigami, K., Horiko, Y.: A high quality voltage divider using optoelectronics for impulse voltage measurements. IEEE Trans. PAS **91**, 494–500 (1972)
82. Kind, D., Arndt, V.: Hochspannungsteiler mit optimierter Abgriffshöhe. Unpublished Paper (1992)
83. Kouno, T., Kato, S., Kikuchi, K., Maruyama, Y.: A new voltage divider covered with metal sheath and improved by frequency division method. In: Proceedings of 2nd ISH Zürich, pp. 222–225 (1975)
84. Groh, H.: Hochspannungsteiler mit 4 Nanosekunden Anstiegszeit. etz-a Bd. **98**, 436–438 (1977)
85. Aro, M., Punkka, K., Huhdanmäki, J.: Fast divider for steep front impulse voltage tests. In: Proceedings of 5th ISH Braunschweig, Paper 73.01 (1987)
86. Denicolai, M., Hällström, J.: A self-balanced, liquid resistive, high impedance HV divider. In: Proceedings of 14th ISH Beijing, Paper J-05 (2005)
87. Burch, F.G.: On potential dividers for cathode-ray oscillographs. Phil. Mag. **13**, 760–774 (1932)
88. Schwab, A.J., Pagel, J.H.W.: Precision capacitive voltage divider for impulse voltage measurements. IEEE Trans. PAS **91**, 2376–2382 (1972)
89. Wolzak, G.G., van der Laan, P.C.T.: A new concept for impulse voltage dividers. In: Proceedings of 4th ISH Athen, Paper 61.11 (1983)
90. Shi, B., Zhang, W., Qiu, Y.: A new type of divider for measuring fast rising high voltage impulses. In: Proceedings 10th ISH Montreal, Paper 3088 (1997)
91. Zaengl, W., Weber, H.J.: A high voltage divider made for education and simulation. In: Proceedings of 6th ISH New Orleans, Paper 41.06 (1989)
92. Malewski, R., Hyltén-Cavallius, N.: A low voltage arm for EHV impulse dividers. IEEE Trans. PAS **93**, 1797–1804 (1974)
93. Harada, T., Aoshima, Y., Harada, M., Hiwa, K.: Development of high-performance low voltage arms for capacitive voltage dividers. In: Proceedings of 3rd ISH Mailand, Paper 42.14 (1979)
94. Li, Y., Rungis, J.: Compensation of step response "creeping" of a damped capacitive divider for switching impulses. In: Proceedings of 11th ISH London, Paper 1.128.P4 (1999)
95. Feser, K.: Ein neuer Spannungsteiler für die Messung hoher Stoß- und Wechselspannungen. Bull. SEV **62**, 929–935 (1971)
96. Feser, K., Rodewald, A.: Die Übertragungseigenschaften von gedämpft kapazitiven Spannungsteilern über 1 MV. In: Proceedings of 1st ISH München (1972)

97. Feser, K.: Transient behaviour of damped capacitive voltage dividers of some million volts. IEEE Trans. **93**, 116–121 (1974)
98. Feser, K.: Ein Beitrag zur Berechnung der Spannungsverteilung von Hochspannungskondensatoren. Bull. SEV **61**, 345–348 (1970)
99. Arndt, V.; Schon, K.: On the uncertainty of the new IEC response parameters. In: Proceedings of 8th ISH Yokohama, Paper 51.02, pp. 293–296 (1993)
100. Harada, T., Aoshima, Y., Okamura, T., Hiwa, K.: Development of a high voltage universal divider. IEEE Trans. PAS **95**, 595–602 (1976)
101. Breilmann, W.: Effects of the leads on the transient behaviour of a coaxial divider for the measurement of high alternating and impulse voltages. In: Proceedings of 3rd ISH Mailand, Paper 42.12 (1979)
102. Meppelink, J., Hofer, P.: Design and calibration of a high voltage divider for measurement of very fast transients in gas insulated switchgear. In: Proceedings of 5th ISH Braunschweig, Paper 71.08 (1987)
103. Gsodam, H., Muhr, M., Pack, S.: Response of the measurement circuit for high frequency transient overvoltages. In: Proceedings of 5th ISH Braunschweig, paper 71.09 (1987)
104. Bradley, D.A.: A voltage sensor for measurement of GIS fast transients. In: Proceedings of 6th ISH New Orleans, Paper 49.01 (1989)
105. Rao, M.M., Jain, H.S., Rengarajan, S., Sheriff, K.R.S., Gupta, S.C.: Measurement of very fast transient overvoltages (VFTO) in a GIS module. In: Proceedings of 11th ISH London, Paper 1.144.P4 (1999)
106. Liu, J.-L., et al.: Coaxial capacitive dividers for high-voltage pulse measurements in intense electron beam accelerator with water pulse-forming line. IEEE Trans. IM **58**, 161–166 (2009)
107. Burow, S., Köhler, W., Tenbohlen, S., Strauchmann, U.: Messung und Dämpfung von sehr schnellen Transienten (VFT) in gasisolierten Schaltanlagen. ETG Fachtagung Kassel (2012)
108. Küchler, A., Dams, J., Dunz, T.H., Schwab, A.: Kapazitive Sensoren zur Messung transienter elektrischer Felder und Spannungen. Arch. E-techn. **68**, 335–344 (1985)
109. Kurrer, R., Feser, K., Krauß, T.: Antenna theory of flat sensors for partial discharge detection at ultra-high frequencies in GIS. In: Proceedings of 9th ISH Graz, Paper 5615 (1995)
110. Rizzi, R., Tronconi, G., Gobbo, R., Pesavento, G.: Determination of the linearity of impulse divider in the light of the revision of IEC 60: comparison among several methods. In: Proceedings of 8th ISH Yokohama, Paper 52.05 (1993)
111. Ishii, M., Li, D., Hojo, J.-I., Liao, W.-W.: A measuring system of high voltage impulse by way of electric field sensing. In: Proceedings of 8th ISH Yokohama, Paper 56.06 (1993)
112. Feser, K., Pfaff, W.: A potential free spherical sensor for the measurement of transient electric fields. IEEE Trans. PAS **103**, 2904–2911 (1984)
113. Pfaff, W.R.: Accuracy of a spherical sensor for the measurement of three dimensional electric fields. In: Proceedings of 5th ISH Braunschweig, Paper 32.05 (1987)
114. Krauß, T., Köhler, W., Feser, K.: High bandwidth potential-free electric field probe for the measurement of three dimensional fields. In: Proceedings of 10th ISH London, Paper 3061 (1997)
115. Feser, K., Pfaff, W., Weyreter, G., Gockenbach, E.: Distortion-free measurement of high impulse voltages. IEEE Trans. PD **3**, 857–866 (1988)
116. Braun, A., Brzostek, E., Kind, D., Richter, H.: Development and calibration of electric field measuring devices. In: Proceedings of 6th ISH New Orleans, Paper 40.09 (1989)

Chapter 5
High Impulse Currents

Abstract Tests with high impulse currents are performed to simulate the stress of power apparatus in the high-voltage grid caused by lightning and short circuits. Various impulse test currents and the corresponding measurement methods are standardized in IEC 62475, together with DC and AC currents. Impulse currents with peak values of up to several 100 kA are generated in the test laboratory, using capacitive energy storage units which are charged slowly and then discharged abruptly to the test object via an RC network. For the conventional measurement of impulse currents up to the highest current magnitudes, low-ohmic measuring resistors or magnetic coils with and without magnetic core are used together with digital recorders, enabling computer-assisted evaluation of the impulse parameters. A well-founded measurement technique is essential, which is achieved mainly by comparison measurement of the current measuring system with a reference system. One of the important properties of a current sensor is its step response, which characterizes the transfer behavior. Numerical convolution can be used for improving the dynamic behavior of a current measuring system.

Tests with high impulse currents are performed to simulate the stress of power apparatus in the high-voltage grid caused by lightning and short circuits. Various impulse currents and the corresponding measurement methods are standardized in IEC 62475, together with DC and AC currents (Ref. [4] of Chap. 2). Impulse currents are generated in the test laboratory with peak values of up to several 100 kA. In order to generate high impulse currents, capacitive energy storage units are usually slowly charged and then discharged abruptly to the test object via an RC network.

For the conventional measurement of impulse currents up to the highest current magnitudes, low-ohmic measuring resistors or magnetic coils with and without magnetic core are used together with digital recorders. In measurement practice, the impulse parameters are predominantly determined by computer-assisted evaluation of the digitally recorded measurement data. One of the important properties of a current sensor is its step response, which characterizes the transfer behavior and thus enables numerical convolution. Current measuring systems can be subject to

K. Schon, *High Voltage Measurement Techniques*, Power Systems,
https://doi.org/10.1007/978-3-030-21770-9_5

interference from magnetic fields, which can be largely reduced by special measures.

Current sensors operating on the basis of the Hall effect are suitable for measuring DC and AC currents as well as for impulse currents up to a few 10 kA (see Sect. 3.5.2). Magneto-optical sensors on the Faraday effect are basically also suitable for DC, AC and impulse current measurements (see Sect. 6.2).

5.1 Definitions and Parameters of Impulse Currents

The shape of impulse currents can be very different depending on the intended test purpose [1]. Basically, impulse currents with exponential and rectangular time characteristics can be distinguished in power engineering, which are standardized in IEC 62475 (Ref. [4] of Chap. 2). Impulse currents are characterized by their *peak value*, which can be up to several 100 kA, and two *time parameters*. In addition, the *impulse charge* and the *Joule integral* of the impulse may be of interest. For impulse currents, tolerances in the generation and uncertainties in the measurement are specified in the test standards.

5.1.1 Exponential Impulse Currents

Exponential impulse currents show a relatively fast, approximately exponential rise to the peak, followed by a rather slow current drop to zero. Depending on the circuit of the generator and the test object, the decrease takes place either exponentially or as a heavily damped sine wave. In the latter case, a transient oscillation of the impulse current below zero is to be expected (Fig. 5.1). The analytical expression of an exponential impulse current in the time and frequency domain is derived in Sect. 8.2. The parameters of an exponential impulse current are, in addition to the

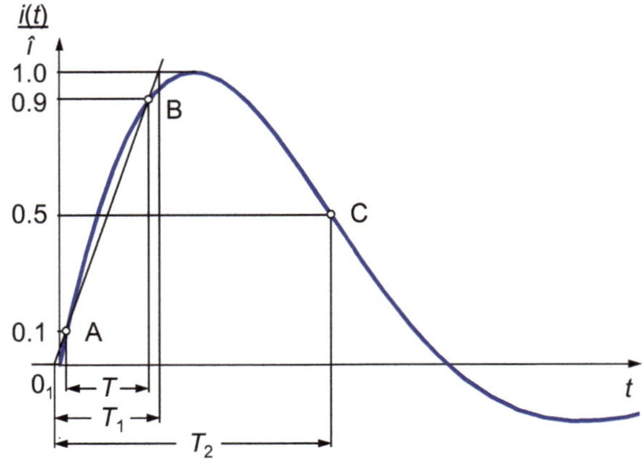

Fig. 5.1 Example of an exponential impulse current with polarity reversal in the back

peak value $\hat{\imath}$ as the *value of the test current*, the *front time* T_1 and the *time to half-value* T_2. Both time parameters are related to the *virtual zero point* O_1, which results from the intersection of the straight line through the front with the zero line. Unlike impulse voltages, the front line of impulse currents passes through the points A at $0.1\hat{\imath}$ and B at $0.9\hat{\imath}$. The front time is calculated as follows:

$$\boxed{T_1 = 1.25\, T_{AB}}, \tag{5.1}$$

where T_{AB} is the time between the two points A and B. The time T_{AB} thus corresponds to the definition in the low-voltage range for the rise time T_a of an impulse (see Sect. 9.6).

The time to half-value T_2 is defined as the time interval between the virtual zero point O_1 and the instant at which the impulse current has decreased to half of the peak value (Fig. 5.1). If oscillations are superimposed on the exponential impulse current in the tail, resulting in two T_2 values, their mean value is taken as a fictitious time to half-value.

An exponential impulse current is characterized by its front time and time to half-value in microseconds. For example, the 8/20 impulse current has a front time T_1 = 8 µs and a time to half-value T_2 = 20 µs. The tolerance limits for generating an 8/20 impulse current are ±10% for the peak value and ±20% for the time parameters. When generating other impulses, other tolerances may be specified (Ref. [4] of Chap. 2; [1]). The expanded measurement uncertainty is limited to 3% for the peak value and to 10% for the time parameters.

In the case of an exponential impulse current with superimposed oscillation, the *polarity reversal* after crossing the zero line must not result in a *reversal peak* which is more than 30% of the peak value. Otherwise, there is a risk that the test object will be damaged by the reversal current. The calculation in Sect. 8.2 shows that the condition for maximum reversal peak in the simple impulse current circuit of Fig. 5.1 is only achieved for T_2 > 20 µs. The reversal peak may need to be limited by an appropriate chopping device. With a *crowbar spark gap*, the polarity reversal can be effectively suppressed, but with the disadvantage of an enlarged time to half-value (see Sect. 5.2.1.2).

The *charge of an impulse current* $i(t)$ is defined as the time integral over the absolute value of the current:

$$\boxed{Q = \int_0^\infty |i(t)|\mathrm{d}t}. \tag{5.2}$$

The upper integration limit is chosen so that the residual contribution of the integral is negligible. Another measurement quantity is the *Joule integral* as the time integral of the square of the impulse current:

$$\boxed{W = \int_0^\infty i^2(t)\mathrm{d}t}, \tag{5.3}$$

with which the maximum permissible energy consumption in the test object or measuring resistor is calculated. The actual values of Q and W during a test shall not be below the values specified in the test standards for this equipment, i.e. the lower tolerance limit is zero.

5.1.2 Rectangular Impulse Currents

The typical time course of a *rectangular impulse current*, formerly referred to as *long-duration impulse current*, is shown in Fig. 5.2. It is characterized by the *value of the test current* $\hat{\imath}$ and the time parameters T_d and T_t (Ref. [4] of Chap. 2). The value of the test current is determined by the maximum value of the impulse current including a superimposed oscillation. Rectangular impulse currents often have a more or less pronounced roof slope. The *duration* T_d is the time in which the current is always greater than $0.9\hat{\imath}$. If the rectangular impulse current is superimposed with oscillations (see Fig. 5.2), T_d is taken as the longest time interval in which the current is constantly higher than $0.9\hat{\imath}$. Rated values for T_d are 500, 1000 and 2000 μs or even longer times up to 3200 μs. Because of the long duration of the peak, the test with rectangular impulse currents represents a heavy load on the test object.

An additional time parameter is the *total duration* T_t, during which the current is greater than $0.1\hat{\imath}$, with the requirement $T_\mathrm{t} \leq 1.5\ T_\mathrm{d}$. Apart from this indirect requirement on the rise time, there are no other requirements. To characterize rectangular impulse currents, the values of $T_\mathrm{d}/T_\mathrm{t}$ are given.

The upper tolerance limit in the generation of rectangular impulse currents is +20% for $\hat{\imath}$ and T_d, the lower limit is zero. The amplitude of a possible polarity reversal below zero must not exceed 10% of the test current value $\hat{\imath}$. For the charge according to Eq. (5.2) and the Joule integral according to Eq. (5.3), the lower

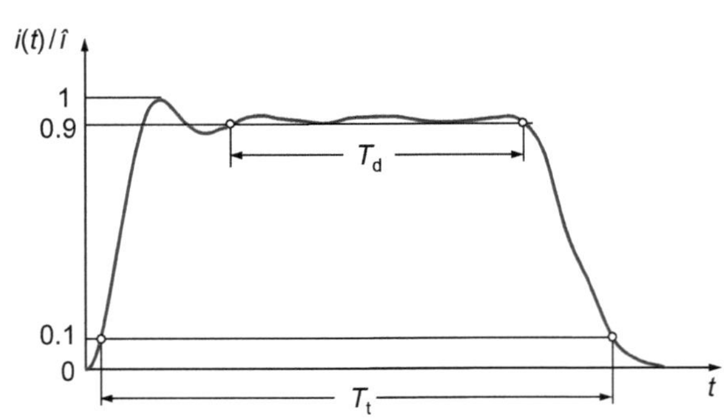

Fig. 5.2 Rectangular impulse current with superimposed oscillation

tolerance limit is also zero. The permissible measurement uncertainties are 3% for the peak value and 10% for the time parameters.

5.2 Generation of High Impulse Currents

In test laboratories, impulse currents with peak values of more than 100 A up to several 100 kA can be generated. Usually, a storage capacitor is slowly charged and then abruptly discharged to the test object via an RC network that determines the impulse shape. Impulse currents can also be generated with exploding current-carrying wires.

5.2.1 Generator Circuit for Exponential Impulse Currents

To generate exponential impulse currents in the test laboratory, mainly circuits with capacitive energy storage are used (Fig. 5.3). The capacitor C is slowly charged to a predetermined voltage U_0 and then abruptly discharged via the resistor R, the inductivity L and the test object P by means of a switch, usually a thyristor or a triggered spark gap. Via the built-in resistor R_m, the measurement voltage $u_m(t)$ proportional to the impulse current $i(t)$ can be tapped. The shape of the generated impulse current depends not only on R, L and C, but also on R_m and the characteristic impedance of the test object (see Sect. 8.2).

Compact impulse current generators are used as table-top devices with peak values of some 10 kA up to spatially extended devices with 200 kA and more. The maximum charging voltage U_0 of table-top devices and larger set-ups ranges from 10 to 200 kV. Impulse current generators for very high current amplitudes are modularly constructed with several capacitors connected in parallel, which are arranged in a partial or full circle (Fig. 5.4). To generate the different impulse shapes according to the specifications, the capacitors and resistors can be exchanged.

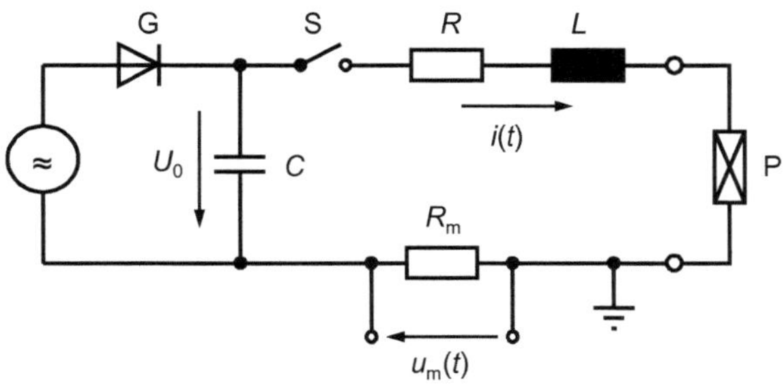

Fig. 5.3 Equivalent circuit diagram of a generator for generating exponential impulse currents

Fig. 5.4 Design of a modular 200 kA impulse current generator (100 kV, 250 kJ) (HIGHVOLT Prüftechnik Dresden GmbH)

The test standards provide for a variety of different impulse shapes. By appropriate selection of the plug-ins in table-top devices or change of the elements in larger units, impulse current generators can be relatively easily adapted to the requirements. The calculation of the desired current and the required components can be done by different methods [2–4]. A method described in [5] applies commercial software with which the circuit elements of a modular impulse current generator can be calculated for a given impulse shape. If the characteristic data of the test object are not known, these can also be determined using the calculation method. The otherwise time-consuming experimental preparation for adapting the circuit elements to the desired impulse shape can thus be avoided.

Note To avoid dangerously high open-circuit voltages, the output terminals of an impulse current generator must be short-circuited via the low-ohmic test object or, if the generator is not in operation, via a short-circuit bar.

In principle, impulse voltage generators can also be modified in such a way that they generate impulse currents in short-circuit operation [6]. However, the achievable current amplitudes are below the subjectively expected values, e.g. 40–70 kA for an 8/20 current impulse that can be generated with a 2 MV impulse voltage generator depending on its capacitors.

5.2.1.1 Influence of the Test Object on the Impulse Shape

The shape and thus the parameters of an exponential impulse current are determined by the circuit impedances including the test object, the measuring system and the high-voltage leads. Figure 5.5 shows the measured time parameters T_1 and T_2 depending on the resistance R_p of the test object P, which is positioned in the discharge circuit of a 20 kA table-top impulse current generator with a charging voltage of 10 kV. As R_p increases, the front time of the 8/20 impulse current decreases and the time to half-value increases. The same effect is also caused by the measuring resistor R_m. In addition, as the load resistance increases, so does the voltage at R_p, and the generator can no longer generate the specified maximum current amplitude. Knowing the values of C and L in the equivalent circuit diagram, the influence of the resistance on T_1 and T_2 can also be calculated (see Sect. 8.2).

5.2.1.2 Crowbar Technique

The tail of exponential impulse currents may have a more or less pronounced oscillation, which can lead to a *polarity reversal* (see Fig. 5.1). For the exponential 8/20 impulse current, the *reversal peak* with opposite polarity is approximately one third of the main peak (see Sect. 8.2). A reversal peak of this magnitude is unacceptable when testing certain power apparatus such as arresters. Increasing the damping resistance in the circuit of the current generator shown in Fig. 5.3 reduces the reversal peak, but also reduces the peak value.

An effective improvement in the case of oscillating impulse currents with polarity reversal brings the *crowbar technique*. With the circuit in Fig. 5.6, very high current impulses can be generated, which fall exponentially in the tail without

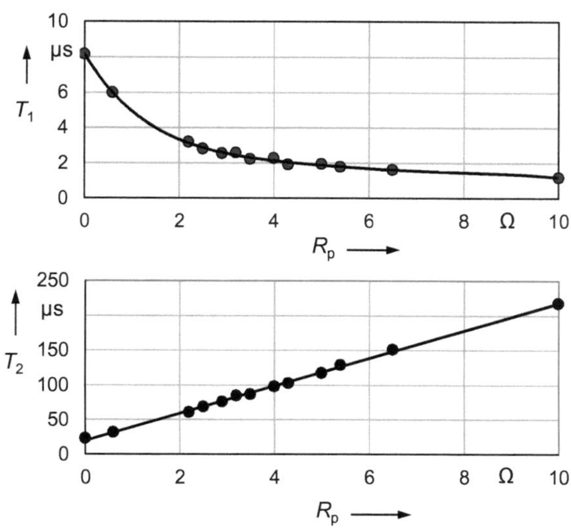

Fig. 5.5 Influence of the load resistance R_p on the time parameters T_1 and T_2 of an 8/20 impulse current, measured in the discharge circuit of the impulse current generator according to Fig. 5.4

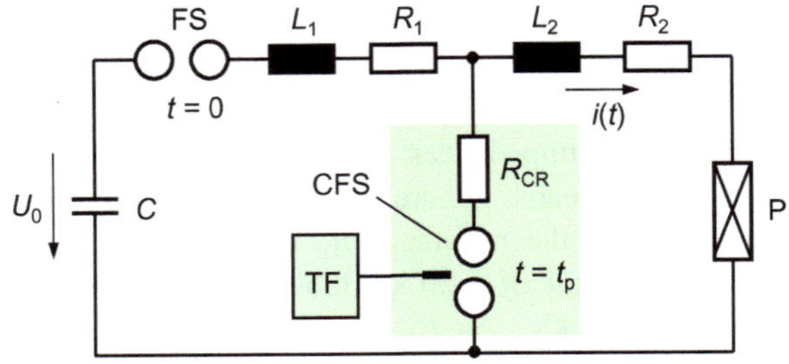

Fig. 5.6 Impulse current generator with crowbar gap CFS to avoid polarity reversal in the tail of impulse currents

oscillation. An essential element of the extended generator circuit is the triggerable crowbar spark gap CFS with the resistance R_{CR} of the spark [7, 8]. The indicated circuit elements L_1, R_1 and L_2, R_2 take into account the inductances and lead resistances of the generator circuit and the test object. The spark gap FS and the crowbar spark gap CFS are initially open and the capacitor C is charged. At time $t = 0$, the charging voltage is U_0, which causes the firing of the spark gap FS, whereby C is discharged via the circuit elements and the test object P (see Fig. 5.3). The initial shape of the impulse current through the test object P is that of an 8/20 impulse current (Fig. 5.7, curve *1*). At the time of the current peak $t = t_p$, the crowbar spark gap is triggered by the spark gap TF. During the firing of the crowbar spark gap, the circuit with L_2, R_2 and the test object P is short-circuited via the spark gap resistance R_{CR}. In addition, at peak time t_p, almost all of the energy previously stored in C is present in the inductance $L_2 \gg L_1$, and which now discharges via the test object. For $t > t_p$, the resulting current through P decreases exponentially to zero with the time constant $L_2/(R_{cr} + R_2)$, i.e. there is no polarity reversal (curve **2** in Fig. 5.7).

Exponential currents can also be generated with inductive energy storage devices. In this case, an inductor is charged with DC current via a charging circuit. Then, an initially closed switch parallel to the test object is opened and the current is commutated abruptly into the test object. Examples of fast commutation switches are circuit-breakers or wires that evaporate explosively at high current amplitudes, whereby the initial discharging circuit is interrupted [9, 10]. In order to simulate

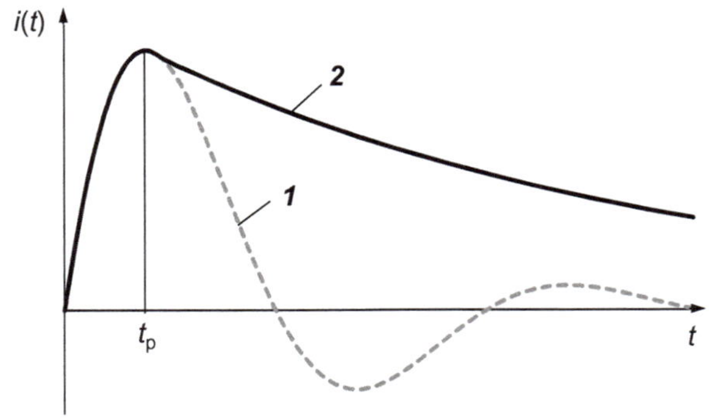

Fig. 5.7 Impulse shape *1* without, impulse shape *2* with crowbar spark gap (schematic)

multiple lightning strokes, current generators are used which can generate a fast sequence of impulse currents with different impulse shapes and both polarities [11, 12].

5.2.2 Generation of Rectangular Impulse Currents

The basic circuit of a generator for generating *rectangular* (*long-duration*) *impulse currents* with duration of more than 1 μs is shown in Fig. 5.8. The series-connected LC elements form an n-stage *ladder network*. The parallel capacitors C' are charged by the rectified AC voltage to the DC voltage U_0 and discharged via a triggered spark gap FS on the terminating resistor R_1 and the test object. The terminating resistance of the homogeneous ladder network is determined by:

$$R_1 = \sqrt{\frac{L}{C}} \tag{5.4}$$

with $L = nL_i$ and $C = nC'$. Wherever required, the resistive part of the test object P is taken into account in Eq. (5.4). The duration T_d of the rectangular impulse current shown in Fig. 5.8 is approximated by (Ref. [1] of Chap. 1):

$$T_d \approx 2\frac{n-1}{n}\sqrt{LC}. \tag{5.5}$$

From Eqs. (5.4) and (5.5), L and C can be calculated for the desired rectangular current of duration T_d. Numerical calculations for a generator with $n = 8$ elements show that an asymmetrical design of the ladder network is more advantageous in order to achieve as rectangular an impulse shape as possible without much overshoot and undershoot at the beginning and end of the impulse. Accordingly, the individual inductances L_1 to L_n are significantly different, while the partial capacitances C' of the ladder network remain constant [13].

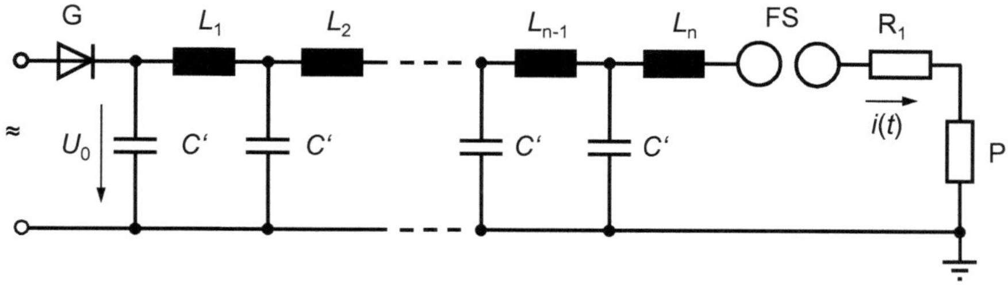

Fig. 5.8 Schematic circuit diagram of a generator for rectangular impulse currents

5.3 Measurement of High Impulse Currents

The conventional measurement of impulse currents up to the highest peak values of some 100 kA is usually carried out with low-ohmic resistors or magnetic coils with and without magnetic core. They are commonly referred to as *current sensors*. Measuring resistors provide an output voltage proportional to the current, measuring coils an output voltage proportional to the time derivative of the current. The measuring instruments at the output of the current sensors are comparable to those used for high-voltage impulse measurements, mainly digital recorders. When using magnetic coils, integration of the output voltage can be achieved in different ways. Further measurement options are provided by current sensors based on the *Hall effect* and *Faraday effect*, which are discussed in detail in Sect. 3.5.2 or Sect. 6.2.

The use of the various types of sensors is associated with advantages and disadvantages of a fundamental nature. Important criteria for the selection of a current sensor are, for example, the electrical insulation from ground and the measurement behavior for DC and high-frequency components. Current measuring systems are exposed to electrical and strong magnetic fields, which may lead to incorrect measurements due to interference. Depending on the sensor type and the measuring circuit, the effect of interference can often be largely avoided.

5.3.1 *Measuring Systems with Low-Ohmic Resistors*

In order to measure high impulse currents, traditionally low-ohmic measuring resistors are used. The measuring resistor R_m is inserted at a suitable location in the circuit, where it is directly flowed through by the current $i(t)$ to be measured (Fig. 5.9). In the preferred position, the resistor is grounded at one end. Measuring resistors for impulse currents are generally constructed as a quadrupole with a

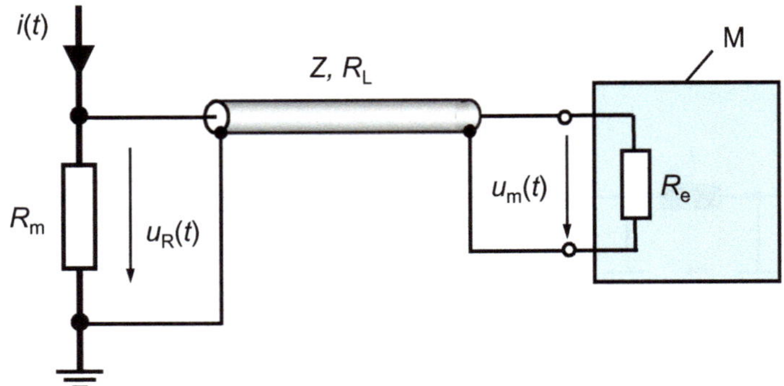

Fig. 5.9 Measuring circuit for impulse currents with low-ohmic resistor R_m. M: Measuring instrument with input resistance R_e

defined voltage tap. Under ideal conditions, the voltage $u_R(t)$ at the measuring resistor R_m is proportional to the current $i(t)$:

$$\boxed{u_R(t) = R_m i(t)}. \tag{5.6}$$

An important prerequisite for the validity of Eq. (5.6) is that R_m behaves purely resistive in the considered frequency range. The resistance of R_m is usually in the range of 50 μΩ to 50 mΩ. A low value of R_m is advantageous for several reasons. The power dissipation in the measuring resistor is limited and a change in resistance or even damage due to excessive self-heating or voltage stress is avoided. Moreover, the shape of the impulse generated by the current generator will remain substantially unchanged in the case of a small resistance. Finally, even at very high currents, the voltage across R_m is limited to values below 2000 V and can be recorded directly with the measuring instruments specially designed for high-voltage impulse.

On the other hand, a low-ohmic measuring resistor has the disadvantage that the output voltage u_R is small for low current amplitudes. The measuring instrument must then have a high sensitivity, and interference voltages can have a greater influence on the measurement results. A low-ohmic resistor also impedes the measurement of the frequency response or the step response, since the available generator circuits are usually designed only for low current amplitudes. Hence, only a small voltage across R_m is available for the measurement.

5.3.1.1 Ground Loops and Coupling Impedance

The impulse voltage $u_R(t)$ at the measuring resistor is fed via a coaxial cable to the measuring instrument M with the input resistance R_e. In order to avoid a *ground loop* in the measuring circuit by double grounding, the measuring instrument M is not connected directly, but via the shield of the coaxial cable to ground. In addition, the mains supply is provided by an isolating transformer. But even without direct grounding of the measuring instrument, the resulting stray capacitance C_e of the housing and of the isolating transformer form a closed circuit for high-frequency signals (Fig. 5.10). The magnetic field H generated by the impulse current $i(t)$ permeates the area shown hatched and induces in the ground loop the voltage u_{in}, which drives the interference current i_{St} (Refs. [66, 67] of Chap. 4).

The interference current i_{St} flows in the cable shield and generates via the *coupling impedance* Z_k the voltage u_{St} on the inner surface of the shield (Fig. 5.11). The coupling impedance of a coaxial cable is defined as $Z_k = u_{St}/i_{St}$, with the inner conductor and the inner side of the cable shield short-circuited at one end. A good shielding effect of the coaxial cable is characterized by low coupling impedance Z_k. The coupling impedance of a coaxial cable with woven metallic braid as the shield initially decreases with increasing frequency, but then increases again due to the inductance of the shielding braid. The interference voltage u_{St} is superimposed on

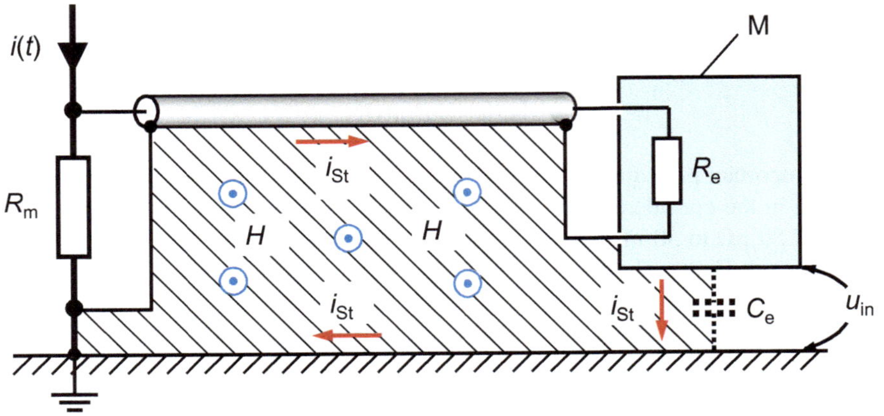

Fig. 5.10 Example for explaining electromagnetic interference in an impulse current measuring system with low-ohmic resistor R_m, coaxial cable and measuring instrument M. H: Magnetic field generated by the impulse current $i(t)$, u_{in}: voltage induced by the magnetic field H permeating the hatched area, i_{St}: interference current in the ground loop via the stray capacitance C_e of the housing of M

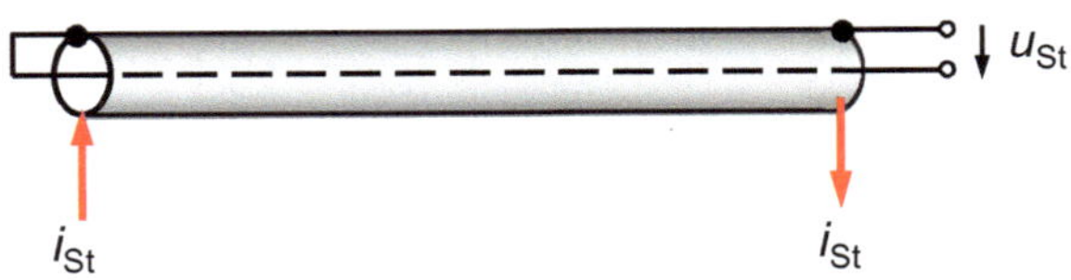

Fig. 5.11 Definition of the coupling impedance $Z_K = u_{St}/i_{St}$ of a coaxial cable. i_{St}: Interference current, u_{St}: generated interference voltage

the voltage measured at R_e. The magnitude of the current also depends on the impedances in the circuit, mainly R_e. With high-ohmic input of the measuring instrument M, this part of the interference current is usually negligible.

Also, the coaxial cable and the measuring instrument positioned at some distance away are directly exposed to the magnetic field generated by the impulse current. On the non-magnetic cable shield and the instrument housing, however, eddy currents are formed, which generate an opposing magnetic field that counteracts the penetration of the magnetic field. Although the shielding may be limited by inhomogeneity of the coaxial cable, the residual interference effect on the measured voltage is small compared to that in the grounded loop arrangement. The interference effect due to the electric field occurring in the generation of the impulse current can usually be considered low.

Avoiding of cable shield currents and housing currents due to ground loops is therefore the main objective to prevent interference on the measuring circuit. An effective means is to additionally shield the coaxial cable and place the measuring instrument in a shielded cabin (Fig. 5.12). Both ends of the outer cable shield and the shielded cabin S are grounded. The interference current i_{st} generated by the external magnetic field now flows via the outer cable shield and the shielded cabin

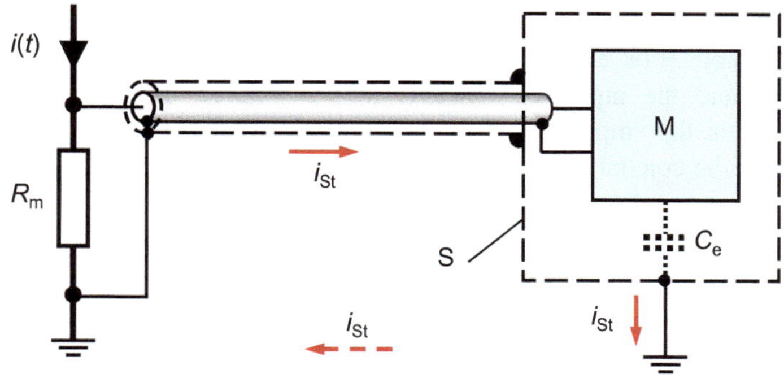

Fig. 5.12 Double shielding of the coaxial cable and measuring instrument to prevent the influence of interference currents on the measurement

to ground. An opposing magnetic field is built up and the interior of the shielding becomes field-free.

Additional shielding of coaxial cables can be achieved by several measures. A double-shielded coaxial cable in turn has the disadvantage that the shielding effect of the two shield braids is still not optimal and decreases at higher frequencies. A better choice is a cable provided with a thin metal foil shield on which the braid is arranged. The most suitable shielding is to place the coaxial cable: i in a flexible corrugated pipe, ii in a metal tube lying on (or under) the floor, or iii in a cable channel below the grounded floor area. A good shielding effect also has a twisted wire pair in a surrounding shield.

5.3.1.2 DC Resistance of a Coaxial Cable

Measurements of impulses with rise times well below 1 μs require that the coaxial cable with impedance Z at the input of the measuring instrument be terminated by a resistance $R_e = Z$. Otherwise, travelling waves propagate along the cable and are superimposed on the signal to be measured. At high currents, impulse voltages of up to 2000 V occur at the current measuring resistor R_m. Therefore, R_e must also have a sufficiently high impulse load capacity.

When using a long coaxial cable, it may be necessary to consider its DC resistance R_L, in particular when the cable is terminated with its characteristic impedance Z, i.e. with low resistance in the range of 50 Ω. The value of R_L per unit length is usually in the range of 15–150 mΩ/m. Since R_L and R_e are connected in series, they form a voltage divider for the measurement voltage. At the input of the measuring instrument, therefore, the voltage is:

$$\boxed{u_m(t) = \frac{R_m R_e}{R_L + R_e} i(t)}, \tag{5.7}$$

which is reduced compared to $u_R(t)$ in Eq. (5.6) due to the voltage across the measuring cable. For example, if the long coaxial cable has a DC resistance $R_L = 0.5\ \Omega$ and the input resistance is $R_e = Z = 50\ \Omega$, then the input voltage $u_m(t)$ and thus the impulse current $i(t)$ is measured by 1% too low. The DC resistance of the coaxial cable affects all types of voltage measurements.

5.3.1.3 Inductances of a Low-Ohmic Resistor

Low-ohmic resistors, which are not specially designed to measure short impulses, no longer show a pure resistive behavior already beyond 1 kHz. The simple equivalent circuit of a low-ohmic resistor shows the resistive component R and in addition a series inductance L (Fig. 5.13). An impulse current $i(t)$ flowing through the resistor therefore causes the output voltage $u(t) = u_R(t) + u_L(t)$ with the two partial voltages $u_R(t) = Ri(t)$ and $u_L(t) = L\mathrm{d}i/\mathrm{d}t$ (Fig. 5.14). Characteristic is the inductive voltage peak of $u_L(t)$, which can be even larger than the resistive partial voltage $u_R(t)$ for a very steep impulse current. The inductance thus prevents a true-to-scale measurement of the impulse current.

Figure 5.15 shows the equivalent representation of the output voltage $u(t)$ in the frequency domain. The amplitude density $F(f)$ increases with frequency and attains at the upper limit frequency f_2, which is characterized by $\omega L = R$, twice the value of the DC component F_0. In order to achieve a wideband behavior, therefore, the inductive component of a resistor must be kept as small as possible by a suitable construction and the materials used. In addition, it should be mentioned that resistors in the equivalent circuit diagram usually also have a parallel capacitance. However, the parallel capacitance has no influence here due to the low resistance and is not considered in Fig. 5.13.

Low-ohmic resistors can have two main inductance components. One of them is determined by the construction of the resistor and its connecting leads. The self-inductance L of a wire or a cylindrical conductor of non-magnetic material ($\mu_r = 1$) with diameter d and length l is calculated as (Ref. [2] of Chap. 9):

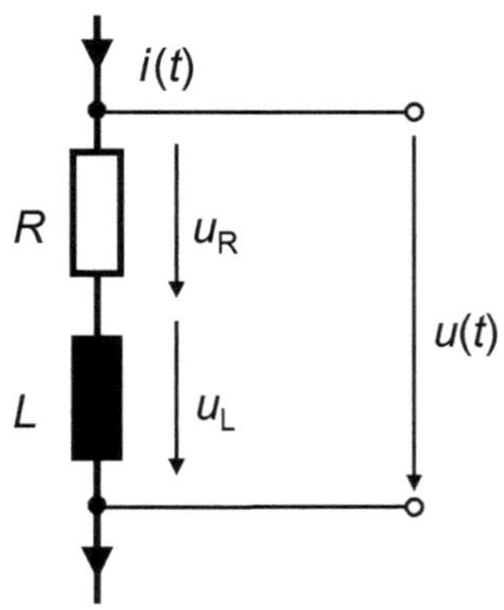

Fig. 5.13 Simplified equivalent circuit diagram of a low-ohmic resistor R with self-inductance L

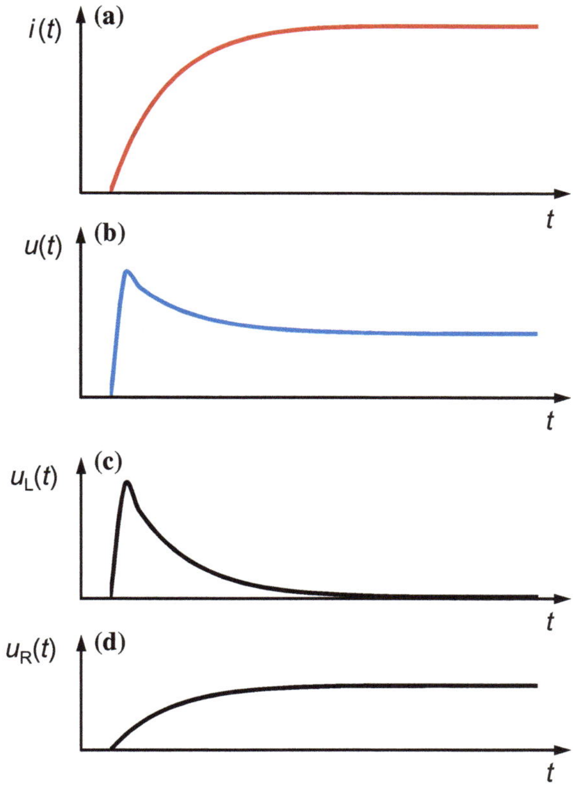

Fig. 5.14 Impulse response of a low-ohmic resistor with self-inductance (schematic). **a** Impulse current $i(t)$, which is fed into the resistor, **b** output voltage $u(t)$ of the resistor (impulse response), **c** inductive and resistive partial voltages u_L or u_R according to the equivalent circuit diagram in Fig. 5.13

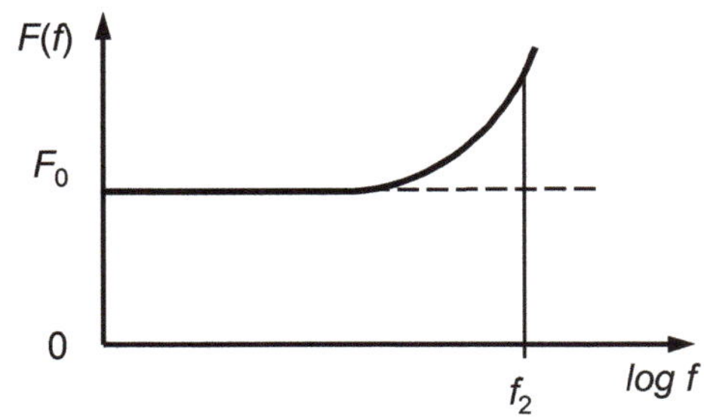

Fig. 5.15 Frequency response of a resistor with self-inductance (schematic)

$$\boxed{L = \frac{\mu_0 l}{8\pi}\left(4\ln\frac{l}{d} - 3\right)}, \tag{5.8}$$

where $\mu_0 = 0.4\pi \times 10^{-6}$ H/m ≈ 1.257 μH/m is the *magnetic field constant*. A cylindrical film resistor with d = 1 cm and l = 5 cm then has a self-inductance of 9 nH. If the resistor is provided with two connecting wires of diameter d = 1 mm and total length l = 2 cm, their inductance is also 9 nH. The total inductance of the

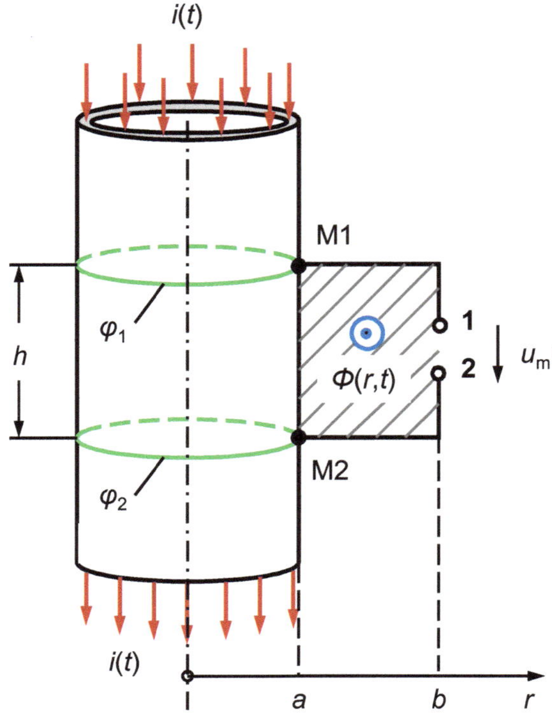

Fig. 5.16 Simple model of a current-carrying thin-walled tubular resistor. M1, M2: Measurement points with potential taps

resistor is then 18 nH. The upper limit frequency f_2 of the resistor, at which the inductive and ohmic components are equal, results from the relationship $\omega L = R$. For a resistance $R = 1$ mΩ and an inductance L = 18 nH, $f_2 = 9$ kHz.

The second inductance component is caused by the type and form of the measurement tap on the resistor. The principle is explained by the simple model of a thin-walled tubular resistor through which the current $i(t)$ flows (Fig. 5.16). The *skin effect* has not yet been taken into account here. With a homogeneous distribution of the resistance and a long tube, the current flow results in circular equipotential lines on the tube wall. If the potentials φ_1 and φ_2 at the measurement points M1 and M2 are tapped, the voltage u_R is obtained under simplifying conditions:

$$u_R = \varphi_1 - \varphi_2 = Ri(t), \tag{5.9}$$

where R represents the effective resistance of the tube between the measurement points M1 and M2 (Ref. [1] of Chap. 1). Further, according to the law of magnetic flux (see Sect. 5.3.2.1), the current generates a tangential magnetic field outside the resistor tube, which decreases with increasing distance $r > a$ from the surface of the resistor. The magnetic field inside the tube is negligible.

In the simple model, the potential taps at M1 and M2 are connected via short leads to the output terminals ***1*** and ***2***, from which the coaxial cable leads to the measuring instrument. This results in a measurement loop which encloses a portion of the magnetic field marked by the hatched area $(b - a)h$. The temporal change of the *magnetic flux* $\Phi(r, t)$ in the measurement loop induces the partial voltage $u_L = -d\Phi/dt$ at the output terminals ***1–2***, which is superimposed on the resistive partial voltage u_R according to Eq. (5.9). Taking into account the polarity rule for the induced voltage, the resulting measurement voltage $u_m(t)$ is given by:

$$\boxed{u_m(t) = u_R - u_L = R\,i(t) + M\frac{di(t)}{dt}}. \tag{5.10}$$

Here, M is the effective *mutual inductance*, which can be calculated from the geometric dimensions of the measurement loop. For a rectangular loop, the mutual inductance is:

$$\boxed{M = \frac{\mu_0 h}{2\pi}\int_a^b \frac{dr}{r} = \frac{\mu_0\, h}{2\pi}\ln\frac{b}{a}}. \tag{5.11}$$

In the equivalent circuit diagram of Fig. 5.13, the mutual inductance M must be added to the self-inductance L of the measuring resistor. Resistors with this type of voltage tap have an increased time constant $(L + M)/R$ and thus a rise time that is probably not much better than 1 µs. The goal of any construction of wideband measuring resistors is therefore to place the voltage tap in the magnetic field-free area.

5.3.1.4 Coaxial Measuring Resistors

The requirement to place the voltage tap of a measuring resistor in the magnetic field-free area can be fulfilled by a coaxial design. In the simplest case, the low-ohmic measuring resistor consists of a parallel connection of equal individual resistors with good high-frequency behavior, which are housed in a metal cylinder. The impulse current $i(t)$ enters the connecting bolt ***1*** and is evenly distributed to the individual resistors ***2*** (Fig. 5.17). The outer metal tube ***3*** serves as the return conductor, whereby the resulting magnetic field within the cylinder disappears (see Sect. 5.3.1.5). The voltage tap on the resistors and the lead ***4*** to the coaxial measurement terminal ***5*** are therefore in field-free space, so that no interference voltage can be induced here. The current-proportional voltage is tapped at the output terminal ***5*** and passes via a coaxial cable to the measuring instrument. Measuring resistors of this type can be produced without much effort for impulse currents of several 10 kA with mostly satisfactory transfer behavior.

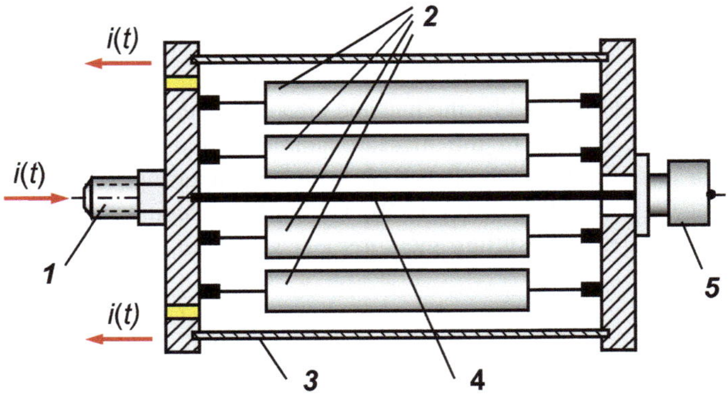

Fig. 5.17 Measuring resistor with parallel connected individual resistors in a metal housing. ***1*** Current terminal, ***2*** individual resistors, ***3*** current return conductor, ***4*** measurement tap and lead to the output terminal, ***5*** coaxial output terminal

The parallel connection of individual resistors reduces the total inductance and improves the frequency behavior. If the requirements regarding the frequency behavior are not so high, bifilar-wound wire resistors are sufficient. They are manufactured with a low temperature coefficient and remain approximately constant even under high current load. Film resistors and chip resistors with ceramic carrier basically show a better frequency behavior. Film resistors should not have a helical groove in the resistive layer, which is particularly common with carbon film resistors to achieve a very precise nominal resistance. However, the helical groove can lead to discharges and flashovers between adjacent resistance paths, which initially cause an irreversible increase in resistance and later the destruction of the resistor.

Optimal transfer behavior is provided by the perfectly homogeneous *coaxial measuring resistor*, also called *coaxial shunt* or *tubular shunt*. The first types of these resistors are reported in [14]. In the construction according to Fig. 5.18, the impulse current $i(t)$ is fed into the coaxial shunt at the terminal ***1***, flows through the thin-walled tubular resistor ***3*** to the right metal plate and back again over the outer metal tube ***4***. The latter is insulated by the insulating ring ***2*** from the current terminal ***1***. The current-proportional impulse voltage at the resistance tube ***3*** is tapped and passed via the lead ***5*** to the coaxial output terminal ***6***. The coaxial arrangement in turn ensures that no magnetic field can form within the coaxial shunt and thus no interference voltage is induced at the measurement tap ***5***. The homogeneous tube resistor ***3*** in Fig. 5.18 consists of a non-magnetic resistance alloy or a graphite-coated insulating foil. As will be shown below, the thickness of the resistor body determines the achievable bandwidth and the thermal load capacity. The temperature coefficient of the tube resistance must be small and the resistance reversible, even at temperatures of 100 °C and more. In conclusion, it can be said that the coaxial shunt according to Fig. 5.18 is very close to the ideal of pure resistance without self-inductance. The lightning current transducer described in [15] has a 60 μm thick NiCr metal foil as forward and return conductors in a coaxial arrangement.

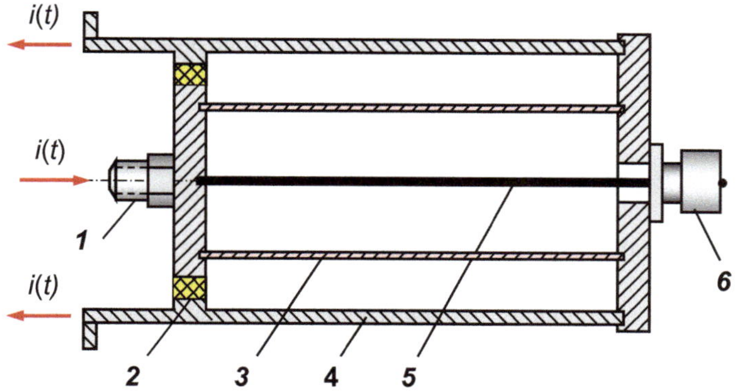

Fig. 5.18 Principle construction of a wide-band coaxial shunt for impulse current measurements. *1* Current terminal, *2* insulating ring, *3* resistance tube, *4* current return conductor, *5* measurement tap with lead to the output terminal, *6* coaxial output terminal

Figure 5.19 shows some examples of low-ohmic measuring resistors. In addition to an AC current shunt on the left side, there are three vertically arranged coaxial shunts designed according to the shunt in Fig. 5.18 for impulse currents of up to

Fig. 5.19 Examples of different measuring resistors for high-current measurements. Left: AC current resistor, right: three coaxial shunts for impulse current measurements (HILO-TEST)

100 kA and bandwidths of more than 100 MHz. As will be shown in the next section, the material and thickness of the resistance tube ***3*** determine the achievable bandwidth and the thermal load capacity for impulse currents.

5.3.1.5 Skin Effect

Although the coaxial shunt shown in Fig. 5.18 is largely homogeneous and low-inductance, its bandwidth cannot be increased infinitely. A time-varying current in a cylindrical conductor generates a magnetic field, which in turn generates currents called *eddy currents*. The eddy currents are superimposed on the original current and thereby influence the current distribution in the conductor. The resulting current is no longer distributed homogeneously over the entire cross-section of the conductor, but is displaced by the *skin effect* with increasing frequency more and more to the outer edge region of the conductor. Furthermore, the eddy currents generate a magnetic opposing field that is superimposed on the original magnetic field, so that the resulting magnetic field within the conductor disappears. The interlinked electric and magnetic fields occurring in the conductor can be determined under idealizing assumptions by Maxwell's equations or by the laws of magnetic flux and induction (Ref. [1] of Chap. 9).

Because of the skin effect, it does not make sense to make a cylindrical conductor or resistor for high-frequency currents out of solid material. Even with the usual tubular constructions, the wall thickness cannot be fully utilized for high-frequency currents. The current density is largest on the outer surface of the resistance tube and decreases radially to the tube interior (Fig. 5.20). The uneven

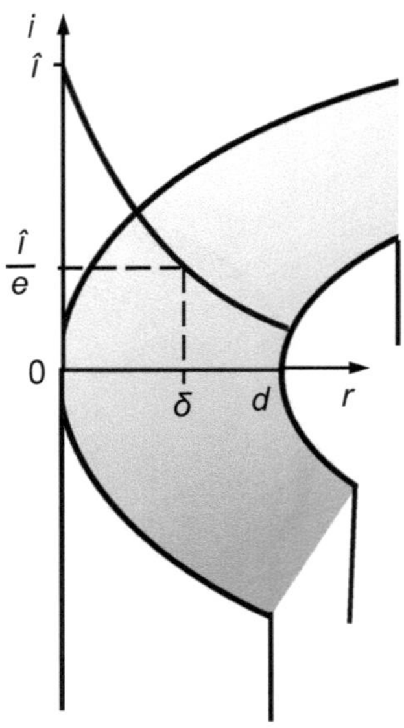

Fig. 5.20 Uneven current distribution in the cross-section of a tubular conductor due to the skin effect

current distribution across the conductor cross-section is equivalent to an increase in resistance compared to the DC resistance.

Characteristic parameter for the skin effect in a conductor with the conductivity σ and the permeability $\mu = \mu_r \mu_0$ is the *penetration depth* (or: *skin depth*):

$$\boxed{\delta = \frac{1}{\sqrt{\pi \mu \sigma f}}}, \tag{5.12}$$

at which the current density with the frequency f has fallen to 1/e of the surface value (to about 37%). This relationship applies exactly to infinitely long plate or cylindrical conductors, which have a very large thickness or a large diameter compared to the penetration depth. With a tube wall thickness $d \ll \delta$, the skin effect has only a minor effect, and the resistance retains nearly its DC value up to the frequency specified for δ. In order to limit the skin effect, the penetration depth δ according to Eq. (5.12) is made as large as possible by using non-magnetic materials with low conductivity σ. Suitable resistance materials with $\mu_r = 1$ are Constantan, Manganin and chromium-nickel alloys in the normal conducting state. Even a thin graphite layer deposited on an insulating body shows good frequency behavior but, depending on the manufacturing process, has only limited stability against thermal overload and mechanical stress.

The transfer behavior of a measuring resistor R_m can be characterized in a similar manner as for an impulse voltage divider by the unit step response $g(t)$ (see Sects. 9.1 and 9.8). It is defined as the voltage at R_m, divided by $I_0 R_m$, when injecting a step current with the amplitude I_0 into R_m. To go into more detail, Maxwell's equations can be used to calculate the distribution of the injected step current in the nonmagnetic resistance tube, whereby the unit step response of the coaxial shunt is obtained as (Ref. [1] of Chap. 1; [16, 17]):

$$\boxed{g(t) = 1 + 2\sum_{k=1}^{\infty} (-1)^k \exp\left(-\frac{k^2 \pi^2}{\mu_0 \, \sigma \, d^2} t\right)}. \tag{5.13}$$

Figure 5.21 shows the calculated step response of a coaxial shunt with a specific conductivity $\sigma = 1$ m/Ωmm^2 for three different wall thicknesses d. The basic form of the step response shows no inductive voltage peak, i.e. in contrast to the simple resistor in Fig. 5.14, the coaxial shunt has no inductive component. The finite rise of $g(t)$ is determined by the ratio of the wall thickness d of the resistance tube to the penetration depth δ according to Eq. (5.12). For example, for $d = 0.1$ mm, the rise time of $g(t)$ is 3 ns (curve ***1*** in Fig. 5.21). As the wall thickness increases, the rise time increases disproportionately (curves ***2*** and ***3*** in Fig. 5.21).

The step response of coaxial shunts according to Eq. (5.13) is, as far as the equation type is concerned, identical to the step response of the high-ohmic voltage divider in Eq. (4.23). The latter is characterized by the RC ladder network or by the simplified equivalent circuit diagram in Fig. 4.23. The formal comparison of the

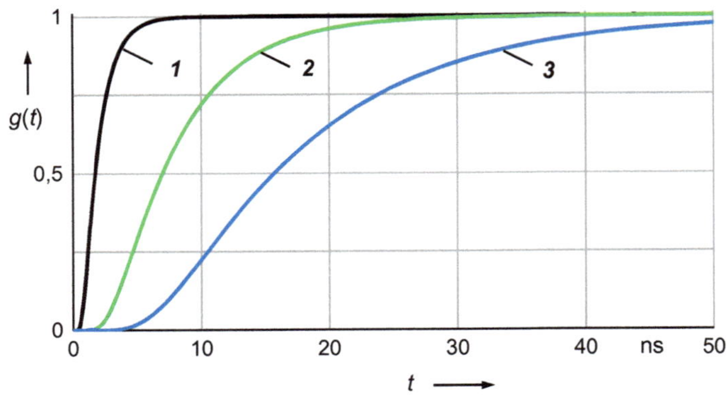

Fig. 5.21 Step response of a coaxial shunt with skin effect, calculated according to Eq. (5.13) for a conductivity $\sigma = 1$ m/Ωmm^2 and three wall thicknesses *d*. ***1*** $d = 0.1$ mm, ***2*** $d = 0.2$ mm, ***3*** $d = 0.3$ mm

expressions in Eqs. (5.13) and (4.23) provides the response time *T* and the bandwidth *B* of the coaxial shunt:

$$\boxed{T = \frac{\mu_0\, \sigma\, d^2}{6}} \tag{5.14a}$$

$$\boxed{B = \frac{1.46}{\mu_0\, \sigma\, d^2}}. \tag{5.14b}$$

Thus, the dynamic behavior of a coaxial shunt can be improved primarily by reducing the thickness *d* and, to a lesser extent, by selecting materials with low conductivity σ. The specific conductivity of suitable alloys is in the range of 0.8–2 m/Ωmm^2 at room temperature.

Very thin resistance foils down to 10 μm can be made with chromium-nickel alloys, the penetration depth δ at 1 GHz being about 17 μm. Short coaxial shunts of about 3 cm in length made of extremely thin resistive foils of Cu (1 μm) and NiCr (10 μm) are discussed in [18]. The measured rise times were below the inherent rise time of 0.4 ns of the then available oscilloscope.

5.3.1.6 The Coaxial Shunt as Ladder Network

In analogy to the resistive voltage divider, the coaxial shunt with the step response according to Eq. (5.13) is represented by the ladder network as equivalent circuit in Fig. 5.22a with *n* distributed longitudinal inductances L' and transverse resistances R' [16, 17]. For the limiting case $n = \infty$, the parallel connection of the *n* resistances $R' = nR_0$ yields the DC resistance R_0 and the series connection of the *n* series

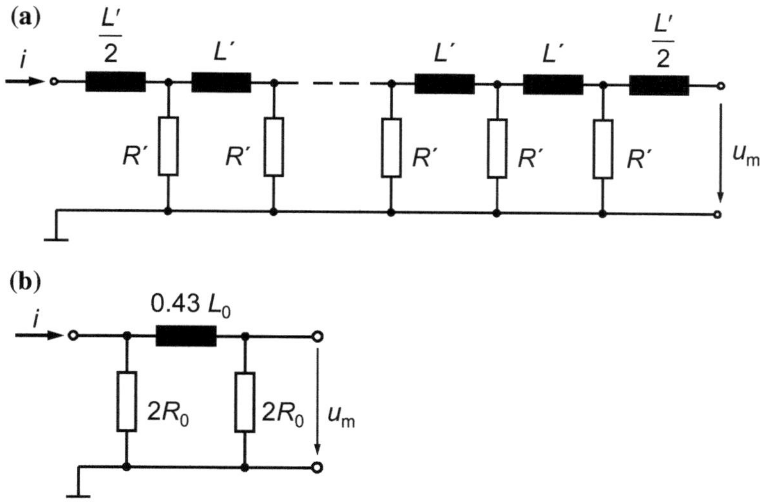

Fig. 5.22 Equivalent circuit diagrams of the coaxial shunt. **a** Ladder network with n elements $L' = L_0/n$ and $R' = nR_0$, **b** simplified equivalent circuit with the same rise time as the ladder network

inductances $L' = L_0/n$ the resulting inductance L_0. For impulse currents with front times of more than 1 μs, a simplified equivalent circuit diagram with discrete elements and a comparable step response can be derived (Fig. 5.22b).

From the requirement that the response times of both equivalent circuit diagrams should be equal, the result for the inductance in the simplified equivalent circuit diagram is initially a value of $0.67L_0$. An improved value of $0.43L_0$ is obtained from the equality of the rise times of both equivalent circuit diagrams. Finally, however, it should be remembered that the coaxial shunt due to its special construction has virtually no inductance and that its transfer behavior is characterized by the skin effect.

5.3.1.7 Experimental Step Response of Measuring Resistors

For generating step currents, the step generators mentioned in Sect. 9.8.4 are available. The low-ohmic measuring resistor is usually connected to the step generator in series with an internal and/or external terminating resistor of 50 Ω. When using mercury wetted reed contacts, step currents with rise times down to 1 ns and amplitudes of up to 3 A can be achieved. The step response of low-ohmic resistors then reaches amplitudes of usually only 1 mV or even less, requiring a preamplifier for the recording instrument. When using a cable generator, higher current amplitude can be achieved by connecting several cables in parallel. In [19], a 500 m long coaxial cable is charged to 300 V and then discharged via a manually operated spark gap to the measuring resistor. The step voltage at the resistor has a rise time of

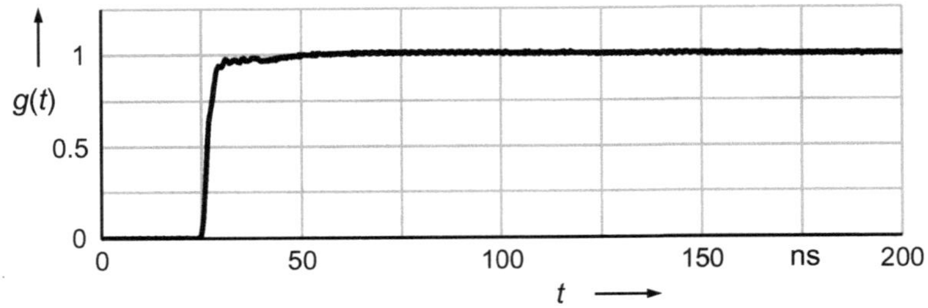

Fig. 5.23 Experimental step response of a 5 kA coaxial shunt

less than 6 ns and generates a step current with a maximum amplitude of 120 A. The electronic step generator in [20] provides an output current of up to 1.2 kA, but with a long rise time of 60 ns. With gas-filled spark gaps, higher voltages can be switched, and thus larger current amplitudes are achievable.

Figure 5.23 shows the experimental, i.e. measured step response of a wideband 5 kA coaxial shunt, consisting of a thin resistance sheet made of chromium-nickel alloy. The rise time of the coaxial shunt is 3 ns, which corresponds to a bandwidth of more than 100 MHz according to Eq. (9.38). This value is comparable to the theoretical rise time of a coaxial shunt with a thickness of $d = 0.1$ mm, whose step response is calculated according to Eq. (5.13) (see Fig. 5.21, curve ***1***). An overshoot of the experimental step response does not occur, i.e. in accordance with the above theoretical investigation, the coaxial shunt under study has no self-inductance. In the first 25 ns, the step response shows a *creeping* to the final level, but this is mainly due to the non-ideal voltage step of the cable generator used. To measure the step response, the current return conductor of the coaxial shunt (***4*** in Fig. 5.18) was connected to the shield of the coaxial output of the step generator via a funnel-shaped adaptor in order to achieve a reflection-free connection.

In principle, the conditions in the measurement of the step response of coaxial shunts can be described as nearly ideal. Compared to high-voltage dividers, coaxial shunts have smaller dimensions, are usually completely shielded and traveling wave phenomena in the measurement circuit can be largely avoided by a terminating resistor equal to the cable impedance. The experimental step response of a coaxial shunt is therefore largely free of interference effects. If the generated current step does not have the expected ideal curve, the effect on the step response can optionally be corrected by a convolution calculation.

These are good prerequisites for using the step response in numerical convolution calculations, e.g. to prove the suitability of a coaxial shunt for measuring a given current impulse. Compliance with defined error limits for the peak value and the time parameters can thus be checked by calculation for arbitrary current waveforms. With the convolution calculation, possibly in conjunction with a check measurement by comparison with a reference system, an effective method for proving the dynamic behavior of coaxial shunts is therefore available, which is basically also accepted in IEC 62475 (Ref. [4] of Chap. 2). It is therefore not necessary to specify response parameters and their limits for current measuring resistors, as is the case with voltage dividers for lightning and switching impulse voltages and which are not always justified (Refs. [30, 44] of Chap. 9).

5.3.1.8 Special Designs of Coaxial Shunts

For rise times below 1 ns or frequencies above 1 GHz, the length of a coaxial shunt can no longer be considered short compared to the wavelength of the measurement signal. The arrangement of the resistor element and the junction from the resistor to the coaxial cable connector must be optimized at high frequencies in order to achieve even shorter rise times. Extremely large bandwidths are achieved with disk-shaped resistors, wherein the resistive material may be a thin graphite layer deposited on an insulating support [21] or a NiCr foil with a thickness between 10 and 50 μm [22]. In the principle arrangement according to Fig. 5.24, the current source is connected via the coaxial conductor ***1*** to the vertically arranged disk resistor ***2***, which is mechanically supported by the insulating disc ***3***. In the coaxial conductor ***1***, a plane wave propagates perpendicularly from the current source to the disk resistor ***2***. From the inner conductor, a current flows radially through the disk resistor ***2*** to the outer conductor and back to the current source. The voltage tap at the disk resistor is designed as a conical transition ***4***, through which the characteristic impedance of the transition ***4*** increases nearly reflection-free from virtually zero to the impedance of the coaxial cable connected at ***5***.

The rise time of the disc-shaped resistor with a 15 μm thick NiCr foil is reported to be less than 0.35 ns, which corresponds to the inherent rise time of the oscilloscope used at that time. Due to their small thickness, disk resistors are suitable for measuring very short current impulses with maximum peak values of 10 kA. The disc resistor is also available as a 50 Ω terminator for oscilloscopes with

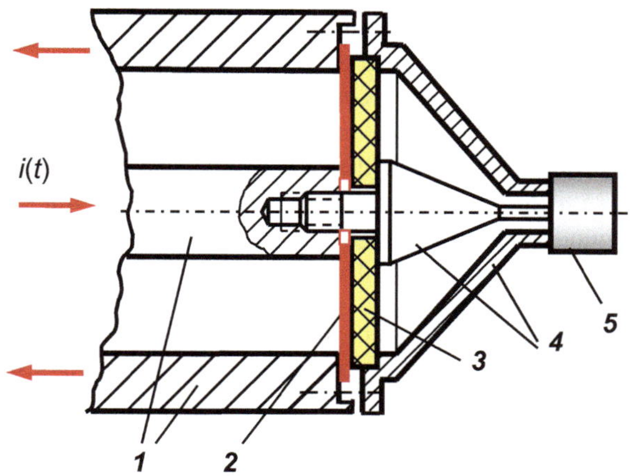

Fig. 5.24 Principle of a disk-shaped measuring resistor. ***1*** Coaxial conductor, ***2*** disk-shaped resistor, ***3*** insulating disc, ***4*** reflexion-free transition, ***5*** output terminal

bandwidths of 500 MHz and more. Theoretical studies on the conduction mechanism in disc resistors lead to the same equations for the step response and the rise time as in Eqs. (5.13) and (5.14a) for the coaxial shunt.

In addition to the dynamic behavior of measuring resistors, the maximum current load is an important criterion. Coaxial shunts for very high current amplitudes and long impulse duration must have a correspondingly large wall thickness (see Fig. 5.18, resistance tube ***3***) in order to limit Joule's self-heating and the associated resistance change. However, a large wall thickness leads to significantly longer rise times due to the skin effect. Various variants for improving the transfer behavior of high current shunts can be found in [23, 24]. In the *squirrel cage resistor*, the resistance tube is replaced by several circularly arranged resistance wires or rods. This allows the magnetic field generated by the measurement current to partially penetrate into the cage arrangement and influence the voltage tap inside. In the measurement loop of the tap thereby a voltage is induced, which increases the output voltage for the higher frequency components. With optimized execution, this is noticeable in the step response by a steeper rise and an overshoot.

The calculation of different cage-type resistors taking into account the skin effect shows that the step response largely depends on the number of rods or wires. By accepting a larger overshoot of the step response, rise times in the range of 1 ns and sometimes also negative response times are achievable. A comparable penetration of the magnetic field into the measurement tap can be obtained in a coaxial shunt, in which one or more longitudinal slits are cut in the outer cylinder serving as the return conductor. But even with optimal dimensioning, the achieved frequency response is generally unsatisfactory and must be equalized by complex electrical compensation circuits (Ref. [6] of Chap. 1).

In another variant of the coaxial shunt for very high current amplitudes, the lead from the measurement tap to the output connector is embedded in the wall of the resistance tube. As a result, a well-defined part of the time- and location-dependent magnetic field in the wall is coupled into the measurement lead with the aim of improving the frequency response of the coaxial shunt. The optimum arrangement of the measurement lead within the tube wall results from the calculation of the variable magnetic field of a step current. In a first simpler prototype, the resistance tube was composed of several layers of a resistance foil. Between two of the foil layers, the measurement lead was embedded isolated, with the optimal position resulting from field calculations. For the multi-layered coaxial shunt with embedded measurement lead to the output connector, the rise time improved from originally 125 ns to 9 ns [25]. Figure 5.25 shows a later implementation of a 250 kA coaxial shunt for short-circuit measurements. The insulated measurement lead ***4*** runs in a calculated groove milled into the thick tube wall ***2***. The magnetic field can act optimally on the lead in the groove and the transfer behavior of the shunt is significantly improved. The measured step response shows a considerable reduction of the rise time from the original 350 μs to 1 μs [26–28].

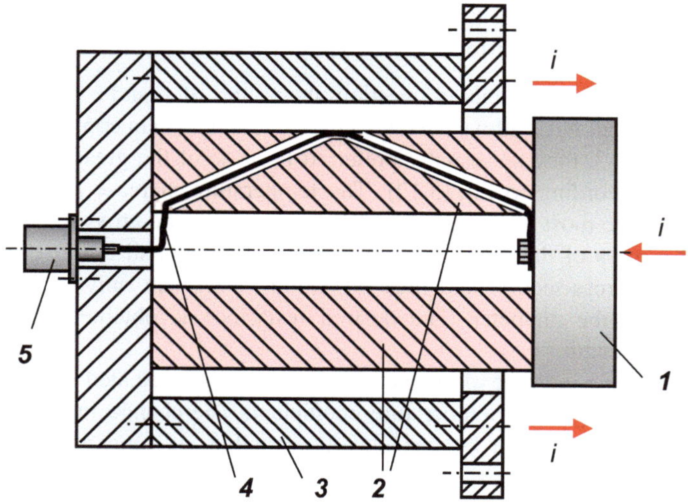

Fig. 5.25 Cross section of a 250 kA coaxial shunt with ideally profiled measurement lead in the resistance tube to improve the dynamic behavior. *1* Current terminal, *2* resistance tube with groove, *3* return conductor, *4* measurement lead, isolated arranged in the groove, *5* coaxial output

5.3.1.9 Limit Load Integral

The heavy load of measuring resistors at high current amplitudes has already been mentioned. The heating of the measuring resistor by a single impulse current is assumed to be approximately adiabatic, since due to the short duration of the current, heat dissipation into the environment is practically zero. Under this condition and neglecting the skin effect, the energy converted into heat by an impulse current $i(t)$ in the measuring resistor R with the mass m and the specific heat capacity c is (Ref. [1] of Chap. 1):

$$\boxed{\int_0^\infty i^2(t) R\,\mathrm{d}t = mc\Delta T}. \tag{5.15}$$

Considering first R to be constant, the temperature increase ΔT of the measuring resistor follows from Eq. (5.15) as:

$$\boxed{\Delta T = \frac{1}{\sigma \rho\, c A^2} \int_0^\infty i^2 \mathrm{d}t} \tag{5.16}$$

where:

σ specific conductivity of the resistance material
ρ density of the resistance material

c specific heat capacity of the resistance material
A cross-section of the resistor

For the resistance materials in question, the values of ρ and c are only slightly different. Although NiCr alloys have lower conductivities than copper or Manganin by a factor of 2–3, they are usually preferred because of the greater current penetration depth according to Eq. (5.12). The temperature increase ΔT according to Eq. (5.16) can be most effectively limited by increasing the cross-section A. Since the wall thickness of the resistance tube should remain small because of the skin effect, a large cross-section and thus limitation of ΔT can only be achieved by increasing the tube diameter. The desired resistance is obtained with a corresponding tube length.

The temperature increase of the measuring resistor is associated with a change in resistance according to its temperature coefficient. The change in resistance by a single impulse current can be considered reversible, as long as the *limit load integral*:

$$\boxed{I_G = \int_0^\infty i^2 \mathrm{d}t} \tag{5.17}$$

does not exceed the limit value specified for the measuring resistor. The limit load integral can be used to calculate the permissible peak value of an impulse current for a given impulse shape. For example, a resistor whose limit load integral is specified by the manufacturer as $I_G = 2 \times 10^4\ \mathrm{A^2s}$, can be used to measure 1 ms long rectangular impulse currents up to a peak value of 4.5 kA. For an impulse current of 10 ms, the permissible peak value is only 1.5 kA. Furthermore, an 8/20 impulse current according to Eq. (8.26a) can be used up to the maximum permissible peak value of 63.6 kA.

The limit load integral according to Eq. (5.17) is given by the manufacturer of an impulse current measuring resistor for a specified temperature increase, usually $\Delta T = 100$ K. Thus, the maximum reversible resistance change due to a single impulse current can be estimated. For a temperature rise $\Delta T = 100$ K and an assumed typical value of the temperature coefficient of $5 \times 10^{-5}\ \mathrm{K^{-1}}$, the resistance increases by 0.5%. This is synonymous with an equally large measurement error in the peak value. Since the change in resistance affects the entire shape of the impulse current, it basically influences the front time as well. This influence on the front time can usually be assumed to be negligible.

The limit load integral does not apply if the resistor is loaded by successive impulse currents, nor during continuous load due to DC or AC currents. During a pulse sequence, the contributions of the individual impulses add up to the temperature increase, but heat is also emitted into the environment. Advantageous for the heat dissipation of a coaxial shunt are a large diameter and a great length. After prolonged impulse loading, provided that the resistance change is reversible, a quasi-steady state for the absorbed and emitted heat is established. With continuous load due to DC or AC currents, the permissible current magnitude is often less than 1% of the maximum impulse current.

5.3.2 *Current Measuring Systems with Coils*

Current measuring systems with coils are well suited for the potential-free measurement of high AC and impulse currents. Their working principle is based on the law of magnetic flux and the law of induction. In use are measuring coils with or without magnetic core, which thus have different properties. Their output voltage after integration is proportional to the measured current, whereby measuring coils with magnetic core are usually equipped with an internal integrating element. The measuring instrument is usually a digital recorder with software-based data evaluation (see Chap. 7).

5.3.2.1 Law of Magnetic Flux and Law of Induction

Every current I in a conductor and every magnetic flux Θ in space generates a magnetic field. According to the *law of magnetic flux* (*Ampere's Law*), the line integral of the magnetic field strength H is proportional to I and Θ (Ref. [2] of Chap. 9):

$$\boxed{\oint H\mathrm{d}s = \Theta = I}. \tag{5.18}$$

For the simple example of a straight, infinitely long conductor, the surrounding field lines lie on circles with the radius r for reasons of symmetry. The line integral in Eq. (5.18) is $2\pi r$, and the magnetic field strength H in simple notation is therefore given by:

$$\boxed{H(r) = \frac{I}{2\pi r}}. \tag{5.19}$$

The law of magnetic flux according to Eq. (5.18) applies to DC, AC and impulse currents. It also applies if the permeability of the considered space has different values. If a measuring loop of the area A is placed perpendicular to a homogeneous magnetic field, it is permeated by the *magnetic flux* Φ:

$$\boxed{\Phi = \int B\mathrm{d}A = AB = \mu A\,H}. \tag{5.20}$$

Here $\mu = \mu_r \mu_0$ is the *permeability*, $\mu_0 = 0.4\pi \times 10^{-6}$ H/m ≈ 1.256 µH/m the *magnetic (field) constant* and μ_r the *permeability number* (relative permeability) of the permeated material. For air and all non-magnetic materials, $\mu_r \approx 1$.

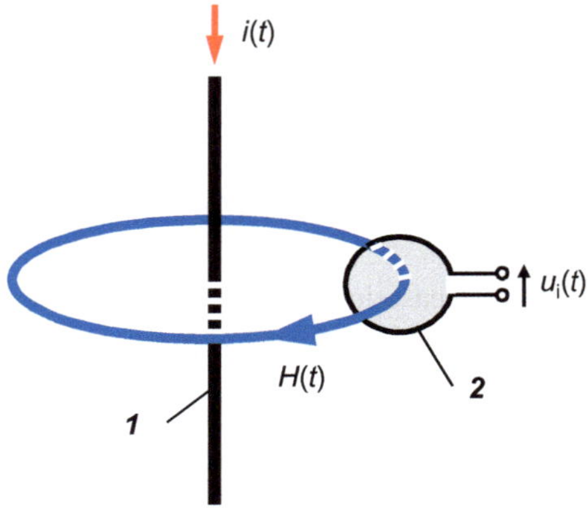

Fig. 5.26 Schematic diagram to explain the law of magnetic flux and law of induction. ***1*** Conductor, ***2*** measuring loop, *H*(*t*): magnetic field strength generated by *i*(*t*), u_i(t): voltage induced by *H*(*t*)

According to the *law of induction*, a time-variable current *i*(*t*) in the conductor ***1*** generates a magnetic field *H*(*t*) that induces in the open measurement loop the voltage (Fig. 5.26):

$$\boxed{u_{\mathrm{i}}(t) = -\frac{\mathrm{d}\Phi(t)}{\mathrm{d}t}}. \tag{5.21}$$

The negative sign in Eq. (5.21) means that the induced voltage u_{i} wants to generate a current in the measurement loop whose magnetic field counteracts the original magnetic field *H*(*t*).

If instead of the simple measurement loop a toroidal coil with *N* turns is arranged around the conductor, the induction effect is amplified *N* times (Fig. 5.27). With Eqs. (5.20) and (5.18), the induced voltage is:

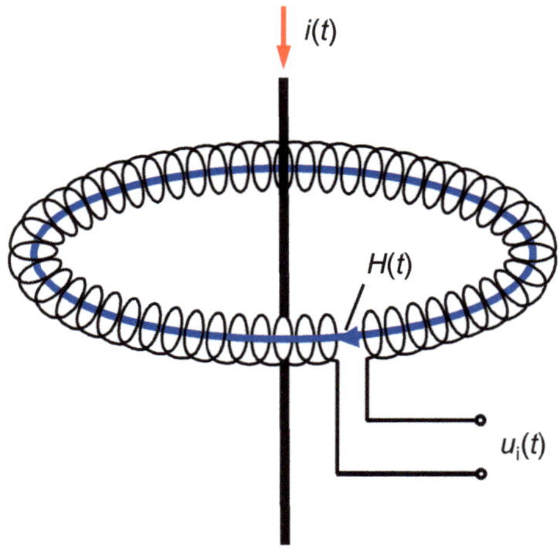

Fig. 5.27 Induced voltage $u_i(t)$ at the output of a toroidal coil due to the magnetic field *H*(*t*) generated by the current *i*(*t*) through the conductor

$$\boxed{u_{\mathrm{i}}(t) = N\frac{\mathrm{d}\Phi}{\mathrm{d}t} = M\frac{\mathrm{d}i}{\mathrm{d}t}}, \tag{5.22}$$

where M is the *mutual inductance* between the measuring coil and the conductor. The indication of the negative sign is not taken into account in Eq. (5.22). Measuring coils are partially provided with a symbol indicating the direction of current flow in the conductor to obtain a positive output voltage. The ideal toroidal coil having the cross-section area A and a mean coil circumference l_{m} equal to the length of the enclosed field line has the mutual inductance:

$$\boxed{M = \frac{\mu N A}{l_{\mathrm{m}}}}. \tag{5.23}$$

According to Eq. (5.22), the induced voltage $u_{\mathrm{i}}(t)$ at the output of the measuring coil is proportional to the temporal change of the current to be measured. The output voltage must therefore be integrated over time to obtain the desired current $i(t)$:

$$\boxed{i(t) = \frac{1}{M}\int_{0}^{\infty} u_{\mathrm{i}}(t)\mathrm{d}t}. \tag{5.24}$$

Basically, a distinction is made between measuring coils with and without magnetic core, on which the value of M largely depends. Since the induction effect exists only for time-variable measurement quantities, DC currents cannot be measured with coils. However, special types of coils with magnetic core, additional auxiliary windings and an electronic module also allow the measurement of slowly varying AC currents and even DC currents (*zero-flux principle*, see Sect. 3.5.3).

In the practical construction of a coil winding, the end of the wire is led back through the coil turns in the opposite direction to the other end of the winding in order to reduce interference from external magnetic fields. Instead of returning the wire, a second winding wound in the opposite direction may be arranged. For shielding against electric fields, the measuring coil is surrounded by a slit toroidal shield, occasionally even by two shields. In this case, the circumferential longitudinal slit enables penetration of the magnetic field associated with the measured current.

Occasionally, the functional principle of the measuring coil is assumed to be comparable to that of the current transformer used in power supply networks for measurement purposes. However, they differ significantly from each other due to their external circuitry (Ref. [2] of Chap. 1). The law of induction according to Eq. (5.22) applies if the measuring coil is unloaded; the induced voltage u_{i} is then proportional to the derivative of the primary current. In contrast, the current transformer is operated virtually in the short-circuit mode, i.e. $u_{\mathrm{i}} \approx 0$, and the secondary current is proportional to the primary current and thus proportional to the turns ratio. The short-circuited current transformer therefore does not require, unlike the no-load induction coil, a subsequent integrating circuit to generate a measurement quantity proportional to the primary current.

5.3.2.2 Integration Methods

The integration of the induced voltage u_i according to Eq. (5.24) is often achieved with passive or active circuits, rarely by numerical integration. The passive integration with an RL element is particularly easy to implement, since the self-inductance L of the measuring coil is used for integration. In the equivalent circuit diagram, the resistor R is between the output terminals at which the output voltage u_m is tapped (Fig. 5.28a). Due to the integration with the RL element, u_m is directly proportional to the current. Frequently, R is already built into the current measuring coil and is equal to the characteristic impedance of the coaxial cable to the measuring instrument.

The passive RL integration according to Fig. 5.28a can be found especially in measuring coils with magnetic core. The high permeability number μ_r ensures that the voltage u_i induced in the measuring coil—and thus also the output voltage u_m obtained after integration—is still sufficiently large in the lower frequency range of a few hertz. They are therefore suitable for measuring low-frequency signals.

In another passive integrating circuit, an RC element is connected to the output of the measuring coil, and the voltage u_m proportional to the measuring current is tapped at the capacitor C (Fig. 5.28b). The resistor R_d damps the high-frequency natural oscillation of the measuring coil.

Measuring coils without magnetic core are mainly operated with active integrator for impulse current measurements. The reason is that passive integrating circuits are only suitable for special coil types that are designed for the measurement of extremely high-frequency currents. Active integrators consist of operational amplifiers with capacitive feedback, which allow the measurement of low-frequency currents of well below 1 Hz, depending on the amplification.

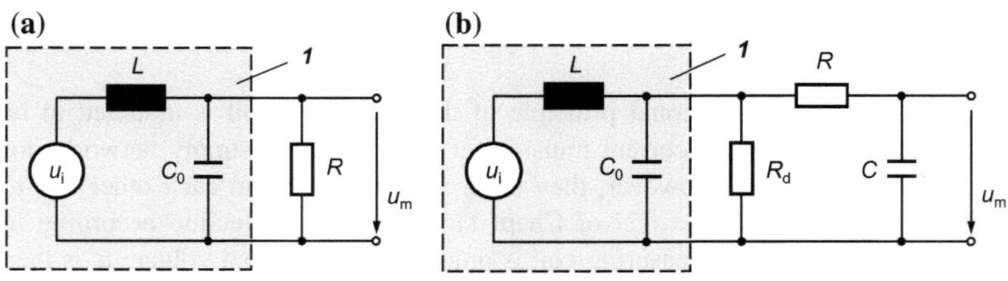

Fig. 5.28 Equivalent circuit diagram of the measuring coil *1* with passive integrating circuit. **a** Integration with self-inductance *L* and resistance *R*, **b** integration with external capacitance *C* and resistance *R*

Numerical integration methods are rarely used, although they have a number of advantages. In this case, the induced output voltage $u_i(t)$ of the measuring coil according to Eq. (5.22) is measured with a digital recorder and stored as a data set for numerical integration with the PC. Since the analog electronic integrator is dispensable, numerical integration provides a cost-effective alternative that can also give more accurate results. The digital recorder and the PC are not additionally required for numerical integration, since they are usually also needed when using an analog integrator for subsequent signal recording and data processing. Examples of numerical integration are can be found in Sect. 5.3.2.5.2.

5.3.2.3 Step Response of Toroidal Coils

The transfer behavior of coils used to measure impulse currents is preferably characterized by the step response. As step generators, similar circuits as for impulse voltage dividers and measuring resistors are used (see Sect. 9.8.4). The simplified measurement arrangement is shown in Fig. 5.29. From the output of the step generator ***1***, a current conductor is led concentrically through the opening of the toroidal coil ***2*** and connected via the wideband resistor $R = 50\ \Omega$ to the copper foil ***3***, which serves as a low-inductive return conductor to the step generator. The DC current that first flows through the coil opening is suddenly interrupted when closing the switch S. This has the effect of a negative step current applied to the toroidal coil.

When using a mercury-wetted reed contact as switch S, the maximum achievable current amplitude is usually limited to 2 A, and the rise time can be less than 1 ns. The output of the toroidal coil, i.e. the step response or its derivative, is fed to the

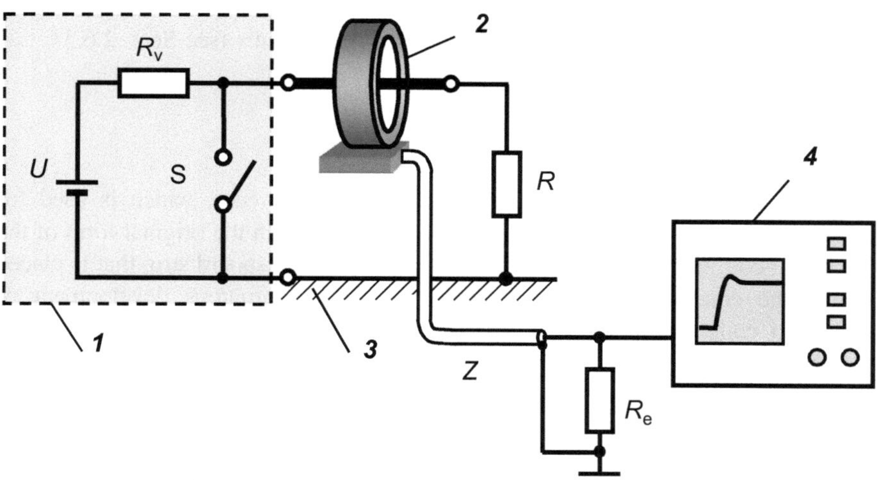

Fig. 5.29 Measurement arrangement for recording the step response of a toroidal coil. ***1*** Step generator, ***2*** measuring coil with integrator, ***3*** surface conductor, ***4*** recorder

input of the recorder ***4*** directly or, in the case of low amplitude resolution, via a preamplifier. The input resistance R_e is equal to the characteristic impedance Z of the measuring cable. A symmetrical arrangement for measuring the step response of extremely wideband measuring coils in a coaxial TEM cell is described in [29]. The step current with a maximum amplitude of 600 A and a rise time of 3 ns is generated here with a cable generator.

For simple characterization of the dynamic behavior of measuring coils, either the rise time of the step response or the bandwidth (or upper limit frequency) calculated therefrom is specified (see Sect. 9.6). If the recording time is long enough, it can be seen that the step response of a coil decreases with time and approaches zero. This corresponds to the limited transfer behavior of measuring coils in the lower frequency range. This disadvantage plays no role in the measurement of exponential impulse currents with time parameters in the range of 1 μs, but must be taken into account in the case of long rectangular impulse currents and short-time currents with superimposed DC component. Manufacturers of current measuring coils typically report the *amplitude droop*, which is defined as the percentage decrease in the step response after 1 μs or 1 s. With the amplitude droop, it is possible to estimate whether the measuring coil is suitable to measure a low-frequency current impulse within the permissible error limits. More precise knowledge is obtained by a frequency response measurement in the lower frequency range. For example, a lower limit frequency of less than 0.2 Hz is considered sufficient for measuring short-circuit currents (Ref. [4] of Chap. 2).

5.3.2.4 Potential-Free Current Measurement at High Voltage

The galvanic insulation of the measuring coil from the primary current circuit enables potential-free measurements in the high-voltage circuit and analog or digital data transfer via optical fiber, comparable with AC currents (see Sect. 2.6.3)

5.3.2.5 Rogowski Coils

The *Rogowski coil* is a toroidal coil without magnetic core, which is used for potential-free measurement of AC and impulse currents. In the original form of the Rogowski coil, the turns are wound around a flexible pressboard strip that is placed around the conductor for measurement [30]. The continuous development of Rogowski coils allows their application for many measurement tasks in power engineering and physics. Depending on the type of construction, smallest currents with rise times in the nanosecond range or largest short circuit currents with power frequency can be measured [31–34]. In addition to rigid Rogowski coils in closed ring form, there are versions that consist of two halves with separate windings (Fig. 5.30a). They can be easily opened, placed around the conductor and closed again without changing the test setup. In another model, the winding is carried out in zigzag form on both sides of a printed board. Very practical are flexible

(a) (b)

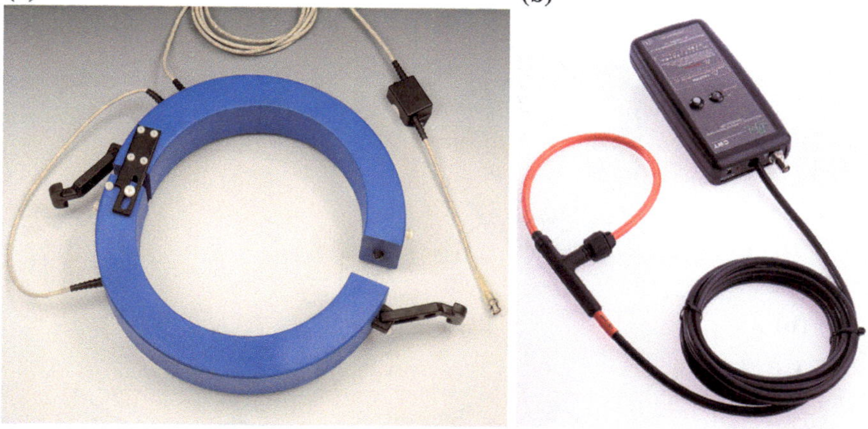

Fig. 5.30 Two types of Rogowski coils. **a** Rogowski coil with two rigid halves of the coil winding (photo: PTB), **b** flexible Rogowski coil with screw plug of the winding and electronic integrator (PEM)

Rogowski coils, which can be easily opened and conveniently positioned as clip-around probe around the conductor. The reproducibility in measurements with this type is about 1%. The flexible Rogowski coil in Fig. 5.30b has a clip-in locking mechanism with screw plug to connect the winding to the coaxial cable which leads to a handy battery-operated integrator. The output voltage of the transducer can be recorded by a digital recorder for further data processing.

Position of the Conductor

Rogowski coils are usually characterized by low non-linearity. Prerequisite for this is that the temporal current change d*i*/d*t* does not exceed the limit specified by the manufacturer and that no deformation of the coil occurs, e.g. due to the magnetic forces of the current to be measured. For the achievable measurement accuracy of a Rogowski coil, the regularity and rigidity of the winding are crucial. A high-precision Rogowski coil with its winding on a stable (non-magnetic) toroidal core shows only a small dependence on the position of the conductor in the coil opening or the return conductor outside the coil. The same statement applies to a Rogowski coil wound on two rigid core halves. The measured deviations δu_i of the induced output voltage with centric and eccentric positioning of the conductor are within ±0.1% (Fig. 5.31). This *position dependence* has practically no significance for impulse current measurements within the scope of the targeted measurement uncertainty in the percentage range. In addition, the reproducibility of impulse current measurements with optimum closure of the two halves is better than 0.1%. In contrast, inexpensive flexible Rogowski coils with a simple screw

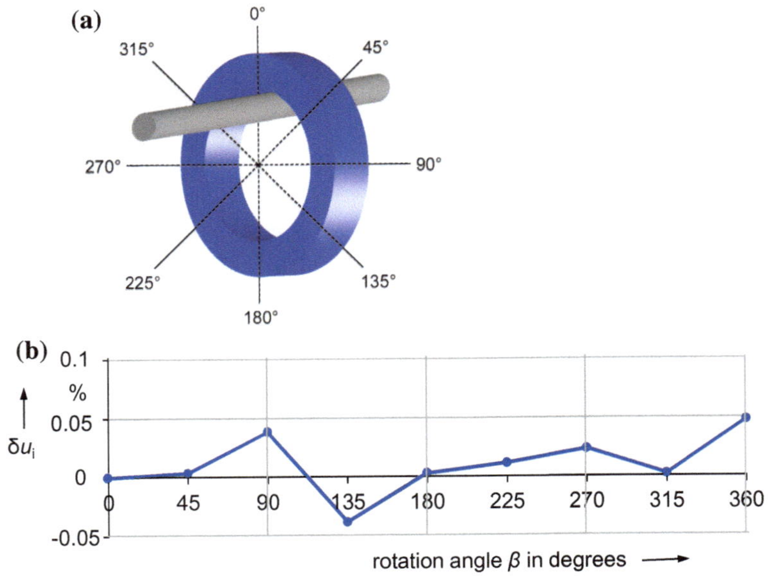

Fig. 5.31 Influence of the position of the conductor in the opening of a Rogowski precision coil. **a** Sketch to explain the rotation angle β of the eccentrically arranged conductor, **b** deviation δu_i of the induced output voltage as a function of the rotation angle β

plug have a significantly higher dependence on the conductor position in the range of 1–2% [35]. Therefore, a centric position of the conductor is always recommended in current measurements with flexible Rogowski coils.

The induced output voltage of a Rogowski coil must be integrated in order to obtain the desired time course of the current according to Eq. (5.24). Passive integration circuits according to Fig. 5.28 are only suitable if the Rogowski coil is specially constructed for the measurement of very short current impulses with frequency components of more than 1 kHz. To measure slowly varying current impulses with frequency components of 1 Hz and even below, the Rogowski coil must be operated with an electronic integrator with high amplification. Passive or electronic integrating circuits are components of the entire current measuring system and contribute to the uncertainty of measurement in addition to the Rogowski coil.

Examples of Numerical Integration

Numerical integration methods are rarely used for impulse current measurement with Rogowski coils, although they have a number of advantages. Numerical integration of the output voltage $u_i(t)$ of the Rogowski coil is usually achieved using the trapezoidal rule that is commonly implemented in commercial software. With a

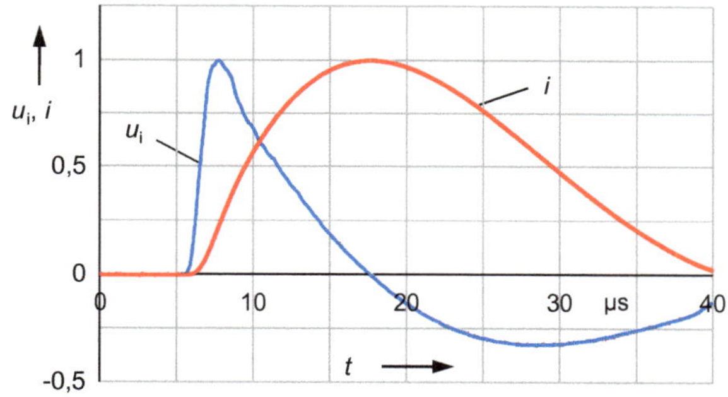

Fig. 5.32 Application example of numerical integration for an 8/20 impulse current. u_i: Induced output voltage of the Rogowski coil, i: impulse current obtained by numerical integration of u_i

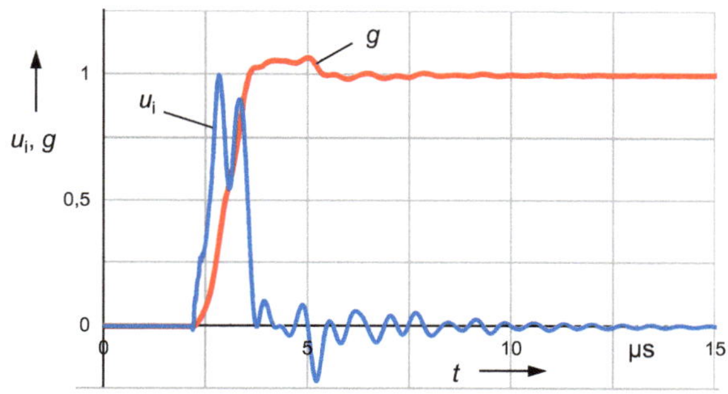

Fig. 5.33 Application example of numerical integration for the step response. u_i: induced output voltage of the Rogowski coil, g: step response obtained by numerical integration of u_i

sufficiently large number of samples, this calculation algorithm can be regarded as virtually error-free within the scope of the desired measurement uncertainty of 0.1–1% for the complete measurement system. As an example, Fig. 5.32 shows the recorded output voltage $u_i(t)$ of a Rogowski coil and the numerically integrated 8/20 impulse current $i(t)$ in a normalized representation [35].

When applying numerical integration, it should be noted that the induced output voltage of the Rogowski coil represents a differentiated signal and thus places higher demands on the sampling rate and bandwidth of the recording digital instrument than the current impulse itself. This applies in particular to the step response (Fig. 5.33). The investigated Rogowski coil is specially designed to measure very large AC currents and short-circuit currents and therefore has an inner diameter of 30 cm. Because of the large dimensions of the Rogowski coil, its step response $g(t)$ in Fig. 5.33 shows a relatively large rise time of 0.8 µs. However, this value is not typical for Rogowski coils. Smaller Rogowski coils have much shorter rise times in the nanosecond range.

Maximum Current

The maximum permissible current that can be measured with a Rogowski coil is determined by the permissible limit $u_{i,max}$ of the induced voltage. To avoid over-stressing of the insulation of turns and leads, $u_{i,max}$ should be limited to 500 V.

For a pure sinusoidal current of amplitude $\hat{\imath}$, the induced voltage according to Eq. (5.22) is determined by:

$$\boxed{u_i(t) = \omega M \hat{\imath} \sin\left(\omega t + \frac{\pi}{2}\right)}. \quad (5.25)$$

The induced voltage $u(t)$ is also sinusoidal with a phase displacement of $\pi/2$ and a frequency-dependent amplitude $\hat{u}_i = \omega M \hat{\imath}$. For a Rogowski coil with mutual inductance $M = 1\ \mu H$ and permissible insulation voltage is $u_{i,max} = 300$ V, the maximum permissible AC current at power frequency is 1 MA.

For impulse currents, significantly lower limits apply. They induce high voltages at relatively low current amplitudes because of the large di/dt values in the impulse front. If di/dt is expressed approximately as the quotient of the peak value $\hat{\imath}$ and the front time T_1, then the maximum permissible peak value $\hat{\imath}_{max}$ results from Eq. (5.22):

$$\hat{\imath}_{max} \approx \frac{T_1}{M} \hat{u}_{i,max} \quad (5.26)$$

where $u_{i,max}$ is the specified limit of the induced voltage. Accordingly, for a Rogowski coil with $M = 1\ \mu H$ and $u_{i,max} = 300 \times \sqrt{2}$ V, the peak value of an 8/20 impulse current must not be greater than $\hat{\imath}_{max} = 3.4$ kA. Rogowski coils for measuring larger impulse currents are therefore produced with significantly smaller mutual inductances than 1 μH.

The mutual inductance M of a Rogowski coil can be determined according to Eq. (5.25) by measuring the voltage induced by a sinusoidal current. The frequency of the AC current must be well above the lower limit frequency f_1 of the measuring coil. If f_1 is less than 1 Hz, M can be determined with sufficient accuracy by comparison with a standard current transformer at power frequency. Since Eq. (5.25) applies only to sinusoidal currents, a falsification of the measurement result due to higher harmonics in the AC test current is to be avoided. Instrument transformer test sets that evaluate only the fundamental oscillation of the AC current are well suited for the comparison measurement in conjunction with a standard instrument transformer.

Fast Varying Impulse Currents, Shielding

With specially constructed Rogowski coils, very fast transient currents can be measured. To avoid unwanted interference from external electric fields, the coil is surrounded by a toroidal metal shield with a slit through which the magnetic field can penetrate. The optimization of the slit shielding, which reduces the bandwidth of the Rogowski coil, is discussed in [36]. For rise times in the nanosecond range, the signal delay time in the coil is comparatively long. The “electrically long” Rogowski coil can then no longer be regarded simply as a concentrated inductance, but is represented in the equivalent circuit diagram as a delay line with distributed elements and a defined signal propagation time. Due to the rapidly changing magnetic field, corresponding partial voltages are induced in the individual coil turns, which occur approximately simultaneously with the same amplitude. They can be represented in the equivalent circuit diagram by distributed voltage or current sources. The capacitive coupling between the coil turns and the surrounding electrical shield is taken into account by transverse capacitances.

For fast varying current impulses, the Rogowski coil together with the shield acts as a traveling wave line with the characteristic impedance Z. Depending on the wiring of the Rogowski coil at the winding ends, different traveling wave phenomena can be proved theoretically and experimentally. Particularly favorable measurement conditions are present when the Rogowski coil is connected to the electrical shield at one end directly and at the other end via a low-ohmic measuring and integrating resistor $R_m \ll Z$. The partial voltages of the individual turns induced by a current step cause two current waves travelling in opposite direction in the Rogowski coil. The waves are reflected at the winding ends and generate a voltage step at R_m proportional to the current step. Because of $R_m \neq 0$, the reflection factor $r < 1$, so that the voltage amplitude decreases step-by-step every two times the travel time in the Rogowski coil. The staircase-shaped voltage curve is consistent with the approach for slow transients. In this case, the "electrically short" Rogowski coil is described by lumped elements, which lead to an exponential decrease in the voltage with the time constant L/R [37, 38].

5.3.2.6 Current Measuring Coils with Magnetic Core

The progress in the development of magnetic materials with excellent frequency behavior of the permeability and low eddy current losses have made it possible for several decades to produce very wide-band current measuring coils with iron or ferrite core [39, 40]. Because of the high permeability of the magnetic core, the mutual inductance M and thus the induced voltage $u_i\ (t)$ according to Eq. (5.22) are much larger than that of the ironless Rogowski coil. The integration of $u_i\ (t)$ is preferably achieved according to Fig. 5.28a by an internal LR element with the self-inductance L and resistance R, which often corresponds to the characteristic impedance Z = 50 Ω of the connecting cable. Since the integrator is located directly in the coil, the erroneously impression may appear that the coil works without an integrating unit. As an example, Fig. 5.34 shows a measuring coil with two halves of winding and magnetic core, which can easily be clamped together around a conductor ("Pearson current monitor"). Such a clamping device allows convenient current measurements without having to open the circuit. Due to the internal RL integrating unit, the output voltage of the measuring coil is proportional to the measured current and is fed via the measuring cable directly or through an attenuator to the digital recorder.

Depending on the type of construction, the rated current of measuring coils with magnetic core can be up to 500 kA for impulse currents. In addition to the peak

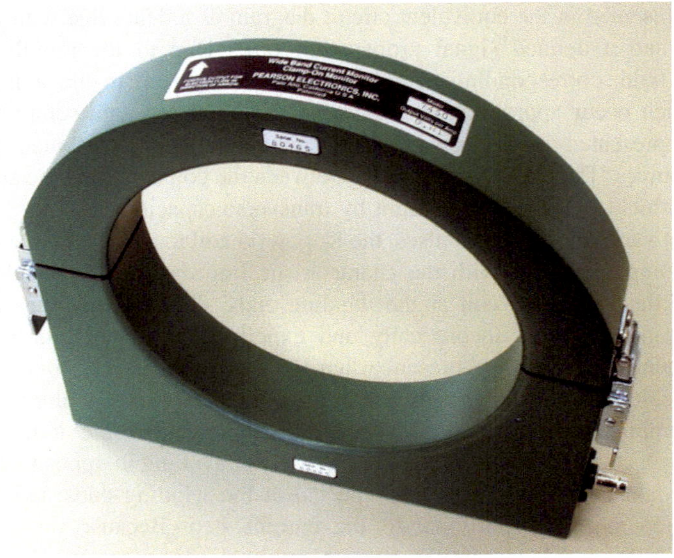

Fig. 5.34 Wideband current measuring coil with magnetic core and internal integrating element as clamp-on unit (Pearson)

value, the manufacturer also specifies a limit for the product of peak value and duration of a rectangular current, i.e. the permissible peak value decreases with increasing impulse duration. The maximum continuous load due to power frequency AC currents is usually only a few percent of the permissible impulse load.

Due to the high permeability of the magnetic core, lower limit frequencies of less than 1 Hz and, secondly, upper limit frequencies of more than 100 MHz can be achieved. Figure 5.35 shows the step response $g(t)$ of a wideband 5 kA measuring coil with magnetic core and internal integrator in three time ranges [41]. The evaluation of the step response up to 150 ns provides a response time of 8 ns and a settling time of 30 ns (Fig. 5.35a). In the further course until 0.5 ms, the step response remains approximately constant with deviations within $\pm 1\%$ (Fig. 5.35b). After 91 ms, the step response has dropped to 70% of the initial value (Fig. 5.35c). The upper limit frequency is 25 MHz resulting from the rise time of the step response. A frequency response measurement gives a lower limit frequency of 1 Hz. Associated with the magnetic core are the well-known disadvantages such as nonlinearity, polarity effect, remanence, core saturation due to DC currents, etc. However, for the intended use and the targeted measurement uncertainty in the percentage range, these deficiencies can be usually neglected or reduced by calibration.

With appropriately designed measuring coils with ferrite cores, impulse currents with frequency components above 1 GHz or with rise times of less than 1 ns can be measured. The measuring coils are shielded like Rogowski coils against external electric fields, while the magnetic field can act on the coil winding through a

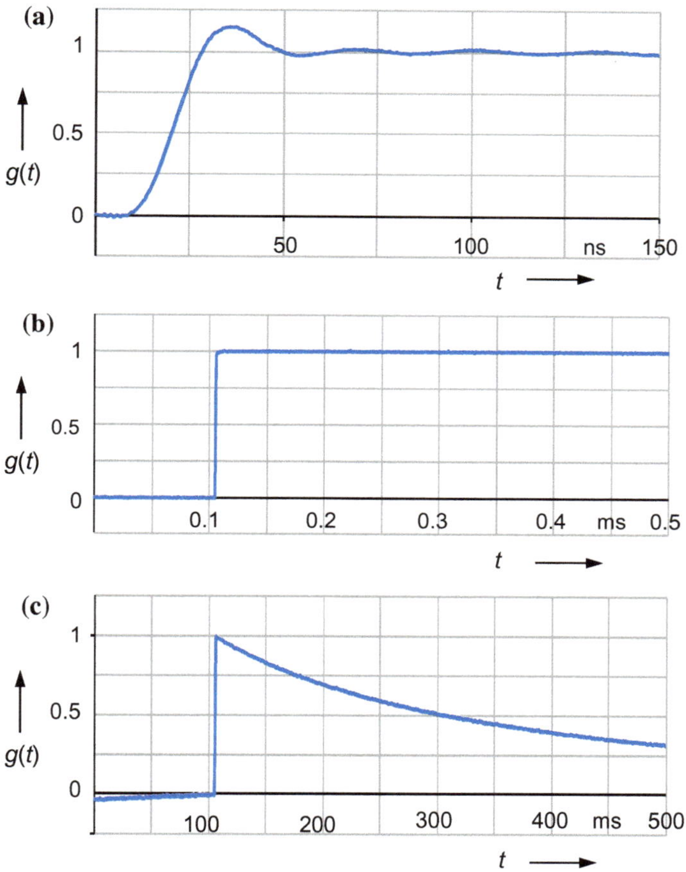

Fig. 5.35 Step response of a wideband 5 kA measuring coil with magnetic core shown for different recording times. **a** Record up to 150 ns, **b** record up to 0.5 ms, **c** record up to 500 ms

longitudinal slit in the shield. Internal damping resistors that are connected between some coil turns and the outer shield serve to damp high-frequency oscillations [39].

5.3.2.7 Magnetic Field Sensor

If the Rogowski coil is reduced to a single turn, the induction loop *2* shown in Fig. 5.26 is obtained. With correspondingly small dimensions, it is particularly suitable as a sensor for measuring very fast varying magnetic fields in a spatially limited environment. The component of the magnetic field perpendicular to the induction loop induces at the output of the sensor a voltage u_i which, according to the law of induction in Eq. (5.21), is proportional to the temporal change of the magnetic field. The magnetic field strength is obtained by integration of this output voltage. By onsite calibration of the sensor at the installation site, the current that generates the magnetic field according to the law of magnetic flux can be determined.

Small induction loops are often installed in gas-insulated switchgear and powerful pulse generators. Figure 5.36 shows schematically three possible installation variants of the induction loop in the outer wall of a gas-insulated coaxial conductor. In addition to the simplest arrangement A, the measuring loop can also be formed symmetrically to ground in order to suppress interference (arrangement B) or arranged in a groove of the tubular conductor (arrangement C). The measuring loop can be designed so that it is largely shielded against the effect of electric fields [38, 42].

The induction loop can be represented in the simplified equivalent circuit diagram by its self-inductance L, a voltage source and a load and measuring resistor R (see Fig. 5.28a). The magnitude of the time constant L/R compared to the duration and rise time of the measurement signal determines the type of integration. In high-frequency magnetic fields and when L/R is relatively large, i.e. R is very small, the measuring circuit has an integrating effect and indicates a measurement quantity proportional to the magnetic field strength. The upper limit frequency is determined here by stray capacitances of the measuring loop and the circuit. For low-frequency magnetic fields and when L/R is relatively small, the measuring circuit has a differentiating effect. In this case, an electronic integration amplifier is required to obtain an indication that is proportional to the magnetic field strength.

An induction loop sensor for the one-dimensional measurement of quasi-stationary and transient magnetic fields in high-voltage test field and in EMP simulators is described in [43]. The analog output voltage of the sensor is fed via optical fiber to the receiver on ground potential for evaluation. In the frequency range from 30 Hz to 10 MHz, the integration takes place electronically with an integration amplifier. At higher frequencies up to 300 MHz, the measurement loop acts self-integrating.

With a spherical sensor, three-dimensional magnetic fields in free space can be measured potential-free [44]. On the metal sphere, three induction loops are aligned orthogonal to the three spatial axes. Inside the sphere is the battery-powered

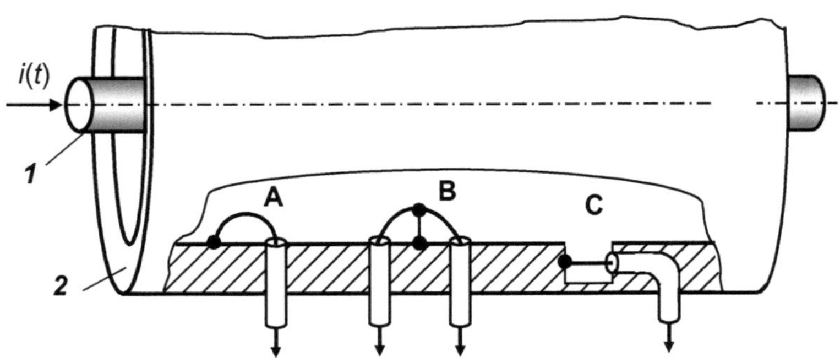

Fig. 5.36 Examples of induction loops for detecting fast transient magnetic fields in a gas-insulated conductor (principle). ***1*** Inner conductor, ***2*** outer conductor with the three examples A, B and C of induction loops

electronics for optoelectronic transmission of the measurement and control signals. Several examples show the broad application of the measuring system with a bandwidth of 50 Hz to 350 MHz. The paper reports on magnetic field measurements in the vicinity of a transformer during switching operations, of a cable bushing of a gas-insulated system during disconnector switching and of an arrester in impulse current tests.

References

1. Gamlin, M.: Impulse current testing. In: Proceedings of 14th ISH Beijing, Paper J-09 (2005)
2. Modrusan, M.: Normierte Berechnung von Stoßstromkreisen für vorgegebene Impulsströme. Bull. SEV **67**, 1237–1242 (1976)
3. Schwab, A., Imo, F.: Berechnung von Stoßstromkreisen für Exponentialströme. Bull. SEV/VSE **68**, 1310–1313 (1977)
4. Kuzhekin, I.P., Hribar, Z.: Imitator of lightning currents with amplitudes of 200 kA. In: Proceedings of 13th ISH Delft (2003)
5. Körbler, B., Pack, S.: Analysis of an impulse current generator. In: Proceedings of 12th ISH Bangalore, Paper 7–22 (2001)
6. Zhao, G., Zang, X.: EMTP analysis of impulse voltage generator circuit. In: Proceedings of 14th ISH Beijing, Paper A-11 (2005)
7. Zischank, W.: A surge current generator with a double-crowbar spark gap for the simulation of direct lightning stroke effects. In: Proceedings of 5th ISH Braunschweig, Paper 61.07 (1987)
8. Pietsch, R., Baronick, M., Kubat, M.: Impulse current test system with crowbar gap extension for surge arrester testing. In: Proceedings of 15th ISH Ljubljana, Paper T10-745 (2007)
9. Salge, J., Peier, D., Brilka, R., Schneider, D.: Application of inductive energy storage for the production of intense magnetic fields. In: 6th Symposium on Fusion Technology, Aachen (1970)
10. Kind, D., Salge, J., Schiweck, L., Newi, G.: Explodierende Drähte zur Erzeugung von Megavolt-Impulsen in Hochspannungsprüfkreisen. ETZ-A **92**, 46–51 (1971)
11. Feser, K., Modrusan, M., Sutter, H.: Simulation of multiple lightning strokes in laboratory. In: Proceedings of 3rd ISH Mailand, Paper 41.05 (1979)
12. Klein, T., Köhler, W., Feser, K.: Exponential current generator for multiple pulses. In: Proceedings of 12th ISH Bangalore, Paper 7-21 (2001)
13. Modrusan, M.: Langzeit-Stoßstromgenerator für die Ableiterprüfung gemäß CEI-Empfehlung. Bull. SEV **68**, 1304–1309 (1977)
14. Park, E.H.: Shunts and inductors for current measurements. NBS J. Res. **39**, 191–212 (1947)
15. Navarro, A., et al.: Transducer for lightning current measurement. In: Proceedings of 14th ISH Beijing, Paper B-29 (2005)
16. Schwab, A.J.: Die Berechnung der Bandbreite und der Anstiegszeit rohrförmiger Messwiderstände unter Berücksichtigung der Stromverdrängung. ETZ A **89**, 604–606 (1968)
17. Schwab, A.J.: Low-resistance shunts for impulse currents. IEEE Trans. PAS **90**, 2251–2257 (1971)
18. Pfeiffer, W.: Aufbau und Überprüfung von koaxialen Rohrwiderständen sehr kurzer Anstiegszeit. ETZ-A **91**, 59–60 (1970)
19. Wakimoto, T., Ishii, M.: Step response and comparison test of impulse current measuring system. In: Proceedings of 18th ISH Seoul, Paper PC-10 (2013)
20. Nagel, A., Kerwer, T.: Design of a 1 kA pulsed current source with 60 ns rise time for the analysis of current probes. In: Proceedings of PCIM Europe Nürnberg, pp. 899–902 (2004)

21. Högberg, L.: 1 Gc/s bandwidth shunt for measurements of current transients in the 10 A–10 kA range. J. Sci. Instrum. **42**, 273–279 (1965)
22. Wesner, F.: Koaxiale Flächenwiderstände zur Messung hoher Stoßströme mit extrem kurzer Anstiegszeit. ETZ-A **91**, 521–524 (1970)
23. Witt, H.: Response of low ohmic resistance shunts for impulse currents. Elteknik **3**, 45–47 (1960)
24. Lappe, F., Westendorf, K.B.: Ein Meß-Widerstand für Hochfrequenz. Z. Angew. Phys. **3**, 29–32 (1951)
25. Malewski, R.: New device for current measurement in exploding wire circuits. Rev. Sci. Instr. **39**, 90–94 (1968)
26. Malewski, R.: Micro-Ohm shunts for precise recording of short-circuit currents. IEEE Trans. PAS **96**, 579–585 (1977)
27. Malewski, R., Nguyen, Chinh T., Feser, K., Hyltén-Cavallius, N.: Elimination of the skin effect error in heavy-current shunts. IEEE Trans. PAS **100**, 1333–1340 (1981)
28. Schwab, A., Imo, F.: Übergangsverhalten koaxialer Strommesswiderstände mit exzentrischem Spannungsabgriff. ETZ A-Arch. **2**, 95–100 (1980)
29. Lu, L., Mou, L., Liu, J., Li, Y.: Development of a standardizing device for Rogowski coil. In: Proceedings of 14th ISH Beijing, Paper J-48 (2005)
30. Rogowski, W.: Über einige Anwendungen des magnetischen Spannungsmessers. Arch. Elektrotech. **1**, 511–527 (1913)
31. Destefan, D.E., Ramboz, J.D.: Advancements in high current measurement and calibration. In: NCSL Intern. Workshop and Symposium, Toronto (2000)
32. Ward, D.A., Exon, J.L.T.: Using Rogowski coils for transient current measurements. IEE Eng. Sci. Educ. J. **2**, 105–113 (1993)
33. Ray, W.F., Hewson, C.R.: High performance Rogowski current transducers. In: Conference Proceedings of IEEE-IAS Conference, Rome (2000)
34. Kojovic, L.: PCB Rogowski coils benefit relay protection. IEEE Comput. Appl. Power **15**, 1–4 (2002)
35. Schon, K., Schuppel, W.: Precision Rogowski coil used with numerical integration. In: Proceedings of 13th ISH Ljubljana, Paper T10-130 (2007)
36. Hewson, C.R., Ray, W.F.: Optimising the high frequency bandwidth and immunity to interference of Rogowski coils in measurement applications with large local dV/dt. In: Conference Proceedings of IEEE-APEC Conference, Palm Springs (2010)
37. Cooper, J.: On the high-frequency response of a Rogowski coil. J. Nucl. Energy Part C **5**, 285–289 (1963)
38. Bellm, H., Küchler, A., Herold, J., Schwab, A.: Rogowski-Spulen und Magnetfeldsensoren zur Messung transienter Ströme im Nanosekundenbereich. Arch. Elektrotechn. **68**, part 1, 63–74, part 2, 69–74 (1985)
39. Anderson, J.M.: Wide frequency range current transformers. Rev. Sci. Instrum. **42**, 915–926 (1971)
40. Waters, C.: Current transformers provide accurate, isolated measurements. In: Conference PCIM (1986)
41. Schon, K., Mohns, E., Gheorghe, A.: Calibration of ferromagnetic coils used for impulse current measurements. In: Proceedings of 12th ISH Bangalore, vol. 5, 166–1169 (2001)
42. Di Capua, M.S.: High speed magnetic field and current measurements. In: NATO ASI Series E Applied Sciences, vol. 1, issue 108, 1, pp. 223–262 (1986)
43. Krauß, T., Köhler, W., Feser, K.: Improved high bandwidth potential-free magnetic field probe for the measurement of transient magnetic fields. In: Proceedings of 11th ISH London, Paper 467 (1999)
44. Kull, M., Krauß, T., Köhler, W., Feser, K.: High bandwidth 3D-magnetic field probe for the measurement of transient magnetic fields. In: Proceedings of 12th ISH Bangalore, Paper 1–16 (2001)

Chapter 6
Electro-optic and Magneto-optic Sensors

Abstract The basics of electro-optic and magneto-optic effects, also named after their discoverers, have been known for more than a century. The Pockels effect and the Kerr effect characterize the optical properties of certain crystals, liquids and gases under the influence of an electric field, whereby the polarization of a light wave propagating in the optical axis of the medium is affected. According to the Faraday effect, a magnetic field also changes the polarization of a passing light wave. For all effects, a rotation of the polarization plane of the light occurs in the medium, which is indicated by a downstream analyzer and photodetector as the corresponding electric or magnetic field strength. The optical processes in the medium take place in the nanosecond range, so that bandwidths from zero to the GHz range can generally be achieved with electro-optic or magneto-optic sensors. The clause discusses the nature and characteristics of various recently developed sensors. In conjunction with fiber optics, there are good prospects for the use of these sensors in the high-voltage area. After calibration of the sensors on site, the voltages or currents that generate the fields can be displayed directly. The technical realization of sensors based on the Pockels or Faraday effect has progressed in the last two decades thanks to the solution of many individual problems, resulting in a variety of electro-optic and magneto-optic transducers in the high-voltage network.

The basics of *electro-optic* and *magneto-optic effects*, also named after their discoverers, have been known for more than a century. The *Pockels effect* and the *Kerr effect* characterize the optical properties of certain crystals, liquids and gases under the influence of an electric field, whereby the *polarization* of a light wave propagating in the optical axis of the medium is affected. According to the *Faraday effect*, a magnetic field also changes the polarization of a passing light wave. For all effects, a rotation of the polarization plane of the light occurs in the medium, which is indicated by a downstream analyzer and photodetector as the corresponding electric or magnetic field strength. The optical processes in the medium take place in the nanosecond range, so that bandwidths from zero to the GHz range can generally be achieved with electro-optic or magneto-optic sensors. In conjunction with fiber optics, there are good prospects for the use of these sensors in the

K. Schon, *High Voltage Measurement Techniques*, Power Systems,
https://doi.org/10.1007/978-3-030-21770-9_6

high-voltage area. After calibration of the sensors on site, the voltages or currents that generated the fields can be displayed directly. The technical realization of sensors based on the Pockels or Faraday effect has progressed in the last two decades thanks to the solution of many individual problems, resulting in a variety of electro-optic and magneto-optic transducers in the high-voltage network.

6.1 Electro-optic Effects

Under the influence of an electric field E, certain crystals, liquids and gases change their optical properties. When a light wave passes through such a medium with the *refractive index n, induced birefringence* occurs:

$$\boxed{n = n_0 + aE + bE^2 + \cdots}, \tag{6.1}$$

where n_0 denotes the *natural refractive index*. While the Pockels effect characterizes the linear relationship between refractive index and field strength and therefore $b = 0$, the Kerr effect describes the quadratic dependence with $a = 0$. The Pockels effect occurs in two variants, which are determined by the direction of the electric field. In the case of the *longitudinal Pockels effect*, the electric field and the light wave have the same direction. The *transversal Pockels effect* (like the *Kerr effect*) is characterized in that the electric field acts perpendicular to the light wave.

6.1.1 Pockels Effect

The basic principle of a measurement arrangement that exploits the longitudinal Pockels effect for field strength or voltage measurements is shown in Fig. 6.1. With the laser ***1***, a light beam of the wavelength λ is generated in the z-direction and linearly polarized by the polarizer **2** in the x-direction. The subsequently arranged $\lambda/4$ wave plate **3** is a thin crystal plate in which there are two mutually perpendicular axes x' and y', along which the refractive indices and hence the propagation velocities differ. The angle of incidence of the polarized light to the optical axis and thus to the "slow" y'-axis is 45°. Upon entering the $\lambda/4$ wave plate, the light clearly splits into two orthogonal partial waves of equal amplitude, which propagate in the z-direction with different velocities due to natural birefringence. The resulting phase shift causes the two partial waves to leave the plate again as *circularly polarized light*. With appropriate thickness and orientation of the plate, the phase shift of the partial waves at the plate output is just $\lambda/4$ corresponding to $\Delta\varphi = \pi/2$. As a result, the operating point of the Pockels cell, as described below, is placed in the linear part of the characteristic curve [1, 2].

The actual Pockels cell ***4*** is likewise an optically uniaxial, birefringent crystal of length l in the z-direction. The propagation direction of the polarized light is

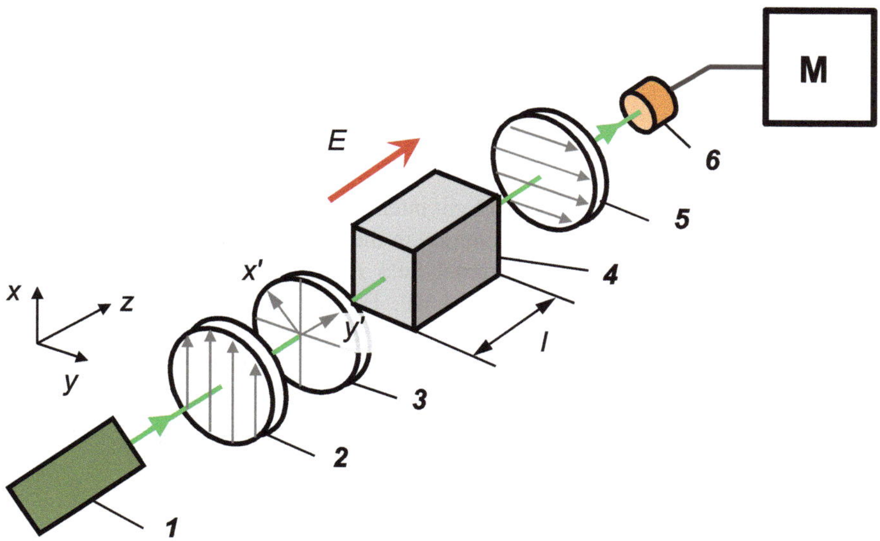

Fig. 6.1 Longitudinal Pockels effect for high-voltage measurement (principle). ***1*** Laser, ***2*** polarizer, ***3*** λ/4 wave plate, ***4*** crystal, ***5*** analyzer, ***6*** photodetector, **M** measuring instrument, e.g. oscilloscope or recorder

placed in the optical axis of the crystal in which the natural birefringence is not effective, i.e. $n_0 = 0$ in Eq. (6.1). An electric field E aligned parallel to the optical axis causes an *induced birefringence*, which leads to a comparable formation of x'- and y'-axes with different propagation velocities of the orthogonal partial waves. At the output of the Pockels cell, the light is elliptically polarized. Depending on the distance z passed through in the crystal, the phase difference between the two partial waves is:

$$\boxed{\Delta\phi(z) = \frac{2\pi\,\Delta n}{\lambda} z}, \tag{6.2}$$

where Δn denotes the difference of refractive indices or propagation velocities for the two partial waves. For Δn, the already mentioned linear relationship with the field strength applies:

$$\boxed{\Delta n = n_0^3\, r_{\mathrm{ij}}\, E}. \tag{6.3}$$

Here, r_{ij} is the effective *electro-optic coefficient* in this arrangement, which depends on the crystal temperature, the wavelength λ of the light and the frequency of the electric field E or the applied voltage.

At the output of the Pockels cell ***4*** is the analyzer ***5***, which is aligned perpendicular to the input polarizer ***2***. The analyzer acts as a polarizer that lets the light through only with the component in its polarization direction. The phase modulation is thereby converted into an intensity modulation. The *light intensity* I at the output of the analyzer is:

$$\boxed{I = I_0 \sin^2\left(\frac{\Delta\phi}{2}\right)}. \tag{6.4}$$

Here, I_0 is the maximum measured light intensity and $\Delta\varphi$ the phase displacement for $z = l$ at the crystal output. For example, in the arrangement of Fig. 6.1 without the λ/4 plate ***3*** and without the external field E, no light can pass through the analyzer ***5*** due to $\Delta\varphi = 0$ and hence $I = 0$. For a total phase displacement of $\Delta\varphi = \pi$, the intensity reaches its maximum, i.e. $I = I_0$.

If the sensor is provided for voltage measurements, the voltage U is applied to the end faces of the crystal ***4*** and the electric field is $E = U/l$. According to Eq. (6.2) with Eq. (6.3), $\Delta\varphi$ is proportional to the field strength and Eq. (6.4) can be rewritten:

$$\boxed{I = I_0 \sin^2\left(\frac{\pi}{4} + \frac{\pi}{2}\frac{U}{U_\pi}\right)}. \tag{6.5}$$

The first term in the bracket of Eq. (6.5) takes into account the phase displacement $\Delta\varphi = \pi/2$ of the light wave after passing through the $\lambda/4$ plate ***3*** (see Fig. 6.1). The term U_π denotes the *half-wave voltage* for which there is a phase difference of $\lambda/2 = \pi$ between the two orthogonal light waves at the crystal output. The U_π values of commonly used crystals are in the range of 3–31 kV.

Because of the ambiguity of sin $2x$ for $x > \pi$, the working area is chosen to lie in the approximately linear part of the sine square of Eq. (6.5) with $U < U_\pi/2$ (Fig. 6.2). Therefore, a bipolar voltage signal with low amplitude leads to an approximately linear modulation of the downstream analyzer. The zero shift obtained with the $\lambda/4$ plate corresponding to an initial phase rotation of $\pi/2$ could alternatively be achieved by an applied DC voltage $U_\pi/4$.

The photodetector ***6*** in Fig. 6.1 detects the voltage-dependent light intensity according to Eq. (6.5) and converts it into an electrical signal for further processing with a digital voltmeter or digital recorder. In the case of the *transversal Pockels effect* with the field strength acting perpendicular to the light wave, similar processes take place in the crystal as in the longitudinal Pockels effect.

6.1.1.1 Pockels Sensors for Field Measurements

The earlier development of electro-optic sensors with Pockels cells for use in high-voltage engineering is summarized in [3, 4] including a large number of references. Basically, there are two main types: miniaturized Pockels cells that can be positioned freely in the electric field for field measurements, and Pockels cells that are usually arranged in series between two electrodes and connected directly to

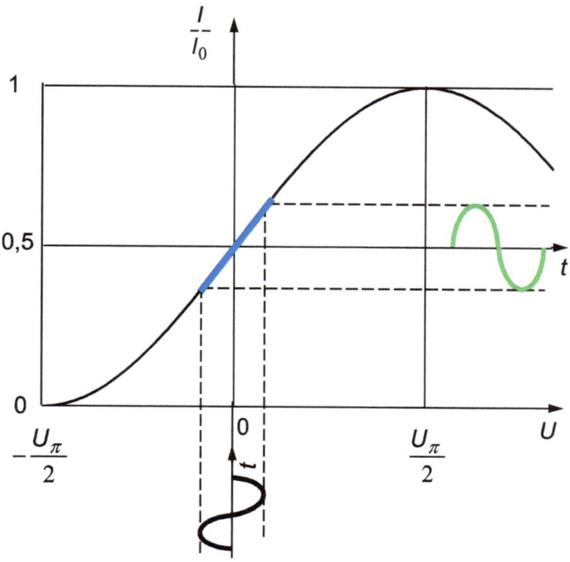

Fig. 6.2 Initial course of the characteristic I/I_0 of the Pockels sensor whose linear part is used to convert an optical signal into an electrical signal

the high voltage to be measured. Instead of a laser, which was mainly used earlier in orientation investigations in the laboratory, today other light sources such as *laser diodes* (LD) or *light-emitting diodes* (LED) are in use. *Fiber optic cables* serve for coupling and decoupling the light wave. Investigations on optical fibers in the high-voltage field show, however, that at higher air humidity flashovers can occur, depending on the coating material of the optical fiber used [5]. Optical fibers are often used in special designs as *polarization-maintaining optical fibers* (PMF).

For Pockels cells, anisotropic crystals such as $Bi_4Ge_3O_{12}$ (BGO), $Bi_{12}SiO_{20}$ (BSO), $Bi_{12}TiO_{20}$ (BTO) or $LiNbO_3$ are used in certain cutting directions. A typical arrangement of an electro-optic sensor with Pockels cell is shown in Fig. 6.3. The light is generated by an LED ***1*** (or LD) and passes through the optical fiber **2** to the sensor in the high voltage field, the beam being coupled through a microlens ***3*** in the fiber. Via a linearly polarizing beam splitter ***4*** and a $\lambda/4$ plate ***5***, the circularly polarized light wave enters the Pockels cell ***6***, from which it exits again under the action of the electric field with elliptical polarization. The light then passes through a second polarizing beam splitter ***4***, the microlens ***3*** and the optical fiber ***2*** before reaching the photodetector *7*. Here, the beam intensity is converted into a proportional voltage, which is detected by the measuring device M [4].

The technical development in the realization of electro-optic sensors with Pockels cells has led to considerable advances, in particular with regard to sensor dimensions and variety of applications. Thus, the individual elements of an electro-optic sensor with two *GRIN lenses* **2** (*gradient index lenses*), polarizer ***3***, $\lambda/4$ plate ***4***, LNO crystal ***5*** and analyzer ***6*** can be glued together to form a unit in the dimensions 10 mm × 75 mm. Together with the connections of the two fiber optics, this results in an easily manageable sensor (Fig. 6.4). The sophisticated

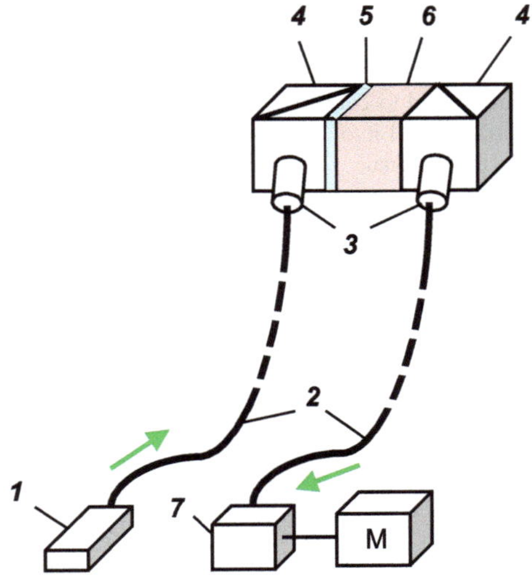

Fig. 6.3 Typical arrangement of an electro-optic sensor with Pockels cell. ***1*** Light source, ***2*** optical fiber, ***3*** microlens, ***4*** polarizing beam splitter, ***5*** $\lambda/4$ plate, ***6*** crystal, ***7*** photodetector, M measuring instrument

bonding technique requires a special mounting device and adjustment unit on an optical bench [6]. Critical points in choosing a suitable adhesive are high light transmission, no stress during curing and minimal elongation with temperature changes.

Extremely small dimensions are achieved by using *optical waveguide technology* in the LNO crystal with diffused Ti and the use of a *dielectric mirror* [4, 7]. The light beam generated by an LED is passed through the optical fiber ***1*** to an optical coupler ***2*** and polarizer ***3***, where it is linearly polarized (Fig. 6.5). The light propagates in the 7 μm wide waveguide ***4***, is reflected at the end face and exits the waveguide. It then returns to the polarizer ***3*** and the coupler ***2***, where a part of it is coupled out and passed through a second optical fiber ***1*** to the analyzer and photodetector. In [8, 9] further improvements and even smaller versions of this sensor type with two parallel optical waveguides are presented. In one of the examples presented, the sensor is used to measure surface discharges.

The arrangement in Fig. 6.6 also shows a sensor with a dielectric mirror for reflecting the light at the end face of the Pockels cell [10]. The linearly polarized light is conducted in a polarization-maintaining fiber ***1*** (PMF) via a GRIN lens ***2*** to

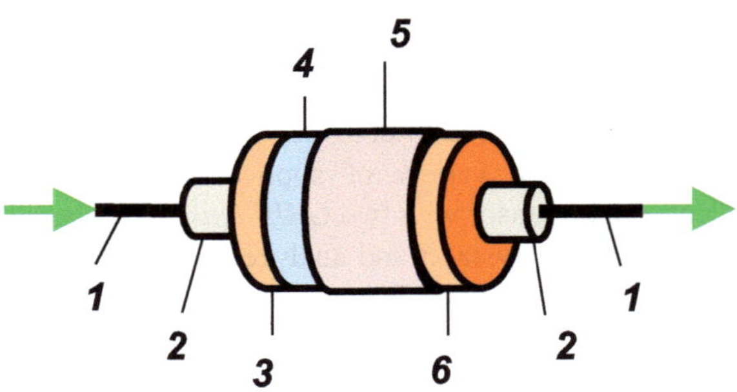

Fig. 6.4 Pockels sensor element in bonded design. ***1*** Optical fiber, ***2*** GRIN microlens, ***3*** polarizer, ***4*** $\lambda/4$ plate, ***5*** LNO crystal, ***6*** analyzer

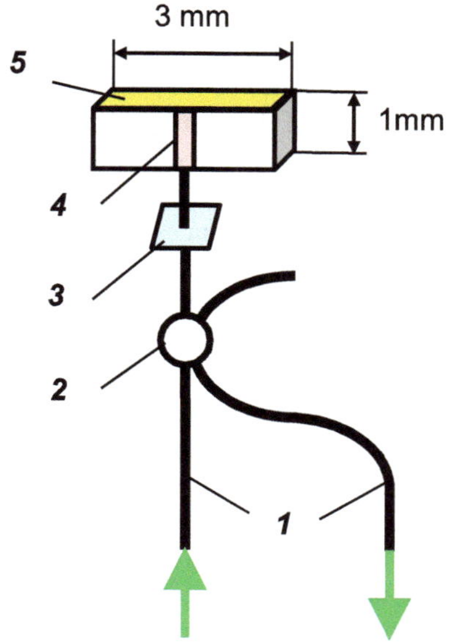

Fig. 6.5 Pockels sensor in miniature version with optical waveguide and dielectric mirror. *1* Optical fiber, *2* optical coupler, *3* polarizer, *4* optical waveguide, *5* dielectric mirror

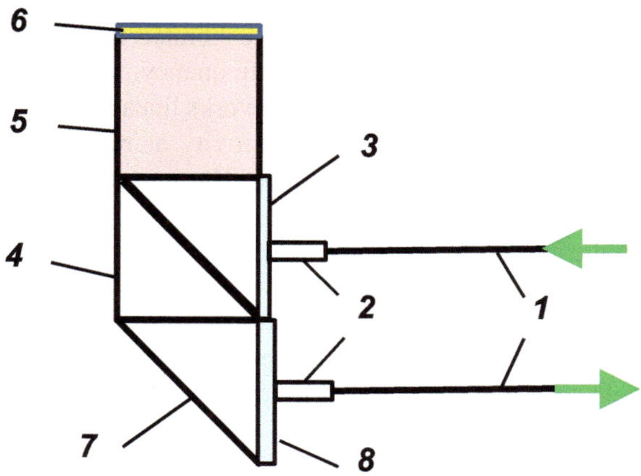

Fig. 6.6 Pockels sensor with dielectric mirror and beam splitter (principle). *1* Optical fiber, *2* optical lens, *3* $\lambda/4$ plate, *4* beam splitter, *5* BTO crystal, *6* dielectric mirror, *7* prism, *8* analyzer

the film *3*, which acts as a λ/4 plate. By means of the beam splitter *4*, a part of the circularly polarized light is directed into the electro-optic BTO crystal *5*. In order to keep the distortion of the electric field low, the crystal has a cylindrical shape with no sharp edges, the diameter and length of each being 5 mm. Under the influence of a longitudinal electric field, the light is elliptically polarized, which is coupled with a phase rotation. The light is reflected by the dielectric mirror *6* and propagates through the crystal *5* a second time. Although the second pass through the crystal causes a doubling of the phase rotation, the advantage of the corresponding intensity gain is lost again by the subsequent beam splitter *4.* The light wave continues to the prism *7* where it is directed to the analyzer *8* and further via the lens 2 and the optical fiber ***1*** to the photodetector for intensity indication. The individual elements of the sensor are glued after careful alignment.

Furthermore, in [10] the influence of different parameters and in particular the interaction of the total 17 mm long sensor with the surrounding field is considered theoretically and experimentally. According to Maxwell's equation of continuity for the conduction and displacement current densities, the introduction of a dielectric in air changes the electric field strength in both the air and the dielectric. Due to the large relative permittivity $\varepsilon_r \approx 50$ of BTO and the high conductivity $\kappa \approx 1.2 \times 10^{-13}\ \Omega\text{m}^{-1}$, the AC field strength in the crystal is mathematically reduced by a factor of 13. This leads to a corresponding reduction in sensor sensitivity. In the DC field, the field reduction in this crystal is even several orders of magnitude due to the very different conductivities. The sensitivity loss is correspondingly high, so that the crystal is not readily suitable for use in the DC field. In addition, the conductivity of the air—and thus the sensitivity of the crystal—is strongly influenced by various parameters such as temperature and humidity.

The different behavior of electro-optic sensors with the BTO crystal for DC and AC fields investigated in [10] is clearly evident in the step response. The initial rapid rise of the electric field is correctly reproduced by the sensor, but in the subsequent range of constant field strength, the step response falls to zero within 35 s. This is comparable to the step response of a measuring coil (see, for example, Fig. 5.35c). The BTO sensor is calibrated in the defined AC field of a large plate electrode arrangement at power frequency, as described in (Ref. [116] of Chap. 4). The result shows that the sensor works linearly for field strengths in the investigated range of ± 150 kV/m. The sensitivity at room temperature is 1.65×10^{-7} m/V, with reproducibility within $\pm 3\%$. The sensor is intended for field measurements on silicone insulators that are tested with AC and impulse voltages. In another application of the sensor, the field distribution on electrical apparatus is measured, e.g. the tangential field of the replica of a stator end winding in a large turbine generator [11]. The experimental results are intended to support numerical field calculations. The aim is to improve the arrangement in order to avoid high local field strengths, which can lead to partial discharges and thus to long-term damage in the winding area.

In [12] a sensor with an LNO crystal is described, which also has a dielectric mirror at the end face for reflecting the light beam and thus has small dimensions. Two identical height-adjustable sensors are installed to the right and left of a 150 kV AC overhead line insulator, which allow the axial and radial field components in the vicinity of the suspension insulator to be measured relatively easily and quickly. Based on the measured spatial field distributions, the presence, location and type of damage in the tested insulators can be clearly identified.

6.1.1.2 Pockels Sensors for Voltage Measurements

The examples in Sect. 6.1.1.1 deal with the use of sensors with Pockels cells in miniature design for field measurements. For the direct measurement of high voltages, sensors of larger dimensions are used, which are usually installed in SF_6-insulated electrode arrangements. A sensor with a BGO crystal ($\varepsilon_r = 16$), which has a diameter of only 1 mm and a length of 40 mm, is described in [13]. The sensor with the fragile crystal is tested for suitability for DC, AC and impulse measurements up to 50 kV. The lightning impulse voltages measured with this sensor have no superimposed oscillations, which occur with other sensors due to the inverse *piezoelectric effect* (see Sect. 6.1.1.3). The rise time of the step response of the sensor is reported as less than 3 ns, which corresponds to a bandwidth of more than 116 MHz.

To measure higher voltages, several electro-optic crystals can be connected in series and arranged one above the other with spacers or glued into a unit. The frequently used BGO crystal is known to have a stable behavior for at least one hour, even in DC operation. A sensor array of eight series-connected BGO crystals ***1***, each 1 mm in length, is described in [14]. Together with the spacers ***2*** between the individual crystals, the sensor reaches a total length of 110 mm (Fig. 6.7). The sensor structure corresponds to the typical arrangement with the polarizer ***3*** or analyzer, $\lambda/4$ retardation plate ***4***, microlens ***5*** and optical fiber ***6***. The sensor is installed between the two electrodes ***7*** in the insulating cylinder ***8*** and housed in a larger SF6-insulated electrode assembly. For the complete sensor and a wavelength

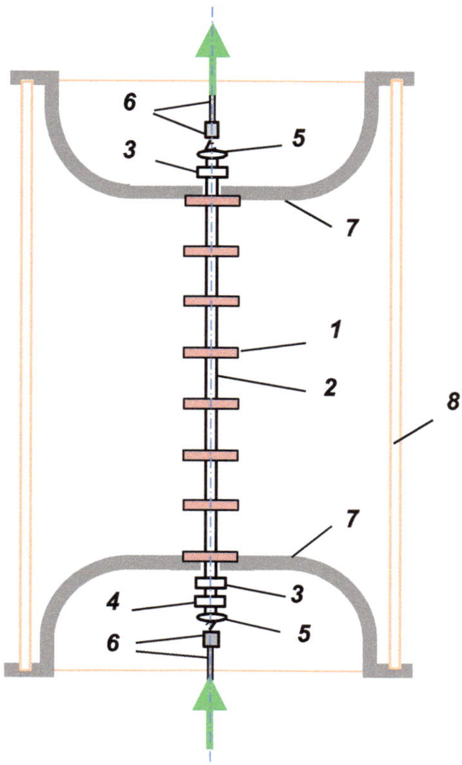

Fig. 6.7 Electro-optic high-voltage measuring system with eight BGO crystals in series. ***1*** BGO crystal, ***2*** spacer, ***3*** polarizer/analyzer, ***4*** λ/4 plate, ***5*** microlens, ***6*** optical fiber, ***7*** electrode, ***8*** insulating cylinder

$\lambda = 1300$ nm, a half-wave voltage $U_\pi = 14$ MV is calculated, which is thus several hundred times larger than the U_π value of a single BGO crystal. The sensor shows excellent linearity, the deviations being within ±0.2% for DC voltages in the range of ±225 kV, within ±0.1% for 50 Hz AC peak voltages up to 250 kV and within ±0.2% for lightning impulse voltages up to 400 kV. The measurement of DC voltage was carried out in each case 5 min after applying the voltage. The data of the recorded lightning impulse voltages were filtered to eliminate the superimposed oscillation of about 160 kHz due to the inverse piezoelectric effect (see Sect. 6.1.1.3).

Another study deals with the behavior of a 1.1 m long sensor, which consists of eight BGO crystals as in [14] and is provided at both ends with a polarizing beam splitter and a transparent electrode [15]. Two light waves with wavelengths of 1300 and 1550 nm are introduced into the grounded base of the sensor via an optical multiplexer and a polarization-maintaining optical fiber. On the high-voltage side, the light waves pass through a demultiplexer, which supplies them to two opto-electronic converters for evaluation. The combination of the two individual characteristic curves (see Fig. 6.2) for the light waves results in a locus curve from which the applied voltage can be determined. In this method, the expectation is that higher voltages than for a single light wave are unambiguously measureable. Due to the Faraday and piezo-gyration effects, however, the characteristic curves and thus also the measured voltages are superimposed by low noise. During the measurement of the natural birefringence and the sensitivity of each crystal at DC voltage, a 50 Hz modulation voltage is applied to one end of the crystal, whereby the above-mentioned decrease in photosensitivity can be avoided for at least 1 h. On the other hand, the results show that each crystal has a different temperature behavior. Due to the inverse piezoelectric effect (see Sect. 6.1.1.3), the measured lightning impulse voltages are superimposed by an 80 kHz oscillation, which can be eliminated by a special filtering method including FFT.

6.1.1.3 Inverse Piezoelectric Effect

Due to the sub-nanosecond processes in the electro-optic crystal, Pockels cells can theoretically achieve bandwidths from zero to the gigahertz range. Several investigations are therefore concerned with wide-band electro-optic measuring systems for impulse voltages [6, 10, 13, 15, 16]. However, due to the inverse piezoelectric effect, the crystal undergoes a periodic change in length in the tail of a lightning impulse voltage, which causes an oscillation of the phase angle $\Delta\varphi$ according to Eq. (6.5). The resulting fluctuation in the intensity of the polarized light wave is converted by the photodetector at the output of the crystal (***6*** in Fig. 6.1) into an electrical oscillation superimposed on the measured lightning impulse voltage (Fig. 6.8). The frequency of the superimposed oscillation lies between 80 kHz and 1 MHz for the examined crystals.

The test results relating to the occurrence of oscillations superimposed on impulse voltages do not give a uniform picture for the electro-optic crystals studied.

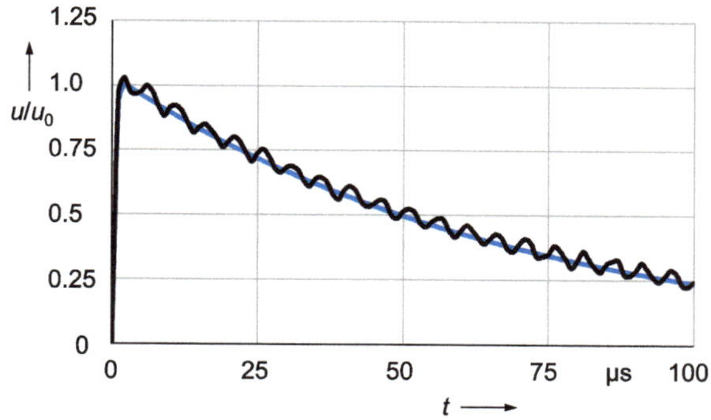

Fig. 6.8 Schematic diagram of a measured lightning impulse voltage with superimposed oscillation, which is generated by the inverse piezoelectric effect of a Pockels cell

For example, the lightning impulses measured with the 40 mm long BGO crystal (diameter 1 mm) presented in [13] have no superimposed oscillations. On the other hand, the majority of sensors with the same or with different crystals show this annoying effect. Furthermore, by reducing the length of an LNO crystal to 2.5 mm, the oscillation frequency increases from a few 100 kHz to almost 1 MHz [6]. The multilayer arrangement with thin crystals is also advantageous in order to shift the oscillation to a higher, no longer disturbing frequency range. The oscillations and noise components superimposed on the lightning impulse voltage can then be largely eliminated by digital filtering of the measurement signal [14–16]. In the case of well reproducible impulse voltages, a significant reduction of the superimposed oscillation is also achievable by multiple recording and averaging of the impulses [10].

6.1.1.4 Electro-optic Voltage Transducers

An important goal in the development of sensors with Pockels cells is their use in *electro-optic voltage transducers* (or: *electro-optic voltage transformers*) as a replacement for the inductive and capacitive voltage transformers, which are used almost exclusively for energy metering in the supply network so far. The replacement of conventional oil-insulated voltage transformers is sought from an ecological point of view, e.g. because of the danger of leaking oil or an explosion. This is also advantageous in view of lower weight and the acquisition and operating costs. Also advantageous is the small output voltage of electro-optic transducers in the range of a few volts, which enables the direct connection of electronic circuits with digital acquisition of the measured data. In addition, the transmission voltages of up to 1 MV planned in Asia place high demands that conventional measuring systems with voltage transformers and dividers can hardly meet due to the required dimensions, the insulation effort and the transmission bandwidth at reasonable costs.

The development of electro-optic voltage transducers—often referred to as "non-conventional transformers"—for measurement and protection purposes in the energy supply network has progressed over the last two decades. They are now proven in practice and are offered commercially. On the other hand, for inductive voltage transformers with appropriate maintenance and monitoring, a lifetime of 40 years can be expected. In the case of electro-optic transducers, comparatively long experiences are not yet available, so that their spread only increases slowly by replacement.

Pockels sensors in electro-optic voltage transducers are either floating in the electric field between the high-voltage and ground electrodes or are directly connected to the voltage to be measured. In the electro-optic voltage transducer presented in [17], the Pockels sensor ***1*** is located in the field of a hemispherical high-voltage electrode **2**, which is supported by a 2 m high insulating tube ***3*** (Fig. 6.9). The sensor is connected via an optical fiber ***4*** to the devices ***5*** at ground potential (light source, measuring instrument, control units, etc.). The experimental studies, in which the position of the sensor in the electric field as well as the position of devices in the vicinity of the transformer was varied, showed satisfactory results. The authors come to the positive conclusion that the electro-optic voltage transducer is suitable for use in the 500 kV supply network after on-site calibration. A variant of the arrangement is shown in [18], in which the high-voltage electrode encloses the spherical ground electrode with built-in field sensor.

In general, GIS and GIL provide good conditions for the installation of the complete electro-optic sensor including polarizer, λ/4 plate and analyzer [19, 20]. The sensor works as a field sensor whose display is converted into a voltage after on-site calibration. The relative measurement uncertainty of the sensor is reported as less than 0.5% for the usual industrial operating temperature range.

Electro-optic voltage transducers whose sensors are directly connected to voltage can be subdivided into two groups. In one group, the sensor is located at the

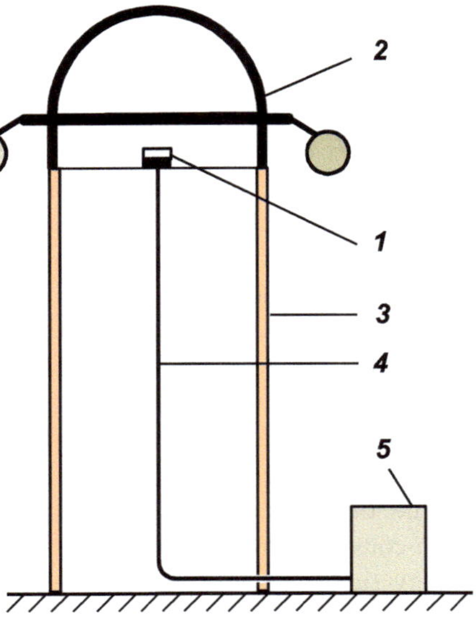

Fig. 6.9 Electro-optic voltage transformer (schematic diagram). ***1*** Pockels sensor in the electric field of the voltage to be measured, ***2*** high-voltage electrode with flange, ***3*** insulating tube, ***4*** optical fiber, ***5*** light source, measuring instrument, control devices, etc.

output of a capacitive voltage divider whose disadvantages must be taken into account as with the conventional capacitive voltage transformer (see Sect. 2.5.6). The sensors of the other group are connected directly to the full high voltage of up to several 100 kV. In [21, 22], an electro-optic voltage transducer with BGO crystals ($Bi_{12}GeO_{20}$ or $Bi_4Ge_3O_{12}$) in SF6 atmosphere is described for use up to the 550 kV level. The introduction of two light waves with a phase shift of $+\lambda/8$ and $-\lambda/8$ proved to be beneficial for the stability of the crystal. The voltage transducer, also designed as an outdoor combined instrument transducer with Faraday sensor for current measurement, meets the requirements of IEC accuracy class 0.2.

The periodic change in length due to the inverse piezoelectric effect (see Sect. 6.1.1.3) is also used for AC voltage measurement [22]. The measuring sensor consists of a 100 mm long cylindrical quartz crystal around which a special *dual-mode optical fiber* with an elliptical core is wound. The resulting differential phase modulation of the two fiber modes is proportional to the voltage applied to the quartz crystal. The piezo-optic voltage transducer is housed in a compact arrangement and can be plugged into a suitably prepared 170 kV GIS. In addition, a 420 kV prototype sensor with four series-connected quartz crystals has been developed and investigated for outdoor operation.

6.1.2 Electro-optic Kerr Effect

Like the transverse Pockels effect, the *electro-optic Kerr effect* is based on the induced birefringence of a light wave in the Kerr medium under the action of a transverse electric field *E*, which in turn causes a rotation of the polarization plane. The basic circuit in Fig. 6.10 consists of a laser ***1***, polarizer **2**, container ***3*** with the mostly liquid or gaseous Kerr medium, an analyzer ***4***, a photodetector **5** and a measuring instrument M. As with the Pockels sensor, the arrangement can be extended by a $\lambda/4$ plate (see Fig. 6.1) or by a mirror for reflection of the light wave (see Fig. 6.5). According to the quadratic dependence of the induced birefringence in the Kerr effect according to Eq. (6.1), the path difference of the two wave components and thus the phase shift is [1–4]:

$$\boxed{\Delta\phi = 2\pi\, l\, K\, E^2}, \tag{6.6}$$

where K is the *Kerr constant* and l is the effective length of the medium traversed by the light wave under the action of the electric field E. The intensity of the light wave after leaving the Kerr medium is again proportional to the square of the sine function according to Eq. (6.4) or Eq. (6.5), but the argument is proportional to the square of the field strength:

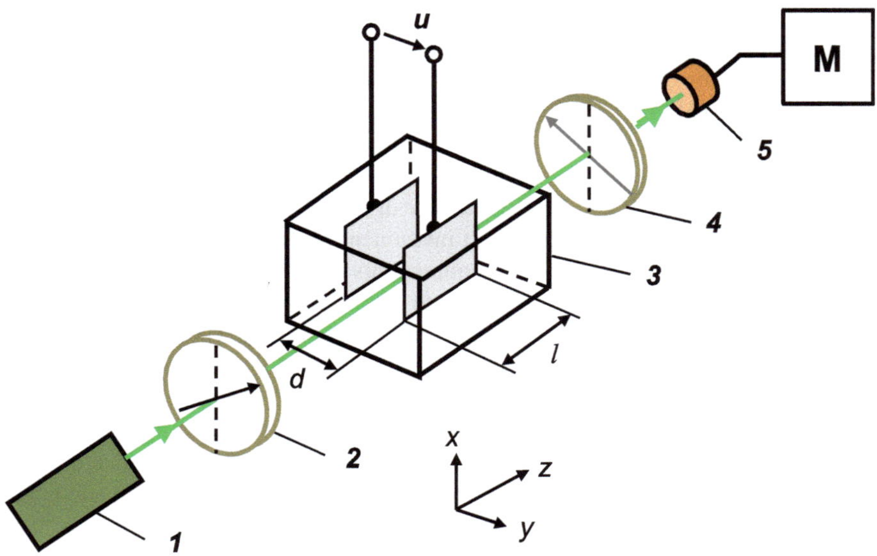

Fig. 6.10 Electro-optic Kerr effect for high-voltage measurement (principle). ***1*** Laser, ***2*** polarizer, ***3*** container filled with liquid Kerr medium, ***4*** analyzer, ***5*** photodetector, **M** measuring instrument, e.g. oscilloscope

$$\boxed{I = I_0 \sin^2 \frac{\pi}{2} (E/E_\mathrm{m})^2}. \tag{6.7}$$

The course of the intensity I/I_0 according to Eq. 6.7 is characterized by maxima and minima, which follow each other with increasing field strength at ever shorter intervals. The first maximum $I = I_0$ for $k = 1$ occurs at the field strength E_m. Equation (6.7) gives the relative field strength E/E_m [23]:

$$\frac{E}{E_m} = \sqrt{k + \frac{2}{\pi} \text{arc sin} \sqrt{\frac{I}{I_0}}}. \tag{6.8}$$

Due to the quadratic dependence of the Kerr effect on the field strength or voltage, the polarity information is lost. Therefore, and because of instability and high temperature dependence, Kerr cells are less well suited for voltage or field strength measurements than Pockels crystals.

When using a Kerr cell for voltage measurement, the applied voltage is $u = (E/E_\mathrm{m})\,(E_\mathrm{m}d)$ with E/E_m according to Eq. (6.8), where d denotes the effective electrode spacing (see Fig. 6.10). The product $E_\mathrm{m}d$ is determined by calibration with a reference divider at DC, AC or impulse voltage [23, 24]. In the early days of Kerr cell measurement, an oscilloscope was used as instrument M and the number

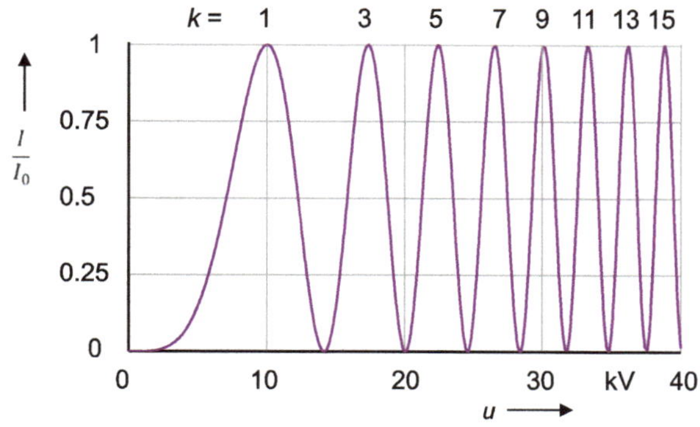

Fig. 6.11 Characteristic of a Kerr cell showing the light intensity I/I_0 versus the applied voltage

of voltage maxima displayed was counted (Fig. 6.11). Basically, the accuracy of this evaluation increases with increasing field strength or voltage, since the number of maxima increases disproportionately.

The Kerr effect has been extensively investigated in a large number of studies concerning its use in high-voltage and field measurement, pulse physics and insulation technology [3, 4]. The materials that were initially preferred because of their large Kerr constant K were flammable, explosive or toxic liquids. So, Kerr sensors with nitrobenzene were manufactured for impulse voltages of up to 300 kV (Ref. [6] of Chap. 1; [4]). Later, less hazardous materials such as water, transformer oil, solids and gases were also used as the Kerr medium. Although nitrobenzene has a large Kerr constant, on the other hand it has a rather large relative permittivity $\varepsilon_r = 36$. If the Kerr cell is operated in a substance of lower relative permittivity, the field strength in the Kerr medium decreases with the permittivity ratio, and the sensitivity of the Kerr cell is reduced. Therefore, not only the value of the Kerr constant K is decisive, but also the quotient K/ε_r [25].

Transformer and silicone oils have small, gases such as O_2, N_2, CO_2 and SF_6 still much smaller Kerr constants. Even at higher field strengths or voltages, only a small indication is achieved with these materials that barely exceed the noise level. A proven remedy is the modulation of the voltage to be measured by an auxiliary voltage whose frequency is between 1 and 50 kHz and amplitude between 200 V and 1 kV [4, 26, 27]. The modulated output signal of the photo detector is then fed via a lock-in amplifier to the PC and the influence of the superimposed auxiliary voltage is eliminated by digital filtering. Further increase in sensitivity is achieved by providing the Kerr cell at the entrance and exit sides of the light wave with translucent concave mirrors. The light wave thus experiences a multiple reflection before it leaves the Kerr cell. Due to this multiple reflection, the entire path of the light wave in the electric field—and thus the phase shift $\Delta\varphi$—is significantly increased [27].

6.2 Faraday Effect

In most solid and liquid dielectric materials that are optically transparent, the *magneto-optic effect* occurs, which according to its discoverer is also referred to as *Faraday effect* or *Faraday rotation*. It describes the interaction of linearly polarized

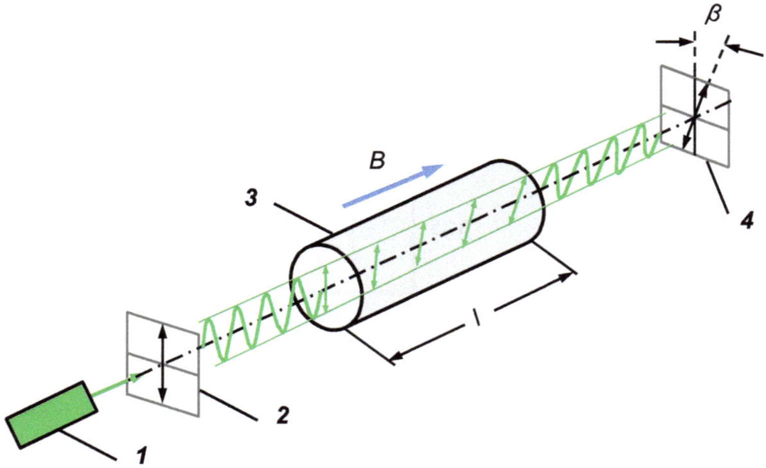

Fig. 6.12 Principle of the magneto-optic Faraday effect for measuring magnetic fields or currents. *1* Laser, *2* polarizer, *3* glass rod, *4* analyzer

light with a magnetic field in a transparent medium. The polarization plane of the light as it passes through the medium is rotated under the influence of a longitudinal magnetic field, which is shown in the example of Fig. 6.12 for a glass rod. The rotation is based on induced circular birefringence, i.e. the linearly polarized light is decomposed into right- and left-circularly polarized waves which propagate in the transparent rod at slightly different speeds due to different refractive indices. At the output of the glass rod with the effective length l, the two partial waves recombine to form a linearly polarized wave. However, its polarization plane is rotated by the angle β with [1–3]:

$$\beta = V\,l\,B = V\,l\,\mu_0 H. \tag{6.9}$$

Here, V is the *Verdet constant* of the medium with $\mu_r = 1$, which can be positive or negative and which depends on the wavelength of the light, the medium and the temperature. Occasionally, instead of V, the product $\mu_0 V$ is given in numerical form. The processes in the medium take place in the range of nanoseconds. With a downstream analyzer, the rotation of the polarization plane is converted into an intensity modulation. Then, a photodetector converts the intensity modulation into a corresponding electrical signal to be measured by an instrument.

6.2.1 *Magneto-optic Current Sensors*

The basic properties of magneto-optic sensors and their use in measuring high DC, AC and impulse currents are discussed in numerous publications [18–22, 28–33].

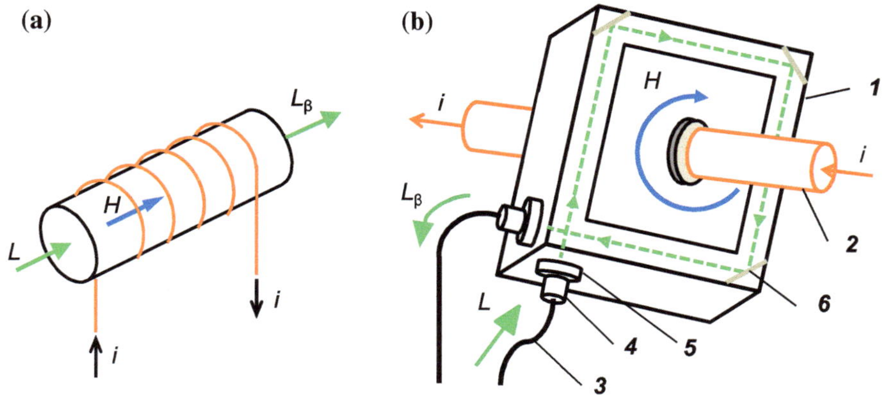

Fig. 6.13 Basic versions of two different magneto-optic current sensors in block form. **a** Faraday sensor with multiple wound conductor, **b** Faraday sensor with a centric opening through which the conductor is inserted. ***1*** Quartz glass transducer, ***2*** conductor, ***3*** optical fiber, ***4*** lens, ***5*** polarizer, ***6*** mirror

According to Eq. (6.9), the angle β and thus the measured value of the instrument is directly proportional to the magnetic flux or current generating this flux:

$$\beta \sim i. \tag{6.10}$$

Basically, a distinction is made between sensors in block form and fiber form. Around the simple cylindrical sensor made of quartz glass in block form, the current carrying conductor is wound with several turns, whereby the longitudinal magnetic field is generated (Fig. 6.13a). The current-proportional rotation of the polarization plane at the output of the quartz glass is converted by an analyzer into an intensity modulation and by a photodetector into an electrical signal for further evaluation. The production of sensors in block form is described as simple and their long-term stability as good. The response time of the complete sensor is a few 100 ns, which corresponds to a bandwidth of several megahertz [34].

A special embodiment of a sensor in block form is the *magneto-optic current transducer* with *glass ring* sensor head (Fig. 6.13b). It consists of a square quartz glass arrangement ***1*** with a centric opening through which the conductor ***2*** is inserted [1, 3, 18, 21]. The current generates in the glass plate ***1*** a magnetic field *H* with field components that are parallel to the edges of the glass plate. Via the optical fiber ***3***, the lens ***4*** and the polarizer ***5***, the light enters as a linearly polarized light wave L in the glass plate parallel to an edge. At each plate corner ***6***, the light wave is mirrored by 90° so that it propagates parallel to the four edges—and thus also approximately parallel to the magnetic field—until it leaves the glass plate near the entrance point again. Under the influence of the magnetic field, the light is circularly polarized as described above. After a complete circulation, the light wave

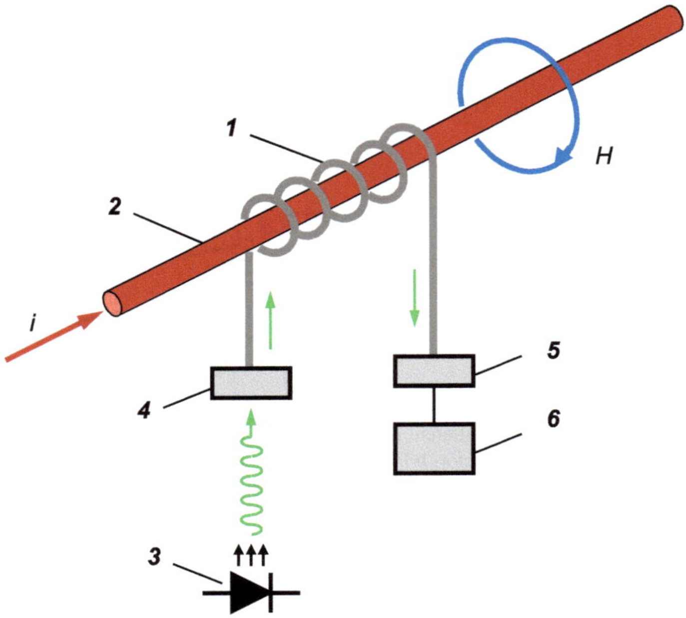

Fig. 6.14 Magneto-optic current sensor in fiber form wound around the conductor (principle). ***1*** Optical fiber, ***2*** conductor, ***3*** laser diode, ***4*** polarizer, ***5*** analyzer, ***6*** photodetector

L_β leaves the quartz glass again with linear polarization, but with a phase rotation β proportional to the current i according to Eq. (6.10). At the output of the glass ring, the phase rotation is converted by an analyzer into an intensity modulation, which is then converted by a photodetector into an electrical signal for further evaluation.

Magneto-optic current sensors in fiber form have recently come to the fore. Optical glass fibers in special versions are used as sensors. In the basic arrangement, the glass fiber ***1*** is wound around the conductor ***2*** in one or more turns (Fig. 6.14). The light source is a laser diode ***3*** with a downstream polarizer ***4*** for generating linearly polarized light. The magnetic field H generated by the current i is parallel to the sensor fiber and causes the Faraday rotation of the polarization plane by the angle $\beta \sim i$. The phase modulation of the linearly polarized light is converted by the analyzer ***5*** into an intensity modulation, which is displayed by the detector ***6*** as an electrical signal for further data processing.

In the practical implementation of the simplified measuring principle according to Fig. 6.14, however, there are great problems. The polarization is changed by minimal thermal and mechanical effects on the glass fiber winding. In addition, optical fibers have slight asymmetries that affect polarization. After decades of research and development in various laboratories, the breakthrough for the practical application of the fiber optic current sensor is considered successful. As a result, for example, the light is reflected at the end of the optical fiber and passes through it a second time in the opposite direction. This doubles the Faraday rotation and disturbing factors such as temperature, pressure, vibration and asymmetry of the tempered optical fiber are largely compensated. In electrolytic metal extraction, e.g. in the aluminum industry, very high DC currents of several 100 kA are needed. The parallel current busbars require an opening diameter of the winding of the sensor coil of more than 1 m. In [35, 36], a sophisticated fiber optic measuring system is described, with which currents of up to 500 kA can be measured, whereby the uncertainty is not more than 0.1%.

6.2.2 *Magneto-optic AC Current Transducers*

The use of magneto-optic sensors as AC current transducers in the high-voltage network instead of oil-filled inductive transformers offers economic and ecological advantages, similar to the use of electro-optic voltage transducers (see Sect. 6.1.1.4). The *magneto-optic transducers* (also: *magneto-optic transformers*) have sensors in block and fiber forms. Together with the polarizer and analyzer, they form a passive unit with small dimensions, which does not require an external power supply. The sensor positioned at high-voltage potential is connected via optical fibers to the light source and the detector at ground potential. In dry air, the magneto-optic sensor with the connecting optical fiber does not significantly affect the electric and magnetic fields and does not require any special insulation effort itself. Electromagnetic interferences as in hybrid transducers, e.g. with Rogowski coil and optical data transmission, do not occur.

The magneto-optic glass ring transducer according to Fig. 6.13b is distinguished by its simple and cost-effective construction. After intensive investigation and testing in the energy supply network, satisfactory experiences are available [18, 21, 37–39]. The best optical glass ring transducers are assigned to the accuracy class 0.2. The commercial implementation of magneto-optic current transducers in fiber form (see Fig. 6.14) was more complex and the first prototypes were later introduced to the market [18, 20]. In contrast to the measuring assembly for very high DC currents in [35, 36] with diameters of the sensing coil in the range of 1 m, the optical fiber wound around the high-voltage conductor has a stronger curvature and is therefore more affected by mechanical stress. Advantageous is the greater possibility of variation of the current measuring range, which can be selected as in inductive current transformers by changing the number of turns of the optical fiber.

Magneto-optic current transducers are used in combination with electro-optic voltage transducers as combined instrument transducers for power measurements in the high-voltage network. Here, the different propagation times of the current and voltage signals must be considered. The extended test specifications for transformers, which have so far mainly prescribed secondary currents of 1 A and secondary voltages of 100 V, now accept much smaller output quantities that are suitable for further data processing with electronic and digital circuits (IEC 60044-7 and -8). The previously predominantly used inductive current and voltage transformers are known to be reliable and long-lasting. Experience gained so far with optical transducers is not yet considered sufficient to assess their accuracy and expected long-term behaviour. This is one of the main reasons that their use in the supply network is not yet widespread.

References

1. Yariv, A., Yeh, P.: Optical Waves in Crystals. Wiley, New York (1984)
2. Pedrotti, F., Pedrotti, L., Bausch, W., Schmidt, H.: Optik für Ingenieure. Springer, Berlin, Heidelberg, New York (2008)
3. Hebner, R.E., Malewski, R.A., Cassidy, E.C.: Optical methods of electrical measurement at high voltage levels. Proc. IEEE Trans. PAS **65**, 1524–1548 (1977)
4. Hidaka, K.: Electric field and voltage measurement by using electro-optic sensor. In: Proceedings of 11th ISH London, Paper 2.1.S2 (1999)
5. Kaluza, K., Peier, D.: The electrical short-time strength of optical fibers. In: Proceedings of 5th ISH Braunschweig, Paper 72.09 (1987)
6. Stolle, D., Niehe, S.: Optische Messung hoher Spannungen im Prüffeld. VDE Berichte Nr. **1530**, 1057–1064 (2000)
7. Takahashi, T., Hidaka, K., Kouno, T.: New optical-waveguide Pockels sensor for measuring electric fields. In: Proceedings of 9th ISH Graz, Paper 8356-1 (1995)
8. Takahashi, T., Okamoto, T., Hidaka, K.: Development of new optical-waveguide Pockels sensors for measuring electric fields. In: Proceedings of 10th ISH Montréal, Paper 3385 (1997)
9. Takahashi, T.: Electric field measurement by optical waveguide Pockels sensors. In: Proceedings of 12th ISH Bangalore, Paper 4-97 (2001)
10. Merte, R.: Measurement of electric fields with an opto-electric miniature probe. In: Proceedings of 15th ISH Ljubljana, Paper T1-507 (2007)
11. Staubach, C., Merte, R.: Direct electrical field strength distribution determination on electrical apparatus by means of an electro-optical miniature field sensor. In: Proceedings of 19th ISH Pilsen, Paper 200 (2015)
12. Pirovano, G., et al.: An innovative electro-optic sensor for AC and DC electric fields detection: experimental results. In: Proceedings of 18th ISH Seoul, Paper PF-24 (2013)
13. Santos, J.C., Taplamacioglu, M.C., Hidaka, K.: Optical high voltage sensors using Pockels fiber crystals. In: Proceedings of 10th ISH Montéal, Paper 3514 (1997)
14. Santos, J.C., Taplamacioglu, M.C., Hidaka, K.: Pockels high-voltage measurement system. In: Proceedings of 11th ISH London, Paper 1.53.S21 (1999)
15. Kumada, A., Hidaka, K.: Directly measuring high voltage measuring system based on Pockels effect. In: Proceedings of 17th ISH Hannover, Paper D-030 (2011)
16. Borowiak, H.: Ein paper zur technischen Realisierung eines optischen Stoßspannungsmess systems. Thesis TU Cottbus (2006)
17. He, Z.-H., Li, J., Ye, Q.-Z.: Electrical field adjustment test on 500 kV optical metering unit. In: Proceedings of 12th ISH Bangalore, Paper 7-13 (2001)
18. Schwarz, H., Honscha, M., Voss, H.-J., Jenau, F.: Optical measurements of high voltages and currents in energy distribution networks. In: Proceedings of 13th ISH Delft, Paper 586 (2003)
19. Mitsui, T., Hosoe, K., Usami, H., Miyamoto, S.: Development of fiber-opic voltage sensors and magnetic field sensors. Trans. IEEE PWRD **2**, 87–93 (1987)
20. Bosselmann, T.: Physikalische Grundagen zur nichtkonventionellen Messung von Spannungen und Strömen auf hohem Potential. In: Proceedings of HIGHVOLT Kolloquium '99 Cottbus, Paper 3.1, pp. 149–152 (1999)

21. Schmitt, O., Lauersdorf, M.: Optische Sensoren und Messwandler im praktischen Einsatz. In: Proceedings of HIGHVOLT Kolloquium '99 Cottbus, Paper 3.5, pp. 175–179 (1999)
22. Bohnert, K., Gabus, P., Brändle, H.: Fiber-optic current and voltage sensors for high-voltage substations. In: Proceedings of 16th Conference on Optical Fiber Sensors, Nara, pp. 752–754 (2003)
23. Cassidy, E.C., Cones, H.N., Wunsch, D.C., Booker, S.R.: Calibration of a Kerr cell system for high-voltage pulse measurements. IEEE Trans. IM **17**, 313–320 (1968)
24. FitzPatrick, G.J., McComb, T.R.: Investigation of the effects of aging on the calibration of a Kerr-cell measuring system for high voltage impulses. In: Proceedings of 8th ISH Yokohama, Paper 54.04 (1993)
25. Kasprzak, W., et al.: Investigations on Kerr cell with various electro-optic liquids. In: Proceedings of 15th ISH Ljubljana, Paper T1-438 (2007)
26. Okubo, H., et al.: Kerr electro-optic measurement of electrical field distribution in silicone liquid insulation systems for transformer. In: Proceedings of 15th ISH Ljubljana, Paper T9-115 (2007)
27. Kumda, A., et al.: High voltage measuring apparatus based on Kerr effect of gas. In: Proceedings of 15th ISH Ljubljana, Paper T10-618 (2007)
28. Kanoi, M., et al.: Optical voltage and current measuring system for electrical power systems. IEEE Trans. PWRD **1**, 91 (1986)
29. Esposti, G.D., Annovazzi-Lodi, Albini, A.: Current measurements on a high voltage apparatus using a fiberoptic sensor. In: Proceedings of 5th ISH Braunschweig, Paper 73.08 (1987)
30. Hirsch, H.: Polarimetrische faseroptische Stromwandler. Thesis TU Dortmund (1991)
31. Menke, P.: Optischer Präzisions-Stromsensor nach dem Faraday-Effekt. Thesis Univ. Kiel (1996)
32. Flerlage, H.: Magnetooptische Messung schnellveränderlicher Ströme. Thesis Univ. Hannover (1999)
33. Silva, R.M., et al.: Optical current sensors for high power systems: a review. J. Appl. Sci. **16**, 6602–6628 (2012)
34. Zhang, G., Luo, C., Pai, S.T.: Magneto-optical sensors for pulsed current measurements. In: Proceedings of 9th ISH Graz, Paper 7851 (1995)
35. Bohnert, K., Guggenbach, P.: Eine Revolution in der Gleichstrommesstechnik. Bull. SEV/VSE **23**, 23–26 (2005)
36. Bohnert, K., Brändle, H., Brunzel, M., Gabus, P., Guggenbach, P.: Highly accurate fiber-optic DC current sensor for the electro-winning industry. In: Proceedings of IEEE 2005 Petroleum and Chemical Industry Committee (PCIC) Technical Conference, Denver, pp. 1–6 (2005)
37. Schwarz, H., Hudasch, M.: Erste Betriebserfahrungen mit optischen Stromwandlern für den Einsatz in 123-kV- bis 420-kV-Freiluftanlagen. PTB-Bericht **E-46**, 97–112 (1994)
38. Luo, C.M., et al.: The study of accuracy of a magneto-optical current transformer. In: Proceedings of 10th ISH Montreál, Paper 3048 (1999)
39. Li, E., et al.: Development of an optical current measuring system for power systems. In: Proceedings of 10th ISH Montreal, Paper 3248 (1999)

Chapter 7
Digital Recorders, Software and Calibrators

Abstract The measurement of high direct, alternating and impulse voltages, as well as the corresponding currents, is carried out mainly with measuring systems, in which digital recorders or other digital instruments are used. Analog measuring devices, e.g. impulse oscilloscopes or impulse voltmeters, have practically no more meaning and are not treated here. An important component of digital measuring instruments is the A/D converter, which digitizes the analog measurement voltage and provides it as a digital data set for further evaluation with the PC. The requirements for A/D converters vary according to the voltage type. For example, the recording of impulse voltages and currents requires high sampling rates, which can only be realized by fast flash converters with limited amplitude resolution. The chapter describes a variety of test and calibration methods to determine the characteristics of digital devices. For the accurate calibration and verification of digital measuring instruments, calibrators are used which generate AC, DC and impulse voltages of several 100 V up to 2000 V. The analysis of the recorded data is carried out with evaluation software, which is also subjected to a comprehensive evaluation using the test data generator (TDG).

The measurement of high DC, AC and impulse voltages, as well as the corresponding currents, is carried out mainly with measuring systems, in which *digital recorders* or other digital instruments are used. Analog measuring devices, e.g. *impulse oscilloscopes* or *impulse voltmeters*, have practically no more meaning and are not treated here. An important component of digital measuring instruments is the *A/D converter*, which digitizes the analog measurement voltage and provides it as a digital data set for further evaluation with the PC. The requirements for A/D converters vary according to the voltage type. For example, the recording of impulse voltages and currents requires high sampling rates that can only be realized with limited amplitude resolution. For the accurate calibration and verification of the measuring instruments, *calibrators* are used which generate AC, DC and impulse voltages of several 100 V up to 2000 V. The analysis of the recorded data according to the test standards is carried out with *evaluation software*, which is also subjected to a comprehensive evaluation.

K. Schon, *High Voltage Measurement Techniques*, Power Systems,
https://doi.org/10.1007/978-3-030-21770-9_7

The requirements for digital measuring instruments and software, as well as the calibration and test methods, are specified in a four-part series of IEC 61083 [1]. This chapter mainly describes the characteristics of digital recorders with fast A/D converters, their calibration and testing of software for the evaluation of the standardized parameters. Some digital circuits developed for precise AC and DC voltage and current measurements are dealt with in Chaps. 2 and 3.

7.1 Construction and Characteristics of Digital Recorders

Digital recorders for measurements in the fields of high voltage and high current were initially used only for measuring impulse voltages and have now largely replaced analog oscilloscope. Other designations are *digital oscilloscope*, *transient recorder* and *digitizer*, which in the past have also been associated with certain constructional and functional principles. Digital recorders are also increasingly used for AC and DC voltage measurements after being included in the test standards [1]. Here, the term "digital recorder" stands for all digital measuring instruments used in high-voltage and high-current applications. The same digital recorder is often used for the measurement of all voltage and current types.

The simplified block diagram of a digital recorder is shown in Fig. 7.1. The input signal $u(t)$ passes through the attenuator ***1*** and preamplifier ***2*** to the *analog-to-digital (A/D) converter* ***3*** and is temporarily stored as digital data set in the semiconductor memory ***4***. From here, the data set can be shifted into a stationary internal or external memory ***5*** for further processing or displayed on a screen as an analogue waveform with the aid of a *digital-to-analog (D/A) converter*. If the trigger condition is satisfied, the contents of the temporary memory ***4*** are overwritten by a new input signal, so that the last recorded data is always stored in the memory. The individual modules of the digital recorder are controlled with the clock generator and the control logic in the correct order [2, 3].

The most important module of the digital recorder is the A/D converter ***3***, which samples the analogue input signal at equidistant time intervals according to its

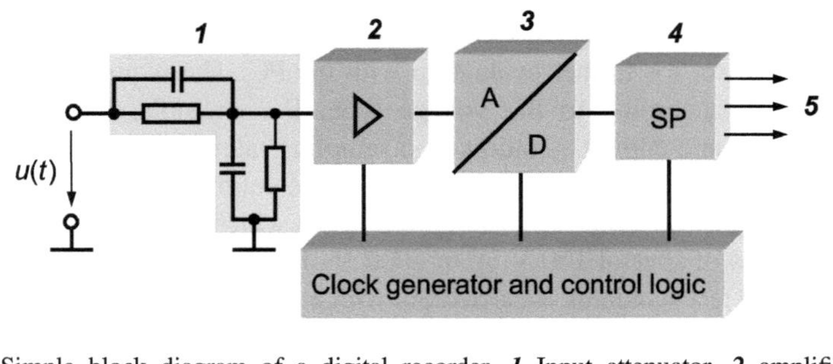

Fig. 7.1 Simple block diagram of a digital recorder. ***1*** Input attenuator, ***2*** amplifier ***3*** A/D converter ***4*** temporary memory ***5*** data output for external equipment (PC, monitor, plotter, etc.)

amplitude resolution. From the beginnings of digital recording of impulse voltages, a series of fundamentally different working principles have emerged for converting a rapidly changing analog signal into a digital data set [4–6]. In a transitional period, attempts have been made to convert the impulse voltage recorded with an analog oscilloscope into a digital data set, e.g. by means of a special cathode ray tube with photosensitive diode matrix memory or a special camera adapter with analog-to-digital image conversion. However, the triumph of the continuously improved digital recorder with electronic A/D converter could not be stopped. For impulse voltage measurement, the A/D converter with parallel conversion has prevailed [7, 8]. With these *flash converters*, recorders with high sampling rate of up to 200 MHz and amplitude resolution between 8 bits and 14 bits are now available. Other versions of A/D converters are used for high AC and DC voltage and current measurements with much higher resolution but lower sampling rates (Chaps. 2 and 3).

7.1.1 Fast A/D Conversion with Flash Converter

The input circuit of a flash converter consists of a multi-stage voltage divider ***1*** with an amplitude resolution of *N* bit corresponding to 2^N quantization values (Fig. 7.2). The divider is connected to the reference voltage U_0 and matched to $2^N - 1$ equal voltage stages. In parallel there is a chain of the same number of comparators 2. The analog input voltage *u*(*t*) is compared with the stage voltages of the voltage divider at the predetermined clock times. As a result, "0" or "1" is simultaneously displayed at the output terminals of the comparator. In the following *encoding unit* **3**, the output signals of the $2^N - 1$ comparators are converted into a binary code with *N*-bit resolution. The maximum achievable speed of the A/D conversion depends mainly on the switching times of the comparators and the delay of the encoding unit.

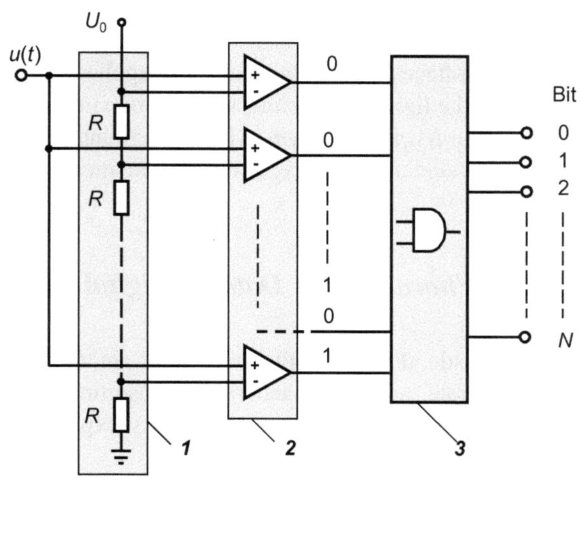

Fig. 7.2 Simplified block diagram of an A/D converter with flash converter ***1*** multi-stage voltage divider with reference voltage U_0 ***2*** comparators ***3*** encoding unit

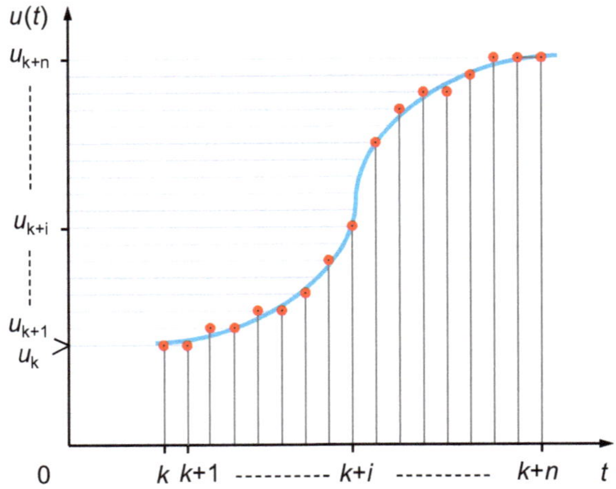

Fig. 7.3 Example of a sampled voltage $u(t)$ (schematic)

Note For recording DC and AC voltages that do not require high sampling rates, other conversion methods are also used to achieve high measurement accuracy. For example, in the case of the *sample-and-hold circuit*, each sampled voltage value is temporarily stored by a capacitor for a short time and then digitized relatively slowly but with high resolution.

The digital recorder is not a linear measuring device like the analog oscilloscope. In the A/D converter, the analog input signal is replaced by the sum of the sampled values at discrete points in time $k\Delta t$, i.e. the information about the measurement signal between adjacent samples is lost. As an example, Fig. 7.3 shows the sampling of an arbitrary voltage at equidistant time intervals with limited amplitude resolution. Depending on the quantization level, the sample values will be below or above the analog curve. The successive sample values u_k, u_{k+1} … u_n are directly connected on the screen, so that the visual impression of a continuous voltage curve is produced, provided that the amplitude resolution and the sampling rate are high. By prescribing sufficiently large minimum values for the sampling rate and the amplitude resolution, the A/D converter can be considered to be quasilinear for the measurement of voltages down to the microsecond time range, i.e. for lightning impulse voltages. This statement also applies to oscillations which are superimposed on the lightning impulse with the maximum frequencies occurring in the test circuit. The *Nyquist theorem*, which determines the sampling rate in relation to the maximum signal frequency, is thus automatically fulfilled.

7.1.2 Characteristic Data of Digital Recorders

Characteristic data of digital recorders are the amplitude resolution, maximum sampling rate, analogue bandwidth or rise time and data memory. The amplitude resolution of a digital recorder is usually expressed by the number N of the A/D converter in bits, from which the corresponding *quantization levels* can be calculated as two to the power of N, i.e. 2^N. Digital recorders developed specifically for use in high voltage and current impulse measurement systems have a rated resolution of 8 to 14 bits. For example, the minimum resolution of $N = 8$ bits corresponds to $2^8 = 256$ quantization levels ranging from 0 to 255, which corresponds to a step height of about 0.4% of full scale. AC and DC measuring instruments with specially developed high-precision A/D converters can have an amplitude resolution of more than 20 bits.

With the input attenuator ***1*** and preamplifier ***2***, the input voltage $u(t)$ is set to the operating range of the A/D converter ***3*** (see Fig. 7.1). High-quality digital recorders have continuously variable input amplification so that the voltage at the A/D converter is increased to its full scale. Then the A/D converter always works with optimum amplitude resolution. The majority of the recorders do not have this feature, so the resolution gets worse with decreasing voltage amplitude. The amplitude resolution can be apparently increased if the input voltage is recorded at a much higher sampling rate than required. With this *oversampling* technique, two or more adjacent samples are combined into a mean value, which is generally between two quantization levels. The measured voltage is then represented by the totality of the mean values, giving the impression of a higher amplitude resolution.

The *sampling rate* indicates the number of samples per second and is expressed in kS/s or MS/s (kilo- or megasamples per second). The formally correct unit "Hz" for the sampling frequency is not used, as this can easily lead to confusion with the bandwidth or the frequency spectrum. The minimum sampling rate required for a standardized impulse voltage test is determined by the time T_{AB} in the front of the impulse voltage or current (see Sect. 7.5). For lightning impulse voltages, sampling rates of several 10 MS/s are required, which are mainly needed for the evaluation of the front time and a possibly superimposed oscillation. The sampling rate of some recorders can be changed after a preset recording time. It is then possible to record an impulse voltage in the front with a high sampling rate and in the tail with a low sampling rate. The advantage is that less memory is required due to the smaller number of samples in the tail. By combining two or more A/D converters that alternately sample the measurement voltage, a multiplication of the sampling rate of up to 200 MS/s or more can be achieved [7, 8].

The *rise time* of a recorder for impulse voltage measurements shall not exceed 3% of T_{AB}. This corresponds to a minimum bandwidth of approx. 25 MHz for recording standardized lightning impulse voltages with the shortest front time $T_1 = 0.84$ µs (see Sect. 9.6). Digital recorders even have bandwidths of up to 100 MHz and sampling rates of up to 200 MS/s and even more with special sampling techniques [7, 8]. Higher bandwidths are difficult to implement because of the internal attenuator for input voltages of up to 2 kV. Low-voltage digital recorders with input voltages of less than 100 V and 8-bit resolution have higher sampling rates and bandwidths in the range of 1 GS/s or 400 MHz. They are thus also well suited for recording the step response of components of high-voltage and high-current impulse measuring systems.

The *temporary data memory*, into which the samples are continuously written, has a limited storage capacity. When the memory is full, further recording is stopped or, depending on the selected recording mode and trigger condition, the memory content is automatically overwritten by a new input voltage. The data of the last recorded input voltage is therefore always stored in the memory. Recording a voltage with a higher sampling rate requires a correspondingly large memory to store the same impulse shape. In the early days of digital measurements, as already mentioned above, some recorders with low memory capacity could record the impulse tail at a reduced sampling rate in order to limit the amount of data.

7.1.3 Further Characteristics of Digital Recorders

For optimum signal recording, there are two *trigger modes*. On the one hand, the value of the trigger pulse and, on the other hand, the partitioning of the memory space before and after reaching the trigger value is set. The phase-locked sampling of a continuous voltage or the time before the beginning of an impulse can thus be conveniently chosen. Unlike analog oscilloscopes, where recording is started by the trigger pulse, the recording of digital recorders is stopped by the trigger pulse and the voltage shape prior to the trigger event is stored in memory. Because of this *pre-trigger mode*, trigger problems, such as with analog oscilloscopes, do not occur with digital recorders.

Most digital recorders enable the repetitive recording of a series of identical impulses and evaluation of the averaged impulse. This is advantageous for the multiple recording of step responses or calibration impulses generated by stable generators. In this case, a smoothing effect and higher accuracy can be achieved since the normally distributed digitization errors are largely eliminated in contrast to the sampling of a single impulse (see Sect. 13.1.3). In addition, some digital recorders use a special scanning technique in which each pulse of a pulse train is sequentially sampled with a very small time shift, and all the samples are then time-correctly combined into a single impulse. The effective sampling rate is thereby increased considerably, but the bandwidth of the recorder will remain unchanged.

A digital recorder usually has at least two measurement channels. This allows the simultaneous recording of current and voltage quantities or carrying out comparative measurements between the measuring system and a reference system (see Sects. 10.3 and 10.4). It also facilitates the performance of the interference test in which the temporal assignment of the generated interference voltage to the measured test voltage is of interest. However, the simultaneous operation of two or more channels in the same recorder may lead to a mutual influencing of the input voltages.

The A/D converter of a digital recorder usually requires an input voltage of several volts for full scale deflection. The voltage to be measured is adapted to this value by means of internal or external *input attenuators* and *preamplifiers*. The attenuators are developed as compensated RC dividers for voltages of up to 1000 V or even 2000 V, which is typically the maximum output voltage of high-voltage dividers. High input voltages are of advantage in suppressing interference voltages generated in the measurement cable by electromagnetic fields, e.g. by the firing of sphere gaps. When the input voltage is attenuated by the internal divider of the recorder, the interference voltage is accordingly reduced. Digital recorders used for low-voltage measurements allow input voltages of not more than 100 V. Therefore, they usually require an external voltage divider for connection to the output of a high-voltage divider [9, 10].

The input impedance of digital recorders shall be at least 1 MΩ with a parallel capacitance of not more than 50 pF. Distortion of the impulse shape, in particular in the tail of switching impulse voltages, is thereby limited. In addition, digital recorders are also equipped with an input resistor equal to the cable impedance of 50 Ω, 60 Ω or 75 Ω. The low-ohmic termination helps prevent reflections when the recorder is operated in conjunction with low-ohmic resistive dividers or measuring shunts. If a recorder, which is not developed specifically for the measurement of high-voltage impulses, is operated with its low-ohmic input resistor, its electrical load capacity must be considered. With high AC or DC input voltages or long impulse duration, there is the danger that the low-ohmic input resistor will change its value or even be destroyed. To protect the input circuit of such a recorder, it is therefore advisable to realize the low-ohmic termination preferably by an external resistor with sufficiently high load capacity.

A variety of options are available for further processing of the data temporarily stored in the recorder. The data can be read from the internal memory of the recorder at a relatively slow repetition frequency, e.g. 1 kHz, and can be repetitively reproduced via a D/A converter as an analog signal on an internal or external screen. Due to the inertia of the human eye, this gives the impression of a stationary display. Furthermore, the data cached in the recorder can be moved to other storage media, such as an internal or external permanent memory, a floppy disc or a CD-ROM drive to be used for later analysis or comparison with other data sets.

In high-voltage and high-power laboratories, strong electromagnetic fields are generated that can adversely affect the function of electronic circuits. Due to their construction and a shielded housing, the digital recorders used there are largely protected against electromagnetic fields and line-conducted interference. When using low-voltage instruments, special precautions must be taken to prevent such interference. Shielding with the help of a *Faraday cage* and a filtered power supply is essential (see Sects. 4.3.1.7 and 5.3.1.1). Disturbances are also coupled via data lines that lead from the measuring instrument to peripheral devices outside the Faraday cage. Data transmission to the external PC and other devices is therefore often carried out via optocouplers with optical fibers.

Digital recorders are available in various versions. In addition to individual devices with an internal or separate computer for control and data processing, there are digital recorders in complex mobile systems. Figure 7.4 shows three versions of digital recorders offered by manufacturers with different amplitude resolution, sampling rates and peripherals.

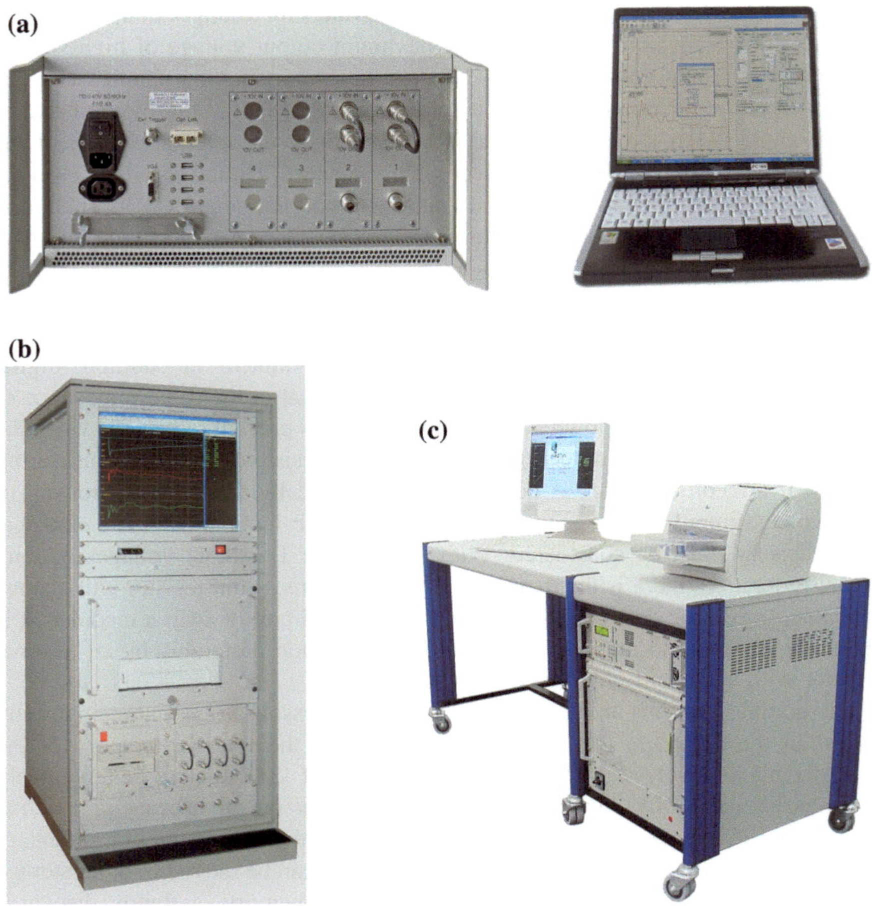

Fig. 7.4 Various types of digital recorders for use in high-voltage measurement systems. **a** Recorder with external notebook and fiber optic data transmission to external PC (HIGHVOLT Prüftechnik Dresden), **b** recorder with built-in PC, printer and CD drive (DR. STRAUSS Messtechnik), **c** Recorder with mobile table frame, monitor and printer (HAEFELY TEST AG)

Digital recorders are used with *software* for the objective and standardized evaluation of the recorded test voltages and currents (see Sect. 7.3). This is particularly important for impulse voltages with superimposed peak oscillations. Also, the filtering of the recorded data to smooth the impulse, the determination of the mean value from a series of data sets, the numerical integration of the stored impulse and the calculation of the spectrum by means of the *Fast Fourier Transform* (FFT) are possible with an internal or external PC. The printout of the stored data on paper enables the manual evaluation of the impulse similar to an analog oscillogram. In case of doubt, the correctness of the software for evaluating the parameters of the measured impulse can thus be easily checked.

Since the introduction of digital recorders in high-voltage and high-current impulse tests in the early 1970s, considerable progress has been achieved in the amplitude and time resolution of A/D converters. Also, the analogue input circuit of recorders, in particular the accuracy and frequency behavior of the input attenuators, has been steadily improved. Further progress in exact measurements is the detailed calibration technique and the introduction of accurate calibration generators. In summary, the uncertainty contribution caused by the digital recorder in high-voltage measurements was significantly reduced.

7.2 Sources of Error in Digital Measurement Technique

The measurement of high voltages and high currents is mainly done with digital measuring instruments whose analog and digital components cause characteristic measurement errors. Even digitizing a voltage with an ideal A/D converter is associated with errors due to the limited amplitude and time resolution. They are called *quantization errors* or *sampling errors* whose maximum values can theoretically be estimated relatively easily. The real A/D converter, in particular the fast flash converter according to Fig. 7.2, causes additional errors due to its technical imperfection. These can only be determined in part by detailed investigations. The error influences can be subdivided into those already present in the sampling of a DC voltage and those which additionally occur with high-frequency voltages. Other sources of error are the input attenuator and amplifier of the recorder, which are already known in principle from the previously used analogue oscilloscopes. This also applies to the effect of interferences caused by the high electric and magnetic fields in the generation of impulse voltages and currents.

7.2.1 *Ideal Quantization*

The basic behavior of an A/D converter is represented by its *quantization characteristic*, which shows the digital *output* in dependence of a DC input voltage. When a DC voltage u_1 is applied to an ideal A/D converter, the corresponding output value is $u_{2,k} = k\Delta u$, where Δu denotes the *step height* of the quantized voltage and k the *code* or *quantization index* (Fig. 7.5, curve 1). The step height is $\Delta u = u_{2,max}/2^N$, where $u_{2,max}$ is the maximum output value at *full scale deflection* of the A/D converter and N is the *amplitude resolution* in bits. When the input voltage u_1 is increased by a small amount, the output value $u_{2,k}$ initially remains unchanged at the level $k\Delta u$. If u_1 exceeds the transition level of the next level, the output value jumps to $u_{2,k+1} = (k + 1)\Delta u$. Altogether, a staircase-like characteristic with the *average code bin width* w_o and step height Δu results. Curve 2 in Fig. 7.5 is an example showing the quantization characteristic of a real, i.e. faulty A/D converter with unequal code bin width w_k [1].

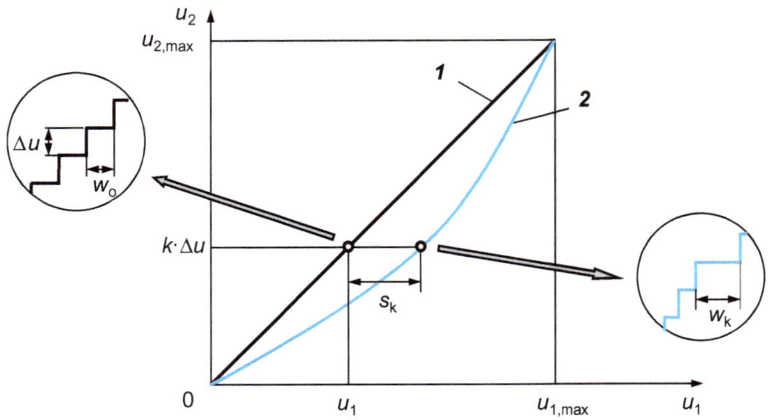

Fig. 7.5 Quantization characteristic of an AD converter for DC voltage ***1*** Ideal A/D conversion with equal step width w_0 and step height Δu ***2*** Faulty AD conversion with unequal step width w_k

Sampling of an arbitrary voltage means that the sampled voltage value is usually between two quantization levels. It is then replaced by the value of the nearest quantization level and stored in the memory. The quantization value therefore deviates from the exact voltage value by the *quantization error* $\delta_{\mathrm{i,k}}$. For an arbitrary voltage, the maximum quantization error $\delta_{\mathrm{i,max}}$ is given by half the difference between two adjacent quantization levels:

$$\delta_{i,max} = 0.5\,\mathrm{LSB} \tag{7.1}$$

where LSB (*Least Significant Bit*) denotes the smallest quantization step. For a resolution $N = 8$ bits, the maximum quantization error at full-scale deflection is $\delta_{\mathrm{i,max}} \approx 0.2\%$. If the input voltage is lower than required for full-scale deflection, the relative quantization error is correspondingly greater.

The discrete quantization errors $\delta_{\mathrm{i,k}}$ in the ideal sampling of a voltage can be characterized by a *rectangular distribution* with the limit values $\pm\delta_{\mathrm{i,max}}$. For a total of m individual values of the quantized voltage, the *standard deviation* σ_{i} of ideal quantization is generally (see Sect. 13.1.3):

$$\boxed{\sigma_{\mathrm{i}} = \sqrt{\frac{1}{m-1}\sum_{k=1}^{m}\delta_{\mathrm{i,k}}^2}} \tag{7.2}$$

and in case of a rectangular distribution with Eq. (7.1):

$$\boxed{\sigma_{\mathrm{i}} = \frac{1}{\sqrt{3}}\left|\delta_{\mathrm{i,max}}\right| = \frac{1}{\sqrt{12}}\,\mathrm{LSB} = 0.289\,\mathrm{LSB} \approx 0.3\,\mathrm{LSB}}. \tag{7.3}$$

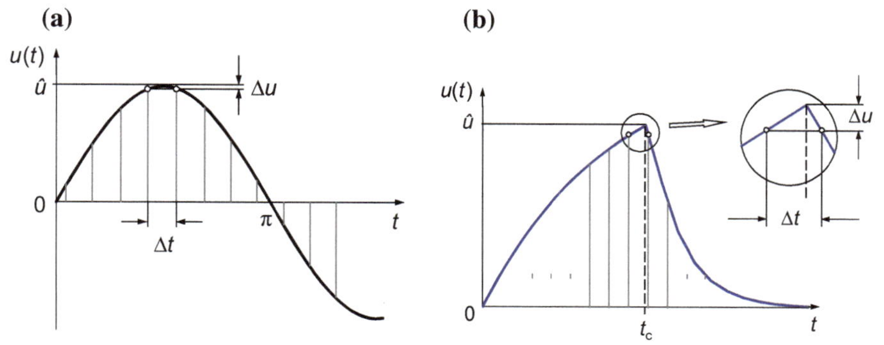

Fig. 7.6 Maximum amplitude error Δu in the case of unfavorable sampling, **a** Sinusoidal voltage, **b** chopped lightning impulse voltage

The standard deviation σ_i according to Eq. (7.3) is the *standard uncertainty* of ideal quantization.

When sampling a voltage, the maximum is generally not exactly captured. The recorded amplitude or peak value is then too small, resulting in a negative sampling error. The most unfavorable case occurs when the two samples adjacent to the maximum value are at the same level [11]. For a pure sinusoidal voltage with the amplitude $\hat{u}$ and frequency f, the negative amplitude error is:

$$\Delta u = -\hat{u}[1 - \cos(\pi \Delta t f)] \tag{7.4}$$

where Δt is the sampling interval, i.e. the reciprocal of the sampling rate (Fig. 7.6a). For an AC voltage with power frequency, the amplitude error according to Eq. (7.4) is negligibly small. Not until a frequency $f = 4.5$ MHz and a sampling rate of 100 MS/s (sampling interval $\Delta t = 10$ ns), a negative amplitude error of up to −1% can occur.

For full lightning impulse voltages and a sampling rate of 100 MS/s, the sampling error of the peak is negligibly small. In the case of a lightning impulse voltage chopped on the front with $T_c = 0.5$ µs, the sampling error at 100 MS/s can theoretically amount to −1% (Fig. 7.6b). In test practice, however, the recorded impulse is rounded in the peak region and the absolute peak value error is therefore smaller.

7.2.2 Static Differential and Integral Non-linearities

The quantization characteristic of a real A/D converter deviates more or less from the ideal characteristic due to the technical imperfection of the converter circuitry. In the example of curve ***2*** in Fig. 7.5, the individual quantization steps have a different width. The relative deviation of the *code bin width* $w(k)$ of the k-th step from the *average code bin width* w_0 in the ideal characteristic ***1*** is the *static* *differential non-linearity d(k)*. The accumulation of steps with too small and too large step widths results in a deformation of the quantization characteristic ***2*** of the real A/D converter compared to the ideal characteristic ***1***. The deformation causes measurement errors not only for the peak value but also for the time parameters of impulse voltages and currents. In the test specifications for digital instruments for measuring DC, AC and impulse voltages, the following limit values are defined: $s \leq 0.5\%$ for the static integral non-linearity and $d \leq 0.8w_0$ for the differential non-linearity [1, part 1 and part 3]. High-quality 8-bit and 10-bit converters have a clearly lower differential non-linearity of $s \leq 0.1\%$ [12].

The standardized determination of the quantization characteristic with DC voltage is quite time-consuming, even with fully automated implementation. The number of applied DC voltages shall be at least five times the number of quantization levels, i.e. approximately 5000 DC voltages ranging from zero to full scale for a 10-bit converter. Furthermore, each voltage shall be evaluated for a sufficiently large number of samples, i.e. $n \geq 1000$. The calibration of a digital recorder with high-resolution A/D converter therefore requires high stability of both the DC voltage source and the digital recorder for several hours. In order to shorten the time for recording the quantization characteristic, the use of ramp voltages or sinusoidal voltages with Fourier analysis is proposed [13, 14]. In this case, the slope or the repetition frequency of the calibration voltage is limited to small values, so that the dynamic behavior of the digital recorder has no additional effect on the quantization. In a sophisticated calibration technique for a 14-bit recorder, a ramp-shaped voltage is applied, which consists of tiny steps corresponding to a resolution of 16 bits. Each voltage level is checked by measurement with a precision digital voltmeter [15].

7.2.3 Differential Non-linearity Under Dynamic Stress

The A/D converter, in particular the one with flash converter according to Fig. 7.2 causes further characteristic errors in the sampling of time-varying voltages, i.e. AC and impulse voltages. With increasing steepness or frequency of the input voltage, it may happen that the A/D converter cannot follow rapid voltage changes due to stray capacitances, inductances and instabilities of the comparators and internal attenuators. Some of the comparator stages respond initially only irregularly and then not at all. The affected codes are therefore present with a lower statistical frequency than adjacent codes or they no longer occur. The relative deviation of the statistical frequency of code k from the average frequency is referred to as *differential non-linearity under dynamic stress d(k)*. The A/D converter in recorders for impulse voltage measurements is tested according to IEC 61083 with symmetrical triangular voltages and those for AC voltage measurements with sinusoidal voltages. The frequency distributions H of code k, which are to be expected for ideal sampling of triangular and sinusoidal voltages, are shown in Fig. 7.7.

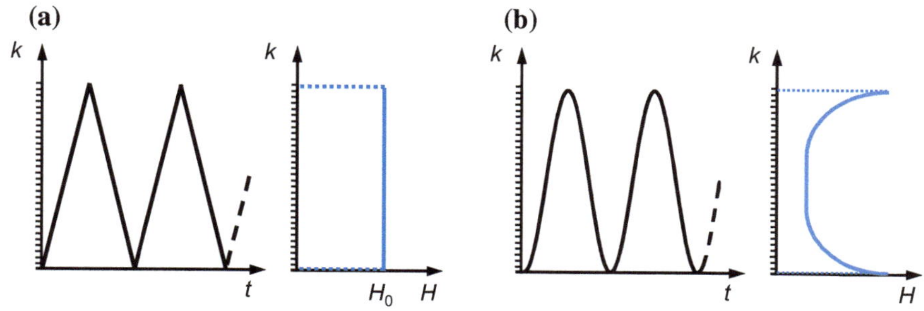

Fig. 7.7 Ideal frequency distribution H of code k for a **a** triangular voltage of constant slope S, **b** sinus voltage of frequency f

The voltages for the test are generated using an analogue signal generator or a digital synthesizer with D/A converter. The amplitude should be close to the value required for full-scale deflection (f.s.d.) of the tested A/D converter. The repetition frequency of the test voltages must not be a harmonic of the sampling frequency. The slope of the triangular voltage must be at least equal to the value f.s.d./0.4T_{AB} definition of T_{AB} (see Sect. 4.1.1). For the reliable evaluation of the histogram, a sufficiently large number of periods of the sampled voltage is necessary. The frequency of occurrence of the discrete codes must be at least 0.8H_0, where H_0 is the mean of the frequency distribution (Fig. 7.7). Further requirements for carrying out the test and the evaluation are specified in [1].

In Sect. 7.2.6, it is shown that the slope of the voltage at the sampling point is decisive for the respective sampling error. This behavior can be illustrated very clearly in the histogram of the codes for linearly increasing test voltages. While at low steepness all codes are approximately equally frequent, certain codes have a decreasing frequency with increasing slope. It may even be that certain codes (*missing codes*) fail completely according to a special pattern. The A/D converter thus loses part of its original amplitude resolution, which increases with signal steepness [16–21]. As an example, Fig. 7.8 shows the measured frequency distribution of the codes of an 8-bit recorder with flash converter for three different slopes S of the test voltage, which is a symmetrical triangular voltage. The frequency distribution H refers to the maximum occurring value H_{max}. For $S = 0.1\ \mathrm{V\mu s^{-1}}$, all codes in the investigated range between 10 and 245 are still present, although not with exactly the same frequency (Fig. 7.7a). With increasing slope, the number of codes decreases more and more (Fig. 7.7b), and for $S = 22\ \mathrm{V\mu s^{-1}}$, only 7 codes are left, which corresponds to a resolution of 4 bits (Fig. 7.7c).

Note The triangular test voltage has the advantage that the dynamic stress of all codes is approximately the same over the entire voltage range. The peaks of the generated triangular voltage are often deformed. It may then be advantageous to set the triangular voltage

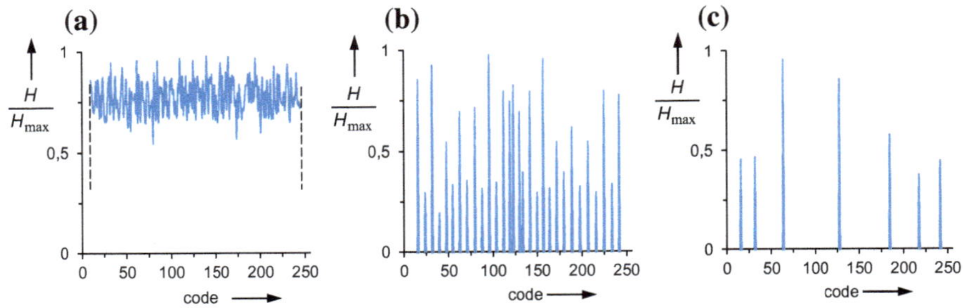

Fig. 7.8 Frequency distribution H/H_{max} of the codes of a digital recorder (8 bits, 10 ns) applying a triangular voltage of the slope S, **a** $S = 0.1\ \text{V}\mu\text{s}^{-1}$, **b** $S = 6\ \text{V}\mu\text{s}^{-1}$, **c** $S = 22\ \text{V}\mu\text{s}^{-1}$

slightly larger than full scale so that the peaks lie outside the range of the A/D converter and will not be evaluated. When testing with sinusoidal voltages, the dynamic stress of the lower and upper codes, i.e. in the region of the negative and positive amplitudes, is rather low and the codes will then show maximum frequency of occurrence. In [20], the frequency distribution of two recorders is determined with impulse voltages with impulse voltages.

7.2.4 Sampling Errors with Sinusoidal Voltages

The quantization characteristic of a flash converter obtained with DC voltage does not automatically apply to fast varying voltages. Above a certain frequency of an AC voltage, the dynamic behavior of the converter deteriorates due to stray capacitances, inductances and instabilities of the internal voltage divider and comparators. For the purpose of metrological testing of A/D converters, sinusoidal voltages that can be generated precisely even in the megahertz range are particularly advantageous. The principle of the evaluation is shown in Fig. 7.9. Sampling of the sinusoidal voltage is done over many periods and provides m stored sample values u_k. These are then approximated by an ideal sinusoid $u(t)$ with respect to frequency, amplitude, phase and offset (Fig. 7.9a). The deviations of the sampled values from the corresponding values of $u(t)$ at the discrete sampling times $k\Delta t$ are the sampling errors $\delta_{r,k}$ (Fig. 7.9b).

In contrast to ideal quantization, the sampling errors of real A/D converters are not limited to 0.5 LSB. They can reach much higher values and have a normal distribution, which is also in contrast to the ideal quantization with uniform distribution. The *empirical standard deviation* σ_r of the sampling errors $\delta_{r,k}$ for the real digitization of a sinusoidal voltage is calculated as [11]:

$$\boxed{\sigma_r = \sqrt{\frac{1}{m-1}\sum_{k=1}^{m}\delta_{r,k}^2}}, \tag{7.5}$$

which is understood as the *standard error* of the A/D converter under study.

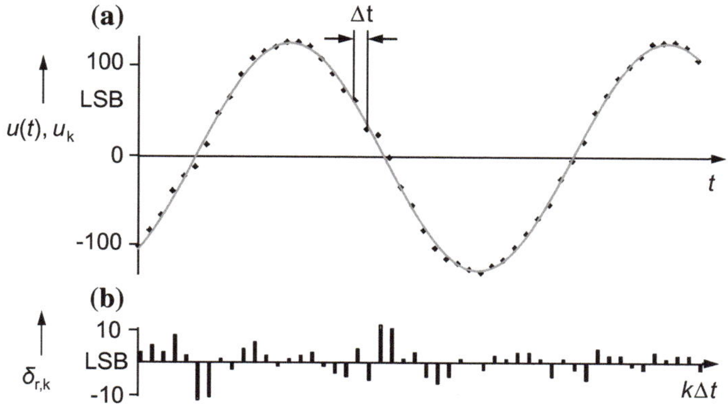

Fig. 7.9 Example of sampling a sinusoidal voltage with an 8-bit converter (Δt = 10 ns). **a** Sample values u_k and best-fit sinusoid $u(t)$, **b** Deviations $\delta_{r,k}$ of the samples u_k from the sinusoid $u(t)$

The plot σ_r as a function of the frequency f provides valuable information about the dynamic behavior of the A/D converter. In general, $\sigma_r(f)$ increases linearly with frequency f of the sine voltage, the initial value being approximately equal to the value σ_i of the ideal quantization according to Eq. (7.3). Figure 7.10 shows the measured curves $\sigma_r(f)$ of different A/D converters in four digital recorders [16–18, 22]. Curve ***1*** denotes the A/D converter of an 8-bit digital recorder, which was used as one of the first ones for high-voltage impulse measurements. Its standard error σ_r according to Eq. (7.5) steeply increases with frequency f, which at first sight indicates the particularly poor dynamic behavior of this A/D converter. However, at low frequencies, the A/D converter achieves approximately the low sampling error σ_i = 0.5 LSB of the ideal quantization. It is therefore well suited for measuring the peak value of AC and impulse voltages where the voltage changes relatively slowly. Curve ***2*** characterizes the A/D converter of a low-cost 8-bit digital oscilloscope with particularly high noise superposition. Therefore, it hardly differs from

Fig. 7.10 Standard deviation σ_r of the sampling errors of digital recorders versus frequency f ***1*** 8-Bit recorder, 100 MS/s ***2*** 8-Bit recorder, 50 MS/s ***3*** 6-Bit recorder, 500 MS/s ***4*** 8-Bit recorder, 200 MS/s

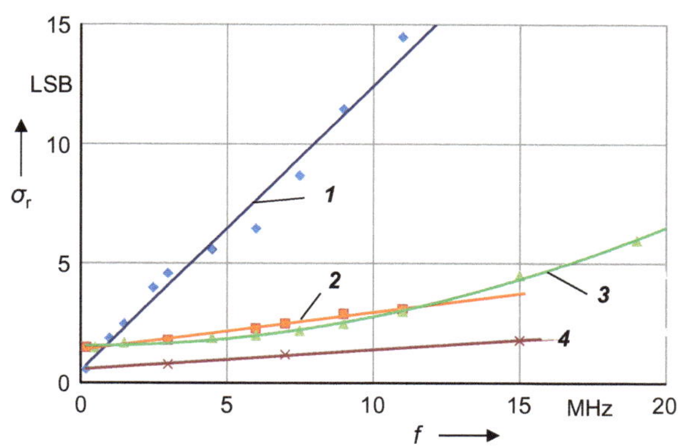

the curve ***3*** obtained for a 6-bit recorder. The best performance shows the A/D converter of a high-quality 8-bit recorder ***4*** which is also characterized by a large analogue bandwidth of 400 MHz.

7.2.5 Effective Number of Bits

Digital recorders are occasionally characterized by the *effective number of bits* according to the following equation [11, 16–22]:

$$EB = N - \log_2 \frac{\sigma_\mathrm{r}}{\sigma_\mathrm{i}} \tag{7.6}$$

with the standard deviations σ_i and σ_r according to Eq. (7.3) or Eq. (7.5). The A/D converter of the recorder is thus compared with an ideal A/D converter via their standard deviations. The plot of the *EB* number versus the logarithm of the sine frequency results in a typical characteristic that is reminiscent of the well-known frequency response of analogue measuring instruments. At low frequencies, the *EB* number of most recorders initially remains approximately constant and is slightly less than the rated resolution *N*. Above a certain frequency limit, the *EB* number decreases significantly with increasing frequency since the sampling error σ_r in Eq. (7.6) increases.

Figure 7.11 shows the *EB* characteristics of the four recorders ***1***, ***2***, ***3*** and ***4*** with their σ_r values plotted in Fig. 7.10. The *EB* values of the two 8-bit recorders ***1*** and ***2*** are noticeable due to their atypical characteristics. The *EB* values of curve ***1*** are approximately ideal at low frequencies up to 0.1 MHz, but curve ***1*** then decreases relatively quickly with increasing frequency and has only slightly less than 6 *EB* at 1 MHz. The *EB* curve ***2*** is also atypical, because the low-cost oscilloscope with a

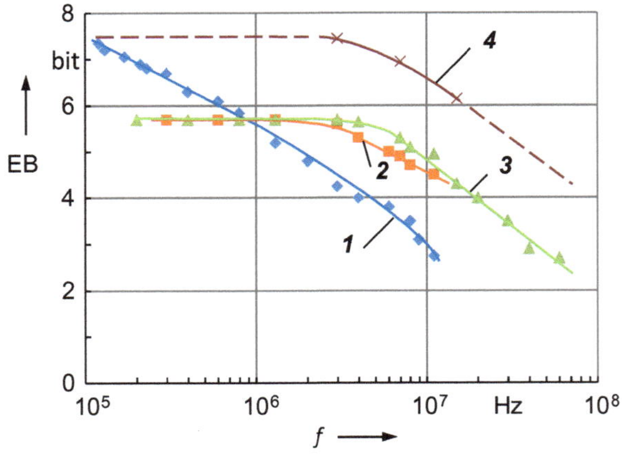

Fig. 7.11 Effective number of bits of different digital recorders versus frequency ***1***, ***2***, ***3***, ***4***: recorder as in Fig. 7.10

specified resolution of 8 bit apparently has an effective resolution of only 6 bits. As in Fig. 7.10, the 8-bit recorder with the *EB* curve ***4*** has the best performance.

The *EB* characteristic as well as the σ_r characteristic provide an informative overview of the dynamic behavior of the investigated A/D converter and allow a limited assessment of different recorders. However, both characteristics do not cover all frequency-dependent parameters that influence digital recorders. The sampling errors $\delta_{r,k}$, with which σ_r and thus also the *EB* number are calculated, refer to the mathematically fitted sine and not to the actual sinusoidal voltage applied to the recorder input. The transfer behavior of the input circuit is thus not taken into account. In addition, the σ_r or *EB* characteristics do not allow quantitative information about the dynamic measurement error of a recorder for an arbitrary signal, as is possible from the frequency response of analog instruments using the numerical convolution algorithm.

7.2.6 Signal Slope and Sampling Error

Looking closer at the sampling errors $\delta_{r,k}$ of most digital recorders with flash converters, it can be found that the largest errors $\delta_{r,k}$ occur during zero-crossing of the sinus voltage and the smallest errors $\delta_{r,k}$ in the amplitude region (see Fig. 7.9). The sampling errors are obviously dependent on the steepness of the input voltage. Sinusoidal voltages with slopes that vary between zero and a maximum are therefore not particularly suitable for testing the dynamic behavior of A/D converters. The basic behavior of A/D converters can be better examined with ramp or triangular voltages whose steepness is approximately constant over the entire voltage range. Due to the statistical evaluation, no particularly high requirement is placed on the linearity of the triangular voltage.

The evaluation of the recorded samples is comparable to that for sinusoidal voltages. The samples in the rising part of the triangular voltage are approximated by a straight line with the slope *S* and their deviations $\delta_{r,k}$ are determined at the sampling times $k\Delta t$. For the two 8-bit recorders ***1*** and ***2*** (see Fig. 7.10), the standard deviations $\sigma_r(S)$ are calculated using Eq. (7.5) and plotted versus slope *S* in Fig. 7.12. Each A/D converter has a characteristic curve $\sigma_r(S)$, which can be regarded as a mean error curve. In this representation, the two 8-bit recorders ***1*** and ***2*** again show a behavior different from that of a good recorder. For comparison, curve ***3*** in Fig. 7.12 represents the frequency-independent standard deviation σ_i 0.3 LSB calculated for ideal quantization according to Eq. (7.3). The advantage of knowing $\sigma_r(S)$ is that the slope-dependent errors of a digital recorder can be determined and corrected for any measured voltage value according to its steepness at the sampling time [17, 18, 22].

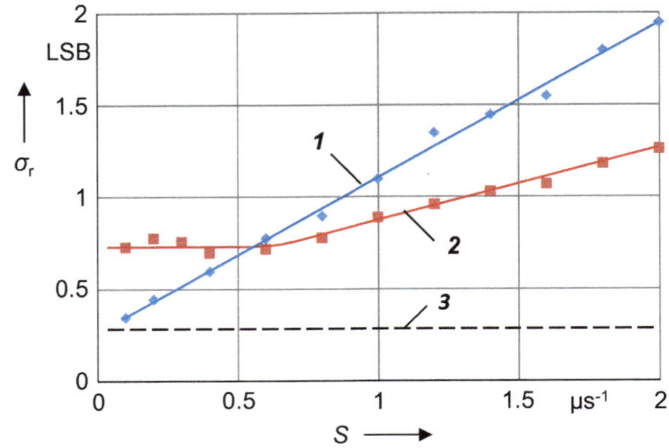

Fig. 7.12 Standard deviation σ_r of the sampling errors of 8-bit recorders versus ramp slope S ***1***, ***2*** recorder as in Figs. 7.10 and 7.11, ***3*** ideal 8-Bit quantization

7.2.7 Internal Noise and Jitter

Digital recorders with flash converters have the typical feature that the sampled values are scattered around the actual voltage curve by one or more quantization levels. The stored raw data is therefore superimposed with *randomly distributed noise* that occurs even during the sampling of a DC voltage, in addition to the *quantization noise* of less than 0.5 LSB with uniform distribution. With greater voltage steepness, the scattering is amplified due to the failure of individual levels (see Fig. 7.8). The randomized noise superimposed on the recorded voltage adversely affects the determination of the characteristic parameters of DC, AC or impulse voltages. An 8-bit recorder causes a typical noise superposition of three or more quantization levels, whereby the maximum value of the recorded voltage appears larger by 1–2%. Recorders with 10 bits and more show a noise superposition of mostly less than 0.5% of the maximum value, due to the smaller quantization steps. Furthermore, the superimposed noise may falsify the evaluation of the front time of impulse voltages. In the case of front-chopped impulse voltages that have only one or two samples at the peak, a reduction of the peak value is also possible.

With various methods also used to reduce superimposed front oscillations, the noisy raw data can be smoothed more or less successfully, thus reducing the influence of noise on the determination of the peak value and front time. Examples are: section-wise approximation of the raw data with parabolas or straight lines, *digital filtering* of the raw data within a certain data "window" or lowering the upper cut-off frequency. The effectiveness of different smoothing methods is investigated theoretically in (Ref. [9] of Chap. 4; 23). When using the standardized filtering method (see Sect. 4.1.1.2) for determining the test voltage value of an impulse voltage with superimposed peak oscillation, the noise is generally completely eliminated. Smoothing a single impulse voltage chopped in the front is particularly problematic because the peak must not be included. The peak will otherwise be reduced or increased depending on the smoothing method used (Ref.

[9] of Chap. 4). The effectiveness of the smoothing method can be checked with test impulses generated with the *Test Data Generator* (*TDG*) (see Sect. 7.3.1).

When calibrating digital recorders and other measuring instruments, a sequence of calibration voltages is applied and the mean is calculated. This averaging results in a significant reduction of the internal noise component. The allowable internal noise level of a recorder, which is determined by DC calibration, is 0.4% of the full scale deflection. Averaging of the records is particularly advantageous when calibrating with impulse or rectangular voltages, whereby, of course, a sufficiently high stability of the calibrator used is a prerequisite. Averaging also reduces the uncertainty in calibrating recorders, which, of course, is not achieved when recording a single impulse voltage.

Note The internal noise of a digital recorder with A/D flash converter seems to have a positive effect on its quantization characteristic. For each DC input voltage u_1, the corresponding output value $u_{2,k}$ is determined as the mean of a large number of samples. Because of the superimposed noise component, $u_{2,k}$ can therefore even have a value between two adjacent quantization levels. The quantization characteristic is then not staircase-like as in Fig. 7.5, but shows a rather steady increase. One gets the impression that there is a quasilinear relationship between the input voltage and the output value of the recorder. In other words, the effective amplitude resolution appears to be greater than the rated value.

The sampling frequency of the A/D converter is generated by an internal oscillator with limited stability. The sampling of a signal then does not always occur exactly at the predetermined times, but is subject to minor deviations. The scattering of the sampling times around their exact value is referred to as *jitter*, which would result in an amplitude error in the sampling of very fast signals. Scattering of digital recorders ranges from 10 ps to 100 ps and therefore need not be considered when measuring standard impulse voltages and currents.

7.2.8 Step Response of Digital Recorders

Digital recorders used in high-voltage measurement systems have input attenuators and preamplifiers that adjust the measurement voltage of up to 2000 V to the input of the A/D converter. The quality of a recorder is seen not only in high rated values for the amplitude and time resolution of the A/D converter but also in the exact adjustment of the analog part. This applies in particular to digital recorders whose input ranges must be balanced over a wide frequency range, including DC voltage. An imperfect balancing of the resistors and capacitors in the individual voltage ranges leads to deviations of the scale and time factors in the case of range switching. The manufacturer usually only specifies a maximum amplitude deviation, e.g. $\leq 1\%$ for DC voltage or AC voltage of 1 kHz. The deviation in linearity can be determined by detailed calibration of the individual voltage ranges. For higher accuracy requirements, the deviation is taken into account by correcting the scale factor.

In the preferred calibration method for recorders in impulse measuring systems, an accurate impulse calibrator is used which generates the standardized impulse voltages according to the test specifications (see Sect. 7.4). For each input range of the recorder, the measurement deviations for the peak value and the time parameters are determined. If an external attenuator is used, it must be adapted to the selected input range of the recorder [24]. Calibrating the relevant input ranges immediately before and after use of the recorder will provide information about the short-term stability. Regular control measurements of the scale factor during a year of operation provide information on the long-term stability.

The dynamic behavior of the individual input ranges of the recorder, which is particularly important for impulse voltage measurements, can be analyzed more precisely by its step response. For the generation of step voltages, circuits with mercury wetted reed contacts are suitable (see Sect. 9.8.4). The maximum step amplitude is, however, limited to 500 V or up to 1000 V for some reed contacts. If each individual input range is optimally balanced, the respectively recorded step response reaches its final values within 1 μs without large over- or undershoots. But even high-quality digital recorders show completely different step responses in the individual input ranges. For example, in the first 10 μs, an over- or undershoot of up to 2% may occur or the step response is creeping and only very slowly reaches its final value [11, 12, 25]. As a result, the peak value and the front time of impulse voltages are erroneously measured.

Figure 7.13 shows the step response of a well-balanced 8-bit recorder in two time ranges using a cable step generator with reed contact. The rise time of the recorder is 0.7 ns, which corresponds to the 400 MHz bandwidth specified by the manufacturer. Apart from minimum oscillations in the first 20 ns, which originate mainly from the step generator and the measuring circuit themselves, the form of

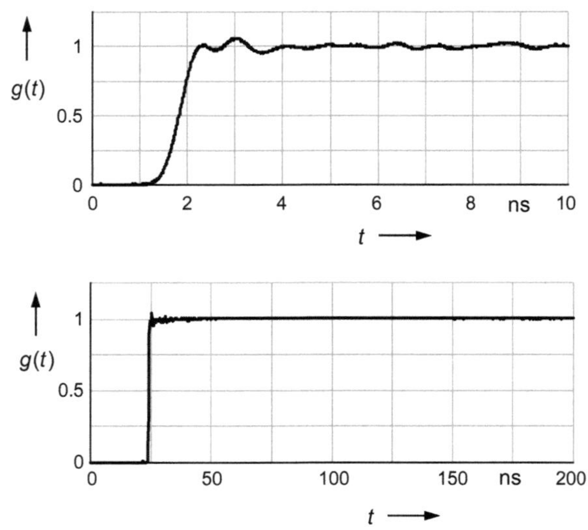

Fig. 7.13 Step response $g(t)$ of a well-balanced 8-bit recorder with 400 MHz bandwidth in two different time ranges

the step response is almost ideal. For input voltages of more than 100 V, the recorder is operated with an external probe, which is matched to the recorder input by means of the recorded step response.

The polarity of the input voltage may also have some influence if the preamplifier of the digital recorder is not properly balanced for positive and negative input voltages [26]. A possible polarity effect is determined by applying positive and negative impulses generated by an impulse calibrator (see Sect. 7.4) or a step generator (see Sect. 9.8.4). If the calibration is performed with step voltages, the direction of the step and not the polarity of the applied voltage is decisive. For example, if a positive DC voltage is applied and short-circuited, a negative voltage step occurs.

7.2.9 Electromagnetic Interference

When firing the spark gaps of an impulse voltage generator, high electromagnetic fields are generated, which can cause interference voltages in an insufficiently shielded measuring circuit [27]. The disturbances are particularly noticeable as high-frequency superimposition in the initial part of the recorded impulse voltage and can thus falsify the front time. In the case of chopped impulse voltages, the region just before and after the chopping instant is additionally affected, so that the peak value is influenced. Furthermore, electric flashovers or breakdowns at AC voltage can cause particularly high current discharges.

Digital recorders and other measuring instruments that have been specially developed for use in high-voltage measurement systems are usually effectively protected against direct electromagnetic interference. Protection includes not only the outer shielding of the housing, but also additional measures already taken care of in the preliminary development phase of the individual components. In the early days of digital high-voltage measurements, mainly recorders developed for the low-voltage range were used, which had to be protected by accommodating them in a shielded cabin and filtering the power supply. The disturbing effect on the recorded impulse voltage can then be demonstrated particularly impressively if the door of the shielded cabin is only slightly opened.

Interference effects on the measurement cable leading from the voltage divider to the recorder are prevented by additional cable shielding, avoiding ground loops and other measures (see Sect. 4.3.1.7). The additional cable shield, which is connected to the shielded housing of the recorder, is indispensable for high-current measurements. The high-frequency magnetic field generated by an AC current or impulse current induces a voltage in the ground loop, which drives a corresponding current through the outer cable shield and the shielded housing of the recorder. This creates a magnetic opposing field that compensates for the primary field and thus prevents interference with the measured voltage (see Sect. 5.3.1.1).

7.3 Software for Data Analysis

Evaluation of the digitally stored data is done with software either supplied by the manufacturer of the measuring instrument or developed by the user himself. The focus here is on the standardized determination of the value of the test voltage and of additional parameters from the stored raw data. The calibration of the measuring instrument with an accurate calibrator for DC, AC or impulse voltages indirectly already includes a certain check of the evaluation software. The parameter values computed with software during calibration should correspond largely to the set calibrator values. However, the time courses of the calibration voltages are generally rather smooth, while the voltages and currents recorded during high-voltage and high-current tests are superimposed by noise and oscillations.

Special attention must therefore be paid to the data processing using software. For example, software for evaluating lightning impulse voltages must recognize the oscillation superimposed on the peak and its frequency and be able to calculate the value of the test voltage effective for the insulation (see Sect. 4.1.1.2). For the evaluation of oscillating impulse voltages during on-site tests, other requirements apply to the evaluation software. During impulse current tests, further parameters such as the impulse charge and the reversal peak in the case of undershoot are to be determined. Not only hardware, i.e. the measuring instrument, but also software has to be thoroughly tested. In every case, the recorded raw data and the result values of a measurement must remain stored in order to be able to verify the data evaluation at a later time if necessary.

7.3.1 *Software Verification with the TDG*

The specifications for high-voltage and high-current tests in (Refs. [1–4] of Chap. 2) as well as software tests in [1] generally have no direct requirements with regard to the use of specific methods, measurement instruments or algorithms for data processing. An exception concerns the evaluation of impulse voltages with superimposed peak oscillations, which is to be carried out using a standardized filtering method (see Sect. 4.1.1.2). With the exception of this filtering method, the user has extensive freedom in the choice of software used for data evaluation. The algorithms of the software implemented in the recorder are not subjected to any more or less complex individual assessment or approval. The validity of the evaluation software is verified with data sets provided by the *Test Data Generator* (*TDG*) for a variety of AC, DC and impulse voltages and currents. The parameter values determined with the evaluation software must correspond within specified limits with the reference values of the TDG.

A first approach to this pragmatic procedure took place in the framework of an international comparison. A floppy disk with invariable data sets of calculated and measured impulse voltages and step responses was distributed to each participating laboratory. Computation of the test data was carried out by each participant with software that was available in the laboratories in the early days of digital measurement technique. The result of the comparison revealed unacceptably large deviations of some of the evaluated parameter values from the reference values [28].

A significant improvement is the TDG, which is part of IEC Publication 61083 [1]. The TDG is software that allows the user to generate the data of selected voltages and currents with the aim of testing the evaluation software of his recorder. These test data originate either from calculated or measured voltages and currents. For example, the first TDG version that has been recently revised generates the test data of full and chopped lightning impulse voltages, switching impulse voltages and impulse currents. The special feature of the TDG is that the user can choose the resolution, sampling rate and superimposed noise according to the characteristics and data format of the recorder he is using.

The revised TDG also contains data sets with impulse voltages that are typical of on-site tests and transformer tests. Furthermore, the TDG data collection contains impulse voltages with superimposed peak oscillations whose test voltage value is to be calculated with the *test voltage function* $k(f)$ (see Sect. 4.1.1.2.3). As digital recorders are increasingly being used also to measure AC and DC voltages and the corresponding currents, the revised TDG is supplemented with the corresponding AC and DC test data [1, Part 4]. The test data are exemplary for AC voltages with harmonics and DC voltages with superimposed ripple [29].

For all TDG test voltages and currents, reference values with upper and lower limits are available for each standardized parameter. The reference values are based either on calculations of analytically given voltages and currents or on mean values resulting from international comparisons with several participating laboratories [30, 31]. It is the responsibility of the user to select those TDG data sets that are typical for the high voltages and currents to be measured in his laboratory and to determine the parameters with the evaluation software implemented in his recorder. If the parameter values agree with the reference values within the specified tolerances, the evaluation software is accepted as passed. From the tolerances for the reference values, the standard uncertainty of the evaluation software can be estimated, which contributes to the measurement uncertainty of the complete measuring system.

7.4 Calibrators

Digital recorders and other instruments for measuring AC, DC and impulse voltages can be calibrated in a number of ways. Analog calibrators with sufficiently high accuracy are preferably used to determine the scale factor and the time parameters. Very accurate *calibrators* for AC and DC voltages up to 1000 V have been available for a long time. The calibration of the previously used analog impulse oscilloscopes was also done with AC voltage. For the calibration of digital recorders for impulse measurements, IEC 61083-1 describes two methods [1]. In the one calibration method, the input ranges of the recorder under test are calibrated with step voltages that can be generated in the circuit of Fig. 9.17. The recorded data of the step response is evaluated within a specified time range. This method has become less importance since impulse calibrators are available that generate double exponential lightning and switching impulse voltages. Even chopped impulse voltages with variable times to chopping down to 0.5 μs, impulse currents and step voltages are available. The time parameters of the double exponential calibration impulses are the specified limits for the shortest front time and the longest time to half-value of full impulse voltages. Depending on the impulse shape and the manufacturer, the peak value of the calibrator can be increased from approx. 100 V in steps of 1 V to a maximum of several 100 V or even 2000 V.

The principle of an analog impulse calibrator is comparable to that of the basic circuit of impulse voltage generators in Fig. 4.10 (or impulse current generators in Fig. 5.3). Often the spark gap is replaced by a thyristor, another electronic component or a mercury-wetted reed relay. The capacitor C_s is charged to the predetermined voltage value and then quickly discharged to the RC element which determines the impulse shape. At the output of the calibrator, a double exponential impulse voltage is generated, whose peak value is determined in a first approximation by the product from the preset DC voltage and the utilization factor of the circuit. In addition to the impulse voltages mentioned above, some standardized current proportional voltages can be generated. Due to the non-linearity of electronic switches, the utilization factor of the calibrator may be voltage dependent, which is particularly noticeable at small peak values of less than 100 V. However, the non-linearity of the calibrator is largely compensated by the internal control unit.

Calibrators of this type are primarily designed to calibrate recorders with an input impedance of at least 1 MΩ and not more than 50 pF. If the load at the calibrator output is larger, e.g. due to the capacitance of a longer coaxial cable to the recorder, oscillations may occur in the beginning of the calibration impulse. This makes the determination of the point at 0.3 $\hat{u}$ and thus the front time of the recorded impulse voltage more difficult. Calibrators adapted to the recorder of the same manufacturer enable fully automatic or at least semi-automatic calibrations in all measurement ranges. High stability of the calibrator is essential, as complete calibration of a recorder can take several hours for all impulse shapes, measurement ranges and amplitudes. Figure 7.14 shows different types of impulse calibrators from three manufacturers.

In addition to the analog calibrators that generate some standardized impulse voltages, programmable *function generators* are also used to generate arbitrary waveforms [32]. The desired waveform is entered as an equation or a data table and converted to the corresponding analog output voltage by a D/A converter. The output resistance of these function generators is usually 50 Ω. It is hence so small in comparison to the input resistance of the recorder to be tested that the voltage shape is practically unaffected. However, due to the low output voltage of the order of 10 V, only the direct input of the recorder without the attenuator can be calibrated. With the programmable function generator, the impulse shape, e.g. the front time of a lightning impulse voltage, can be varied as desired. This allows the recorder's dynamic behavior to be studied in the area of interest in more detail.

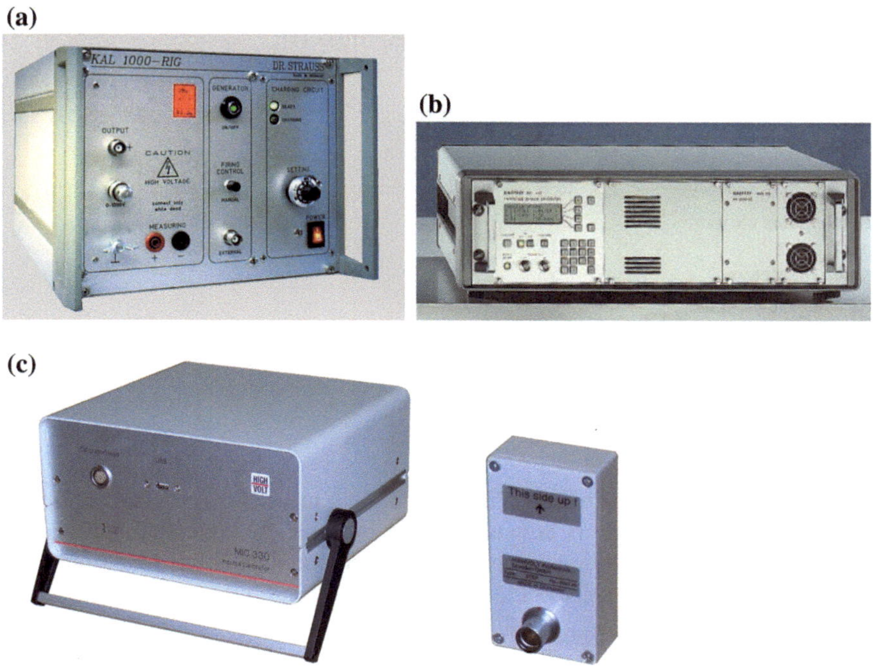

Fig. 7.14 Impulse calibrators from different manufacturers, **a** step voltage calibrator up to 1000 V (DR. STRAUSS Messtechnik GmbH), **b** impulse calibrator up to 1600 V for all standardized impulse shapes (HAEFELY TEST AG), **c** impulse calibrator up to 330 V (basic unit and individual calibrator head for an impulse shape) (HIGHVOLT Prüftechnik Dresden GmbH)

The development of digital recorders with ever greater amplitude and time resolution require calibrators with correspondingly high accuracy [33]. A very accurate impulse calibrator up to 300 V consists of a mercury-wetted reed contact as a switch and precisely dimensioned circuit elements for impulse shaping [34, 35]. The output voltage of the circuit with high-quality components is calculated taking into account the stray capacitances and line inductances. The input impedance of the digital recorder and the capacitance of the coaxial cable to the calibrator are also included in the calculation. This *"calculable" impulse calibrator* promises uncertainties of 0.05–0.14% for the peak value of the generated calibration impulses and less than 0.5% for the time parameters. A powerful variant of the calculable impulse calibrator with MOSFET switch and low output impedance is suitable for accurate calibration of recorders up to 1000 V and even impulse voltage dividers. Further circuit variants with reed contacts and MOSFET switches are dealt with in [36, 37].

The calibrator is an important link in the *chain of traceability* of high DC, AC and impulse voltage and current measurements to the internationally defined *SI units* for voltage and time. The verification of the calibrator properties in an accredited calibration laboratory ensures traceability to the national measurement

Table 7.1 Compilation of the permissible uncertainties ($k = 2$) for the calibration of measuring instruments and reference calibrators according to IEC 61083 [1]

Type of voltage/parameter	Instrument in approved system	Instrument in reference system	Reference calibrator
Impulse voltage:[a]			
peak value	2% (3%)[b]	0.7% (2%)[b]	0.5% (1%)[b]
time parameters	5%	3%	2%
Rectangular current:			
peak value	2%	0.7%	2%
time parameters	5%	3%	2%
AC and DC voltages:[a]			
value of the test voltage	1%	0.7%	0.5%
time and frequency parameters	1%	0.7%	0.5%

[a]values also apply to currents
[b]front-chopped lightning voltage

standards with low uncertainty. The permissible measurement uncertainties of accurate reference calibrators are summarized in Table 7.1. International comparison measurements on calibrators provide the basis for standardized measurements of high voltages and currents [38, 39].

7.5 Uncertainty Requirements

The calibration procedures and requirements for digital measuring instruments, evaluation software and *reference calibrators* used in high-voltage and high-current testing are laid down in a series of IEC 61083 [1]. Part 1 of the test specifications deals with the calibration of digital recorders and Part 2 with the test of the evaluation software for impulse voltages and currents using the TDG (see Sect. 7.3.1). For AC and DC voltages, Part 3 describes the calibration of the digital measuring instruments and Part 4 the software test with the TDG. From the results of the calibrations, the uncertainties of the digital instruments are estimated in a similar manner as shown for the complete high-voltage measuring systems in IEC 60060-2. The maximum permissible uncertainties of the measuring instruments used in approved measuring systems and reference systems are summarized in Table 7.1.

Note The permissible measurement uncertainties for the complete high-voltage measuring system are given for tests with AC voltage in Table 2.1, with DC voltage in Table 3.1 and with impulse voltage in Sect. 4.1.1.1.

For instruments in approved impulse measuring systems, additional requirements apply to the allowable deviations of the measured time parameters from the default nominal values of the calibrator. For most time parameters of impulse voltages and currents, the permissible deviations are ±3%, with the exception of the time to half-value (±2%) and time to chopping (±5% for $T_c < \mu s$).

Note Analog measuring instruments are no longer included in IEC 61083.

In addition to the uncertainties given in Table 7.1, additional requirements on individual parameters must be met. For example, the amplitude resolution of digital recorders for impulse voltage measurements shall be at least $N = 8$ bits. For tests which involve spectrum analysis or waveform transformation, $N = 9$ bits or more are recommended. In test practice, the input voltage of digital recorders is generally lower than the voltage required for full scale deflection. The actual resolution is then less than the rated resolution N. Therefore, a minimum input voltage is required which must not be less than $4/N$ for a recorder in an approved measuring system ($6/N$ for a recorder in a reference system). For an 8-bit recorder, this means that the input voltage in the selected range shall be at least 50% (75%) of full scale deflection.

The required sampling rate depends on the time T_{AB} of the impulse front (see Fig. 4.1) and shall not be less than $30/T_{AB}$. For lightning impulse voltages with the shortest front time $T_1 = 0.84$ µs, the required sampling rate is at least 60 MS/s. This rather high sampling rate is necessary in particular for the evaluation of the front time and a possibly superimposed oscillation.

The amplitude resolution of digital measuring devices for DC and AC voltages depends on the required uncertainty of the respective measurement quantity. The minimum sampling rate of digital recorders for AC voltage measurements results from the quotient n/T, where n is the number of samples per period of the highest harmonic to be measured and T is the period duration. The sampling rate also depends on the required measurement uncertainty u_{SR} for the highest harmonic. The requirement $n \geq \pi/\arccos(1 - u_{SR})$ applies to the number n of samples. For example, the required sampling rate of a 50 Hz voltage including the 10th harmonic is therefore at least 12 kS/s.

The rise time of measuring devices for AC and DC voltages shall not exceed $1/(18\,f_{max})$, where f_{max} is the highest frequency of the harmonic to be measured with an uncertainty of not more than 1%. Since the value of a parameter obtained by calibration is given as a mean value and the required sampling rates are not very high, the requirements for AC and DC voltages will pose no problems.

References

1. IEC 61083: Instruments and software used for measurements in high-voltage and high-current tests—Part 1: Requirements for hardware for impulse tests (2001)—Part 2: Requirements for software for impulse tests (2013)—Part 3: Requirements for digital instruments for tests with alternating and direct voltages and currents (draft)—Part 4: Requirements for software for tests with alternating and direct voltages and currents (draft)
2. Tränkler, H.-R.: Messtechnik und Messsignalverarbeitung – Teil C: Digitale Messtechnik. tm **53**, 433–438 and 470–474 (1986)
3. Tietze, U., Schenk, C., Gamm, E.: Electronic Circuits. Springer, Berlin (2008)
4. Myamoto, T., Takami, T., Tanaka, T., Sakaguchi, I., Kawaguchi, F.: Real time central data acquisition and analysis system for high voltage transients. IEEE Trans. IM **24**, 379 (1975)

5. Wiesendanger, P.: Automatische, digitale Aufzeichnung und Auswertung von transienten Signalen in der Hochspannungstechnik. Thesis ETH Zürich (1977)
6. Malewski, R.: Digital techniques in high-voltage measurements. IEEE Trans. PAS **101**, 4508–4517 (1982)
7. Strauss, W.: Performance progress and calibration of digital recorder for impulse voltage tests. In: Proceedings of 13th ISH Delft, paper 620 (2003)
8. Steiner, T., Böhme, F.: Applications and necessary requirements for digital recorders in the field of high voltage and high current testing. In: Proceedings 15th ISH Ljubljana, paper T10–741 (2007)
9. Cerqueira, W.R., Oliveira, O.B., Nerves, A.S., Chagas, F.A.: Capacitive and resistive attenuators for HV impulse measuring systems. In: Proceedings of 9th ISH Graz, paper 4461 (1995)
10. Mannikoff, A., Bergmann, A.: High impedance passive impulse voltage attenuator for 4 kV and 4 ns. In: Proceedings of 14th ISH Beijing, paper B-66 (2005)
11. Schon, K., Korff, H., Malewski, R.: On the dynamic performance of digital recorders for HV impulse measurement. In: Proceedings of 4th ISH Athen, paper 65.05 (1983)
12. Gobbo, R., Pesavento, G., Cherbaucich, C., Rizzi, G.: Digitizers for impulse voltage reference measuring systems. In: Proceedings of 9th ISH Graz, paper 4516 (1995)
13. Ihlenfeld, G.: Messung der integralen dynamischen Nichtlinearität hochauflösender Analog-Digital-Umsetzer. PTB-Jahresbericht 2004, Forschungsnachrichten der Abt. 2
14. Gobbo, R., Pesavento, G.: Procedure for the check of integral nonlinearity of digitizers. In: Proceedings of 12th ISH Bangalore, paper 7–8 (2001)
15. Lucas, W.: Modernisierte Messeinrichtung für Hochspannungsimpulse. PTB-Jahresbericht 2004, Forschungsnachrichten der Abt. 2. www.ptb.de
16. Korff, H., Schon, K.: Digitalisierungsfehler von Transientenrecordern mit hoher Abtastrate. PTB-Bericht E-28, 27–32 (1986)
17. Korff, H., Schon, K.: Digitization errors of fast digital recorders. IEEE Trans. IM **36**, 423–427 (1987)
18. McComb, T.R., Kuffel, J., Malewski, R.: Measuring characteristics of the fastest commercially-available digitizers. IEEE Trans. PWRD **2**, 661–670 (1987)
19. Kuffel, J., Malewski, R., van Heeswijk, R.G.: Modeling of the dynamic performance of transient recorders used for high voltage impulse tests. IEEE Trans. PWRD **6**, 507–511 (1991)
20. Germain, A.M., Ribot, J.J., Bertin, D., Spangenberg, E.: Practical tests to evaluate the accuracy of digitizer used for H.V. measurement. In: Proceedings of 5th ISH Braunschweig, paper 72.04 (1987)
21. Malewski, R.A., McComb, T.R., Collins, M.M.C.: Measuring properties of fast digitizers employed for recording HV impulses. IEEE Trans. IM **32**, 17–22 (1983)
22. Schon, K.: Test methods for the dynamic performance of fast digital recorders. Proc. NATO ASI Series E Appl. Sci. Fast Electr. Opt. Meas. **1**, 453–466 (1986)
23. McComb, T.R., Li, Y.: The contribution of software to the uncertainty of calibrations of calibration pulse generators. In: Proceedings of 12th ISH Bangalore, paper 7–7 (2001)
24. Schneider, G.A., Ladwig, G.: Design of a high voltage recorder to IEC Publication 1063. In: Proceedings 9th ISH Graz, paper 4931 (1995)
25. Gobbo, R., Pesavento, G., Bolognesi, F., Rizzi, G.: Accuracy assessment of digitizers under impulse conditions. In: International Symposium of Digital Techniques in HV Measurements, Toronto, pp. 1-9–1-13 (1991)
26. Rungis, J., et al.: Use of low voltage calibrators in impulse voltage measurement. Electra **189**, 83–109 (2000)
27. Malewski, R., Dechamplain, A.: Digital impulse recorder for high-voltage laboratories. IEEE Trans. PAS **99**, 636–649 (1980)
28. Schon, K., Lucas, W.: Intercomparison of software for evaluating high voltage impulses and step responses. BCR Report EUR 15260 EN, Brüssel (1993)

29. Rahimbakhsh, M., Gockenbach, E., Werle, P.: Verification of a new developed algorithm for the evaluation of AC and DC signals using the IEC Test Data Generator. In: Proceedings of 19th ISH Pilsen, paper 276 (2015)
30. Schon, K. et al.: International comparison of software for evaluating HV impulses and step responses. In: Proceedings of 8th ISH Yokohama, paper 51.01, pp. 289–292 (1993)
31. Cherbaucich, C. et al.: IEC Test data generator for testing software used to evaluate the parameters of HV impulses. In: Proceedings of 9th ISH Graz, paper 4494 (1995)
32. Beyer, M., Schon, K.: Calibration of digital recorders for hv impulse measurement. In: Proceedings of 7th ISH Dresden, paper 62.02 (1991)
33. Li, Y., Sheehy, R., Rungis, J.: The calibration of a calculable impulse voltage calibrator. In: Proceedings of 10th ISH Montreal, pp. 45–49 (1997)
34. Hällström, J.: A calculable impulse voltage calibrator. Diss. Acta Polytech. Scand. Elect. Eng. Ser. No. 109 (2002)
35. Hällström, J., Chekurov, Y., Aro, M.: A calculable impulse voltage calibrator for calibration of impulse digitizers. IEEE Trans. IM **52**, 400–403 (2003)
36. Sheehy, R., Li, Y.: An impulse voltage calibrator of low output impedance. IEEE Trans. PD **20**, 15–22 (2005)
37. Sato, S., Harada, T.: Evaluation of potential divider's performance using lightning impulse voltage calibrator for low-impedance load. In: Proceedings of 14th ISH Beijing, paper J-15 (2005)
38. Hällström, J., Li, Y., Lucas, W.: High accuracy comparison measurement of impulse parameters at low voltage levels. In: Proceedings of 13th ISH Delft, paper 432 (2003)
39. Wakimoto, T., Hällström, J., Chekurov, Y., Ishii, M., Lucas, W., Piiroinen, J., Shimizu, H.: High-accuracy comparison of lightning and switching impulse calibrators. IEEE Trans. IM **56**, 619–623 (2007)

Chapter 8
Representation of Impulses in the Time and Frequency Domain

Abstract Single and continuous signals can be represented by their shape in the time domain or by their spectrum in the frequency domain. Both forms of representation are equivalent. Which form is preferred in a particular case depends on the measurement task and the specified target. Both the waveform and the spectrum can be used to derive requirements for the correct measurement of a signal. High-voltage impulses and high-current impulses are defined by their waveforms, which are characterized by the value of the test voltage – usually the peak value – and two time parameters. On the other hand, low-voltage measuring instruments, including analog oscilloscopes and digital recorders that are not specifically built to measure impulse voltages and currents, are more characterized by parameters in the frequency domain such as frequency response and bandwidth. In this chapter, the spectra as well as the time courses of some ideal impulse voltages and currents are calculated using the Laplace transform. This allows a statement as to whether the transfer behavior of the measuring instrument is suitable for the measurement task.

Single and continuous signals can be represented by their shape in the *time domain* or by their spectrum in the *frequency domain*. Both forms of representation are equivalent. Which form is preferred in a particular case depends on the measurement task and the specified target. Both the waveform and the spectrum can be used to derive requirements for the correct measurement of a signal. High-voltage impulses and high-current impulses are defined by their waveforms, which are characterized by the *value of the test voltage*—usually the peak value—and two time parameters. For the measurement of these parameters, uncertainties are prescribed which must be observed by an approved measuring system, the suitability of which has preferably been proven by comparison with a *reference system* in the time domain. On the other hand, low-voltage meters, including analog oscilloscopes and digital recorders that are not specifically built to measure impulse voltages and currents, are more characterized by parameters in the frequency domain such as frequency response and bandwidth. For this reason, the spectra of some impulse voltages and currents are also treated in this chapter in addition to the time courses. This allows a statement as to whether the transfer behavior of the measuring instrument is suitable for the measurement task.

K. Schon, *High Voltage Measurement Techniques*, Power Systems,
https://doi.org/10.1007/978-3-030-21770-9_8

8.1 Analytical Representation of Impulse Voltages

Lightning and switching impulse voltages generated by the two basic circuits in Fig. 4.10 can be approximately represented in analytical form. In this case, the stray capacitances and lead inductances are ignored and the spark gaps are assumed to be ideal switches. For the generator circuit A (see Fig. 4.10a), two basic calculation methods are given. After firing of the spark gap FS at time $t = 0$, the capacitor C_s charged to U_0 discharges through the circuit consisting of R_d, R_e and C_b. The discharge current i_d is divided into $i_{e.}$ and i_b:

$$i_d(t) = i_e(t) + i_b(t).$$

The loop equation established for the generator circuit A is:

$$U_0 - \frac{1}{C_s}\int_0^t i_d \mathrm{d}t = i_d R_d + u(t). \tag{8.1}$$

This equation leads to a homogeneous second order differential equation that can then be solved with the known solution approaches.

The *Laplace transform* provides an alternative approach (see Appendix A.2). Applying the rules and correspondences in Tables A.1 and A.2 to Eq. (8.1) yields the following equation in the *complex variable domain*:

$$\frac{U_0}{s} - \frac{1}{sC_s} I_d = I_d R_d + U, \tag{8.2}$$

where $s = \sigma + j\omega$ is a complex number. With:

$$I_d = I_e + I_b = U/R_e + C_b U s$$

and the time constants τ_1 and τ_2, Eq. (8.2) can be solved for $U = U(s)$:

$$U(s) = \frac{U_0}{R_d C_b} \frac{1}{(1 + s/\tau_1)\,(1 + s/\tau_2)}. \tag{8.3}$$

After inverse transform into the time domain with $s = j\omega$ and the correspondences in Table A.2, the impulse voltage $u(t)$ is obtained as the difference of two exponential functions:

$$\boxed{u(t) = \frac{U_0}{R_d C_b} \frac{\tau_1 \tau_2}{\tau_1 - \tau_2} \left(e^{-t/\tau_1} - e^{-t/\tau_2}\right).} \tag{8.4}$$

The two time constants τ_1 and τ_2 are roots of a quadratic equation with:

$$\frac{1}{\tau_{1,2}} = \frac{B_0}{2} \mp \sqrt{\left(\frac{B_0}{2}\right)^2 - B_1} \quad (8.5a)$$

$$B_0 = \frac{1}{R_d C_b} + \frac{1}{R_d C_s} + \frac{1}{R_e C_b} \quad (8.5b)$$

$$B_1 = \frac{1}{R_d C_b R_e C_s}. \quad (8.5c)$$

In the same way, the output voltage of circuit B in Fig. 4.10b can also be calculated. For $u(t)$, we obtain a largely identical expression as in Eq. (8.4) with the only difference that C_s replaces C_b in the third term on the right side of Eq. (8.5b).

For the standardized lightning and switching impulse voltages, $R_e C_s << R_d C_b$. Then, the time constants τ_1, τ_2 and the efficiency η of the circuit A (index "A") can be approximately calculated as follows (Refs. [1, 6] of Chap. 1, Refs. [16, 17] of Chap. 4):

$$\tau_{1,A} \approx (R_d + R_e)(C_b + C_s), \quad (8.6a)$$

$$\tau_{2,A} \approx \frac{R_d R_e}{R_d + R_e} \frac{C_b C_s}{C_b + C_s}, \quad (8.6b)$$

$$\eta_A = \frac{\hat{u}}{U_0} \approx \frac{R_e}{R_d + R_e} \frac{C_s}{C_b + C_s}. \quad (8.6c)$$

Accordingly, for the time constants and the efficiency of the impulse voltage generator in circuit B (index "B") we have:

$$\tau_{1,B} \approx R_e(C_b + C_s), \quad (8.7a)$$

$$\tau_{2,B} \approx R_d \frac{C_b C_s}{C_b + C_s}, \quad (8.7b)$$

$$\eta_B = \frac{\hat{u}}{U_0} \approx \frac{C_s}{C_b + C_s}. \quad (8.7c)$$

The inadequacy of the two idealized basic generator circuits in Fig. 4.10 has already been pointed out in Sect. 4.2.1. Taking into account the inductances and stray capacitances of the impulse voltage generator and the test object, an analytical calculation of the complete equivalent circuit is practically impossible. More suitable for this purpose is software that is either generally available for the calculation of electrical circuits or especially for the optimization of impulse voltage generators (Ref. [18, 22] of Chap. 4).

For theoretical investigations regarding the shape and spectrum of lightning and switching impulse voltages or the transfer behavior of voltage dividers, the following equation derived from Eq. (8.4) is better suited:

$$\boxed{u(t) = \hat{u}A\left(e^{-t/\tau_1} - e^{-t/\tau_2}\right).} \tag{8.8}$$

Here, $A \geq 1$ is a factor with which the two exponential functions are to be multiplied so that the impulse voltage reaches its peak value $\hat{u}$ at the time to peak t_p (Fig. 8.1). For lightning and switching impulse voltages, $\tau_1 \gg \tau_2$ applies. This means that the waveform of the impulse voltage in the tail is mainly determined by the exponential term with τ_1 in Eq. (8.8).

Setting the first derivative of Eq. (8.8) to zero gives the time to peak t_p:

$$\boxed{t_p = \frac{\tau_1 \tau_2}{\tau_1 - \tau_2} \ln \frac{\tau_1}{\tau_2}.} \tag{8.9}$$

The symbols for the time to peak and the other time parameters that can be calculated from Eq. (8.8) are written in lower case. This is in contrast to the capitalization of the standardized time parameters in IEC 60060 (Refs. [1, 2] of Chap. 2), which are referred to the virtual origin of lightning impulses (Fig. 4.1). With the time to peak t_p according to Eq. (8.9), the factor A can be calculated as:

$$\boxed{A = \frac{1}{e^{-t_p/\tau_1} - e^{-t_p/\tau_2}}.} \tag{8.10}$$

For the analytical representation of an impulse voltage with the time parameters T_1 and T_2, the corresponding time constants τ_1 and τ_2 in Eq. (8.8) must be known. The relationship between the time constants and the time parameters of the impulse voltage is determined numerically today, whereas previously τ_1 and τ_2 were determined by means of diagrams. The procedure for determining the time

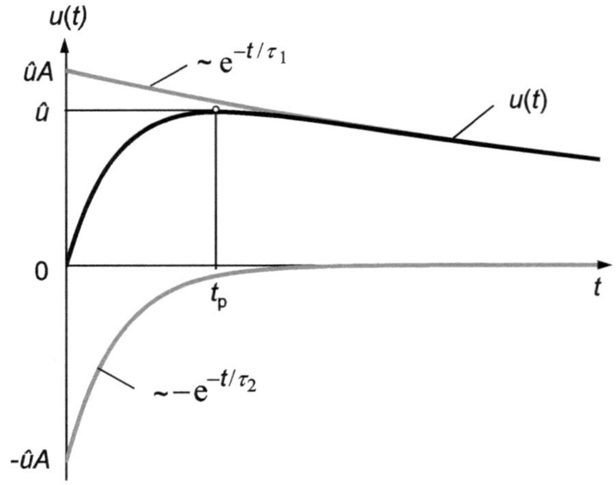

Fig. 8.1 Representation of an impulse voltage $u(t)$ as superposition of two exponentials with the time constants τ_1 and τ_2 according to Eq. (8.8)

constants τ_1 and τ_2 of an impulse voltage according to Eq. (8.8) for a predetermined value pair of the time parameters T_1 und $T_2 \gg T_1$ will be briefly outlined below. The first step is to estimate an initial value for τ_1. Since in the tail of lightning and switching impulse voltages the second exponential term in Eq. (8.8) with the time constant τ_2 is negligible, Eq. (8.8) simplifies for the time to half-value $t_2 \approx T_2$:

$$u(t = T_2) = \hat{u}A\mathrm{e}^{-T_2/\tau_1} = 0,5\hat{u},$$

from which with $A \approx 1$ a first approximate value is obtained:

$$\tau_1 \approx \frac{T_2}{\ln(2A)}. \tag{8.11}$$

With the estimate $\tau_2 \approx T_1/3$, a first data set for the impulse voltage can be calculated using Eq. (8.8). In further iterative steps, improved values are calculated with Eqs. (8.9)–(8.11) and inserted in Eq. (8.8) until the time parameters T_1 and T_2 reach the prescribed nominal values.

8.1.1 Mathematical and Virtual Origin

The standardized virtual origin O_1 of a lightning impulse voltage (see Sect. 4.1.1) obviously differs from the mathematical origin O of the impulse defined by Eq. (8.8). For higher accuracy requirements, this difference may need to be considered e.g. when comparing a measured impulse voltage with the result of the convolution calculation (see Sect. 9.7). The difference between O and O_1, which of course depends on the time course of the considered lightning impulse voltage, is designated here by t_0. It is calculated according to the intercept theorem for the straight line through the points at $0.3\hat{u}$ and $0.9\hat{u}$ (Fig. 8.2a):

$$\boxed{t_0 = \mathrm{O} - \mathrm{O}_1 = 0.3\hat{u}\frac{T_1}{\hat{u}} - t_{30}(\mathrm{O}) = 0.5T_{\mathrm{AB}} - t_{30}(\mathrm{O})} \tag{8.12}$$

where $t_{30}(\mathrm{O})$ is referred to the mathematical origin of the lightning impulse voltage. For the 1.2/50 lightning impulse voltage, the virtual origin O_1 is about t_0 = 0.22 µs before O. Therefore, the time to half-value T_2, which is referred to O_1 and has the nominal value 50 µs, is around 0.5% greater than the value calculated for O. This small difference is certainly negligible in most cases. The time difference t_0 has no influence on the front time defined by Eq. (4.2).

For lightning impulse voltages chopped in the front, it may be necessary to consider the time difference between the origins O and O_1. For the 1.2/50 lightning impulse voltage chopped after $t_c(\mathrm{O})$ = 500 ns, the virtual origin O_1 is t_0 = 61 ns before O (Fig. 8.2b). The time to chopping relative to the virtual origin O_1 is therefore T_c = 561 ns, which is about 12% larger than 500 ns. On the other hand, if

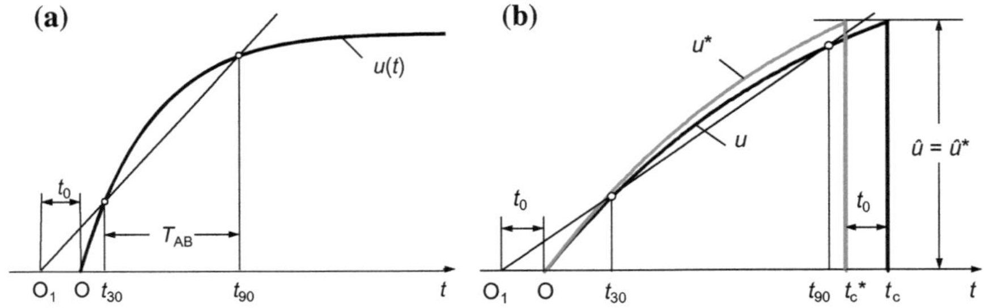

Fig. 8.2 Mathematical origin O and virtual origin O_1 of lightning impulse voltages. **a** Full lightning impulse voltage, **b** Front-chopped lightning impulse voltage

$T_c = 500$ ns is set and the lightning impulse voltage is calculated until $t = T_c - t_0 = 439$ ns according to Eq. (8.8), the peak value is only about 93% of the original value $\hat{u}$. To get the original peak value $\hat{u}$ again, the impulse voltage must increase more steeply (curve u^* in Fig. 8.2b). This, in turn, changes the position of O_1 and thus also of t_0 and T_c. Further iterative steps are necessary, but this will not be discussed further here. If necessary, it must be checked whether the steeper front of u^* and the shorter time to chopping can affect the measurement result, since the requirements on the dynamic behavior of the measuring system are higher.

8.1.2 *Variants of the Impulse Voltage*

The impulse calculated with Eq. (8.8) shows an abrupt rise from the zero line at $t = 0$, whereas in test practice, the impulse voltage generated by the voltage generator rather shows a gradual increase. To better approximate the gradual onset of the impulse voltage, another function is occasionally added in Eq. (8.8), which leads to a rounding of the initial curve. Also for the time $t > t_c$, i.e. after the collapse of a chopped lightning impulse, an additional function with a third time constant τ_3 can be inserted:

$$\boxed{u_c(t - t_c) = u(t_c)e^{-(t-t_c)/\tau_3},} \tag{8.13}$$

where by an exponential decay of the voltage is achieved.

Furthermore, the impulse voltage chopped in the front, which is often represented by a wedge-shaped voltage, can be approximated by a triangular function with the slope S:

$$\begin{aligned} \boxed{u(t) = \frac{\hat{u}}{t_c}t = St} \quad & \text{for } 0 \le t \le t_c \\ = 0 \quad & \text{for } t > t_c \end{aligned} \tag{8.14}$$

The triangular function has the advantage that it is easier to handle for basic calculations than the equation for the front-chopped lightning impulse voltage, with no significant differences in the final result (see Sect. 8.1.4).

8.1.3 Parameter Values of Impulse Voltages

For some of the most common high-voltage impulses, the numerically calculated parameter values are listed in Table 8.1. The time to peak t_p of lightning impulse voltages (LI) refers to the mathematical origin O, which lies after the virtual origin O_1 by the time t_0. The lightning impulse voltage chopped on the front (LIC) has a time to chopping $t_c = t_p = 0.5$ µs, based on the mathematical origin O. The quotient $\hat{u}_{LIC}/\hat{u}_{LI}$ indicates the peak value of the chopped lightning impulse voltage (LIC) relative to that of the corresponding full 1.2/50 lightning impulse voltage (LI). For switching impulse voltages (SI), the obvious origin corresponds approximately to the mathematical origin, i.e. $O_1 = O$. Therefore, $t_0 = 0$ and $T_p = t_p$ can be set.

More and more frequently, the task is to find a closed mathematical expression for the recorded data set in order to further process it using special computational algorithms. In principle, the synthesis or analysis of any signal can be carried out with sinusoidal oscillations according to Fourier. For the synthesis of impulse voltages, parabola segments or exponential functions, which are fitted in sections to the recorded impulse, are more suitable. The sum of these functions then represents the complete data set in a good approximation. For example, the *Test Data Generator* (TDG, see Sect. 7.3) comprises, in addition to the analytically given test impulses, the series representation of measured impulses. With the TDG, the software for evaluation of typical high-voltage and high-current parameters is verified. Both types of test impulses can thereby be processed in the same manner by the evaluation software, taking into account the characteristics of the digital

Table 8.1 Parameter values of some high-voltage impulses calculated according to Eq. (8.8) in comparison to the standardized parameters in IEC 60060 (Ref. [1, 2] of Chap. 2)

	Lightning impulse voltage				Switching impulse voltage
Parameter	1.2/50		0.84/60		250/2500
Symbol	LI	LIC (0.5 µs)	LI	LIC (0.5 µs)	SI
τ_1 µs	68.217		83.666		3155
τ_2 µs	0.405		0.2746		62.487
A	1.037		1.022		1.104
t_p^a µs	2.089	0.5	1.576	0.5	250
t_0 µs	0.2210	0.061	0.1563	0.074	0
T_1 µs	1.2	0.524	0.84	0.509	165.1
T_c µs	–	0.561	–	0.574	–
$\hat{u}_{LIC}/\hat{u}_{LI}$	–	0.728	–	0.851	–

[a]Referred to the mathematical origin O

recorder used (quantization, superposition of noise, etc.). An example for the representation of the data set of a TDG test impulse with superimposed oscillation by a series of complex exponential functions is described in (Ref. [10] of Chap. 4). The series representation offers, inter alia, the possibility to determine the mean curve through the oscillation by reducing the number of exponential functions.

8.1.4 Spectrum of Impulse Voltages

To calculate the spectrum of impulse voltages or currents, the complex or real Fourier integral and the Laplace transform are available (see Appendix A). In the following, the Laplace transform is preferred for calculating some spectra. Using the Laplace correspondences in Table A.2, the Laplace transform of the double exponential impulse voltage according to Eq. (8.8) is obtained:

$$F(s) = \hat{u}A\left[\frac{1}{s+\frac{1}{\tau_1}} - \frac{1}{s+\frac{1}{\tau_2}}\right]. \tag{8.15}$$

With $s = \mathrm{j}\omega$ and after multiplying the fractions in Eq. (8.15) with their conjugate complex denominators, the Laplace transform can be separated into its real and imaginary parts. The absolute value, referred to $F(\omega = 0)$, is equal to the normalized amplitude density of the impulse voltage:

$$\begin{aligned} F(\omega) &= \left|\frac{F(\mathrm{j}\omega)}{F(0)}\right| \\ &= \frac{1}{\tau_1 - \tau_2}\sqrt{\left[\frac{\tau_1}{1+(\omega\tau_1)^2} - \frac{\tau_2}{1+(\omega\tau_2)^2}\right]^2 + \left[\frac{\omega\tau_1^2}{1+(\omega\tau_1)^2} - \frac{\omega\tau_2^2}{1+(\omega\tau_2)^2}\right]^2}. \end{aligned} \tag{8.16}$$

The spectrum of chopped impulse voltages can, in principle, also be calculated with the help of the Laplace transform, but the calculation is quite extensive because of the numerous expressions. A lightning impulse voltage chopped on the front can be approximated by the triangular function of Eq. (8.14), whose spectrum is easier to calculate. The discontinuity at the instant of chopping $t = t_c$ is overcome in that the triangular function $u(t)$ is expressed by the sum of three functions which are unlimited in time:

$$u(t) = u_a + u_b + u_c \tag{8.17}$$

that cancel each other out for $t > t_c$ (Fig. 8.3). Here, u_a is a ramp with the steepness $\hat{u}/t_c$, u_b a negative step starting at t_c with the amplitude $-\hat{u}$ and u_c a negative ramp starting delayed at $t = t_c$ with the slope $-\hat{u}/t_c$. The sum of the three functions, together with the calculation rules and correspondences in Tables A.1 and A.2, leads to the Laplace transform of the triangular function:

$$F(s) = \hat{u}\left[\frac{1}{t_c s^2} - \frac{1}{s}e^{-st_c} - \frac{1}{t_c s^2}e^{-st_c}\right]. \tag{8.18}$$

After separating Eq. (8.18) into real and imaginary parts, we have:

$$F(s) = \hat{u}t_c\left[\frac{\sin\omega t_c}{\omega t_c} + \frac{\cos\omega t_c}{(\omega t_c)^2} - \frac{1}{(\omega t_c)^2} + \mathrm{j}\left(\frac{\cos\omega t_c}{\omega t_c} - \frac{\sin\omega t_c}{(\omega t_c)^2}\right)\right]. \tag{8.19}$$

Considering $s = \mathrm{j}\omega$ and $F(\omega = 0) = \hat{u}t_c/2$, the normalized amplitude density of the triangular function is the absolute value of Eq. (8.19):

$$\boxed{F(\omega) = \left|\frac{F(\mathrm{j}\omega)}{F(0)}\right| = \frac{2}{(\omega t_c)^2}\sqrt{(\omega t_c)^2 + 2(1 - \cos\omega t_c - \omega t_c \sin\omega t_c)}.} \tag{8.20}$$

The expression in Eq. (8.20) are undefined for $\omega = 0$, since both the numerator and denominator are zero. For small ω, Eq. (8.20) also becomes unusable, since the smallest inaccuracies in the computation cause strong oscillations of $F(\omega)$. By series expansion of $\cos\omega t_c$ and $\sin\omega t_c$, we obtain:

$$\boxed{F(\omega) = \sqrt{1 + (-1)^{\frac{n}{2}} \cdot \frac{8(n-1)}{n!}(\omega t_c)^{n-4}}} \quad \text{for } n = 6, 8, 10 \ldots \tag{8.21}$$

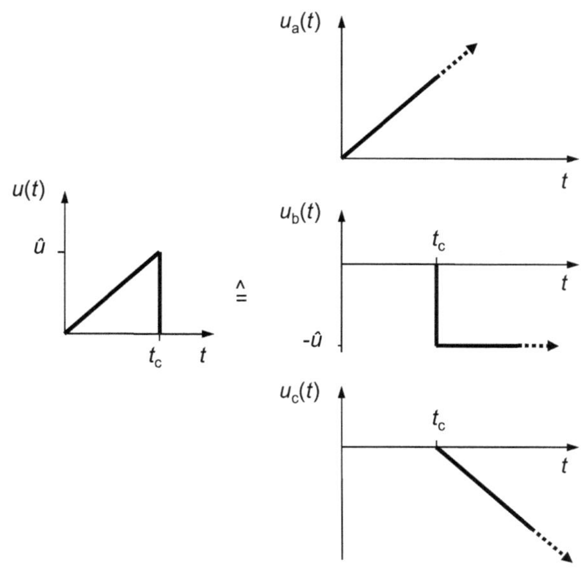

Fig. 8.3 Equivalent representation of a triangular function (left) by the sum of three functions u_a, u_b and u_c (right) which are unlimited in time

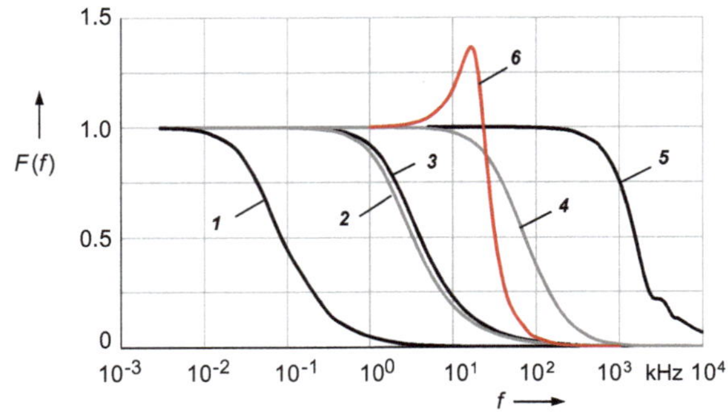

Fig. 8.4 Amplitude density *F(f)* of some impulse voltages and an impulse current ***1*** 250/2500 switching impulse voltage ***2*** 0.84/60 lightning impulse voltage ***3*** 1.2/50 lightning impulse voltage ***4*** 1.2/5 lightning impulse voltage ***5*** triangular function (t_c = 0.5 μs) ***6*** 8/20 impulse current

The series in Eq. (8.21) for the triangular function converges rapidly for $\omega t_c < 1$. The normalized amplitude density is practically equal to 1 for $\omega t_c < 0.1$. This corresponds to a frequency $f < 32$ kHz at a time to chopping t_c = 0.5 μs. Above this frequency, Eq. (8.20) can be used to calculate the amplitude density.

Figure 8.4 shows the amplitude density of some impulse voltages (curves ***1–5***) normalized to the respective DC component $F(f = 0)$. In the semi-logarithmic representation, the amplitude density *F(f)* of each impulse voltage is approximately constant up to a limit frequency determined by the impulse shape and then decreases more or less rapidly as the frequency increases. From the frequency characteristic, the upper 3 dB limit frequency f_2 can be determined, at which the normalized amplitude density has fallen to $1/\sqrt{2} \approx 0.7$. For comparison, the amplitude density of the 8/20 impulse current is also shown (curve ***6***). The oscillation in the course of time appears as a resonance peak in the frequency characteristic before the steep decay in the spectrum occurs.

The 250/2500 switching impulse voltage (curve ***1*** in Fig. 8.4) has an upper limit frequency $f_2 \approx 50$ Hz, which is comparable to the power frequency. For the 1.2/50 lightning impulse voltage (curve ***3***), f_2 = 2.4 kHz. Above 200 kHz, i.e. approximately 100 times the value of f_2, the amplitude density has decreased to less than 1%. In comparison, the amplitude density of the 0.84/60 impulse voltage, which is used to calibrate digital recorders and impulse voltmeters, differs only insignificantly (curve ***2***).

The triangular function used as a wedge-shaped impulse with the time to chopping t_c = 0.5 μs has the highest limit frequency f_2 = 1.1 MHz. The amplitude density remains approximately constant up to 100 MHz (curve ***5*** in Fig. 8.4). Basically, the shorter and steeper the impulse, the more its spectrum extends to higher frequencies. For the ideal *Dirac impulse* with infinitely narrow impulse width, the spectrum is known to be constant up to infinitely high frequencies.

8.2 Analytical Representation of Impulse Currents

Assuming linear elements in the RLC circuit for generating exponential impulse currents, *Kirchhoff's loop equation* applies to the voltages (see Fig. 5.3):

$$u_{\mathrm{C}} = u_{\mathrm{L}} + u_R \tag{8.22}$$

and *Kirchhoff's junction equation* to the currents:

$$U_0 - \frac{1}{C}\int i\mathrm{d}t = L\frac{\mathrm{d}i}{\mathrm{d}t} + Ri, \tag{8.23}$$

where U_0 is the charging voltage at the capacitor C at the beginning of the discharge at t = 0. Applying the Laplace transform with its correspondences in Table A.1 to Eq. (8.23), the equivalent equation in the complex variable domain is:

$$\frac{U_0}{s} - \frac{1}{Cs} = Lsi + Ri, \tag{8.24}$$

which is dissolved for the current:

$$i(s) = \frac{U_0}{L}\frac{1}{s^2 + \frac{R}{L}s + \frac{1}{LC}}. \tag{8.25}$$

The form of Eq. (8.25) is comparable to that of Eq. (8.3) for impulse voltages, whose solution is given by two exponential functions corresponding to Eqs. (8.4) and (8.8). In contrast to impulse voltages, however, the values for the front time and the time to half-value are not very different for the 8/20 impulse current and some other impulse currents. The solution of Eq. (8.25) thus differs from Eq. (8.8) with the two exponential terms. In this case, the inverse transform of Eq. (8.25) with the correspondences in Table A.2 offers two further solutions, one for the damped oscillating impulse current:

$$\boxed{i(t) = \frac{U_0}{\omega_{\mathrm{d}}L}\mathrm{e}^{-\delta t}\sin(\omega_{\mathrm{d}}t),} \tag{8.26a}$$

and the other for the aperiodic impulse current:

$$\boxed{i(t) = \frac{U_0}{\omega_{\mathrm{d}}^{*}L}\mathrm{e}^{-\delta t}\sinh\left(\omega_{\mathrm{d}}^{*}t\right)} \tag{8.26b}$$

where:

$$\delta = \frac{R}{2L}, \tag{8.27}$$

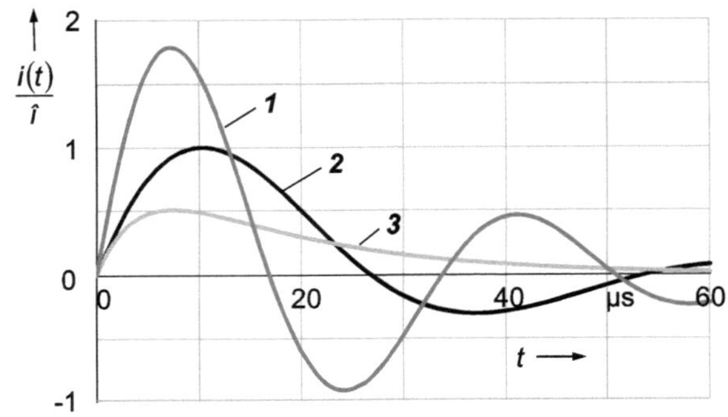

Fig. 8.5 Exponential impulse currents calculated according to Eq. (2.26a, b) *1* weakly damped impulse current *2* 8/20 impulse current *3* aperiodic impulse current

$$\omega_{\mathrm{d}} = \sqrt{\frac{1}{LC} - \left(\frac{R}{2L}\right)^2}, \tag{8.28a}$$

$$\omega_{\mathrm{d}}^{*} = \sqrt{\left(\frac{R}{2L}\right)^2 - \frac{1}{LC}}. \tag{8.28b}$$

In Fig. 8.5, the calculated time courses of exponential impulse currents are plotted for various damping resistors $R_1 < R_2 < R_3$. Curve ***1*** shows a weakly damped impulse current according to Eq. (8.26a), curve ***2*** the 8/20 impulse current, also calculated according to Eq. (8.26a), and curve ***3*** shows an aperiodic impulse current according to Eq. (8.26b). For $R = 0$, we obtain from Eq. (8.26a) the theoretical special case of an undamped sinusoidal oscillation.

By setting the first derivative of Eq. (8.26a) to zero, the times of occurrence of the positive and negative maxima of a damped oscillating impulse current are obtained as:

$$\boxed{t_{\mathrm{max,k}} = \frac{1}{\omega_{\mathrm{d}}}\left[\arctan\left(\frac{\omega_{\mathrm{d}}}{\delta}\right) + k\pi\right]} \quad \text{for } k = 0, 1, 2, \ldots. \tag{8.29}$$

The time to peak $t_{\mathrm{p}} = t_{\mathrm{max,0}}$ is determined for $k = 0$ and the time $t_{\mathrm{max,1}}$ of the first (negative) undershoot for $k = 1$. The amplitude of the first undershoot is calculated by setting $t = t_{\mathrm{max,1}}$ in Eq. (8.26a). For tests, limit values are specified for undershoot, as a higher current amplitude with opposite polarity could destroy the test object.

It should be noted that the impulse currents calculated according to Eq. (8.26a) refer to the mathematical origin O. The difference between O and the standardized virtual origin O_1 is determined in analogy to the impulse voltages according to Fig. 8.2a, if t_{30} is replaced by t_{10}:

$$\boxed{t_0 = \mathrm{O} - \mathrm{O}_1 = 0.125\, T_{\mathrm{AB}} - t_{10}(\mathrm{O}).} \tag{8.30}$$

In Eq. (8.30), $t_{10}(\mathrm{O})$ is the time between O and the time when the current reaches the value $0.1\hat{\imath}$, and T_{AB} is the time between the values at $0.1\hat{\imath}$ and $0.9\hat{\imath}$. For the 8/20 impulse current, O_1 is about $t_0 \approx 0.28$ μs before O. This has no effect on the front time T_1 calculated from Eq. (5.1), but the time to half-value T_2 referred to O_1 is greater by 1.4% than the value referred to O. In test practice, the difference between O and O_1 is tolerable due to the large permissible tolerance for T_2, thus a correction can be omitted. For higher accuracy requirements, however, the exponential impulse current is calculated according to Eq. (8.26a) with a value T_2 that is smaller by t_0, e.g. $T_2 \approx 19.72$ μs for the 8/20 impulse current.

8.2.1 Determination of the Circuit Elements

When an impulse current defined by its time parameters T_1 and T_2 is to be generated, the required circuit elements *R*, *L* and *C* in Fig. 5.3 can be calculated from Eq. (8.26a) by iteration with numerical calculation methods (Ref. [5] of Chap. 5). They replace the previously proposed graphical methods with the help of diagrams (Ref. [1] of Chap. 1, Refs. [2, 3] of Chap. 5). Usually, the capacitance *C* of the charging capacitor, which is a considerable cost factor of an impulse current generator, is prescribed and the two other circuit elements are calculated to obtain the desired impulse shape. As an example, Fig. 8.6 shows three 8/20 impulse currents calculated with the circuit elements according to Eq. (8.26a) for various charging capacitors. It can be seen that for the same charging voltage U_0 the peak value $\hat{\imath}$ becomes larger with *C*. However, the impulse shapes computed for 50, 40 and 30 μF are identical when normalized to the respective peak value $\hat{\imath}$. The 8/20 impulse current referred to the virtual origin O_1 attains its peak after $t_p = 10.6$ μs. The maximum undershoot occurs after 36.8 μs and is 33.5% of the main peak $\hat{\imath}$.

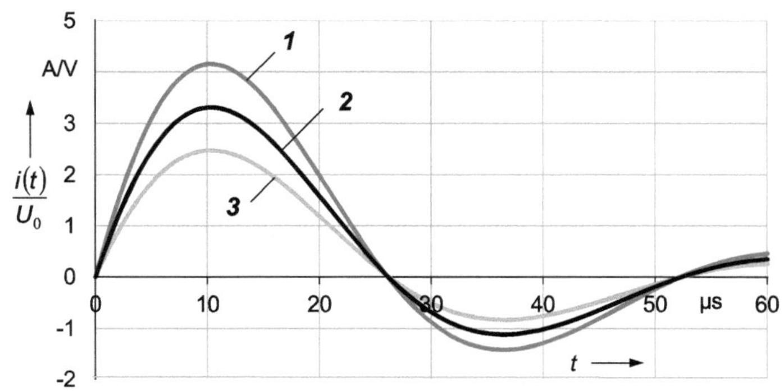

Fig. 8.6 8/20 impulse current calculated using Eq. (8.26a) for various charging capacitors *C* ***1*** *C* = 50 μF ***2*** *C* = 40 μF ***3*** *C* = 30 μF

Table 8.2 Circuit elements and current peaks for the 8/20 impulse current, calculated using Eq. (8.26a) for the virtual origin O_1 or mathematical origin O

Origin	C (μF)	R (Ω)	L (μH)	$\hat{\imath}/U_0$ (A/V)
O_1	50	0.102	1.235	4.15
O_1	40	0.128	1.547	3.31
O_1	30	0.173	2.067	2.48
O	30	0.188	2.13	2.38

Table 8.2 compiles the values of the circuit elements used to calculate the 8/20 impulse currents for the three charging capacitors from Eq. (8.26a). The values in the first three table rows refer to the virtual origin O_1. For comparison, the values referred to the mathematical origin O are also given for C = 30 μF in the last table row.

8.2.2 *Spectrum of Exponential Impulse Currents*

The spectrum of exponential impulse currents results from Eq. (8.26a) with the correspondences of the Laplace transform in Table A.2 as:

$$I(s) = \hat{\imath}\frac{1}{s^2 + \frac{R}{L}s + \frac{1}{LC}}. \tag{8.31}$$

After separation into real and imaginary parts, the normalized amplitude density is obtained with $s = \mathrm{j}\omega$:

$$\boxed{F(\omega) = \frac{|F(\mathrm{j}\omega)|}{F(0)} = \frac{1}{\sqrt{(1 - LC\omega^2)^2 + (RC\omega)^2}}.} \tag{8.32}$$

As an example, curve ***6*** in Fig. 8.4 shows the normalized amplitude density of the 8/20 impulse current compared to different impulse voltages. At a frequency of 16.4 kHz, which corresponds to that of the damped oscillation in the time domain (Fig. 8.5, curve **2**), it has an increase of 36%. The upper 3 dB limit frequency is f_2 = 27 kHz. The drop in the amplitude density is steeper than that of the impulse voltages shown in Fig. 8.4.

8.3 Analytical Representation of Short-Time AC Currents

On the basis of the simple equivalent circuit diagram for the *short-circuit AC current* in Fig. 2.8, the following differential equation can be set up after closing the switch S:

$$u(t) = \hat{u}\sin(\omega t + \psi) = L\frac{\mathrm{d}i(t)}{\mathrm{d}t} + Ri(t), \tag{8.33}$$

where ψ is the switching angle at the time of closing S. The solution of the differential equation yields the short-time AC current:

$$\boxed{i(t) = \hat{i}\left[\sin(\omega t + \psi - \varphi) - \sin(\psi - \varphi)\cdot \mathrm{e}^{-t/\tau}\right]} \tag{8.34}$$

with the amplitude $\hat{i}$, the phase angle φ and the time constant τ:

$$\hat{i} = \frac{\hat{u}}{\sqrt{R^2 + (\omega L)^2}}, \tag{8.35}$$

$$\varphi = \arctan\frac{\omega L}{R}, \tag{8.36}$$

$$\tau = \frac{L}{R}. \tag{8.37}$$

Equation (8.34) consists of two parts. The first part represents the steady-state AC component of the short-time current and the second part describes the exponential decrease of the DC component of the current (see Fig. 2.2a). The AC current, i.e. the first term in the square bracket of Eq. (8.34), consists of an ohmic and an inductive component which is in phase with the voltage or lags behind by 90°. The resulting AC component has the phase angle φ corresponding to Eq. (8.36). The second term in the square bracket represents a DC current that decreases exponentially with the time constant τ.

The initial form of the short-time AC current depends on the instant t_s or the switching angle ψ at which the AC voltage is switched to the RL circuit. Under unfavorable switching and phase conditions, the steady-state and transient currents can superimpose such that at low values of R, the first peak value of the resulting short-time AC current reaches almost double the amplitude of the steady-state current (see Fig. 2.2a). This means, of course, a considerable load on the test object. If the switching is in the zero-axis crossing of the AC voltage and R is very small, the DC current practically disappears and the short-time AC current is almost sinusoidal (see Fig. 2.2b).

Chapter 9
Transfer Behavior of Linear Systems, Convolution and Deconvolution

Abstract The transfer behavior of a linear system can be described by its input and output signals and expressed in the time domain or frequency domain. In order to analytically determine the transfer behavior of a system, the Laplace transform of a time function into the complex variable domain (spectral domain, frequency domain) and the subsequent inverse transform into the time domain is a very effective tool. The step responses of simple RC and RCL circuits, which represent the basic elements of voltage dividers, shunts and measuring coils, are calculated using the Laplace transform. With the convolution integral, the output signals of the RC and RCL circuits are calculated for some characteristic input signals. This opens up a variety of possibilities to thoroughly analyze and optimize impulse voltage and current measuring systems and their components without the need for extensive experimental investigations. For calculations with experimental step responses, numerical convolution is applicable due to the high computing power of the PC as well as the significantly improved properties of digital recorders.

The transfer behavior of a *linear system* can be described by its input and output signals and expressed in the *time domain* or *frequency domain*. Which domain will be preferred depends on whether the input signal is also predetermined in the time or frequency domain. The *Laplace transform* of a time function into the *complex variable domain* (*spectral domain*, *frequency domain*) and the subsequent *inverse transform* into the time domain is a very effective way to analytically determine the transfer behavior. The relationships derived for linear systems can be applied to the high-voltage measurement systems used in practice, in particular for impulse voltages and impulse currents. Voltage dividers, shunts and measuring coils, which are provided for measuring impulse voltages or impulse currents in the time domain, are preferably characterized by their step response. In contrast, in the low-voltage range and in the measurement technique of high AC voltages and currents, it is customary to characterize electrical measuring instruments by their frequency response. For oscilloscopes and digital recorders, which are used to measure impulse voltages as well as AC voltages, both forms of representation in the frequency or time domain are common.

K. Schon, *High Voltage Measurement Techniques*, Power Systems,
https://doi.org/10.1007/978-3-030-21770-9_9

If the transfer behavior of a linear system is known in the time or frequency domain, its output signal can be calculated for any input signals with the convolution integral. This opens up a variety of possibilities to thoroughly analyze and optimize impulse voltage and current measuring systems and their components without the need for extensive experimental investigations. For the calculations, both analytical and numerical approaches are available. The enormously high computing power of the PC as well as the significantly improved measurement properties of digital recorders now offer good technical prerequisites for the application of convolution using numerical algorithms. The following principles for voltage measurement systems apply accordingly to current measurement systems.

9.1 Step Response of a System

Figure 9.1 schematically shows an ideal system with the input voltage $u_1(t)$ and the output voltage $u_2(t)$. The transfer behavior of the system is characterized in the time domain by the *step response* $g(t)$ or in the frequency domain by the *complex transfer function* $H(\mathrm{j}\omega)$ (see Sect. 9.3). The relationships for linear systems can be applied to measuring systems for high voltages and currents, but the influence of electromagnetic interference on the most unshielded measuring systems often adversely affects the result. The transfer behavior of impulse voltage and current measuring systems is preferably expressed by the step response $g(t)$ and that of AC voltage and current measuring systems by the complex transfer function $H(\mathrm{j}\omega)$.

When the input of a system is the step voltage:

$$u_1(t) = U_{10} s_0(t), \tag{9.1}$$

the step response is obtained at the system output:

$$u_2(t) = U_{20} g(t). \tag{9.2}$$

Here, $s_0(t)$ is the *unit step function* with the time-dependent form:

$$\begin{aligned} s_0(t) &= 0 \quad \text{for}\, t \leq 0, \\ &= 1 \quad \text{for}\, t > 0, \end{aligned}$$

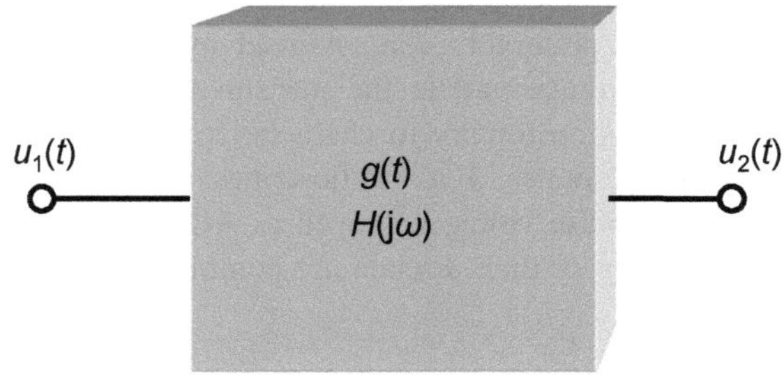

Fig. 9.1 Characterization of a linear system by the step response $g(t)$ or the complex transfer function $H(\mathrm{j}\omega)$

Fig. 9.2 Unit step $s_0(t)$ and unit step response $g(t)$ of a damped oscillating system with the partial response times T_α, $-T_\beta$, T_γ, $-T_\delta$, …

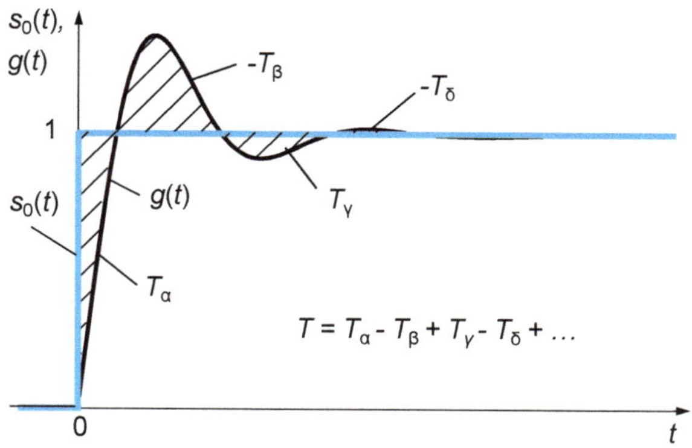

and $g(t)$ the *unit step response*. The quotient U_{10}/U_{20} denotes the nominal division ratio or the transformation ratio of the system. For the sake of simplicity, unless otherwise pointed out in this book, the term "step response" is understood as the unit step response $g(t)$. Figure 9.2 shows an typical example of the step function $s_0(t)$ and the step response $g(t)$ of a damped oscillating system which also has an inductive component in addition to resistive and capacitive components.

The step response of linear systems can be characterized by a set of different *response parameters*. The hatched partial areas of the step response mark the deviations from the step function and are referred to as partial response times T_α, T_β, T_γ, etc. Their sum yields the response time T, with the areas above the unit line having a negative sign. In the test standards for high-voltage impulse dividers, the response parameters are slightly differently defined, which takes into account the imperfections in the recording and evaluating of the experimentally obtained step response. Compliance with specified limit values serves as proof of the adequate dynamic behavior of the voltage divider. The significance of the response parameters is discussed in more detail in Sects. 9.8.1 and 9.8.2.

9.2 Convolution Integral and Convolution Algorithm

If the step response $g(t)$ of a system is known, the *convolution integral*, also known as *Duhamel integral*, provides the possibility to calculate the output signal $u_2(t)$ for an arbitrary input signal $u_1(t)$ [1–4]:

$$u_2(t) = \frac{\mathrm{d}}{\mathrm{d}t}\left[\int_0^t u_1(\tau)\cdot g(t-\tau)\mathrm{d}\tau\right] = \frac{\mathrm{d}}{\mathrm{d}t}\left[\int_0^t u_1(t-\tau)\cdot g(\tau)\mathrm{d}\tau\right]. \tag{9.3}$$

Both versions of Eq. (9.3) are identical due to the commutative law. One of the four possible forms of the convolution integral is given by:

$$\boxed{u_2(t) = u_1(0)g(t) + \int_0^t u_1(\tau)\frac{\mathrm{d}g(t-\tau)}{\mathrm{d}t}\mathrm{d}\tau}. \tag{9.4}$$

The following example may serve to understand the convolution principle. The input signal $u_1(t)$ is considered as the superimposition of many small time-displaced single steps (Fig. 9.3). Each step $\Delta u_{2,\mathrm{i}}$ produces the corresponding small response $\Delta u_{2,\mathrm{i}}$ at the output of the system. The superposition of the individual responses yields the time-discrete output signal $u_2(k\Delta t)$, which turns for infinitely small time intervals Δt into $u_2(t)$. For impulse voltages, we have $u_1(0) = 0$, so that on the right side of Eq. (9.4) only the integral remains. However, the convolution integral can be solved analytically only for a few step responses and input signals. The high computing power of the PC and the improved measurement properties of digital recorders now offer good technical prerequisites for the application of numerical convolution algorithms.

The time courses of $g(t)$ and $u_1(t)$ measured with digital recorders are available as time-discrete samples and the convolution integral in Eq. (9.4) is calculated numerically [5–8]. For the numerical integration, the simple trapezoidal rule is suitable, and with an adequate number of samples, it enables a sufficiently accurate calculation of the convolution integral. With the abbreviation for the derivative (or the differential quotients) of the step response:

$$g' = \frac{\Delta g}{\Delta t}$$

and assuming $u_1(0) = 0$, the following *convolution algorithm* is obtained to calculate the output voltage $u_2(t)$:

$$\boxed{u_2(k\Delta t) = u_{2,\mathrm{k}} = \Delta t\left[\frac{u_{1,\mathrm{k}}g'_0}{2} + \sum_{\mathrm{i}=1}^{\mathrm{k}-1} u_{1,\mathrm{i}}g'_{\mathrm{k-i}}\right]} \tag{9.5}$$

for $k = 2, 3, 4, \ldots N - 1$.

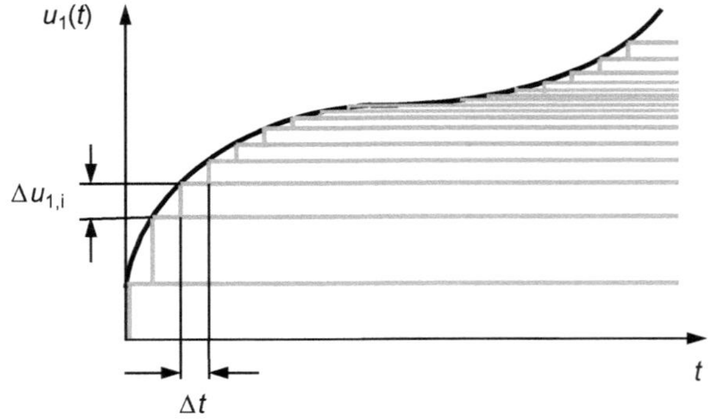

Fig. 9.3 Representation of a signal $u_1(t)$ by time-displaced single steps $\Delta u_{1,\mathrm{i}}$, which generate the corresponding step responses $\Delta u_{2,\mathrm{i}}$ at the system output

The initial value of the output signal for $k = 1$ is:

$$u_{2,1} = \frac{\Delta t}{2} u_{1,1} g_0' \tag{9.6}$$

where N is the number and Δt the equidistant time interval of the discrete samples of $g(t)$ und $u_1(t)$. The sampling frequency for recording both signals must be the same. For the numerical differentiation of the step response, the calculation of the central mean value from the two neighboring values is recommended:

$$g_\mathrm{k}' = \frac{(g_{\mathrm{k}+1} - g_{\mathrm{k}-1})}{2\Delta t} \tag{9.7}$$

for $k = 1, 2, 3, \ldots N$. The first value g_0' at the initial time $t = 0$ is not defined by Eq. (9.7). It is either set equal to the first calculable value g_1' or, if the derivative of the step response at the beginning changes very much, is preferably determined by extrapolation of the first five values $g_1' \ldots g_5'$.

With convolution, the dynamic behavior of impulse voltage and current measuring systems, which are characterized by their step response, can be proved without the need for complex experimental investigations. For example, deviations of the measured peak value from the correct value can thus be calculated as a function of the front time or time-to-chopping. Convolution is particularly useful for detecting or confirming the quality of wide-band measurement systems with very good characteristics, maintained in metrology institutes or highly qualified calibration laboratories. The reason is that even more accurate reference systems are rarely available for comparative measurements. The conditions and limits for the application of convolution are dealt with in Sect. 9.7.

At first glance, it seems easy to rearrange the convolution algorithm in Eq. (9.5) in such a way that the input voltage $u_1(k\Delta t)$ can be calculated from the output voltage $u_2(k\Delta t)$ and the derivative of the step response $g'(k\Delta t)$. However, the inverse of the convolution, the *deconvolution*, provides only a few satisfactory results in measurement practice. Small disturbances and sampling errors that are inevitably superimposed on the recorded data $u_2(k\Delta t)$ and $g'(k\Delta t)$, even the smallest computational errors of the PC, result in a rapid amplification of the disturbances due to the recursive calculation formula. Without any special countermeasure, the disturbances add up to $u_1(k\Delta t)$ after each computed series expansion, so that the final result of the numerical deconvolution becomes unusable. Smoothing methods and an iterative procedure, in which every intermediate result for u_1 is verified by convolution, help to a limited extent. The corresponding deconvolution in the frequency domain, briefly discussed in Sect. 9.3, is often more promising [9–21].

9.3 Fourier Transform and Transfer Function

As an alternative to the step response, the transfer behavior of a system can be characterized by its transfer function. According to *Fourier*, every signal can be decomposed into a series of partial sine waves $u(\omega t) = \hat{u}\sin(\omega t + \varphi)$ with the amplitude $\hat{u}$, the angular frequency ω and the phase angle φ (see Appendix A). A partial sine wave $u_{1,i}$ applied to the input of a linear system produces at its output a sine wave $u_{2,i}$ of the same frequency, but generally of different amplitude and deviating phase angle (Fig. 9.4a). Between the two partial sine waves $u_{1,i}$ and $u_{2,i}$ with the angular frequency ω, there exists a time delay t_0 due to the finite signal propagation time:

$$t_0 = \frac{\varphi_1 - \varphi_2}{\omega} = \frac{b}{\omega}, \tag{9.8}$$

where $b = \varphi_1 - \varphi_2$. The entirety of the partial sine waves at the output, which are superimposed according to their amplitudes and phase angles, results in the output signal $u_2(t)$.

The quotient of the *Fourier transforms* of the output and input signals of a system yields the complex *transfer function* [3]:

$$\boxed{H(\mathrm{j}\omega) = \frac{U_2(\mathrm{j}\omega)}{U_1(\mathrm{j}\omega)} = \frac{\hat{u}_2 \mathrm{e}^{\mathrm{j}(\omega t + \varphi_2)}}{\hat{u}_1 \mathrm{e}^{\mathrm{j}(\omega t + \varphi_1)}} = |H(\mathrm{j}\omega)|\mathrm{e}^{-\mathrm{j}(\varphi_1 - \varphi_2)} = H(\omega)\mathrm{e}^{-\mathrm{j}b}}. \tag{9.9}$$

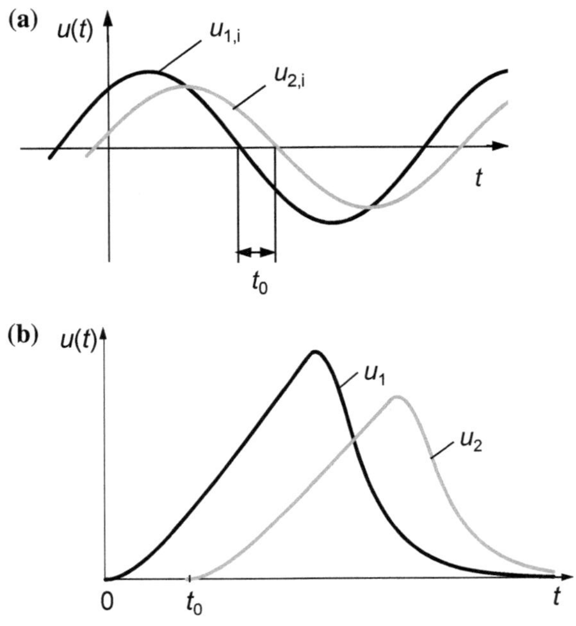

Fig. 9.4 Decomposition and synthesis of a signal after Fourier. **a** Partial sine wave $u_{1,i}$ of the decomposed input signal $u_1(t)$ and corresponding sine wave $u_{2,i}$ at the output of a linear system; **b** input signal $u_1(t)$ of the system and output signal $u_2(t)$ obtained by synthesizing the sine waves $u_{2,i}$

The absolute value is called the *transfer factor* $H(\omega)$ and the phase difference $b(\omega) = \varphi_1 - \varphi_2$ the *transfer angle.* $H(\omega)$ and $b(\omega)$ can be graphically represented as the *amplitude response* or *phase response* of the system versus frequency. The absolute value $H(\omega)$ is synonymous with the reciprocal of the division ratio of a voltage divider, the transformation ratio of a transformer or the amplification factor of an amplifier.

The special case of ideal signal transfer is given by $H = 1$ and $b = 0$. Input and output signals of the system are then identical. However, due to the finite velocity of signal propagation in passive or active systems, this particular case cannot occur in the strict sense. The distortion-free transfer with $H = H_0 = \text{const.}$ and $b/\omega = t_0 = \text{const.}$ is theoretically possible and highly desirable in practice. This means that the output signal u_2 is increased or decreased relative to u_1 and delayed by the time t_0 (Fig. 9.4b). The output signal then represents a true-to-scale reproduction of the input signal.

If the complex transfer function $H(\mathrm{j}\omega)$ of a system is known, the output signal $u_2(t)$ can be calculated for any arbitrary input signal $u_1(t)$. Calculation is made in the frequency domain according to the relationship:

$$U_2(\mathrm{j}\omega) = H(\mathrm{j}\omega) \cdot U_1(\mathrm{j}\omega) \tag{9.10}$$

which results from Eq. (9.9). First, the *spectral function* $U_1(\mathrm{j}\omega)$ of $u_1(t)$ is calculated with the Fourier integral according to Eq. (A.1), i.e. the input signal is decomposed into its frequency components. Then, each frequency component of U_1 is multiplied by the corresponding frequency component of the transfer function H, resulting in the spectral function U_2. The inverse transform of $U_2(\mathrm{j}\omega)$ is carried out according to Eq. (A.4) and leads to the desired output signal $u_2(t)$ in the time domain:

$$\boxed{u_2(t) = \frac{1}{2\pi} \int_{-\infty}^{\infty} U_2(\mathrm{j}\omega) \mathrm{e}^{\mathrm{j}\omega t} \mathrm{d}\omega.} \tag{9.11}$$

In Appendix A.1, it is pointed out that the complex Fourier integral in Eq. (A.1) can only be solved analytically for a few elementary cases. In measurement practice, therefore, the frequency analysis is performed predominantly numerically using the *Fast Fourier Transform* (*FFT*). This evaluation option is offered by most of the measuring instruments with digital storage of the measured data. The number of samples in the FFT must be a power of 2. For higher accuracy requirements and arbitrary number of samples, the *Discrete Fourier Transform* (*DFT*) is more suitable, but their application requires a considerably longer computation time.

The calculation of $u_2(t)$ in Eq. (9.11) with the spectral functions $H(\mathrm{j}\omega)$ and $U_1(\mathrm{j}\omega)$ corresponds to the convolution in the time domain. In general, the amplitude and phase response of a shielded measuring system, which is connected reflection-free to the frequency generator via a coaxial cable, can be measured very well even at higher frequencies. For high-voltage dividers, which are usually large and unshielded, the measurement of the frequency response is not customary. Because of the limited signal amplitude of the frequency generator and the large division ratio, the divider output signal is quite small, so that strong electromagnetic interference is to be expected. The transfer behavior of high-voltage dividers is therefore preferably characterized by their step response $g(t)$.

Transition from the time domain into the frequency domain may be advantageous in order to obtain the transfer function $H(\mathrm{j}\omega)$ according to Eq. (9.10). The transition is possible since between the transfer function $H(\mathrm{j}\omega)$ and the *Dirac impulse*, i.e. the time derivative of the step response $g(t)$, the following relation exists:

$$\boxed{H(\mathrm{j}\omega) = \int_{-\infty}^{\infty} \frac{\mathrm{d}g}{\mathrm{d}t} \mathrm{e}^{-\mathrm{j}\omega t} \mathrm{d}t} \tag{9.12}$$

which is also derived from Eq. (9.14) with the Laplace transforms $H(s)$ and $G(s)$ with $s = \mathrm{j}\omega$. Once again, the integral in Eq. (9.12) can only be solved for a few analytical functions. The step response $g(t)$ usually exists as digital data set, which can be numerically differentiated and transformed into the frequency domain by FFT.

According to Eq. (9.9), the transfer function $H(\mathrm{j}\omega)$ is defined as the quotient $U_2(\mathrm{j}\omega)/U_1(\mathrm{j}\omega)$ in the frequency domain. If the input and output signals are given in the time domain, the corresponding spectral functions can be formed using Eq. (A.1). When forming the quotient $H(\mathrm{j}\omega)$, care must be taken to ensure that there is no zero in the denominator $U_1(\mathrm{j}\omega)$, so that a division by zero is excluded.

In high-voltage testing, the voltage applied to the test object is primarily of interest. However, fast high-voltage impulses cannot always be measured correctly, e.g. due to the limited response of the measuring system. By rewriting Eq. (9.10), we obtain the spectral function $U_1(\mathrm{j}\omega)$ as the quotient $U_2(\mathrm{j}\omega)/H(\mathrm{j}\omega)$. This operation corresponds to the *deconvolution* in the time domain. The time function $u_1(t)$ corresponding to $U_1(\mathrm{j}\omega)$ is calculated using Eq. (A.4). This method of deconvolution using the spectral functions is often more successful than the deconvolution in the time domain.

9.4 Laplace Transform

The *Laplace transform* of a function from the original *time domain* to the *complex variable domain* (or *s domain*) and the *inverse Laplace transform* back to the time domain provide another, very convenient and comprehensive possibility to determine the transfer behavior of a system analytically [4]. The transfer function $H(s)$ of a linear system according to Eq. (9.9) can be formally expressed as the quotient of the Laplace transforms of the output and input functions:

$$\boxed{H(s) = \frac{U_2(s)}{U_1(s)}} \tag{9.13}$$

where s is generally a *complex variable*, e.g. $s = \sigma + j\omega$. For example, when a system is characterized by a passive network of resistors, capacitors and inductors, *Kirchhoff's laws*, along with the *Heaviside operator s* for differentiation ($1/s$ for integration), provide the conditional equation for determining the output voltage $U_2(j\omega)$, when $U_1(j\omega)$ is the voltage applied to the system input. For a series of input and output functions, there are appropriate correspondences of the Laplace transforms (see Appendix A) that can be directly inserted into Eq. (9.13). The advantage of using the Laplace transform is that the well-known calculation rules and correspondences can be applied very effectively, which greatly simplifies the calculation.

The step function $s(t)$ in the time domain corresponds to 1/s in the complex variable domain (see Table A.2). If the step function is applied to the input of a system, then $U_1(s) = 1/s$ and the output function $U_2(s)$ is identical to the step response $G(s)$. From Eq. (9.13), we obtain a simple relationship between the Laplace transforms of the transfer function $H(s)$ and the step response $G(s)$ as:

$$\boxed{G(s) = \frac{1}{s} \cdot H(s)}. \tag{9.14}$$

For the inverse transform of $G(s)$ into the time domain, the calculation rules and correspondences are again available in Tables A.1 and A.2. Application examples of the Laplace transform and inverse transform to RC and RLC elements are given in Sect. 9.5.

In another specific task, let the step response $G(s)$ be known and the output function $U_2(s)$ of a system for an arbitrary input function $U_1(s)$ to be determined. With Eqs. (9.13) and (9.14), the output voltage is obtained in the complex variable domain:

$$\boxed{U_2(s) = s \cdot U_1(s) \cdot G(s)}. \tag{9.15}$$

The inverse transform of $U_2(s)$ into the time domain using the correspondences in Table A.2 yields the desired output voltage $u_2(t)$. This method corresponds to the analytical convolution in the time domain (see Sect. 9.2). The convolution integral in Eq. (9.4) is directly obtained from Eq. (9.15) using the convolution theorem of the Laplace transforms in Table A.1. Even in the case of convolution, calculation with the Laplace transform is simpler than evaluation of the convolution integral in the time domain. The same applies to the solution of linear differential equations, which are set up for the purpose of network analysis using Kirchhoff's laws for voltages and currents.

In a similar manner, the complex transfer function $H(s)$ can also be defined as the quotient of the output current $I_2(s)$ to the input voltage $U_1(s)$:

$$\boxed{H(s) = \frac{I_2(s)}{U_1(s)}} \tag{9.16}$$

or other combinations of input and output quantities. The transfer function in Eq. (9.16) has the dimension of admittance. An important field of application is the verification of the integrity of the windings of power transformers. The test is often carried out with impulse voltages, with the simultaneously recorded current and voltage impulses being transformed by the Fourier transform into the frequency domain. Depending on the type and condition of the power transformer, a characteristic form of the transfer function versus frequency is obtained. Any deviation from the initial form found during later inspections should be regarded as serious indication for a possible damage of the transformer winding. This method can also be used during *on-line monitoring* in the power supply network, where transients caused under operational conditions are used as the exciting input voltage [22–27].

9.5 Characteristics of RC and RLC Elements

RC and RLC elements can be found in the simplified equivalent circuits of voltage and current measuring systems. For a basic understanding of the transfer behavior of high-voltage dividers, current shunts and other components of a measuring system, it is therefore very instructive to know the transfer behavior of RC and RLC elements. Their step responses are defined by relatively simple analytic expressions, which can be determined with the aid of the Laplace transform and inverse transform. The step response and the transfer function of an element can be converted into one another in the frequency domain. The amplitude responses of different RC and RLC elements are discussed in the following sections.

9.5.1 Step Response of an RC Low-Pass Filter

Figure 9.5a shows an RC element that is set up as a low-pass filter, at whose input the unit step $s_0(t)$ is applied and the unit step response $g(t)$ is tapped at the capacitor C. According to Kirchhoff's loop rule, we obtain in the time domain:

$$s_0(t) = Ri + \frac{1}{C}\int i\mathrm{d}t \tag{9.17}$$

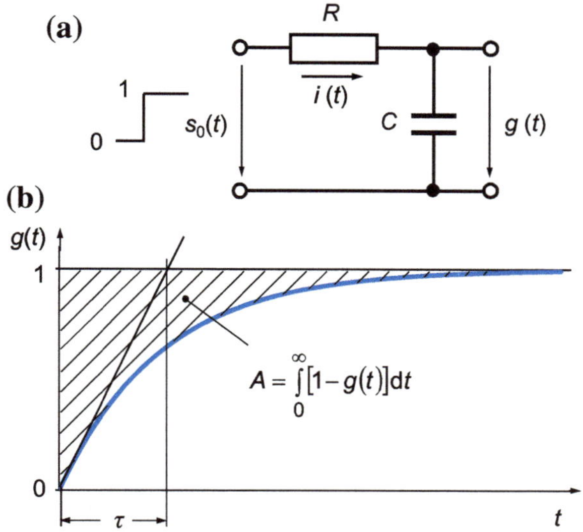

Fig. 9.5 Transfer behavior of an RC element set up as a low-pass filter. **a** RC element with unit step $s_0(t)$ at the input; **b** unit step response $g(t)$ at the output of the RC element with response time $T = \tau = RC = A$

and in the complex variable domain (see Appendix A):

$$\frac{1}{s} = I(s)\left(R + \frac{1}{Cs}\right). \tag{9.18}$$

With $I(s) = sCG(s)$, we obtain the unit step response in the complex variable domain:

$$G(s) = \frac{1}{s(1 + RCs)}. \tag{9.19}$$

The inverse transform into the time domain using Rule 5 in Table A.2 gives the unit step response $g(t)$ of the RC element (Fig. 9.5b):

$$\boxed{g(t) = 1 - \mathrm{e}^{-\frac{t}{\tau}}}, \tag{9.20}$$

where $\tau = RC$ is the time constant. It also results graphically as the point of intersection of the tangent at $g(t = 0) = 0$ with the horizontal through the value 1 or additionally from the area A between $g(t)$ and the horizontal through 1.

9.5.2 *Step Response of an Oscillatory RLC Element*

For the RLC element in Fig. 9.6a, we obtain from Kirchhoff's loop equation the Laplace transform:

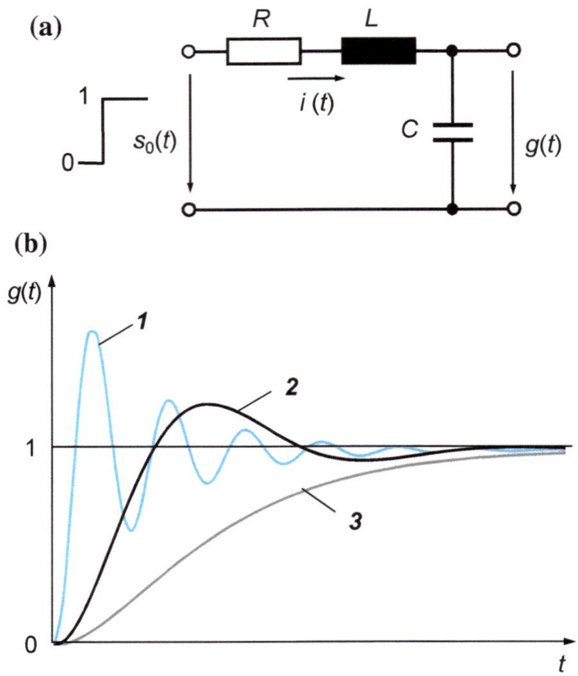

Fig. 9.6 Transfer behavior of an oscillatory RLC element. **a** RLC element with step $s_0(t)$ at the input; **b** unit step response $g(t)$ at the output. ***1*** Weakly damped oscillation, ***2*** normally damped oscillation, ***3*** aperiodic limiting case

$$\frac{1}{s} = I\left(R + Ls + \frac{1}{Cs}\right). \tag{9.21}$$

Further calculation is analogous to that for the RC low-pass filter and yields the unit step response at the capacitor C in the complex variable domain:

$$G(s) = \frac{1}{s}\frac{1}{LCs^2 + RCs + 1} = \frac{1}{s}F(s). \tag{9.22}$$

Inverse transform of $G(s)$ into the time domain is favorably performed in two steps. First, Rule 1 in Table A.1 is applied, which gives the integral of the quotient in Eq. (9.22), expressed in the time domain, and, secondly, the solution of the integral for the here interesting case $R < 2(L/C)^{1/2}$ results in the step response of the oscillatory RLC element:

$$\boxed{g(t) = 1 - \mathrm{e}^{-\delta t}\left(\cos\omega_0 t + \frac{\delta}{\omega_0}\sin\omega_0 t\right)} \tag{9.23}$$

where δ is the decay constant and ω_0 the natural angular frequency:

$$\delta = \frac{R}{2L} \tag{9.24a}$$

$$\omega_0 = \sqrt{\frac{1}{LC} - \left(\frac{R}{2L}\right)^2}. \tag{9.24b}$$

For very small resistances R, the decay constant is $\delta \ll \omega_0$, and the step response results from Eq. (9.23) with $\omega_0^* = (LC)^{-1/2}$ approximately as:

$$\boxed{g(t) \approx 1 - \mathrm{e}^{-\delta t} \cdot \cos\omega_0^* t}. \tag{9.25}$$

The amplitude of the strongly oscillating step response in Eq. (9.25) can reach almost twice the value of the exciting step voltage.

For the aperiodic limiting case with $\omega_0 = 0$ and since $\sin x/x = \mathrm{si}(x) = 1$, Eq. (9.23) reduces to:

$$\boxed{g(t) = 1 - \mathrm{e}^{-\delta t}(1 + \delta t)}. \tag{9.26}$$

Figure 9.6b shows three step responses of the RLC element for different values of δ and ω_0. Curve ***1*** represents the step response of a weakly damped resonant circuit and curve ***2*** that of a normally damped resonant circuit according to Eq. (9.23). Curve ***3*** characterizes the aperiodic limiting case given by Eq. (9.26).

9.5.3 Transfer Functions of RC and RLC Elements

According to Eq. (9.14), the transfer function $H(s)$ and the step response $G(s)$ of a system are linked with each other in the complex variable domain. For the RC low-pass filter with $G(s)$ according to Eq. (9.19), we obtain the simple expression:

$$H(s) = sG(s) = \frac{1}{1 + s\tau} \tag{9.27}$$

where $\tau = RC$. The absolute value of Eq. (9.27) with $s = \mathrm{j}\omega$ yields the amplitude response $H(\omega)$ of the RC element in the time domain:

$$H(\omega) = |H(\mathrm{j}\omega)| = |H(s)| = \frac{1}{\sqrt{1 + (\omega\tau)^2}}. \tag{9.28}$$

For the oscillatory RLC element with $G(s)$ according to Eq. (9.22), the transfer function is:

$$H(s) = sG(s) = \frac{1}{LCs^2 + RCs + 1} \tag{9.29}$$

and the amplitude response:

$$H(\omega) = |H(\mathrm{j}\omega)| = \frac{1}{\sqrt{(1-\omega^2 LC)^2 + (\omega RC)^2}}. \tag{9.30}$$

Figure 9.7 shows the amplitude response *H*(*f*) in double-logarithmic representation calculated for two RC elements (curves **2** and **3**) and one RLC element (curve **4**). For comparison, curve **1** is the amplitude density *F*(*f*) of a 1.2/50 lightning impulse voltage, which can also be seen as curve **3** in Fig. 8.4.

The time constant of the RC element with *H*(*f*) corresponding to curve **2** is $\tau = RC = 436$. This value is chosen so that the rise time $T_a = 2.2RC = 960$ ns calculated with Eq. (9.35) equals that of the lightning impulse voltage according to Eq. (9.34). If the rise time is reduced to one tenth, i.e. $T_a = 96$ ns and thus $\tau = 43.6$ ns, the amplitude response according to curve **3** is obtained with a 3 dB limit frequency of 3.7 MHz. The amplitude response of the RLC element with $RC = 0.436$ µs and $LC = 0.275 \times 10^{-12}$ s (curve **4**) shows a slight overshoot at 220 kHz, corresponding to the oscillation in the step response. The 3 dB limit frequency is 410 kHz.

The amplitude responses of the three elements investigated extend to significantly higher frequencies than the amplitude density of the 1.2/50 lightning impulse voltage. The question of which limit frequency a measuring system must have for the true-to-scale recording of impulse voltages is discussed in Sect. 9.8.

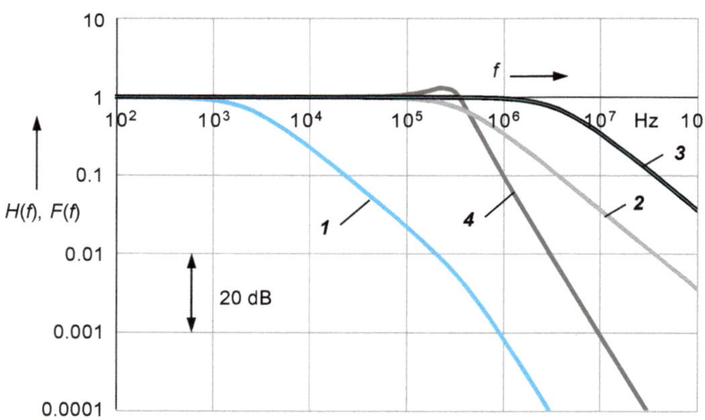

Fig. 9.7 Amplitude response *H*(*f*) of RC and RLC elements in comparison to the amplitude density *F*(*f*) of the 1.2/50 lightning impulse voltage. **1** 1.2/50 lightning impulse voltage, **2** RC element with *RC* = 436 ns, **3** RC element with *RC* = 43.6 ns, **4** RLC element with *RC* = 436 ns

9.6 Response Time, Rise Time and Bandwidth

An important parameter of the unit step response $g(t)$ is the *response time T*. The general mathematical definition is:

$$\boxed{T = \int_0^\infty [1 - g(t)]\mathrm{d}t}. \tag{9.31}$$

For the RC element in Fig. 9.5a with the time constant $\tau = RC$ and the step response $g(t)$ according to Eq. (9.20), the response time is:

$$\boxed{T = \int_0^\infty e^{-\frac{t}{\tau}}\mathrm{d}t = \tau = RC \equiv A} \tag{9.32}$$

where $A = RC$ denotes the hatched area in Fig. 9.5b. The time constant τ of the RC element, as well as its response time T, are obtained graphically as the time at which the tangent to the beginning of the step response intersects the horizontal with the value of 1 (see Fig. 9.5b).

For the RLC element in Fig. 9.6a with the step response according to Eq. (9.23), the response time is calculated as follows:

$$\boxed{T = \int_0^\infty \mathrm{e}^{-\delta t}\left(\cos\omega_0 t + \frac{\delta}{\omega_0}\sin\omega_0 t\right)\mathrm{d}t = RC}. \tag{9.33}$$

The response time T can also be determined graphically as the sum of the partial areas of the step response shown hatched in Fig. 9.2, wherein the partial areas with $g(t) > 1$ are entered with negative sign:

$$T = T_\alpha - T_\beta + T_\gamma - T_\delta + \ldots.$$

In general low-voltage engineering, it is customary to characterize current and voltage impulses by their *rise time* T_a. It is defined as the time between the two points in the front at 10 and 90% of the impulse peak or final value (Fig. 9.8). High impulse voltages and currents, on the other hand, are characterized by the front time T_1 (see Sects. 4.1 and 5.1). For the idealized double exponential impulse voltage according to Eq. (8.8), the rise time is approximately:

$$\boxed{T_\mathrm{a} \approx \frac{4}{5}T_1 = \frac{4}{3}T_\mathrm{AB}} \tag{9.34}$$

where the time T_{AB} is defined in Eq. 4.2. The rise time T_a assigned to a measuring instrument is the rise time of its step response. For a measuring instrument with RC

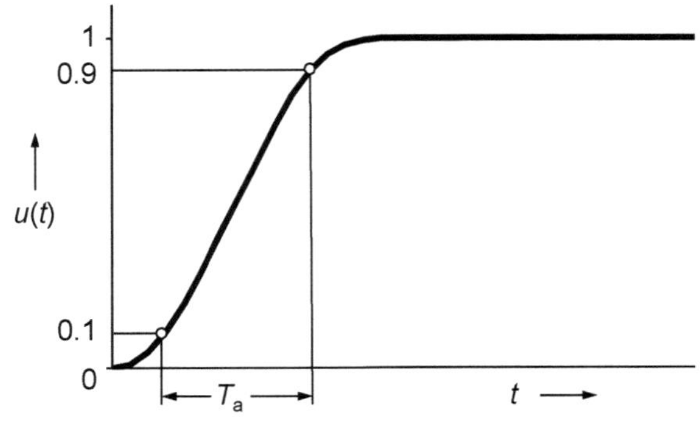

Fig. 9.8 General definition of the rise time T_a of a low-voltage impulse

characteristic according to Fig. 9.5, the following relationship exists between the rise time T_a of the step response and its response time T:

$$\boxed{T_a \approx 2.2RC = 2.2T}. \tag{9.35}$$

Sometimes, measuring instruments are used whose inherent rise time $T_{a,e}$ cannot be ignored. When evaluating a recorded impulse, the measured value $T_{a,m}$ is then larger than the true impulse rise time T_a. Under certain conditions, which are generally given in the measurement practice, the following relationship is valid:

$$T_{a,m}^2 = T_a^2 + T_{a,e}^2, \tag{9.36}$$

from which the true rise time of the impulse is calculated as follows:

$$T_a = \sqrt{T_{a,m}^2 - T_{a,e}^2}. \tag{9.37}$$

If the rise time $T_{a,e}$ of the measuring instrument is significantly smaller than that of the impulse, conversion according to Eq. (9.37) can be dispensed with. If $T_{a,e}$ is less than one fifth of the impulse rise time, i.e. $T_{a,e} < 0.2\, T_a$, then we have $T_a \approx T_{a,m}$ with an error of less than 2%.

The statement of the rise time in the time domain is equivalent to the statement of the *bandwidth B* in the frequency domain with 3 dB limit frequency. For all wideband systems whose bandwidth is practically equal to the upper 3 dB limit frequency, the following simple relationship applies (Ref. [6] of Chap. 1):

$$B = \frac{0.35\ldots0.45}{T_a}. \tag{9.38}$$

The factor 0.35 in Eq. (9.38) applies to a system whose step response attains the final value without overshoot, e.g. an RC network. A bandwidth of 10 MHz then corresponds to a rise time of T_a = 35 ns. The factor 0.45 in Eq. (9.38) applies to a system with about 10% overshoot in the step response.

9.7 Examples of Convolution

In the preceding sections, the basic concept of convolution to calculate the output signal of a linear system for arbitrary input signals has been presented. The analytical calculation with the convolution integral according to Eq. (9.3) leads to a solution only for a few step responses and input signals. Comparable with the analytical convolution in the time domain is the calculation of the output signal by means of the Laplace transform, which can be applied to a large number of systems and signals thanks to the correspondences in the time and complex variable domains. Using the Laplace transform, the basic properties of linear systems are derived using examples of simple RC and RLC elements. The results can serve as model and are useful for understanding the measurement behavior of impulse voltage dividers, current sensors and other components of measuring systems. Examples of numerical convolution are given below for three fictitious voltage dividers. The response errors calculated for impulse voltages with different rise times can be graphically displayed in error diagrams.

9.7.1 Wedge-Shaped Impulse Applied to RC Element

As a first example, the response of the RC element in Fig. 9.5a to a wedge-shaped impulse with an infinitely steep tail at time t_c is investigated. According to Fig. 8.3, the wedge-shaped impulse $u_1(t)$ is composed of three functions. The Laplace transforms of $u_1(t)$ and the step response $g(t)$ of the RC element with $\tau = RC$ are represented by Eqs. (9.15), (8.18) and (9.19). The Laplace transform of the output voltage is given by:

$$U_2(s) = s \cdot U_1(s) \cdot G(s) = \frac{\hat{u}_1}{t_c} \frac{1 - t_c s e^{-st_c} - e^{-st_c}}{s^2(1 + RCs)}. \tag{9.39}$$

The inverse transform into the time domain using Table A.2 provides the output voltage:

$$\boxed{u_2(t) = \frac{\hat{u}_1}{t_c} \left\{ \left[t - \tau\left(1 - e^{-t/\tau}\right) \right] - \left[(t - t_c) - (\tau - t_c)\left(1 - e^{-(t-t_c)/\tau}\right) \right] \right\}}. \tag{9.40}$$

Here, the term in the second square bracket on the right side of the equation contributes only for $t > t_c$.

Figure 9.9a shows the response of the RC element according to Eq. (9.40). At the beginning, the output voltage $u_2(t)$ follows the rise of the input voltage $u_1(t)$ with a delay and, after a certain settling time, runs parallel to $u_1(t)$ but displaced in time. The constant time delay of $u_2(t)$ in the steady-state condition is equal to the response time $T = \tau = RC$. The term in the second square bracket in

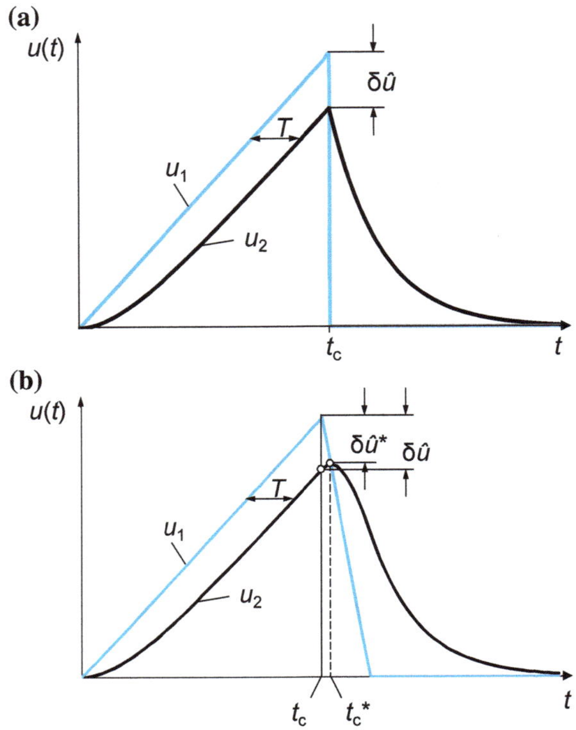

Fig. 9.9 Wedge-shaped impulse u_1 at the input of an RC element and output voltage u_2 calculated by analytical convolution. **a** Infinitely steep tail at t_c; **b** finitely steep tail at t_c

Eq. (9.40) is not considered here. This means, on the other hand, that at the time to chopping t_c with $t_c \gg \tau$, the output voltage is lower than the input voltage by the difference $\delta\hat{u}$. For an infinitely steep collapse of $u_1(t)$ at time t_c, there is a peak value error $\delta\hat{u}$ of the output voltage:

$$\delta\hat{u} = u_2(t_c) - u_1(t_c) = -\hat{u}_1 T/t_c = -ST \tag{9.41}$$

where $S = \hat{u}_1/t_c$ is the steepness of the slope of the wedge-shaped impulse voltage and $T = RC$ the response time of the RC element according to Eq. (9.32). For example, for a time to chopping t_c = 500 ns and a response time T = 25 ns, the relative peak value error is $\delta\hat{u}/\hat{u}_1 = -5\%$.

The wedge-shaped impulse with finitely steep chopping can again, as in Fig. 8.3, be composed of several, in this case five time-displaced functions. The corresponding five output functions of the RC element can be easily calculated and their sum gives the output voltage $u_2(t)$. In contrast to the infinitely steep chopping of $u_1(t)$, the output voltage overshoots the value $u_2(t = t_c)$ and reaches its peak value $\hat{u}_2$ only later at time $t_c^* > t_c$ (Fig. 9.9b). When evaluating the recorded output voltage, therefore, t_c^* appears to be the time to chopping. Due to the overshoot of $u_2(t)$, a smaller peak value error $\delta\hat{u}^* < \delta\hat{u}$ results [28, 29]. In the case of a recorded impulse, $\delta\hat{u}^*$ can be estimated from the slopes of the output voltage shortly before and after the voltage collapse (Ref. [1] of Chap. 4). A further feature is that the tail of the wedge-shaped impulse of any slope intersects the output voltage $u_2(t)$ in the peak [12]. The results obtained for the wedge-shaped impulse also apply approximately to an impulse chopped on the front, whose waveform is quite close to the idealized wedge-shaped impulse (see Fig. 4.1c).

It should be pointed out that the lagging of $u_2(t)$ behind $u_1(t)$ by the response time T must be clearly distinguished from the time delay t_0 in Eq. (9.8) due to the limited signal propagation time. In both examples shown in Fig. 9.9, the time delay is not shown, i.e. $t_0 \equiv 0$. Significant signal delay occurs, for example, when the output of the RC element is connected to the recording instrument via a longer coaxial cable. Impulse voltage dividers have a more or less large signal delay time due to their dimensions, the high-voltage lead and the coaxial cable connected to the divider output. In the steady state, the recorded output voltage $u_2(t)$ of the voltage divider is displaced from the input voltage $u_1(t)$ by the sum of the time delay and the response time, i.e. by $t_0 + T$ (Fig. 9.10).

The contributions of the high-voltage divider with its lead and its coaxial cable to the time delay can be estimated, if necessary, from the divider and lead dimensions and the cable datasheet. In most cases, however, the time delay is of minor interest in impulse voltage tests.

The response time T of a voltage divider can also be determined from simultaneous recordings of the input and output voltages. In addition to the output voltage $u_2(t)$, the input voltage $u_1(t)$ is recorded with a second measuring system (or channel) that triggers the first channel. At a favorable low input voltage level, the voltage dividers or probes can be used with short cables and negligible propagation times. The time difference between u_1 and u_2 in the steady state, as shown in Fig. 9.9, directly indicates the response time T of the measuring system. For a wedge-shaped input voltage $u_1(t)$ with a steep voltage collapse, the response time T can be calculated from the steepness S and the difference δu of the peak values according to Eq. (9.41). Both evaluation methods are used only occasionally in the measurement practice, since the necessary prerequisites are often not fulfilled.

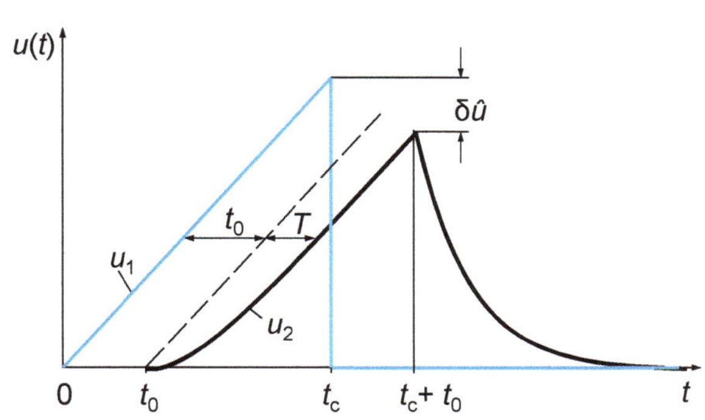

Fig. 9.10 Time delay of the output voltage u_2 with respect to the input voltage u_1 by the signal propagation time t_0

9.7.2 Wedge-Shaped Impulse Applied to RLC Element

Apart from the aperiodic limiting case, the transfer behavior of an RLC element is characterized by a more or less violently oscillating step response (see Fig. 9.6b). In the case of a pronounced overshoot, the response time T in Eq. (9.31) becomes negative. As a special case, $T = 0$ is also possible. The following two examples will illustrate the basic influence of oscillating step responses on the output voltage. The analytical calculation is done in the same manner as for the RC element by using the Laplace transforms of the step response in Eq. (9.22) and the wedge-shaped impulse voltage with infinitely steep tail in Eq. (8.18). The derivation and reproduction of the extensive equations is omitted here.

The results of the convolution calculations for the two RLC elements with oscillating step responses (curves ***1*** and ***2*** in Fig. 9.6b) are presented graphically in Fig. 9.11. After a short settling time, the front of the output voltage $u_2(t)$ of the RLC element with $T = 0$ almost coincides with the input voltage $u_1(t)$ (Fig. 9.11a). For the RLC element with the longer response time, $u_2(t)$ shows a significant time delay (Fig. 9.11b). In the steady state, this time delay is again equal to the response time T of the oscillating step response, comparable to the RC element.

Furthermore, it is evident that the oscillation of the step response has a different effect on $u_2(t)$, depending on the steepness of $u_1(t)$. In the comparatively slowly rising front of $u_2(t)$, the oscillation is barely detectable, while in the steep collapse of $u_2(t)$ for $t > t_c$ it is clearly recognizable. The infinitely steep collapse of $u_1(t)$ in

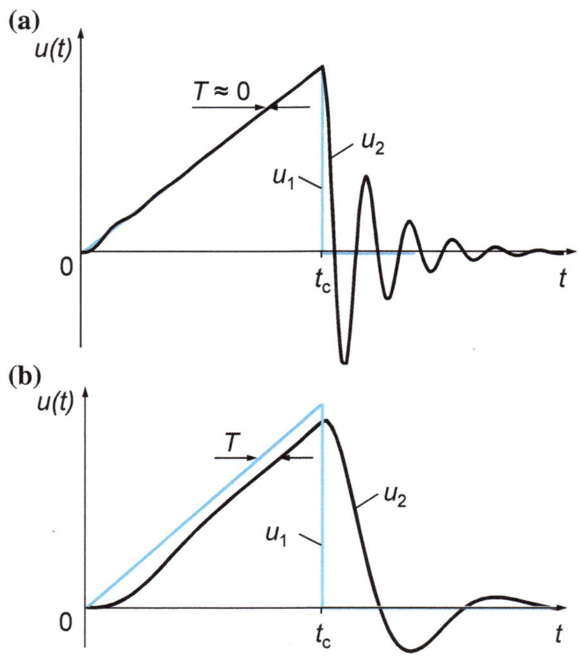

Fig. 9.11 Wedge-type impulse u_1 at the input of an RLC element and output impulse u_2 calculated by analytical convolution for two different response times T of the step response. **a** $T \approx 0$ (step response according to curve ***1*** in Fig. 9.6b); **b** $T > 0$ (step response according to curve ***2*** in Fig. 9.6b)

Fig. 9.11 appears to be a negative step voltage, but the actual response $u_2(t)$ for $t > t_c$ is not exactly the step response. The explanation is that the tail of $u_2(t)$ is also affected by the shape of $u_1(t)$ before chopping, i.e. $u_2(t)$ is defined by the sum of the response signals for the three components u_a, u_b and u_c of the triangular function as shown in Fig. 8.3.

The example in Fig. 9.11a shows that a measuring system optimized for $T \approx 0$ can correctly measure the peak value, but not automatically the front time of the applied impulse voltage in addition. If the point at $0.3\hat{u}_2$, which determines the calculation of the front time T_1, is not located in the steady-state region due to the superimposed oscillation, significant deviations from the correct value of T_1 may occur. If the input voltage of the RLC element drops in finite time for $t > t_c$, the output voltage has an overshoot similar to that of the RC element and reaches a higher peak value than in the case of the infinitely steep chopping (see Fig. 9.9b). For optimum dimensioning of an impulse voltage divider, therefore, the response time T is not the only relevant parameter.

9.7.3 Double Exponential Impulse Applied to RC Element

When a double exponential impulse voltage with the Laplace transform in Eq. (8.15) is applied to the input of the RC element with the Laplace transform in Eq. (9.19), we obtain the output voltage in the complex variable domain according to Eq. (9.15):

$$U_2(s) = s \cdot U_1(s) \cdot G(s) = \frac{\hat{u}A}{RC}\left[\frac{1}{s + 1/\tau_2} - \frac{1}{s + 1/\tau_2}\right]\frac{1}{(s + 1/RC)} \tag{9.42}$$

where τ_1 and τ_2 are the time constants of the double exponential impulse voltage defined by Eq. (8.6). With the correspondences in Table (A.2), the inverse transform provides the output voltage in the time domain:

$$\boxed{u_2(t) = \frac{\hat{u}A}{RC}\left[\frac{e^{-t/RC} - e^{-t/\tau_1}}{1/\tau_1 - 1/RC} - \frac{e^{-t/RC} - e^{-t/\tau_2}}{1/\tau_2 - 1/RC}\right]}. \tag{9.43}$$

Curve ***1*** in Fig. 9.12 shows the initial form of a 1.2/50 lightning impulse voltage $u_1(t)$ applied to an RC element. The curves ***2***, ***3*** and ***4*** represent the output voltages $u_2(t)$ calculated with Eq. (9.43) for three time constants $\tau = RC$. The RC element with the time constant $\tau = RC = T = 436$ ns (curve ***3*** in Fig. 9.12) is characterized in that the rise time T_a of its step response with $T_a = 2.2T = 960$ ns equals the rise time of the 1.2/50 lightning impulse voltage according to Eq. (9.34).

As expected, the larger the time constant RC or rise time T_a, the greater the deviation of the output voltage from the input voltage. It should also be noted that each output voltage (curves ***2*** to ***4***) is intersected at its peak by the input voltage

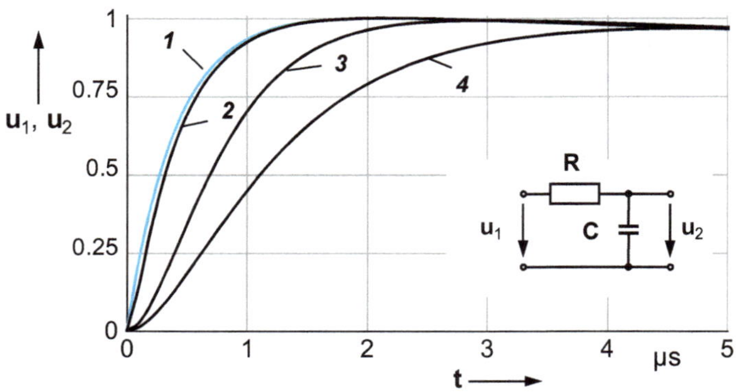

Fig. 9.12 Input and output voltages of an RC element with different time constants $\tau = RC$. ***1*** 1.2/50 lightning impulse voltage u_1, ***2*** output voltage u_2 for $RC = 43.6$ ns, ***3*** output voltage u_2 for $RC = 436$ ns, ***4*** output voltage u_2 for $RC = 1$ µs

(curve ***1***). The tail of $u_2(t)$ is shifted approximately parallel to $u_1(t)$, the time delay being equal to the response time and time constant of the RC element.

9.7.4 Response Errors and Error Diagram

The example of convolution in Sect. 9.7.3 deals with the calculation of the output voltage $u_2(t)$ of an RC element, to whose input terminals a 1.2/50 impulse is applied. By comparing $u_2(t)$ with $u_1(t)$, the deviations $\delta\hat{u}$ for the peak value and δT_1 for the front time can be easily calculated. These deviations are also referred to as *transfer errors* or *response errors* of the system. For the three RC elements investigated in Sect. 9.7.3, which can be considered as fictitious voltage dividers with RC behavior, the relative response errors $\delta\hat{u}$ and δT_1 are summarized in Table 9.1. Both response errors are negligibly small for $RC = 43.6$ ns. For $RC = 436$ ns, the permissible error of 10% for δT_1 is clearly exceeded, while for even larger RC values, $\delta\hat{u}$ also exceeds the limit of 3%.

The response errors can also be represented graphically in *error diagrams* as a function of the front time or another time parameter. This results in a comprehensive characterization of the dynamic behavior of a measuring system. In the following example, the behavior of three fictitious voltage dividers is represented by the step responses of a first and second order system as well as an oscillating system (Fig. 9.13). The selected response times are $T = 50$ ns (curve ***1***), 100 ns (curve ***2***) and 19.9 ns (curve ***3***). Since each response time is greater than the limit of 15 ns recommended in (Ref. [2] of Chap. 2) for a reference divider, none of the three fictitious voltage dividers would be suitable as a reference divider for the measurement of lightning impulse voltages.

Table 9.1 Response errors $\delta\hat{u}$ und δT_1 of three RC elements calculated for a 1.2/50 impulse

RC in ns (curve in Fig. 9.12)	43.6 (2)	436 (3)	1000 (4)
$\delta\hat{u}$ in %	−0.03	−0.8	−3.6
δT_1 in %	0.25	53.7	221

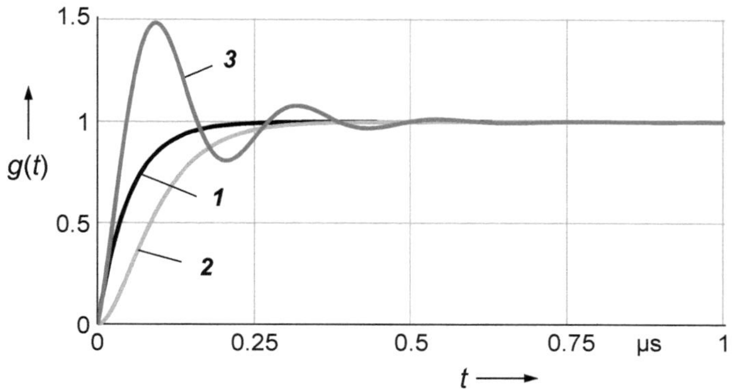

Fig. 9.13 Step responses of different linear systems with the response time *T*. ***1*** First order system (T = 50 ns), ***2*** second order system (T = 100 ns), ***3*** oscillating system (T = 19.9 ns)

However, the results of numerical convolution with the algorithm in Eq. (9.5) lead to a different assessment. The response errors $\delta\hat{u}$ und δT_1 of the three fictitious voltage dividers with the step responses according to Fig. 9.13 are plotted in Fig. 9.14a versus the front time T_1 of full lightning impulse voltages. In summary, it can be concluded that the response errors for the peak value are within ±0.02% and for the front time within ±3%. This means that the transfer behavior of these voltage dividers is adequate to use them as reference dividers for the measurement of impulse voltages with front times in the tolerance range T_1 = 1.2 µs ± 30% [5, 30].

For chopped lightning impulse voltages, Fig. 9.14b shows the calculated response errors $\delta\hat{u}$ and δT_1 of the same fictitious voltage dividers as a function of the time to chopping T_c. The voltage collapse after chopping is exponential with a time constant of 50 ns. As expected, the same response errors are obtained for $T_c \geq 2$ µs as for full lightning impulse voltages. With decreasing time to chopping, the response errors become increasingly negative, i.e. the peak values and front times of the chopped lightning impulse voltages are measured too low. In summary, it can be said that the fictitious voltage dividers could only be used in a very restricted range of T_c as approved impulse voltage dividers or even reference dividers. In testing practice, however, chopped impulse voltages have a rounded peak. The dynamic stressing of the voltage dividers is therefore lower and the response errors are smaller than those calculated for a sharp peak as shown in Fig. 9.14b.

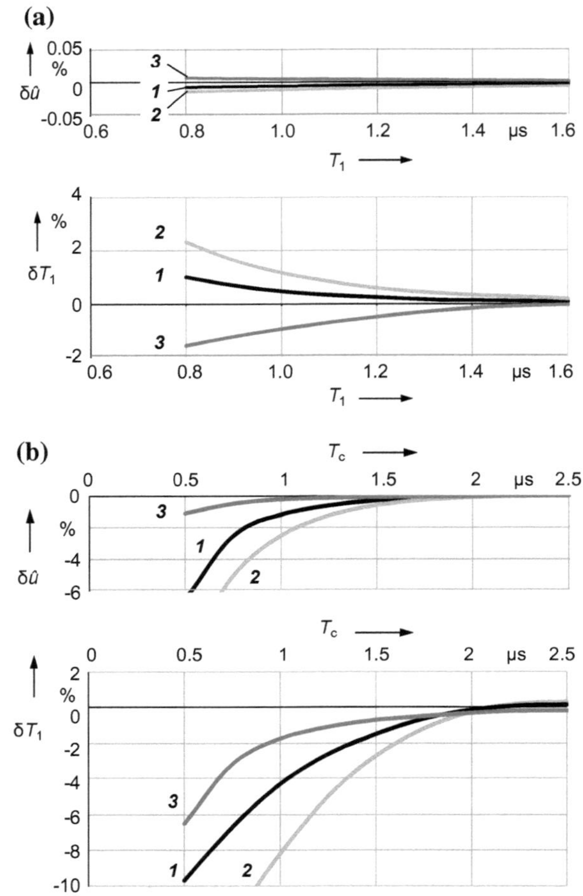

Fig. 9.14 Calculated response errors $\delta\hat{u}$ und δT_1 of three fictitious voltage dividers with the step responses ***1***, ***2*** and ***3*** according to Fig. 9.13. **a** Input voltage: 1.2/50 lightning impulse, top: $\delta\hat{u} = \mathrm{f}(T_1)$, below: $\delta T_1 = \mathrm{f}(T_1)$; **b** input voltage: chopped lightning impulse, top: $\delta\hat{u} = \mathrm{f}(T_c)$, below: $\delta T_1 = \mathrm{f}(T_c)$

It should be taken into account that the data of $g(t)$ and $u_1(t)$ recorded in practice are affected by small measurement errors, which, of course, is noticeable in the result of numerical convolution. The step response of high-voltage dividers cannot be accurately measured due to the well-known insufficiencies of the measurement and test circuits (see Sect. 9.8). Apart from the finite steepness of the generated step voltage and the sampling errors of the digital recorder, the spatially extended, unshielded high-voltage impulse divider cannot be connected to the step voltage generator without reflections and disturbances. The other function in numerical convolution, the impulse voltage $u_1(t)$ applied to the input of the voltage divider, is not known exactly as a rule. In general, $u_1(t)$ can be measured by a high-voltage divider only with limited accuracy, since even reference dividers as defined in (Ref. [2] of Chap. 2) cannot be regarded as error-free due to the permissible errors of $\pm 1\%$ for the peak value and $\pm 5\%$ for the time parameters.

In summary, the numerical convolution method can be very helpful in assessing the dynamic behavior of impulse voltage dividers. However, due to the insufficiencies in the measurement of $g(t)$ and $u_1(t)$, it is advisable to check the results of the convolution calculation by means of a metrological verification. For this purpose, a comparison measurement of the impulse voltages with an accurate reference divider is suitable, with at least one value of the relevant time parameter. This combination of calculation and measurement significantly reduces the effort required to determine the dynamic behavior of an impulse voltage divider. The difference between the calculated and measured parameter values contributes to the uncertainty budget (see Chap. 13).

9.8 Experimental Step Response

The *experimental step response* of voltage or current measuring systems and their components is recorded with a digital recorder under defined measurement conditions and evaluated with software. In contrast to the mathematically exact step response in Sect. 9.1, the experimentally obtained step response will have some imperfections and interferences that affect the determination of the origin, the final value and thus the response parameters. This makes it more difficult to accurately evaluate the dynamic behavior of a measuring system on the basis of its step response, especially for large voltage dividers. In contrast to high-voltage impulse dividers, current sensors are often shielded, resulting in minor disturbances of the experimental step response. In principle, the step response can reproduce transient events in the range of nanoseconds, in particular the time range in which the imperfections and interferences occur. A significant influence on the measurement of lightning impulse voltages and exponential impulse currents due to these interfering effects can therefore usually be excluded.

The experimental step response provides valuable information about the transfer behavior of a measuring system and its components. It reveals a non-optimal design of the measuring system, which can then be improved by means of simulation calculation with commercial or self-developed software. Requirements for the course of the experimental step response are given in the relevant test standards (Refs. [2, 4] of Chap. 2). From the experimental step response, several response parameters can be derived, which also serve as a criterion for the adequate dynamic behavior of the measuring system. Furthermore, the step response may be stored as a *fingerprint* of the measuring system for its identification and proof of long-term stability. Finally, with some limitations, the experimental step response can be used for the numerical convolution calculation in order to obtain the response errors of a measuring system for predefined input impulses.

9.8.1 Analysis of the Experimental Step Response

High-voltage and high-current measuring systems have very different step responses, depending on their construction and dimensions. The analysis of the experimentally determined step response can be affected by several insufficiencies of the measuring circuit and by interference as follows. The generated step voltage may not have the ideal rectangular shape, and measurement errors of the digital recorder may affect the recording of the step response. Impulse voltage dividers with high-voltage lead, which are usually not or only insufficiently shielded, receive electromagnetic interference of all kinds due to their large dimensions and construction like an antenna. On the other hand, they themselves radiate the high-frequency signals of the step voltage to the environment [31–33]. Due to mismatch between the voltage divider and the step voltage generator, reflection phenomena occur, which lead to damped travelling waves on the connecting leads. Frequently, the voltage divider is provided with a different lead in the step response measurement as in the high-voltage test.

The inadequacies mentioned intensify with increasing divider height as shown on the example of a damped capacitive 6 MV voltage divider [34]. Far better measurement conditions exist for coaxial shunts which allow a reflection-free connection to the step generator via a coaxial cable due to their smaller dimensions and their shielded construction. Also, the step response of a current measuring coil, through whose opening the conductor is led and connected reflection-free to the step generator, can generally be recorded without any problem.

The beginning of the recorded step response and thus also the zero point can often not be clearly determined. This may be due to the insufficient bandwidth of the recorder and to electromagnetic interference. In the test specifications, the origin O_1 is therefore defined as the point at which the step response first rises monotonically from zero (Refs. [2, 4] of Chap. 2). The evaluation of the step response at the end of the record is equally problematic due to superimposed disturbances, but is necessary to fix the amplitude value "1" (Fig. 9.15). The *reference level* to which the value "1" is assigned is calculated from the mean of the samples in the *reference level epoch* extending from $0.5t_{min}$ to $2t_{max}$. The times t_{min} und t_{max} denote the limits of the permissible front time of the impulse to be measured. The step response of an approved measuring system may only vary by ±2% in the reference

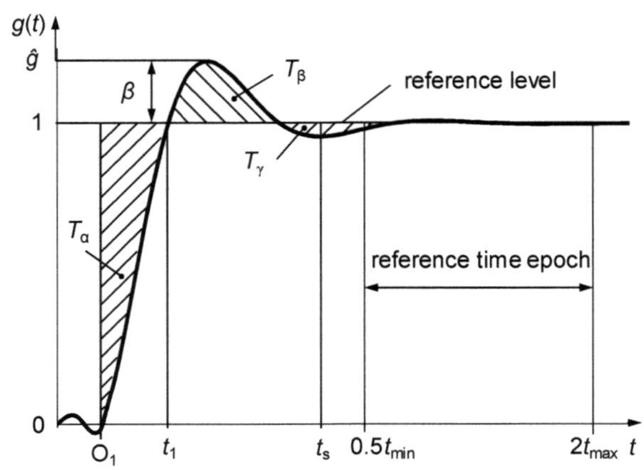

Fig. 9.15 Experimental step response $g(t)$ of a measuring system with defined origin O_1, reference level, reference time epoch and response parameter

level epoch and by ±5% in the extended range from $2t_{max}$ to $2T_{2max}$ (see Sect. 10.3.8). Experiences and suggestions for computer-aided evaluation of step responses can be found in ([32–43]; Ref. [99] of Chap. 4).

9.8.2 Response Parameters of the Step Response

The step response of impulse voltage dividers can be characterized by four *response parameters* for which maximum limit values are prescribed in international standards. Compliance with the limit values is considered as proof that the dynamic behavior of the measuring system is adequate to meet the prescribed error limits for the peak value and the time parameters. However, the significance of the response parameters has been reduced in the revised test standards and recommended limit values for the response parameters refer only to reference systems for full and chopped impulse voltages (Ref. [2] of Chap. 2).

For impulse current measuring systems, the response parameters with prescribed limit values have not gained acceptance (Ref. [4] of Chap. 2). On the one hand, there is a wide variety of standardized impulse currents, for which reliable limit values should have been determined. On the other hand, the direct evaluation of the step response according to Sect. 10.3.8 or the calculation of the measurement deviations with the convolution algorithm according to Sect. 9.2 offer far more effective alternatives.

The four response parameters of the step response $g(t)$ according to IEC 60060-2 (Ref. [2] of Chap. 2) are:

$$\textit{Experimental response time } T_N : \boxed{T_N = \int_{O_1}^{2t_{max}} [1 - g(t)]\mathrm{d}t} \tag{9.44}$$

$$\textit{Partial response time } T_\alpha : \boxed{T_\alpha = \int_{O_1}^{t_1} [1 - g(t)]\mathrm{d}t} \tag{9.45}$$

$$\textit{Settling time } t_s : \boxed{\left| T_N - \int_{O_1}^{t} [1 - g(t)]\mathrm{d}t \right| \leq 0.02 t_s} \tag{9.46}$$

$$\textit{Overshoot } \beta : \boxed{\beta = \hat{g} - 1}. \tag{9.47}$$

The *experimental response time* T_N according to Eq. (9.44) allows to a limited extent an estimation of the peak value error of the measured impulse voltage. It differs from the mathematical definition of the response time T given in Eq. (9.31) by the lower and upper integration limits.

The *partial response time* T_α in Eq. (9.45) is used to judge the correctness of a measuring system during fast voltage changes in the front of an impulse voltage, e.g. in the case of superimposed oscillations.

The *settling time* t_s defined in Eq. (9.46) is the shortest time that satisfies the given inequality for all times $t \geq t_s$ of the step response up to the upper integration limit $2t_{max}$. The settling time t_s can be interpreted such that for $t > t_s$ the remaining contribution of the step response to the response time is not more than 2% of t_s. Besides experimental response time T_N, the settling time t_s is considered the most important parameter of the step response.

The *overshoot* β is given by the amount by which the maximum value of the oscillating step response exceeds the reference level. If the step response of a voltage divider shows a large overshoot, the front of the measured impulse voltage will be reproduced incorrectly. A diagram shows the permissible range of β as a function of the ratio T_α/T_1 resulting from simulation calculations with various forms of the step response and the impulse voltage [44].

Note Due to the modified definition of the origin O_1 of the step response in (Ref. [2] of Chap. 2), the previously defined *initial distortion time* T_0 has practically no more meaning and is not treated here.

The characterization of the step response by the response parameters and the setting of corresponding limit values may seem practical, but there is no clear quantitative relationship between these response parameters and the measurement deviations of an impulse voltage divider. The examples in Sect. 9.7.4 calculated with the convolution algorithm show that the limit values of the response parameters set for impulse voltage dividers are not always justified.

The response parameters are usually calculated using software from the data of the recorded step response. The correctness of the software can be checked with analytically calculable step responses of ideal systems (see Sect. 9.5). However, the partly different definitions of the response parameters for the experimental and analytical step responses have to be taken into account, e.g. with regard to the origin and the integration limits [45].

The response parameters can basically also be determined graphically from the partial areas between the step response $g(t)$ and the reference level (see Fig. 9.15). The experimental response time T_N results from the sum of the partial areas, whereby those above the reference level being are added with a negative sign:

$$T_N = T_\alpha - T_\beta + T_\gamma - \cdots . \tag{9.48}$$

However, graphical determination of the response parameters is now rarely done since the previously used analog oscilloscopes, for which a graphical evaluation was required, have largely been replaced by digital recorders with computer-aided data evaluation.

Note In earlier test specifications and publications before issuing (Refs. [2, 4] of Chap. 2), the virtual origin O_1 of the step response was defined as the intersection of the tangent to the steepest part of the step response with the time axis. Furthermore, the tangent replaced the recorded initial course of the step response below its steepest part. However, the prescribed evaluation often resulted in a significant change in the initial step response, which of course affected the calculation of the response parameters. The modified determination of the zero point O_1 in (Refs. [2, 4] of Chap. 2) can lead to deviations of the response parameters, in particular in the case of T_α und T_N, compared with earlier evaluations.

9.8.3 Measuring Circuits for the Step Response

Three arrangements for recording the step response of a voltage divider are standardized. The circuit according to Fig. 9.16a with horizontal lead between the step generator ***1***, damping resistor ***2*** and voltage divider ***3*** corresponds most closely to the usual arrangement of the voltage divider in the high voltage test circuit and is therefore recommended in the test specifications (Ref. [2] of Chap. 2). The step generator is located at the same height as the divider top, with the lead length approximately equal to the divider height. An at least 0.5 m wide low-inductance copper foil ***5*** serves as the ground return conductor between the step generator and the divider bottom.

The other two arrangements in Fig. 9.16b and c are also standardized (Ref. [2] of Chap. 2), although the position and length of the leads are not comparable to those in the high-voltage test. Therefore, step responses and response parameters are

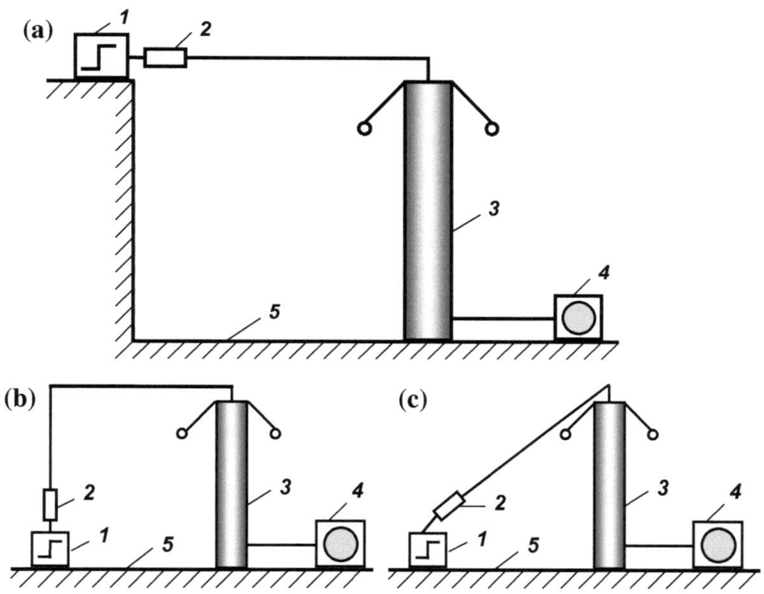

Fig. 9.16 Arrangements for the step response measurement of high-voltage dividers (schematic). **a** Recommended set-up of the voltage divider with horizontal lead; **b** voltage divider with rectangular lead; **c** voltage divider with inclined lead. ***1*** Step voltage generator, ***2*** damping resistor R_d, ***3*** impulse voltage divider, ***4*** measuring instrument, ***5*** earth return conductor

expected to deviate from those obtained in the arrangement of Fig. 9.16a. If the step response is only used as a fingerprint to identify a voltage divider and to verify its long-term stability, i.e. without further evaluation of the response parameters, the simple arrangement in Fig. 9.16c is often preferred.

For the step response measurement, the voltage divider should be assembled as far as possible as for the high-voltage test. The voltage divider, high-voltage lead, external damping resistor and coaxial cable to the recorder always form a unit. Any change of the divider position and spatial arrangement of the lead, the distance to the hall wall or to other objects and even the hall size can affect the step response (see Fig. 4.33). Changes in the measurement setup can be simulated and their influence on the step response can be calculated using special software [46]. In general, the recorder used for the high-voltage test must be replaced by another recorder with a higher sampling rate and sensitivity. The bandwidth of the recorder for the step response measurement also has an influence on the step response and the response parameters.

A small part of the step voltage is radiated from the step generator and the unshielded lead and reaches the unshielded voltage divider as an electromagnetic wave. An interfering signal is thus coupled into the unshielded lower part of the voltage divider, which arrives there earlier than the step voltage conducted via the lead and through the voltage divider. The interfering signal is superimposed on the step response at its beginning, whereby its origin is partially hidden. Interference is stronger in the two circuits with vertical and inclined leads (Fig. 9.16b and c), since the distance of the step generator and the lead to the lower part of the divider is less than in the arrangement according to Fig. 9.16a.

The infinitely long horizontal lead to the voltage divider is characterized by its *characteristic impedance* Z (or: *surge impedance*) (see Sect. 4.3.1.2.1). For example, $Z = 360\ \Omega$ for a lead with a diameter $d = 2$ cm and a height $h = 2$ m above the hall floor according to Eq. (4.12). In the case of a finitely long horizontal lead, similar to the arrangement in Fig. 9.16a, it is common practice to approximately calculate Z also by Eq. (4.12). The external damping resistor R_d at the beginning of the lead has two functions. For $R_d = Z$, the finitely long lead is approximately terminated with its impedance, so that travelling waves on the lead are largely suppressed. With the matched damping resistor, the influence of different lead lengths on the step response and thus on the measurement behavior of the voltage divider is also greatly reduced. This is particularly advantageous for reference dividers that are used with variable lead length for the calibration of differently sized voltage dividers by comparison measurement. The size difference is compensated with the corresponding lead length of the reference divider.

9.8.4 Generation of Step Voltages

The principle of the circuit in Fig. 9.17 is suitable for generating step voltages with short rise times. When the switch S is opened, the DC voltage U_0 is applied to the

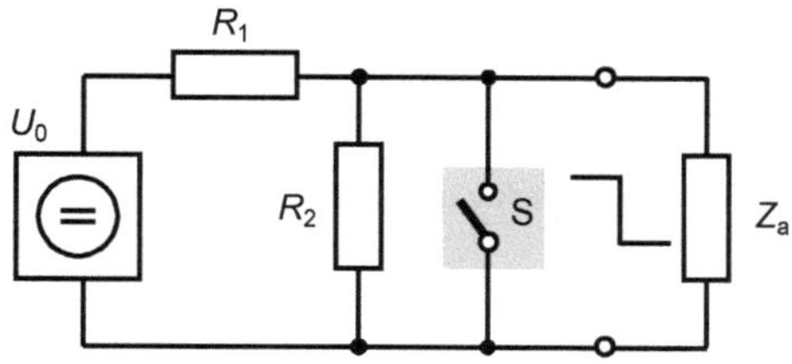

Fig. 9.17 Circuit principle of a generator with reed contact S for generating step voltages

terminating impedance Z_a of the test object via the high-ohmic limiting resistor R_1. As a result, the voltage at Z_a increases relatively slowly up to a final value determined by U_0, Z_a, R_1 and R_2. The charging time is additionally determined by the inductance of the lead and the parallel capacitance of the components. When the switch S is closed, depending on the polarity of the applied direct voltage U_0, a negative or positive voltage step against ground occurs. The steepness of the step depends on the switch itself and the inductance of the leads in the discharge circuit and is usually much greater than that of the initial increase of the charging voltage. Only this steep flank of the shorted voltage is used for the step response measurement.

As a fast switch, a *reed relay* with mercury-wetted contact is preferably used, since in contrast to simple mechanical switches, bounce-free switching of voltages up to a maximum of 1000 V or currents up to 2 A is possible. The contact resistance of the short-circuited reed contact—and thus the output resistance of the step generator—lies below 10 mΩ. The movable reed contact is actuated once or periodically by means of an exciter coil, which is not shown in Fig. 9.17. The period duration of the generated rectangular voltage can be selected almost as long as desired. Thus, with decreasing exciter frequency of the relay coil, the step responses can be recorded in the range of seconds. Restrictions on the use of mercury due to ecological constraints have severely limited the availability of more powerful reed contacts.

The circuit principle in Fig. 9.17 is basically suitable for all components of an impulse voltage or current measuring system. If Z_a is a capacitive voltage divider without or with only a small damping resistor, attention must be paid to the maximum permissible current load of the reed contact during short-circuiting. Currents above the rated value lead to contact disturbances and shorten the life of the reed contact considerably. If Z_a is a recorder or a low-ohmic coaxial shunt connected to the step generator via a long coaxial cable, it is recommended to connect a series resistor equal to the impedance of the cable at the output of the step generator to avoid reflection phenomena.

Another possibility to generate step voltages with short rise times offers a cable generator (Fig. 9.18). The coaxial cable with the wave impedance Z is charged to the voltage U_0 through the series resistor R_1. When the switch S is closed, a steep voltage wave enters the cable. It is reflected at the open end of the cable and runs back to the cable input with the same amplitude but opposite polarity. At the terminating impedance Z_a, a voltage step is produced whose amplitude is

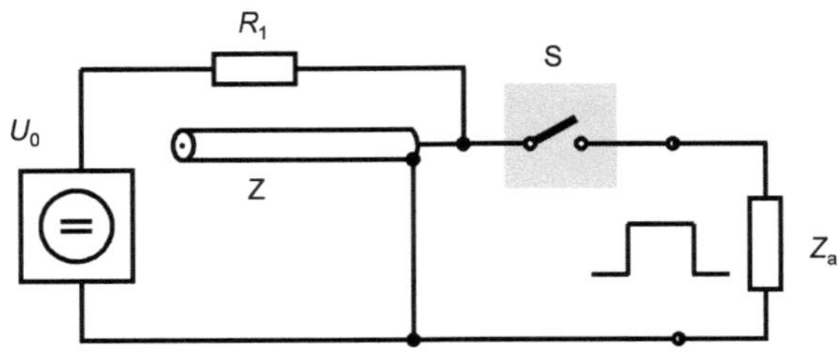

Fig. 9.18 Schematic diagram of a cable generator for generating rectangular voltages with duration of twice the cable delay time and small rise time

determined by the ratio of the impedances Z_a and Z. For $Z_a \neq Z$, reflection phenomena occur after twice the cable delay time. The duration of the rectangular impulse generated depends on the cable length L and the cable delay time $t = 2L/v$ with $v = 1/\sqrt{L_0 C_0} \approx 0.2$ m/s. The cable length is usually limited so that the duration of the rectangular impulse voltage is less than 500 ns in order to limit the drooping of the impulse caused by the attenuation losses of the cable. With a mercury-wetted reed contact as switch S and by optimal matching of the test object to the cable generator, step amplitudes of up to 500 V and rise times of less than 0.5 ns can be achieved.

Step voltage generators with electronic switches, e.g. with thyristors or avalanche transistors, are also used [47, 48]. The tuning of the electronic components must be carefully done in order to avoid drooping of the rectangular impulse. For thyristor circuits, the amplitude of the step voltage is up to 1000 V and the rise time is more than 10 ns. For circuits with avalanche transistors, the maximum step amplitude is a few tens of volts and the rise time is less than 1 ns. Electronic step voltage generators often have an output resistance of the order of the impedance of coaxial cables.

Step voltages of considerably more than 1000 V can be generated with a sphere gap as switch [32, 49]. While with oil-immersed sphere gaps rise times down to 10 ns are available, small compressed-gas insulated sphere gaps enable rise times in the range of a few nanoseconds. The sphere gap is operated with a rectified AC voltage or with an impulse voltage of longer front time. The sphere gap fires in the region of the peak, where the voltage change is practically zero. Sphere gaps are also used together with a high-voltage cable as a cable generator [49, 50].

High step amplitudes have the advantage that the step response of the complete high-voltage measuring system, i.e. including the recorder and its internal attenuator, can be measured. On the other hand, the voltage $u_1(t)$ generated with the sphere gap usually differs significantly from the ideal shape of a step voltage. Consequently, the voltage $u_2(t)$ recorded at the output of the measuring system is not the exact step response. An improved expression for the step response can be obtained by performing the FFT of the voltages $u_2(t)$ and $u_1(t)$ and calculating their quotient in the frequency domain, resulting in the transfer function $H(\mathrm{j}\omega)$ of the system according to Eq. (9.9). With Eq. (9.14) and after inverse transformation into the time domain, a good approximation of the step response $g(t)$ is obtained. However, if the high signal frequencies are missing in the input and output voltages

$u_1(t)$ and $u_2(t)$, the step response calculated from them cannot provide information about the behavior of the measuring system at these high frequencies.

The step voltage generators presented so far are rather small compared to the large dimensions of a high-voltage divider. From the point source, a spherical wave is electromagnetically radiated, which reaches the upper and lower part of the unshielded voltage divider at different times. The portion radiated into the lower part makes itself noticeable as a disturbance signal which is superimposed on the beginning of the recorded step response. Thus, the true origin of the step response cannot be detected (see Sect. 9.8.1). In order to avoid this interference effect, the use of a step generator with distributed sources is suggested in [28, 52]. This pulse generator consists of ten electronic step generators arranged one above the other, which are simultaneously triggered by optical fibers. The spherical waves generated by each of the individual step generators superimpose, so that an approximately cylindrical wave is radiated. The resulting E-field propagates nearly parallel to the multi-stage step generator and reaches the upper and lower part of the voltage divider at the same time. The generated step voltage has a rise time of 10 ns, the maximum amplitude is 500 V. However, the oscillograms shown for the step response of a 3.8 MV impulse voltage divider differ only minimally when using the multi-stage step generator or a conventional single-stage step generator.

The step voltage generators can even be used to calibrate impulse current measuring systems and their components (Refs. [33, 36, 37] of Chap. 5). In this case, the measuring shunt—or the conductor through the measuring coil—is connected to the step voltage generator via a terminating resistor to avoid reflections. If reflections are unavoidable, the step response will only be evaluated until the first reflection occurs. With mercury-wetted reed contacts, current amplitudes of a maximum of 1–2 A can be generated. Due to the usually very low resistance of shunts or equivalent resistance of measuring coils, amplitudes of the step response of only 1 mV and below are obtained, so that a preamplifier for the recorder is required. The use of electronic circuits and sphere gaps for generating higher step currents is often associated with the disadvantage of a larger rise time, drooping on the tail or superimposed oscillations. The step response of the test object cannot be then directly determined experimentally, but must be calculated from the input and output voltages using the Fourier transform [51].

9.9 Supplementary Considerations on the Transfer Behavior

The task of a measuring system is the error-free measurement of test voltages or test currents. An important prerequisite for this is a frequency-independent transfer factor $H(f)$ of the measuring system up to the upper limit frequency f_2, which must lie above the limit frequency f_2' of the signal spectrum $F(f)$ (Fig. 9.19). The lower limit frequency of the measuring system is generally $f_1 = 0$, but may also have a value above zero. At the upper limit frequency f_2, the transfer factor $H(f_2)$ has fallen

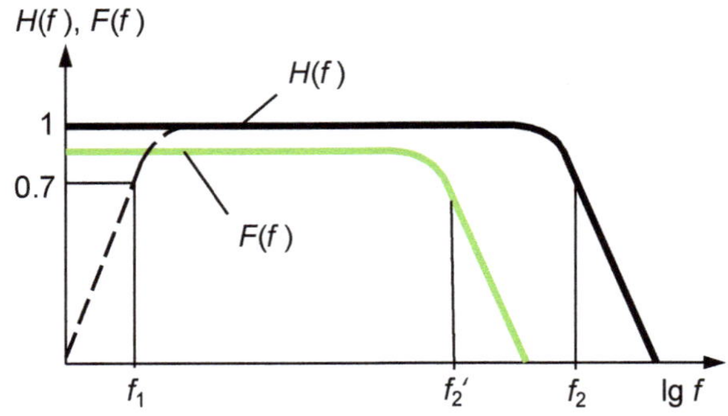

Fig. 9.19 Transfer factor $H(f)$ of a measuring system with the limit frequencies f_1 and f_2 and amplitude density $F(f)$ of an impulse with the upper limit frequency f_2' (schematic)

by 3 dB, i.e. $H(f_2)$ is then only $1/\sqrt{2}$ corresponding to about 70% of the original value. For measurements with accuracy requirements in the percentage range, this amplitude drop is generally too large. The numerical value given for f_2 often leads to a subjective misinterpretation of the transfer behavior of a measuring system. The question arises as to which upper limit frequency f_2 a measuring system must have for accurately measuring impulse voltages. There are several approaches to answer this question.

In impulse measurement technique, it is well-known that the rise time of a measuring system should not be more than one tenth of the rise time of the signal to be measured. For the 1.2/50 lightning impulse voltage with the front time T_1 = 1.2 μs, which according to Eq. (9.34) corresponds to a rise time T_a = 960 ns, this requirement is just fulfilled by the RC element with the time constant $RC = T$ = 43.6 ns and the rise time $T_a = 2.2T$ = 96 ns (curve 3 in Fig. 9.7). The upper limit frequency of the RC element is f_2 = 3.7 MHz and is thus 1500 times greater than the limit frequency f_2' = 2.4 kHz in the spectrum of the lightning impulse voltage (curve 1 in Fig. 9.7). An alternative criterion is that the range of the approximately constant amplitude response $H(f)$ of the measuring system (curve 3 in Fig. 9.7) must extend at least to the frequency at which the amplitude density $F(f)$ of the signal (curve 1 in Fig. 9.7) has fallen by more than 60 dB, corresponding to a factor of 1000.

The step response of a measuring system being known, its output voltage can be calculated for any input voltage by means of convolution. The comparison of the output and input voltages gives the transfer error, i.e. the measurement deviations for the peak value and the time parameters. Model calculations for systems with different frequency responses can be used to find the measurement system with the bandwidth B—or upper limit frequency f_2—for which the transfer errors are smaller than the predefined limit values (see Sect. 9.7). When looking at the results of the numerical convolution in Sect. 9.7.4, the following rule of thumb can be established for the required upper limit frequency f_2 of the measuring system:

$$f_2 \geq (100 \ldots 1000) f_2'$$

where the factor 1000 applies to the measurement of the front time and the factor 100 to that of the peak value, the measurement deviation being within ±1%. The 1.2/50 lightning impulse voltage with $f_2' = 2.4$ kHz (see curve 1 in Fig. 9.7) therefore requires a bandwidth of at least 240 kHz for measuring the peak value and 2.4 MHz for measuring the front time. Requirements related to the upper limit frequency for the measurement of oscillations which may be superimposed on the peak or the front of the impulse voltage, have not been considered here.

The amplitude response of the wedge-shaped voltage, which is represented by curve 5 in Fig. 8.4 and is approximately equal to front-chopped impulse voltages, has a limit frequency $f_2' = 1.1$ MHz. Following the rule of thumb, the upper limit frequency of the measuring system should be more than 110 MHz for the peak value measurement. In test practice, however, the peak area of front-chopped impulse voltages is most likely rounded. This means that the amplitude spectrum has a lower limit frequency, so that the bandwidth requirement is reduced.

Comparable considerations are to be made if the lower limit frequency f_1 of the measurement system is greater than zero. The system then acts as a high-pass filter and is not able to detect the DC component of a measurement signal. Even slowly varying voltages or currents are not measured true to original. In particular, the tail of switching impulse voltages, the roof of rectangular impulse currents and short-circuit currents with slowly decreasing DC component are affected. For example, the impulse tail recorded by a measuring system with high-pass filter characteristic shows an apparently faster drop, so that the time to half-value T_2 of a switching impulse voltage is measured too short. Depending on the type of test, measuring systems for short-circuit currents shall have a lower limit frequency of at least 0.2 Hz in order to correctly measure the transient DC component (Ref. [4] of Chap. 2).

The distortion of an impulse by a measuring system with insufficient limit frequency or rise time is treated here by two examples with rectangular impulses. A measuring system with the lower limit frequency $f_1 = 0$ is referred to as a low-pass filter or, if the upper limit frequency f_2 is very large, as a wide-band system. An example of a simple low-pass filter is the RC element as shown in Fig. 9.20a in which the resistor lies between input and output. Due to its integrating effect on the input signal, the RC low-pass filter is also used as an integrator. If a rectangular voltage $u_1(t)$ with the duration t_d is applied to the input of the RC low-pass filter, the capacitor C will be charged via the resistor R with the time constant $\tau = RC$. An almost true-to-original output voltage $u_2(t) \approx u_1(t)$ is obtained for $RC \ll t_d$. With increasing time constant or decreasing limit frequency of the RC low-pass filter, an increasing deformation of the rising and falling flanks occurs (Fig. 9.20b). For $RC \gg t_d$, the output voltage is strongly distorted (Fig. 9.20c).

A measuring system with the lower limit frequency $f_1 > 0$ acts as a high-pass filter or, if the upper limit frequency f_2 is limited, as a bandpass. The simplest high-pass filter is, in turn, an RC element in which the capacitor C is connected between the input and output and the output voltage is across the resistor R (Fig. 9.21a). Due to its differentiating effect on the input signal, the RC high-pass

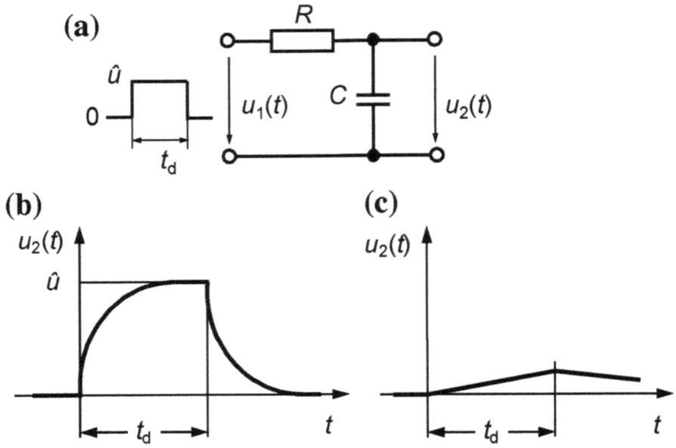

Fig. 9.20 Distortion of a rectangular signal of duration t_d by a low-pass filter. **a** RC low-pass filter with rectangular signal $u_1(t)$ at the input; **b** output signal $u_2(t)$ for $RC < t_d$; **c** output signal $u_2(t)$ for $RC \gg t_d$

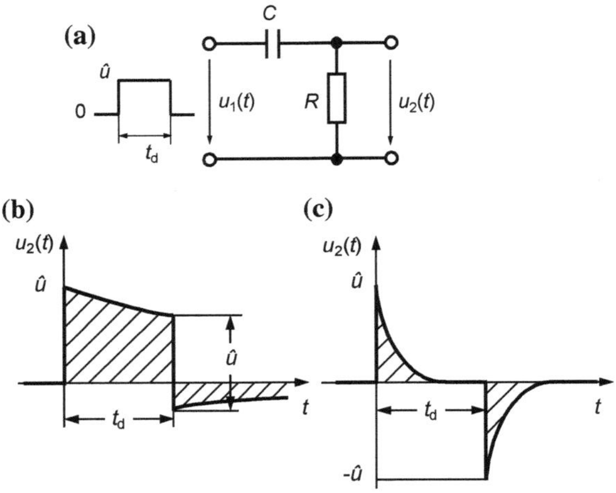

Fig. 9.21 Distortion of a rectangle signal of duration t_d by a high-pass filter. **a** RC high-pass filter with rectangular signal at the input; **b** output signal for $RC > t_d$; **c** output signal for $RC \ll t_d$

filter is also referred to as differentiator. If a rectangular voltage $u_1(t)$ is applied to the input of the RC high-pass filter, the output voltage $u_2(t)$ is more or less deformed due to the differentiating effect (Fig. 2.21b and c). Characteristic is the voltage droop in Fig. 9.21b. The differentiator with $RC \ll t_d$ produces positive and negative needle pulses at the flanks of the rectangular voltage (Fig. 9.21c). Since the high-pass filter cannot transmit the DC component, the hatched positive and negative partial areas of the output pulse are equal.

References

1. Oppenheim, A.V., Willsky, A.S., Young, I.T.: Signals and Systems, 2nd edn. Prentice-Hall, Englewood Cliffs (1996)
2. Beerends, R.J., et al.: Fourier and Laplace Transforms. University Press, Cambridge (2003)
3. Küpfmüller, K.: Die Systemtheorie der elektrischen Nachrichtenübertragung. S. Hirzel Verlag, Stuttgart (1974)

4. Harris, J.W., Stöcker, H.: Handbook of Mathematics and Computational Science. Springer, New York (1998)
5. Glinka, M., Schon, K.: Numerical convolution technique for qualifying hv impulse dividers. In: Proceedings of 10th ISH, Montreal, vol. 4, pp. 71–74 (1997)
6. Li, Y., Rungis, J., Pfeffer, A.: The voltage and time parameter measurement uncertainties of a large damped capacitor divider due to its non-ideal step response. In: Proceedings of 15th ISH, Ljubljana, Paper T10-499 (2007)
7. Gobbo, R., Pesavento, G.: Evaluation of convolution to assess the performance of impulse dividers. In: Proceedings of 14th ISH, Beijing, Paper J-08 (2005)
8. Sato, S., Harada, T., Wakimoto, T., Saeki, M.: Numerical convolution for impulse voltage measuring systems. In: Proceedings of 13th ISH, Delft, Paper 195 (2003)
9. Darveniza, M., Holcombe, B.C.: A fast Fourier Transform Technique for correcting impulse voltage divider measurements. In: Proceedings of 3rd ISH, Mailand, Paper 42.15 (1979)
10. Kiersztyn, S.E.: Numerical correction of HV impulse deformed by the measuring system. IEEE Trans. PAS **99**, 1984–1995 (1980)
11. McKnight, R.H., Lagnese, J.E., Yi, X.Z.: Characterizing transient measurements by use of the step response and the convolution integral. IEEE Trans. IM **39**, 346–352 (1990)
12. Schon, K., Gitt, W.: Reconstruction of high impulse voltages considering the step response of the measuring system. IEEE Trans. PAS **101**, 4147–4155 (1982)
13. Charrat, O., Demoment, G., Segalen, A.: High voltage impulse restoration with a fast deconvolution techniques. IEEE Trans. PAS **103**, 841–848 (1984)
14. Nikolopoulos, P.N., Topalis, F.V.: Error free registration of high impulse voltages. In: Proceedings of 5th ISH, Braunschweig, Paper 71.03 (1987)
15. Xu, W., Kwan, S.-K.: Numerical correction of HV impulses by the method of time-frequency-time domain transform. In: Proceedings of 5th ISH, Braunschweig, Paper 71.04 (1987)
16. Kollár, I., Osvath, P., Zaengl, W.S.: Numerical correction and deconvolution of noisy impulses by means of Kalman filtering. In: Proceedings of IEEE International Symposium on Electrical Insulation, Boston, pp. 359–363 (1988)
17. Narduzzi, C., Zingales, G.: An inverse filtering approach to reconstruction of HV impulses. In: Proceedings of 6th ISH, New Orleans, Paper 47.09 (1989)
18. Bak-Jensen, B., Bak-Jensen, J.: Some problems arising from the use of Fourier Transform to reconstruction of deformed high voltage impulses. In: Proceedings of 6th ISH, New Orleans, Paper 42.29 (1989)
19. Younan, N.H., Kopp, A.B., Miller, D.B., Taylor, C.D.: On correcting HV impulse measurements by means of adaptive filtering and convolution. IEEE Trans. PWRD **6**, 501–506 (1991)
20. Matyas, Z., Aro, M., Damstra, G.C.: Frequency-domain analysis contribution to HV measurement quality. In: Proceedings of 13th ISH, Delft, Paper 807 (2003)
21. Yan, W., Li, Y.: Improvement of measurement uncertainty of electrical fast transient burst parameters using step response measurement and deconvolution. In: Proceedings of 19th ISH, Pilsen, Paper 310 (2015)
22. Poulain, B., Malewski, R.: Impulse testing of power transformers using the transfer function method. IEEE Trans. PWRD **3**, 476–489 (1988)
23. Leibfried, T., Feser, K.: Monitoring of power transformers using the transfer function method. IEEE Trans. PWRD **14**, 1333–1341 (1999)
24. Rahimpour, E., Christian, J., Feser, K., Mohseni, H.: Calculation of the transfer function to diagnose axial displacement and radial deformation of transformer windings. In: Proceedings of 12th ISH, Bangalore, Paper 3519 (2001)
25. Wimmer, R., Feser, K., Christian, J.: Reproducibility of transfer function results. In: Proceedings of 13th ISH, Delft, Paper 165 (2003)
26. Christian, J., Feser, K.: Procedures for detecting winding displacements in power transformers by the transfer function method. IEEE Trans. PWRD **19**, 214–220 (2004)
27. Wimmer, R., Feser, K.: Berechnung der Übertragungsfunktion aus Online-Messdaten. In: ETG-Tagung "Diagnostik elektrischer Betriebsmittel", Köln, Paper 3.13 (2004)
28. Schwab, A., Bellm, H., Sautter, D.: Peak-error correction for front-chopped impulse voltages. In: Proceedings of 3rd ISH, Mailand, Paper 42.13 (1979)

29. Gobbo, R., Pesavento, G.: Analysis of errors in the measurement of chopped impulses. In: Proceedings of 11th ISH, London, Paper 467 (1999)
30. Schon, K.: What is new in IEC 60060-2: uncertainty and convolution. In: HIGHVOLT Kolloquium ‘07, Dresden, Paper 1.2 (2007)
31. Gonschorek, K.H.: The electromagnetic behavior of widely extended high voltage test circuits. In: Proceedings of 3rd ISH, Mailand, Paper 42.02 (1979)
32. Schwab, A., Maier, R., Bellm, H.: Calculation of transient electromagnetic fields with the transient elementary dipole method. In: EMC-Symposium, Zürich, No. 29 F1 (1987)
33. Maier, R., Schwab, A.: Transient radiation-field interaction with impulse voltage dividers. In: Proceedings of 5th ISH, Braunschweig, Paper 71.12 (1987)
34. Palva, V.: Facing UHV measuring problems. ELECTRA **35**, 157–254 (1974)
35. Berger, K., Ašner, A.: Neue Erkenntnisse über das Verhalten und die Prüfung von Spannungsteilern zur Messung sehr hoher, rasch veränderlicher Stoßspannungen. Bull. SEV **51**, 769–783 (1960)
36. Dharmalingam, K., Gururaj, B.I.: Step response of UHV impulse voltage dividers using IEC square loop lead arrangement. In: Proceedings of 3rd ISH, Mailand, Paper 42.03 (1979)
37. Hyltén-Cavallius, N., Chagas, F.A., Appelda Silva, M.: Response errors of impulse test circuits. IEEE Trans. PAS **103**, 3277–3285 (1984)
38. Rungis, J., Schon, K.: The evaluation of impulse divider response parameters. IEEE Trans. PWRD **3**, 88–95 (1988)
39. Bolognesei, F., Bombonato, M., Cherbaucich, C., Rizzi, G.: Digital measurement of a unit step response: analysis of the computer determination of the response parameters. In: Proceedings of 6th ISH, New Orleans, Paper 50.11 (1989)
40. McComb, T.R., Chagas, F.A., Feser, K., Gururaj, B.I., Hughes, R.C., Rizzi, G.: Comparative measurements of HV impulses to evaluate different sets of response parameters. IEEE Trans. PWRD **6**, 70–77 (1991)
41. Oliveira, O.B., Junqueira, A.J.S.: Digitized step response of HV measuring systems and the calculation of their parameters. In: Proceedings of 8th ISH, Yokohama, Paper 53.04 (1993)
42. Kumar, O.R., Kanyakumari, M., Kini, N.K., Priya, S., Nambudiri, P.V.V., Srinivasan, K.N.: Software evaluation of step response parameters of high voltage dividers. In: Proceedings of 8th ISH, Yokohama, Paper 51.01 (1993)
43. Li, Y., Rungis, J., Sheehy, R.: Impulse divider unit step response evaluation. In: Proceedings of 9th ISH, Graz, Paper 4452 (1995)
44. Qi, Q.-C., Zaengl, W.S.: Investigation of errors related to the measured virtual front time T_1 of lightning impulses. IEEE Trans. PAS **102**, 2379–2390 (1983)
45. Schon, K., Lucas, W., Arndt, V., Cherbaucich, C., Rizzi, G., Deschamps, F., Ribot, J.J., Garnacho, F., Perez, J., Gomes, N., Dias, C., Aro, M., Valve, P., Claudi, A., Lehmann, K., Strauss, W., Notkonen, E.: International comparison of software for evaluating HV impulses and step responses. In: Proceedings of 8th ISH, Yokohama, Paper 51.01 (1993)
46. Ishii, M., Liao, W.: Numerical electromagnetic analysis of impulse voltage measuring systems. In: Proceedings of 10th ISH, Montréal, Paper 3519 (1997)
47. Pfeiffer, W.: Erzeugung von Rechteckimpulsen mit Avalanche-Transistoren. Internat. Elektron. Rdsch. **24**, 178–180 (1970)
48. Schoenwetter, H.K.: A programmable precision voltage-step generator for testing waveform recorders. IEEE Trans. IM **33**, 196–200 (1984)
49. Creed, F.C., Kawamura, T., Newi, G.: Step response of measuring systems for high impulse voltages. IEEE Trans. PAS **86**, 1408–1419 (1967)
50. Liu, R., Zou, X., Wang, X., Zeng, N., Han, M.: A high voltage nanosecond pulse generator for calibrating voltage dividers. In: Proceedings of 14th ISH, Beijing, Paper J-27 (2005)
51. FGH: Übertragungsverhalten von Hochstrom-Messeinrichtungen für transiente Ströme. FGH Technischer Bericht Nr. 292 (2000)
52. Bellm, H., Maier, R., Schwab, A.: Impulse voltage divider equivalent circuit considering transient radiation effects. In: Proceedings of 4th ISH, Athen, Paper 61.08 (1983)

Chapter 10
Calibration of the Measuring Systems

Abstract For standardized tests of power apparatus used in the electrical power supply, approved measuring systems are used to measure the relevant test voltages and test currents. “Approved” in this case means that the measuring system meets the requirements of the test and calibration standards. The approval of a measuring system is achieved by a traceable calibration, i.e. a calibration that can be traced back to national or international measurement standards. After an introductory discussion on standardization, accreditation and traceability, the following sections deal with the mainly applied calibration methods, with emphasis on impulse voltage and current measuring systems. These are, for example, the determination of the scale factor by comparison with a reference system, the linearity test and the evaluation of the experimental step response. The principle of measurement methods and the requirements on the measuring systems are largely comparable for the various types of voltage and current, with few exceptions, which mainly concern on-site tests.

For standardized testing of power apparatus in the electrical power supply, *approved measuring systems* are used to measure test voltages and test currents. “Approved” means in this case that the measuring system meets the requirements specified in the test and calibration standards. The approval of a measuring system is achieved by a *traceable calibration*, i.e. by a calibration that can be traced back to national or international measurement standards. Here, on the one hand, the higher-level “horizontal” test and calibration standards (Refs. [1–4] of Chap. 2) and on the other, the special test standards for the individual apparatus like power transformers, cables, gas-insulated switchgear, surge arresters, etc. must be observed.

After an introductory discussion on the subjects of standardization, accreditation and *traceability*, the following sections deal with the mainly applied calibration methods for conventional voltage and current measuring systems, with emphasis on impulse voltage and current measuring systems. The principle of measurement methods and the requirements on the measuring systems are broadly comparable for the various types of voltage and current with a few exceptions, which primarily concern *on-site tests*.

K. Schon, *High Voltage Measurement Techniques*, Power Systems,
https://doi.org/10.1007/978-3-030-21770-9_10

10.1 Standardization, Accreditation and Traceability

The internationally accepted test and calibration standards in electrical engineering are primarily issued by the *International Electrotechnical Commission* (*IEC*) as the result of the collaboration between international experts [1]. The *Technical Committee* (*TC*) 42 of the IEC is responsible for standardization in the high voltage and high current areas. After translation of the *IEC Publications*, they are accepted as harmonized, i.e. technically unchanged standards, by many countries. For example, the *Deutsche Elektrotechnische Kommission* (*DKE*), an association of *VDE* and *DIN*, is responsible for standardization in Germany in the field of electrical engineering [2]. The German mirror committee to TC 42 is the *Komitee K 124* of the DKE. The scientific and technological foundations of standardization in the field of electrical power engineering are worked out by the experts in various organizations, mainly the *Conseil International des Grands Réseaux Électriques* (*CIGRE*) [3].

The qualification and approval of a voltage or current measuring system and its components is carried out in the course of several tests and calibrations according to IEC 60060-2 (Ref. [2] of Chap. 2) or IEC 62475 (Ref. [4] of Chap. 2). The manufacturer of the measuring system first performs an *acceptance test*, if necessary in consultation with the customer, consisting of a general *type test* and a *routine test*. As a rule, the determination of the scale factor and the limit frequencies or time parameters is also included in the first *performance test*. *Performance tests* and *performance checks* are repeated at regular intervals, in which the user verifies the stability of the measuring system in the course of its use. The characteristic data, the results of the tests and calibrations as well as any changes or repairs of the measuring system are documented in the *record of performance*.

Of particular importance is the *traceability of a measurement* to the corresponding national or international standards of measurement maintained in every industrial country by a *National Metrology Institute*. For example, in Germany the *Physikalisch-Technische Bundesanstalt* (*PTB*), in the United Kingdom the *National Physical Laboratory* (*NPL*) and in the USA the *National Institute of Standards and Technology* (*NIST*) are responsible for this task. They develop and maintain the measurement standards required in physics and technology with the lowest possible uncertainties of measurement [4]. Regular intercomparisons among the metrology institutes of all industrialized countries ensure the uniformity of the measurements all over the world, including the area of high-voltage engineering [5–7]. The measurement capabilities of the metrology institutes are listed in the database *Calibration and Measurement Capability* (*CMC*) [8].

In many industrialized countries, testing and calibration laboratories are *accredited* by a national *accreditation body*, e.g. the *Deutsche Akkreditierungsstelle* (*DAKKS*) *in Germany*. Together with the National Metrology Institutes, they ensure traceability in selected measurement ranges, i.e. the dissemination of the units of measurement down to the working standards used in a factory (Ref. [9, 10] of Chap. 7). *Accreditation* means the internationally valid recognition that the general requirements on the management system and the specific requirements on the technical competence of the laboratory, including the employees, are met [9–12]. Accreditation also includes regular supervision and audit of the calibration and testing laboratories every twelve to eighteen months.

In principle, each national accreditation body enjoys regional protection in its own country, so that foreign accreditation bodies can be active there only in justified exceptional cases. The formal accreditation of a calibration or test laboratory is not explicitly demanded in [11]. However, the laboratory must have the competence required for its activities and must be able to prove it at any time. The same criteria apply as for regular accreditation, in particular a comprehensive quality management system, traceable and regularly recalibrated reference measuring systems, suitable laboratories and test equipment, appropriately trained employees for the tasks and documentation of all results.

Important basis for the quality of a manufactured product is the correctness of the tests and measurements carried out within the specified measurement uncertainties. Special attention should therefore be given to the measuring instruments used. In the major industrial nations with a large number of manufacturers and test institutes, the required number of traceable calibrations of the measuring instruments cannot be managed by the *National Metrology Institute* alone, but only in close cooperation with accredited calibration laboratories. An overview of the hierarchy of measuring laboratories and instruments with regard to their accuracy is given in Fig. 10.1, with the *national measurement standard* maintained in the metrology institute of the respective country on the top.

Tests and calibrations of a product according to recognized standards and rules as well as the documentation of the results are an important part of the quality management. Many manufacturers of power apparatus and measuring systems have their own accredited test or calibration laboratory with measuring systems that are traced back to national or international measurement standards. This allows the manufacturer to strengthen confidence in the quality of his products, so that tests and calibrations need not be repeated at the customer. By international agreements, mutual recognition of the calibration and testing laboratories operating in various industrial countries and accredited according to [11], is thoroughly regulated. This

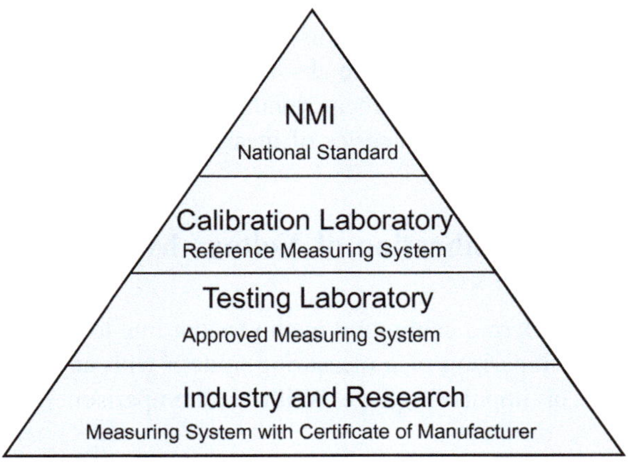

Fig. 10.1 Hierarchy of measuring laboratories and measuring instruments in an industrial country

is an important prerequisite for the international trade of goods free of technical trade barriers, as requested by the *World Trade Organization* (*WTO*) [15]. Of course, the buyer of a product retains his right to perform additional tests and calibrations in his own or a neutral laboratory.

10.2 About the Calibration of Measuring Systems

Calibration of a measuring system means that one or more *measurement quantities* are determined by applying standardized methods or procedural instructions. As a result of calibration, a numerical value for the desired quantity is obtained, e.g. for the scale factor, which is associated with an *uncertainty of measurement*. This uncertainty generally consists of several individual *standard uncertainties* resulting from the uncertainty contributions of various *influence quantities* of the measurement (see Chap. 13). The measurement uncertainty is regarded as a measure of the quality of the measurement. Details of the calibration, the measured value and the uncertainty of measurement are stated in the *calibration certificate*.

The preferred calibration method for determining the *assigned scale factor* and other measurement quantities is the *comparison measurement* of the complete measuring system with an accurate *reference system*. In this case, the comparison measurement is carried out with the voltage or current shape and the values, which correspond to the actual operating conditions. An alternative is the *component calibration*, with which the individual components of a measuring system are calibrated separately. Measurement methods at low voltage, which are customary in low-voltage technique, can also be applied. The linearity of the complete measuring system up to the rated value must then be confirmed by a suitable method.

The reference system of an accredited calibration laboratory plays a central role. It is usually calibrated with minimal uncertainty directly against the national measurement standard. Reference systems are mainly used for the traceable calibration of approved measuring systems used in test laboratories. Calibration laboratories can also calibrate other reference systems with their own reference systems, but with reduced accuracy according to the resulting uncertainty budget. In industry, research and development, both measuring instruments with and without traceable calibrations are used whose measurement accuracy can be extremely high or moderate. In part, the absolute accuracy obtained by a traceable calibration with the national measurement standard is not of interest, but only the sensitivity of the indication during a series of measurements.

10.3 Calibration of Voltage Measuring Systems

The preferred calibration method according to IEC 60060-2 (Ref. [2] of Chap. 2) is the comparison of a measuring system with an accurate reference system for AC, DC or impulse voltage. With the comparison measurement, the *assigned scale factor*, the *linearity* and the *dynamic behavior* of a measuring system are determined directly in relation to the reference system. Alternative methods are also permissible. Furthermore, the influence of the ambient temperature, the short-term and long-term behavior, the proximity effect of neighboring objects and the coupling of interferences must be determined by applying other methods. The calibration of the individual measurement quantities is associated with uncertainties, which are determined according to the rules of the GUM, taking into account the measurement uncertainty of the reference system (see Chap. 13). The uncertainty of the calibration of an approved measuring system shall be sufficiently low to comply with the permissible IEC limits when testing power apparatus. The long-term stability of a voltage measuring system is checked by regular performance tests and performance checks and documented in the record of performance.

10.3.1 Comparison Measurement with Reference System

Figure 10.2 shows schematically the arrangement for the comparison measurement of the impulse voltage measuring system X to be calibrated and the reference system N. The two voltage dividers including their damping resistors $R_{d,X}$ or $R_{d,N}$ and the high-voltage leads to the load capacitor C_b form a "Y" when viewed from

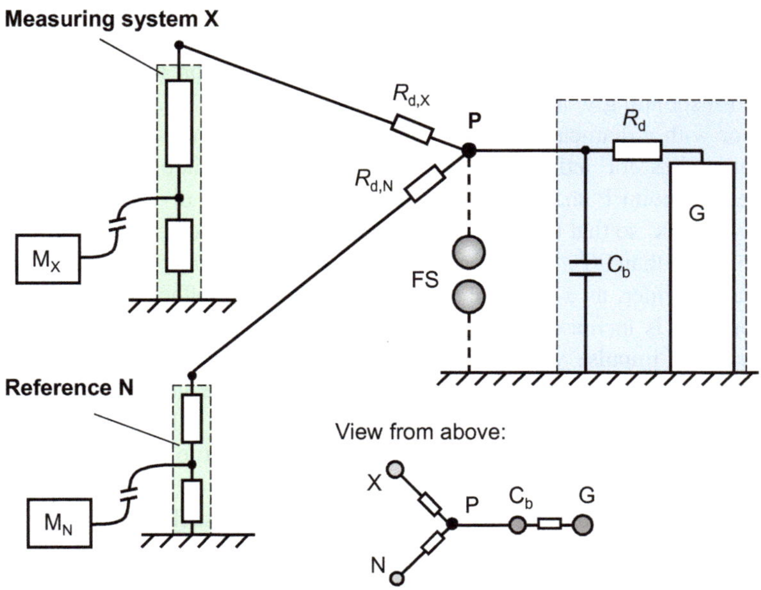

Fig. 10.2 Spatial arrangement for the comparison measurement in the Y circuit (principle), X measuring system, N reference system, M_X, M_N measuring instrument (recorder), P measurement point, C_b load capacitor, G impulse voltage generator, FS chopping gap

above. The common measurement point P is generally realized by a sphere electrode on top of an insulating support. The largely symmetrical Y circuit has several advantages. The influence of the voltage dividers with each other and from neighboring walls (*proximity effect*) as well as the interference effect caused by the firing of the spark gaps of the impulse voltage generator G is a minimum, or both measuring systems are equally affected. Also suitable is the T circuit with the corresponding T-shaped arrangement of the two voltage dividers and the impulse voltage generator.

To carry out comparison measurements with chopped impulse voltages, the insulating support is replaced by a sphere spark gap FS with connection to the common point P (Fig. 10.2). A triggered sphere gap is required for the exact chopping at the peak or on the tail. Chopping on the front is achieved by reducing the spacing between the spheres. Irradiation of the sphere gap with very short-wave ultraviolet light (UVC) increases the number of free charge carriers, thereby reducing the spread of the time to chopping. The time to chopping can be varied by changing the spacing of the spheres. However, the breakdown voltage is also affected and thus the peak value of front-chopped impulse voltage. This is compensated by corresponding readjustment of the charging voltage of the impulse voltage generator. At small spacing of the sphere gap and despite UVC irradiation, a larger spread of the time to chopping and thus also of the peak value of chopped impulse voltages is to be expected.

Reference dividers for lightning impulse voltages are generally designed for rated voltages not exceeding 500 kV. With a divider height of less than 2 m, they have good transfer characteristics and are convenient to transport for *on-site calibrations*. For conducting comparison measurements with tall voltage dividers, the smaller reference divider must be connected to the common measurement point P via a correspondingly long high-voltage lead (see Fig. 10.2). A stable tubular conductor with a diameter of a few centimeters, which consists of several plug-in telescopic tubes for individual length adjustment, is suitable as the lead. The measurement point P shall be at least at the height of the divider top of the measuring system X, so that the angle between the voltage divider and the high-voltage lead is not less than 90°. With smaller angles, the high-voltage lead comes closer to the voltage divider, as a result of which the interference effect due to electromagnetic coupling is increased. As low-inductance ground returns of the two voltage dividers to the impulse voltage generator, strips of copper foil with a width of at least 0.5 m can be used. For safety reasons, the impulse voltage generator itself must be connected to the ground rod of the high-voltage hall.

The coaxial cable connecting the measuring instrument to the voltage divider X shall be the same during calibration and voltage testing, or at least equal in terms of quality and length. The position of the coaxial cable is such that the coupling of electromagnetic interference is minimized. Optimal is the laying of the coaxial cable in grounded pipes or channels in the hall floor. The same applies to the coaxial cable of the reference system N.

The measuring instrument is either specifically built for use with high-voltage dividers and is electromagnetically shielded, or it is protected from interference by operation in a *Faraday cage* with filtered power supply (see Sect. 5.3.1.1). Digital recorders and other measuring instruments, even if calibrated together with the voltage divider during the comparison measurement, must undergo a separate calibration in all measurement ranges. The correctness of the evaluation software is verified with data supplied by the test data generator (see Sect. 7.3).

10.3.2 Special Features of the Comparison Measurement

The calibration of a large measuring system X is preferably carried out as on-site calibration in the test laboratory, i.e. the place where the measuring system X is used. On the one hand, the transport of the reference divider to the large high-voltage system X is generally easier than vice versa—if system X is transportable at all—and on the other, the measurement conditions for the on-site calibration are more or less identical to the use of the measuring system X in high-voltage tests of power apparatus. This concerns in particular the premises of the test hall, installation site of the measuring system, grounding conditions and, in the case of impulse voltage measurements, the shape of the generated impulses. The measurement uncertainty determined during the calibration can then be directly taken over for the voltage measurement during tests. If necessary, additional contributions to uncertainty must be taken into account, e.g. the long-term behavior of the measuring system or another ambient temperature (see Annex B.2).

Large test laboratories often have their own reference systems to regularly calibrating their own approved measuring systems during performance tests and performance checks. If, however, the calibration of the measuring system X is carried out at the premises of the calibration laboratory commissioned for this purpose, the measurement conditions generally deviate more or less from those in the test laboratory. When using the calibrated system X in the test laboratory, further uncertainty contributions must then be considered in addition to the measurement uncertainty stated in the calibration certificate.

In the comparison measurement, an overshoot or oscillation superimposed on the peak of the lightning impulse voltage must be avoided. Peak oscillations occur in particular when the impulse voltage generator is set up for short front times and inductances are effective in the test and measurement circuit. Lightning impulse voltages with superimposed peak oscillations are generally measured correctly by reference voltage dividers with good transfer behavior, but not by large voltage dividers with rated voltages in the megavolt range. The consequence is that the scale factor and time parameters are incorrectly determined during the comparison measurement. The calibration specifications in (Ref. [2] of Chap. 2) do not provide a solution for this case.

Oscillations in the test and measurement circuit can be eliminated by various compensation circuits, but these circuits often result in longer front times (see Sect. 4.2.1.3). A calibration with short front times therefore only seems to be possible if the inductances in the test and measurement circuit are reduced. The case is completely different when the peak oscillation recorded with the measuring system X is greater than what is indicated by the reference system N [16]. This suggests that the transfer behavior of the voltage divider in the measuring system X has a resonance point in the corresponding frequency range. When measuring system X is used in tests, an existing peak oscillation would be erroneously recorded too large and thus misinterpreted as an increased stress on the power apparatus being tested.

Basically, the evaluation of the measured data differs when calibrating a voltage measuring system and when using it during tests. During calibration, the result, such as the scale factor or time parameter of a voltage or current measuring system, is usually given as the mean value of a large number of repetitive measurements. As a result of this averaging, the scattering of the individual measurement values, caused by the limited resolution and internal noise of the digital recorder, is largely eliminated. During voltage tests, however, only a single value for the peak value or a time parameter is measured in general, which can be significantly different if the test is repeated. For example, if the measuring system with digital recorder has an 8-bit resolution, the individual peak values in consecutive measurements may differ by 1% and more. The maximum possible deviation is then to be considered in the uncertainty budget for the test voltage value by an additional uncertainty contribution.

10.3.3 Calibration of a Single Voltage Divider

Frequently, the task is to calibrate not the complete measuring system, but only the high-voltage divider as a single component. For this purpose, the output of the voltage divider is connected to a calibrated measuring instrument of the calibration laboratory, so that the calibration in the arrangement according to Fig. 10.2 can take place again. From a metrological point of view, it is problematic to use the second channel of the recorder in the reference system, in particular in the case of impulse voltage measurements. Thereby, the shields of the long coaxial cables of the two voltage dividers are set to the same potential at the recorder input so that compensating currents can flow and influence the measurement signals. Furthermore, there is the danger of crosstalk, in particular if the input voltages of the two recorder channels are of different sizes. The larger input voltage on the one channel would couple a noticeable interfering voltage into the other channel.

There are two possible ways to check whether interference voltages are coupled in. When the resolution of the two recorder channels is approximately equal, the comparison measurement is repeated with interchanged channels and the two measured values are averaged to eliminate interference. In order to examine interference, it may also be helpful to perform the comparison measurement in two successive measurement series, whereby the digital recorder is connected in each series to only one of the two voltage dividers. Prerequisite for the serial measurement is a good reproducibility of the impulse voltage generator.

10.3.4 Assigned Scale Factor

The most important goal in the calibration of a measuring system is the determination of the *assigned scale factor*. It is generally defined as the factor by which the display of the measuring system is to be multiplied in order to obtain the applied test voltage. The assigned scale factor is preferably determined during the performance test by comparison with a reference system. In this case, the measuring system X to be calibrated and the reference system N are generally of the same type, i.e. they each consist of a high-voltage divider with a digital recorder (Fig. 10.3). When the test voltage $u(t)$ is applied, the output voltages of the two voltage dividers are recorded simultaneously, from which the values of the voltage are evaluated. For example, for impulse voltages without overshoot, the values of the voltage are the peak values $\hat{u}_X$ and $\hat{u}_N$ multiplied with the scale factors F_X or F_N. For the peak value $\hat{u}$ of the applied high-voltage impulse, we then have the equation:

$$\hat{u} = \hat{u}_N F_N = \hat{u}_X F_X, \tag{10.1}$$

from which the desired scale factor F_X of the measuring system X is determined:

$$\boxed{F_X = \frac{\hat{u}_N}{\hat{u}_X} F_N}. \tag{10.2}$$

Frequently, the reference system N directly displays the peak value $\hat{u}$, i.e. the product $\hat{u}_N F_N$.

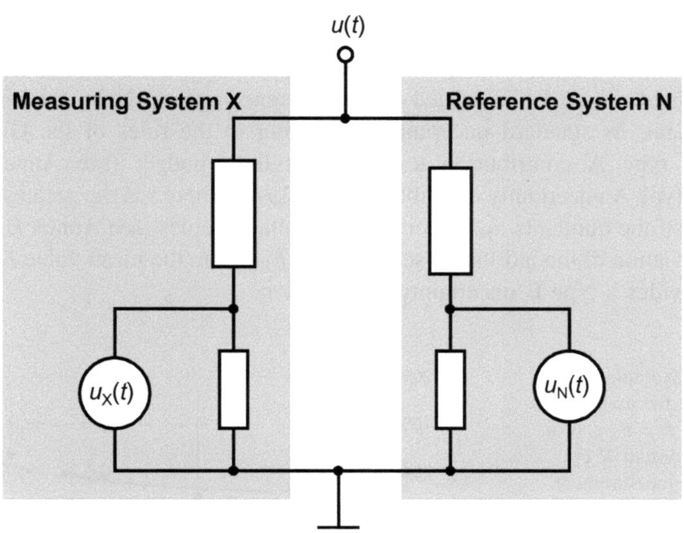

Fig. 10.3 Model of the comparison measurement of the measuring system X with the reference system N at impulse voltage

The comparison measurement is preferably carried out over the entire voltage range in which the measuring system X is to be used. If, however, the reference system N has a smaller rated voltage than system X, the comparison measurement can only be performed up to this voltage. Then a linearity test is required.

10.3.4.1 Comparison Over the Entire Voltage Range (Reference Method)

Performance and evaluation of the comparative measurement for determining the scale factor will be explained on the example of the standard 1.2/50 lightning impulse voltage. As front time, a mean value T_{1cal} within the permissible tolerance range from T_{1min} = 0.84 μs to T_{1max} = 1.56 μs is selected. Furthermore, the time to half-value is set to the upper tolerance value T_{2max} = 60 μs. At first, a series of $n \geq 10$ measurements is carried out with at least 20% of the rated voltage U_0 of the measuring system X, the divider output voltages $u_X(t)$ und $u_N(t)$ being simultaneously recorded. For each of the n impulse voltages, the peak values $\hat{u}_X$ und $\hat{u}_N$ are determined and the quotient $\hat{u}_N/\hat{u}_X$ is calculated. Due to the simultaneous recording, the instability of the impulse voltage generator is eliminated in the quotient formation. The mean of the n quotients, multiplied by the scale factor F_N of the reference system N according to Eq. (10.2), gives the scale factor $F_{X,1}$ of the measuring system X for the lowest voltage value (Fig. 10.4).

The comparison measurements with the reference divider are performed for at least four additional voltage values in steps of 20% up to 100% of U_0. Thus, at least five mean scale factors $F_{X,1}$ to $F_{X,5}$ are obtained for the entire voltage range. Their common mean $F_X = F = F_{X,m} = 2040$, as indicated by the dashed horizontal line in Fig. 10.4, is the assigned scale factor of the measuring system X, which is valid for the entire voltage range.

Equation (10.2) is the basic form of the *model equation* with which not only the scale factor F_X of the measuring system X but also the uncertainty of measurement is determined. Since F_X is related to the reference system N by the comparison measurement, its standard uncertainty according to the rules of the GUM is an important type A contribution to the uncertainty budget (see Annex B.1.1). A further type A uncertainty contribution is $s \cdot F_N/\sqrt{n}$, where s is the greatest standard deviation of the quotients $\hat{u}_N/\hat{u}_X$ for the five voltage values (see Annex B.1.2). The largest deviation of the individual scale factors $F_{X,i}$ from the mean value F_X divided by $\sqrt{3}$ provides a type B uncertainty contribution.

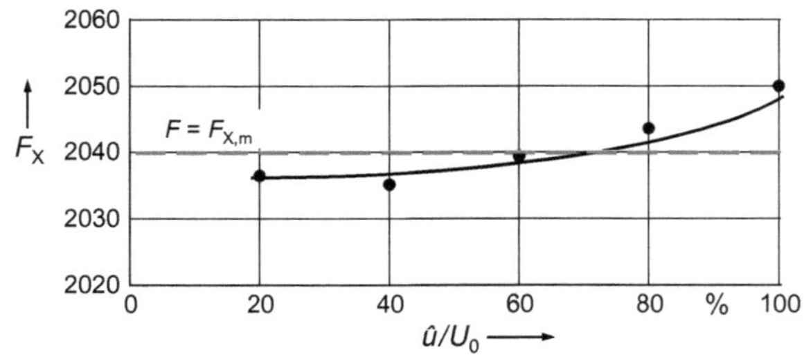

Fig. 10.4 Example for determining the assigned scale factor $F = F_{X,m}$ of a measuring system X by comparison measurement with a reference system N over the entire voltage range

10.3.4.2 Comparison Measurement Over a Limited Voltage Range

Frequently, the measuring system X to be calibrated has a higher rated voltage U_0 than the reference system N. The assigned scale factor $F_X = F_{X,m}$ can then be determined by comparison measurements only up to a reduced voltage level. In this case, the comparison should be made with at least two voltage values. The mean scale factor $F_{X,m}$ calculated therefrom applies to the entire voltage range of the measuring system X up to U_0. As an example, Fig. 10.5 shows the three scale factors $F_{X,1}$, $F_{X,2}$ and $F_{X,3}$, resulting from the comparison at 20%, 40% and 60% of U_0. Their mean is the assigned scale factor $F_X = F = F_{X,m} = 2038$ of the measuring system X, which is valid up to U_0. The uncertainty of F_X is estimated in a manner similar to that in Sect. 10.3.4.1, with an additional uncertainty contribution resulting from the linearity test as described below.

The linearity of the scale factor up to U_0 is to be proved by a method for which there are several alternatives (see Sect. 10.3.5). The following measurement with the alternative system starts at the highest voltage of the comparison with the reference system, i.e. in the example of Fig. 10.5 at $0.6U_0$, and is performed up to U_0. Since it depends on linearity rather than absolute accuracy, the mean scale factor F_X* determined with the alternative system may deviate more or less from F_X. The largest deviation of the individual values from their mean F_X* is taken into account as type B uncertainty contribution in the uncertainty budget of the scale factor F_X of the measuring system X (Ref. [2] of Chap. 2).

10.3.4.3 Calibration of the Components (Alternative Method)

In addition to the preferred comparison of the complete high-voltage measuring system with an accurate reference system, alternative calibration methods for determining the assigned scale factor are also permitted according to IEC 60060-2 (Ref. [2] of Chap. 2). The main components of a voltage measuring system are the high-voltage divider, possibly with damping resistor, and the measuring instrument on the low-voltage side. Further components can be an external attenuator at the input of the measuring instrument, the coaxial cable or another signal transmission device between the voltage divider and the instrument. In the alternative method,

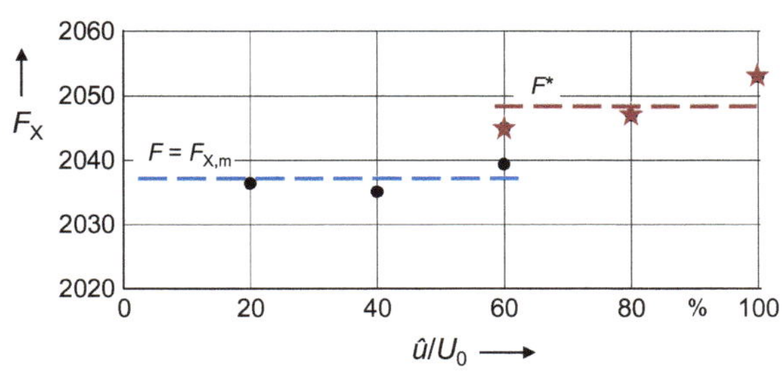

Fig. 10.5 Example for determining the assigned scale factor $F = F_{X,m}$ of a measuring system X by comparison measurement with a reference system N over a limited voltage range

the scale factors of the individual components are determined separately, whereby it should be ensured that the load on the components corresponds to that in the complete system. For calibrating the components, it is also possible to use measuring devices and methods that are customary in the low-voltage range. The product of the scale factors of the individual components gives the assigned scale factor of the complete measuring system. In addition, proof of the linearity of the complete measuring system up to the maximum operating voltage is required. This linearity test can be performed using one of the methods mentioned in Sect. 10.3.5.

The calibration of digital instruments is carried out according to IEC 61083 (Ref. [1] of Chap. 7). Calibrators for AC, DC or impulse voltages are on the market, with output voltages being up to 2000 V. Calibrators with low-ohmic output are also suitable for determining the scale factor of voltage dividers (Ref. [26] of Chap. 7). As the output voltage of the divider is very small, it is advisable not to connect the measurement circuit to the general laboratory ground due to superimposed disturbances. Another alternative method is to measure the resistances or capacitances on the high- and low-voltage side of the voltage divider with a low-voltage bridge and to calculate the scale factor from this. The influence of stray capacitances has to be considered.

Each of the above low-voltage methods is suitable for checking the long-term stability of the scale factor. For regularly repeated performance checks, the scale factor of each component may not deviate by more than 1% from its previous value, so that the assigned scale factor of the complete measuring system is still valid. If the performance check is carried out with the complete measuring system, the verified scale factor must not deviate by more than 3% from the assigned scale factor.

10.3.5 Linearity Test

The linearity test is intended to verify the constancy of the scale factor of a measuring system over the entire voltage range. It is performed up to the required voltage level either by comparison with a reference system or, if its rated voltage is too low, by one of the alternative methods described below. For the linearity test, it is not the absolute accuracy of the voltage measurement that is decisive but the smallest measurable deviation from linearity. Depending on the type of voltage, sufficient linearity can be expected from the following alternative systems (Ref. [2] of Chap. 2) [17–21]:

- approved measuring system with a sufficiently large rated voltage
- primary voltage of an AC transformer, control voltage of a DC voltage generator, charging voltage of an impulse voltage generator
- field measuring probe
- sphere gap, rod-rod gap.

In IEC 60052 (Ref. [5] of Chap. 2), the values of the disruptive discharge voltage of sphere gaps are stated with an uncertainty of 3%, which is mainly due to the influence of ambient temperature, atmospheric pressure and humidity in air. However, due to the short measurement time during the linearity test, the ambient conditions remain approximately constant and therefore have practically no influence on the test result. In the case of short-wave UVC irradiation of a sphere gap, a sufficiently large number of free electrons is generated for rapid initiation of the disruptive discharge mechanism, i.e. even in the case of lightning impulse voltages. Experience has shown that then the scattering of the individual values of the disruptive discharge voltage during a measurement series is within ±1% (see Sect. 2.5.5 and Fig. 3.4.9).

If larger deviations from linearity occur in the test with an alternative system, the cause need not necessarily lie with the measuring system X. A further linearity test with another alternative system is then possible for clarification. The deviation determined during the linearity test with the alternative system does not enter into the calculation of the mean scale factor F_m (see Sect. 10.3.4.2) but is taken into account as Type B uncertainty contribution in the uncertainty budget of the measuring system X (Ref. [8] of Chap. 1).

10.3.6 Time Parameters of Impulse Voltages

Impulse voltage measuring systems shall measure the time parameters of the test voltage with an expanded uncertainty of not more than 10%. The correctness of the time parameter measurement is preferably verified by comparison of the measuring system X with a reference system N. To evaluate the time parameters, the same records of the impulse voltage can be used, from which the scale factor is calculated. From the simultaneous measurement with both systems, the difference of the time parameter T_X to the value T_N of the reference system N results:

$$\boxed{\delta T = T_X - T_N}. \tag{10.3}$$

A total of $n \geq 10$ records are evaluated and the mean difference and the standard deviation are calculated.

As an example, the front time of a lightning impulse voltage is considered, which is formally given by $T_1 = T_{AB}/0.6$ according to Eq. (4.2). The comparison measurement is carried out for the two tolerance limit values $T_{1min} = 0.94$ μs and $T_{1max} = 1.56$ μs, preferably also with a third value lying between them. From at least ten measurements for each value of T_1, the differences δT_1 are calculated according to Eq. (10.3) and from which their mean value is determined. The largest standard deviation of the values δT_1, divided by $\sqrt{n}$, contributes to the type A measurement uncertainty. The greatest difference of the individual values δT_1 from their mean, divided by $\sqrt{3}$, gives a type B contribution to the uncertainty budget.

The differences of the time parameters of chopped and switching impulse voltages are determined in a comparable manner.

The nature of the mean difference δT_1 is systematic. In general, the reference system N itself is not completely error-free, i.e. the front time indicated for an impulse voltage may differ from the correct value, but is still within the permissible limits. If this systematic difference is known, it can be subtracted from the indicated value T_{1N}, so that the error in the front time theoretically becomes zero. Similarly, the measured difference δT_1 of the system X can also be used as correction. The front time of impulse voltages measured during tests on power apparatus can thus be improved. The prerequisite in this case is that the front time of the test voltage is approximately the same as in the calibration.

When calibrating the front time by comparison measurement, there is no requirement for the magnitude of the impulse voltage. However, if the scale factor of the measuring system X is voltage dependent, the front time T_1 and thus δT_1 are also slightly influenced by the voltage level. The reason for this is that the non-linearity of the scale factor is more effective at $0.9\hat{u}$ of the impulse voltage than at $0.3\hat{u}$, the points for calculating the front time.

The correct time parameter measurement of an impulse voltage measuring system can also be checked by means of the component method, preferably using an impulse generator with variable front time (see Sect. 10.3.4.3). Further possibilities are offered by the evaluation of the experimental step response of the voltage divider and the numerical convolution method (see Sect. 10.3.8).

10.3.7 Dynamic Behavior

As a result of the comparison measurement according to Sect. 10.3.4, the scale factor of an AC or impulse measuring system X (or a component) is initially obtained only for a single frequency or a single time parameter. However, an approved measuring system (or component) must also be able to correctly measure AC voltages of other frequencies or impulse voltages with deviating time parameters. It is therefore necessary to proof that the measuring system meets the relevant requirements within the prescribed tolerances and error limits. The dynamic behavior is again preferably examined by comparing the measuring system X with a reference system N for AC or impulse voltages whose frequency or time parameters are varied within the predetermined limits.

For example, the dynamic behavior of a measuring system for lightning impulse voltages is investigated for the lower and upper tolerance values of the front time $T_{1\min} = 0.94$ µs and $T_{1\max} = 1.56$ µs, with the time to half-value being set as the upper tolerance value $T_{2\max} = 60$ µs. In the comparison measurement, the output voltages of both measuring systems are recorded again ten times simultaneously for $T_{1\min}$ and $T_{1\max}$. The measuring system X usually displays a peak value $\hat{u}_X$ deviating from that of the reference system N. The voltage difference $\delta\hat{u}$ between $\hat{u}_X$ and $\hat{u}_N$ for the respective front time is:

$$\boxed{\delta\hat{u} = \hat{u}_X - \hat{u}_N}. \tag{10.4}$$

If the comparison measurement is performed for other front times, the average measurement deviations for the peak value may be listed in tables or graphically in error diagrams (see Sect. 10.3.9) [22–24]. Alternatively, the result of the comparison measurement can also be expressed by the scale factor in dependence on the time parameter.

Besides the comparison with a reference system at high voltage, the dynamic behavior of an impulse voltage divider can also be comprehensively determined using an impulse generator of a few hundred volts. The calibrator has low-ohmic output impedance and generates double-exponential impulse voltages according to Eq. (8.8) with variable front time. The impulse voltages at the input and output of the voltage divider are recorded simultaneously by digital recorders with adequate amplitude resolution. The difference of the peak values $\delta\hat{u}$ is evaluated and can, in turn, be shown in an error diagram in dependence on the front time (see Sect. 10.3.9).

10.3.8 Evaluation of the Step Response

As an alternative to the comparison method in the case of impulse voltages with different time parameters (see Sect. 10.3.7), the dynamic behavior of an impulse voltage divider can be judged on the basis of its step response. In addition, a single comparison measurement with a reference system is recommended. The recorded step response $g(t)$ is first slightly filtered to eliminate the internal noise of the digital recorder and small high-frequency oscillations. In the next step, the *reference level* is determined as the mean of the step response in the *reference level epoch* between $0.5t_{min}$ and $2t_{max}$ and is set equal to one. In the example of Fig. 10.6, the times $t_{min} = T_{1min} = 0.84$ µs and $t_{max} = T_{1max} = 1.56$ µs denote the tolerance limits for the front time of a 1.2/50 lightning impulse voltage. The permissible deviation of the step response from the reference level is 2% in the reference level epoch and 5% in the following time interval between $2t_{max}$ and $2T_{2max}$ (Ref. [2] of Chap. 2).

The evaluation of the step response is supplemented by a comparison measurement with a reference system in order to determine the assigned scale factor. In this case, the lightning impulse voltage has a front time T_{1cal} between the tolerance limits T_{1min} and T_{1max} as well as a time to half-value equal to the upper tolerance value $T_{2max} = 60$ µs. Further, the step response at the time T_{1cal}, may not deviate by more than 1% from the reference level.

If the requirements for the step response and the comparison measurement are met, the dynamic behavior of the voltage divider is recognized as adequate for measuring lightning impulse voltages within the tolerance range of the front time $T_1 = 1.2$ µs ± 30%. The maximum deviation of the step response from the reference level within the reference level epoch may be used to calculate a

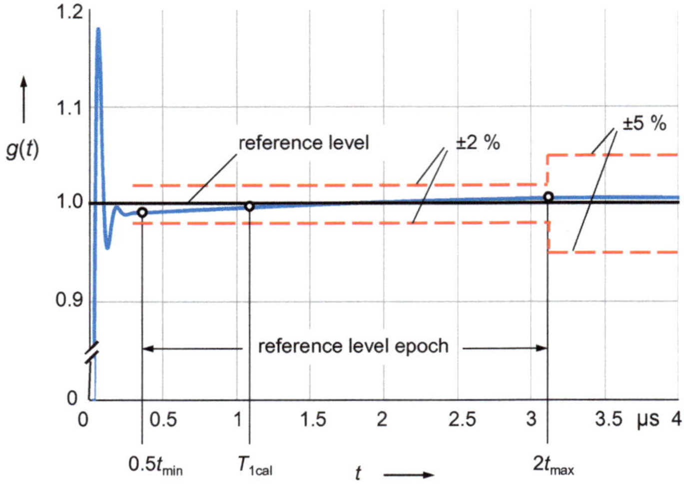

Fig. 10.6 Section of the step response $g(t)$ of an impulse voltage divider with reference level and reference level epoch (schematic)

corresponding type B uncertainty contribution to be taken into account in the uncertainty budget of the scale factor (see Sect. 13.1.4). In the example of Fig. 10.6, the maximum deviation of the step response from the reference level is approximately 0.5%, resulting in a standard uncertainty of 0.3%.

Finally, reference is made to the convolution method with which the output voltage of an impulse voltage divider with known step response can be calculated for arbitrary input voltages (see Sect. 9.2 and 9.7). It is recommended to check the result of the calculation by a single comparison measurement with a reference system. The dynamic suitability of an impulse voltage divider used in a reference system can also be demonstrated by its response parameters being below the specified limit values (Ref. [2] of Chap. 2).

10.3.9 Error Diagram for Peak Value and Time Parameters

The dynamic behavior of an impulse voltage divider can be well represented in an *error diagram*. As an example, the behavior of a resistive 2 MV voltage divider is shown in the error diagram of Fig. 10.7. The 12 kΩ voltage divider was calibrated by comparison with a reference divider at chopped lightning impulse voltages. The error diagram shows the mean measurement deviations $\delta\hat{u}$ for the peak value (Fig. 10.7a) and δT_1 for the front time (Fig. 10.7b) as a function of the time to chopping T_c. Here, $\delta\hat{u}$ and δT_1 are related to the values obtained for full lightning impulse voltages. The front time T_1 defined in Eq. (4.2) is chosen as the quantity to be measured, as it can be unambiguously determined for chopped impulse voltages as well as for full lightning impulse voltages and those chopped in the tail.

A typical feature of the investigated resistive voltage divider of this size is that the measurement deviations $\delta\hat{u}$ and δT_1 become increasingly negative as the time to chopping decreases, i.e. the peak value $\hat{u}$ and the front time T_1 of front-chopped lightning impulse voltages are measured too small. This behavior is caused by the stray capacitances of the voltage divider, through which a frequency-dependent part of the measurement current flows to ground and thus does not contribute to the displayed measurement value (see Sect. 4.3.1.4). Knowledge of the systematic deviations $\delta\hat{u}$ and δT_1 in Fig. 10.7 allows a corresponding correction of the values $\hat{u}$ and T_1 measured with the impulse voltage divider. The entire range of the time to chopping T_c in Fig. 10.7 may also be subdivided into two or more sub-ranges. For each sub-range, a corresponding mean value $\delta\hat{u}$ and δT_1 is determined and used as correction. As an alternative to correcting the measured peak value $\hat{u}$, the scale factor of the measuring system may be set anew for the entire range or a sub-range of T_c, so that the measurement deviation $\delta\hat{u}$ is zero on average.

As another example, Fig. 10.8 shows the error diagram of a resistive 700 kV voltage divider in dependence of the front time T_1 of double exponential impulse voltages. The peak value is 400 V and the front time ranges from 7 μs down to 0.5 μs. With decreasing front time T_1, the peak value $\hat{u}$ is reduced only slightly, but the front time T_1 becomes significantly larger (Ref. [26] of Chap. 7). The increase of δT_1 for low T_1 values can be explained by the step response of the voltage divider showing an overshoot in this time range.

In the example of Fig. 10.8a, the deviation $\delta\hat{u}$ remains within the specified error limits of ±1%, nevertheless, it is recommended to increase the scale factor of the

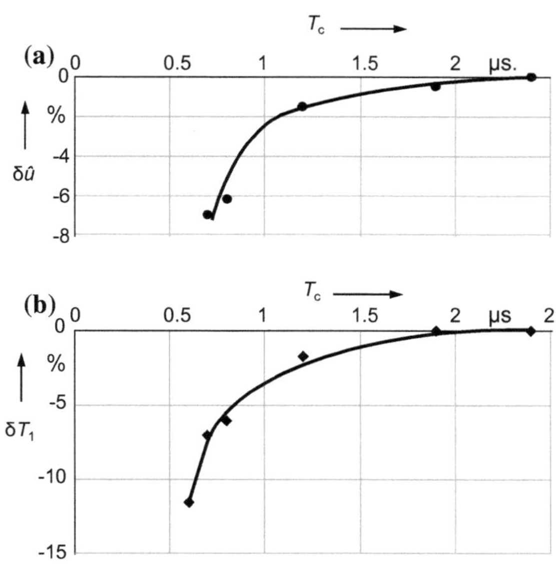

Fig. 10.7 Error diagram of a 2 MV resistive voltage divider for chopped lightning impulses as a function of the time to chopping T_c (**a**) Relative deviation $\delta\hat{u}$ of the peak value (**b**) Relative deviation δT_1 of the front time

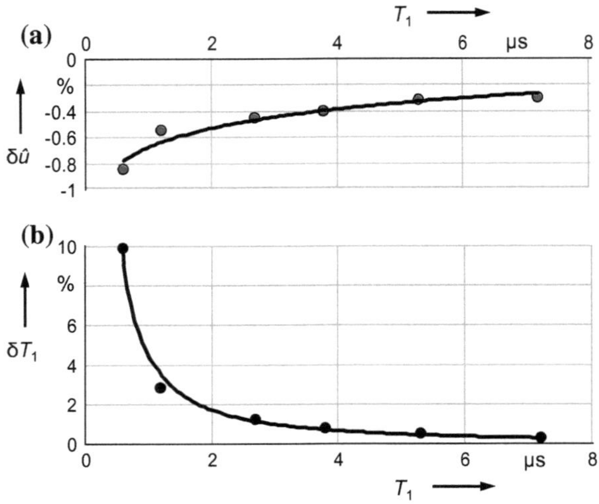

Fig. 10.8 Error diagram of a 700 kV resistive divider as a function of the front time T_1 of lightning impulse voltages at 400 V (**a**) relative deviation $\delta\hat{u}$ for the peak value (**b**) relative deviation δT_1 for the front time

voltage divider by 0.6%. This is equivalent to a corresponding increase in the measured peak values, which causes the error curve in Fig. 10.8a to be shifted upwards to smaller absolute values $\delta\hat{u}|$. In the tolerance range of the front time T_1 = 1.2 µs ± 30% of a 1.2/50 lightning impulse voltage, the deviation $\delta\hat{u}$ is then only within ±0.1%. For longer front times up to 7 µs, then $\delta\hat{u} \leq 0.3\%$. The measurement deviation δT_1 as shown in the error diagram of Fig. 10.8b can be used to correct the front time T_1 measured with this voltage divider (see Sect. 10.3.6).

Further investigations on the resistive 700 kV voltage divider confirm and supplement the results of the calibration with impulse voltages at 400 V. The scale factor values determined with DC voltage, AC voltage and resistance measurements agree within ±0.2%. Measurements of the input and output voltages of the voltage divider at AC voltage with frequencies of up to 150 kHz show a decrease in the output voltage with increasing frequency. This corresponds to the trend of the measurement deviation $\delta\hat{u}$ with decreasing front time in Fig. 10.8a. Finally, the scale factor determined by comparison with a reference system for high-voltage impulses largely confirms the value obtained by the low-voltage methods (Ref. [26] of Chap. 7).

10.3.10 Proximity Effect

The scale factor of an AC or impulse voltage divider is affected by its clearance to neighboring objects and walls. This *proximity effect* can be measured without much effort using a digital voltmeter or a measuring bridge at low voltage. An AC voltage of a few 100 V and preferably 1 kHz is applied to the voltage divider, and its output voltage is measured for different wall clearances. The reason for affecting the scale

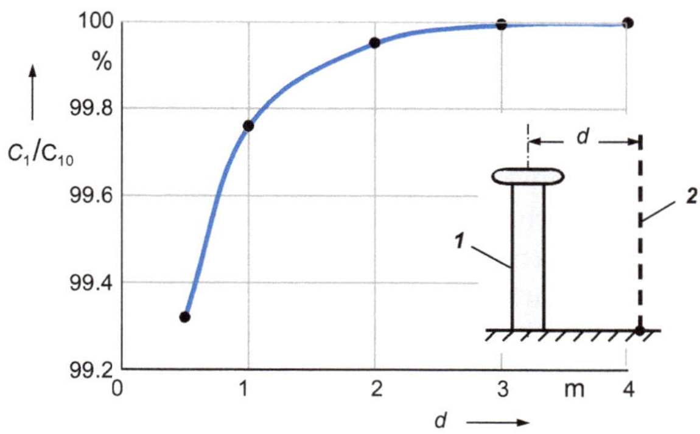

Fig. 10.9 High-voltage capacitance C_1 of a damped capacitive 500 kV voltage divider *1* depending on the clearance to a grounded metallic grid *2* (C_{10} = 150 pF)

factor is the stray capacitance C_e of the voltage divider, through which a part of the measurement current flows directly to ground or grounded objects. This stray current is then lost in the measurement of the divider output voltage (see Sect. 4.3.1.4). For a capacitive voltage divider, this is equivalent to a reduction in the effective capacitance on the high-voltage side, which can be measured directly with a capacitance measuring instrument.

As an example, Fig. 10.9 shows the effective high-voltage capacitance C_1 of a 500 kV damped capacitive voltage divider as a function of the clearance d to a grounded metal grid. At a distance d = 1.4 m, which is equal to the height of the voltage divider, the capacitance decrease is 0.14% of the rated value C_{10} = 150 pF. Accordingly, the voltage at the divider output is too low. To compensate for this, the scale factor can be correspondingly increased. The correction of the scale factor, however, is not absolutely necessary in this case, since the decrease in the high-voltage capacitance is so small. The test result serves as an orientation in the estimation of the proximity effect caused by the influence of surrounding objects such as the impulse voltage generator or other voltage dividers. In addition, the test result is helpful to estimate the appropriate uncertainty contribution to the scale factor. Because of the danger of electric flashover to walls, clearances that are smaller than the divider height need not be considered in the test.

10.3.11 Short-Term and Long-Term Stability

The *short-term stability* characterizes the change in the scale factor of a measuring system (or component) during a typical period of operation. The test is performed at maximum voltage that is applied continuously or at maximum impulse rate. According to IEC 60060-2 (Ref. [2] of Chap. 2), the scale factor shall be measured as soon as the prescribed maximum voltage has been reached and then again just before the voltage is reduced. In measurement practice, the scale factor is measured with a low-voltage method before applying the voltage and after switching off the voltage and grounding the set-up. The difference of the scale factors measured before and after the voltage application, divided by $\sqrt{3}$, gives a type B uncertainty contribution for the short-term stability of the scale factor. The stability test is part of the routine test and includes the *self-heating effect* of the voltage divider or another converting device. Self-heating of a voltage divider causes a mostly reversible change in the scale factor according to its temperature coefficient. Manufacturers of electronic devices generally guarantee that the operating parameters are reached after a short warm-up period within 1, 8 or 24 h.

The *long-term stability* provides information about the constancy of a measuring system (or component) over a longer period, usually one year. The reason for the scale factor deviation may be of random or systematic nature. A small spread of the scale factor within $\pm 1\%$ is acceptable since the measurement conditions are rarely identical in the case of repeated measurements. A systematic change in the scale factor may occur due to ageing of the components in the voltage divider or measuring instrument and can last for many years. This change, also referred to as (*long-term*) *drift*, is usually exponential in the initial time after commissioning of the measuring system, until after a longer operating time a nearly constant value for the scale factor is reached.

Knowledge about the long-term stability of a measuring system is obtained by regularly measuring the scale factor, as required in the course of prescribed performance tests and performance checks. For a new measuring system, it is recommended that the time interval between the initial scale factor checks be less than one year, i.e. six or even three months. During the regular performance checks, the measuring system is compared with an approved measuring system at the prescribed voltage difference between the currently measured voltage values and the previous value is within $\pm 3\%$, the scale factor is accepted as still valid. If the difference is larger, then a new assigned scale factor shall be determined in a performance test (see Sect. 10.3.4). Alternatively, the scale factors of the individual components of a measuring system can be checked. If the change in one of the scale factors is outside $\pm 1\%$, a new assigned scale factor shall be determined.

When a new measuring system is used for the first time, the extent of possible ageing and the resulting change in the scale factor are often not exactly known. In this case, an initial value of the change in the scale factor can first be obtained from a reliable source, e.g. manufacturer data for one year. In general, however, it is not known whether the given deviation is based on a random dispersion of the measured values or a systematic change of the measuring system. Extrapolation to a time other than the specified time does not appear to be justified. The difference of two successive performance tests, divided by $\sqrt{3}$, results in a type B uncertainty contribution for the long-term stability of the scale factor.

10.4 Calibration of Current Measuring Systems

The basic scheme of the measurement methods and requirements for AC, DC and impulse current measuring systems according to IEC 62475 (Ref. [4] of Chap. 2) is largely comparable with that for voltage measuring systems in Sect. 10.3. For the user of a current measuring system, the scale factor and the frequency response or, in the case of an impulse current measuring system, the measurement deviations for the time parameters are most important. In conventional systems for measuring high DC, AC and impulse currents, coaxial resistors are mainly used as converting device (see Sect. 5.3.1). Time-variable currents are also measured with coils with or without magnetic core (see Sect. 5.3.2). Both sensor types provide output voltages which are proportional to the current either directly or after integration. To measure these output voltages, mainly digital recorders or other measuring instruments with A/D converters that meet the requirements of IEC 61083 are used (Ref. [1] of Chap. 7). The preferred calibration method, i.e. the comparison measurement with a reference system, and the alternative methods are described below using the example of an impulse current measuring system. The long-term stability of a current measuring system is checked by regular performance tests and performance checks and documented in the record of performance.

10.4.1 Comparison with a Current Reference System

The preferred calibration method is the comparison measurement of the complete current measuring system X with a reference system N. In this case, the current sensors, i.e. shunts or coils, of both measuring systems are connected in series and the same current $i(t)$ flows through them. Figure 10.10 shows the basic arrangement for the calibration of an impulse current measuring system X by comparison with the reference system N. In the example, the current sensor of system X is a measuring coil with a magnetic core. The current conductor is guided centrally through the coil opening. The measuring coil has an internal integration element and its output voltage is thus proportional to the current $i(t)$. The potential-free output voltage is recorded and evaluated as $u_X(t)$ by a digital recorder suitable for measuring impulse currents. The current sensor of the reference system N with the scale factor F_N is a coaxial measuring resistor with the output voltage $u_N(t)$. Many of the impulse currents standardized in (Ref. [4] of Chap. 2) have a relatively slow temporal change. In this case, it is generally not absolutely necessary to terminate the input of the digital recorder with the impedance of the coaxial cable.

If the recorder of the reference system N is not specifically designed to measure high-current impulses, it is additionally shielded together with the measuring cable to minimize the effects of electromagnetic interference. The interference current induced by the magnetic field of the current to be measured is thus conducted to

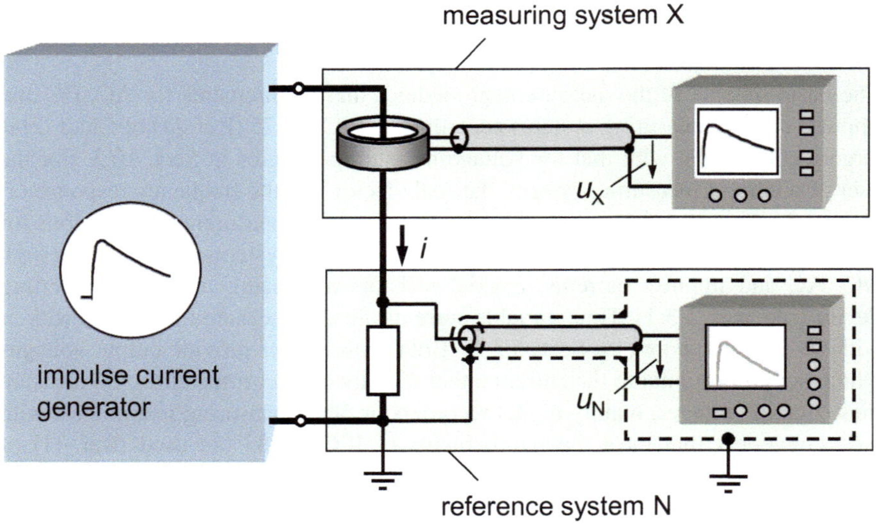

Fig. 10.10 Comparison measurement between an impulse current measuring system X with measuring coil and internal integration element and the reference system N with coaxial shunt

ground via the external shield and does not influence the measurement (see Sect. 5.3.1.1). If the measuring coil of system X has a magnetic core, additional measurements are recommended in order to detect a possible polarity influence. In this case, the coil is reversed so that the current flows in the opposite direction through the coil opening. A well-suited current measuring system X generally also has a double shielding.

If the current sensor of the measuring system X is a resistor, it is arranged on the grounded side of the calibration circuit in Fig. 10.10. Then the use of a potential-free reference system N, e.g. based on a potential-free measuring coil, is then compulsory. In this case, the measuring coil is usually calibrated against another reference system with a measuring resistor. However, this inevitably involves a loss of accuracy of the coil system.

In the comparison measurement, the output voltages $u_X(t)$ and $u_N(t)$ of the two current sensors are recorded simultaneously with the digital recorders and evaluated by software with regard to the peak value, the time parameters and, if necessary, further measurement quantities. The details of the comparison measurement and evaluation of the measured values, including the calculation of type A and B uncertainties, are comparable to those for a voltage calibration (see Sect. 10.3).

In order to determine the scale factor of the current measuring system, the comparison measurement with the reference system is carried out at five or more current magnitudes between 5 and 100% of the maximum operating current. For each current magnitude, the mean scale factor $F_{X,i}$ is calculated from at least ten records analogously to Eq. (10.2). The assigned scale factor F_X is the mean of the individual values $F_{X,i}$ for all current magnitudes. For an approved current measuring system, the assigned scale factor shall be constant within $\pm 1\%$. From the same records of the two systems, the difference of the time parameters is evaluated.

10.4.2 Linearity Test

The preferred linearity test of a current measuring system consists of a comparison measurement with a reference system. The procedure for the linearity test is comparable to that for voltage measuring systems (see Sect. 10.3.5). Often, the comparison measurement with the reference system N cannot be performed up to the full rated current of the measuring system X, since the reference system has a smaller rated current. Above this voltage, the linearity test of X is then carried out by comparison with an alternative system of sufficient linearity up to the maximum current magnitude, e.g. with an approved measuring system. Rigid Rogowski coils, i.e. coils without magnetic core, whose output voltage after integration is proportional to the current, are basically suitable for the linearity test with AC and impulse currents (see Sect. 5.3.2.5).

If the current sensor of an impulse or short-time current measuring system X is a resistor, the linearity proof can also be provided by calculating the temperature increase. A single current impulse causes a nearly adiabatic temperature increase ΔT of the measuring resistor R (see Sect. 5.3.1.9). Depending on the temperature coefficient K_{TK}, the increase ΔT results in a resistance change $\Delta R = RK_{TK}\Delta T$ and thus in a corresponding change in the scale factor. For the desired time course of the impulse or short-time current, the resistance change can be calculated from the manufacturer's specifications for the limit load integral and the maximum temperature increase.

10.4.3 Alternative Calibration of Current Measuring Systems

In the alternative methods for determining the scale factor, the components of a high-current measuring system are individually calibrated. As in the case of high-voltage components, the measurement methods known in the low-current range can also be used. The scale factor of a component can be determined, for example, by simultaneous measurements of the input and output variables, comparison with a reference component, use of calibrators, measurement of the impedances in a bridge circuit or calculation of the measured impedances. The wiring and loading of the components at the input and output should be comparable to those of the complete measuring system. The product of the scale factors of the individual components gives the scale factor of the complete measuring system. In addition, its linearity must be verified with the appropriate currents up to the maximum current magnitude (see Sect. 10.4.2).

10.4.4 Dynamic Behavior

The dynamic behavior of DC and AC current measuring systems is determined by measuring the frequency response. For impulse current measuring systems, the preferred method is the comparison with a reference system (Ref. [4] of Chap. 2). The impulse currents used in the comparison shall have the longest and shortest time parameters of the nominal epoch, e.g. $T_{1\min}$ and $T_{1\max}$ for the front time of exponential currents. In this range, the scale factor may not vary by more than 1%.

The dynamic suitability of a measuring system (or a component) for impulse currents can also be proved by means of numerical convolution on the basis of the measured step response. Due to the large number of impulse currents, this standardized calculation method is believed to be very effective. By convolution, the output currents of the measuring system are calculated for selected impulse currents with variable time parameters. The deviations between the input and output currents result in the errors for the peak value and the time parameters for which error limits are specified in the standards. In addition to the calculation, a measurement of the scale factor and a linearity test are to be performed.

The prerequisites for applying the convolution method to coaxial measuring resistors are almost optimal. Due to their shielded construction and the possibility to feed the step current into the resistor via a coaxial cable without reflection, the step response can be measured correctly. The result of the convolution calculation for a coaxial measuring resistor can therefore generally be regarded as reliable. A check measurement with an impulse current whose time parameter is within the nominal epoch of the step response is recommended.

10.5 Calibration of Digital Recorders

Digital recorders used in the measuring systems for AC, DC and impulse voltages or the corresponding currents require comprehensive calibration (Ref. [1] of Chap. 7). In the foreground is the determination of the scale factor and the dynamic behavior in all measurement ranges of the recorder. Other requirements include amplitude and time resolution, non-linearity, rise time and long-term stability. The software used to evaluate the recorded data is subject to a separate validation test using the test data of the *test data generator* for typical AC, DC and impulse voltages (see Sect. 7.3). While the calibration of recorders with accurate DC or AC calibrators and a frequency generator is generally relatively simple, the calibration of recorders used in impulse voltage and current measuring systems requires considerable effort and will be discussed in more detail below.

10.5.1 Impulse Calibration

The preferred calibration of digital recorders used in high-voltage and high-current impulse measuring systems is carried out with impulse voltages that are comparable with the standardized test voltages and currents (Ref. [1] of Chap. 7). During calibration, the scale factor and the deviation of the recorded time parameters are measured for all ranges of the recorder. *Impulse calibrators* are available for practical use, which generate impulse voltages of both polarities with peak values of up to 2000 V (see Sect. 7.5). The accuracy of the calibrators themselves is proved by *traceable calibrations*. The advantage of the calibration impulses is obvious: The digital recorder records similar impulse shapes during calibration and evaluates them with the same software used in the high-voltage or high-current tests. This also applies to the algorithm by which the recorded data of the impulses are smoothed before their evaluation and the internal noise of the recorder or superimposed oscillations are reduced. The calibration of the recorder thus indirectly also includes a check of the evaluation software for the selected calibration impulses.

When calibrating a digital recorder with standard impulses, the peak value and the time parameters are each determined as the mean value from at least ten records for each input range at different amplitudes. By comparing the mean values with the respective preset calibrator values, the scale factor and the measurement deviations of the time parameters are obtained for the input range examined. The complete calibration of a digital recorder with standard impulses, such as full and chopped impulse voltages, in all measurement ranges with different input levels often comprises more than one million individual records. If the calibrator and the recorder are from the same manufacturer, the automated control of the calibrator as well as the recording and evaluation of the measured data are usually supported. If the digital recorder has several input channels, these can be calibrated in parallel with an impulse calibrator at the same time. In this case, however, the influence of the parallel connected input resistances and capacitances of the recorder channels on the generated calibration pulses must be taken into account.

10.5.2 Calibration with Step Voltages

In the alternative calibration method for digital recorders used in impulse tests, the recorder is calibrated with step voltages (Ref. [7] of Chap. 1). They can be generated by means of a DC voltage source and a mercury-wetted reed contact of up to 500 V (see Sect. 9.8.4). The scale factor is obtained as the quotient of the DC voltage applied to the recorder input and the mean value of the repeatedly recorded amplitude of the step response in a given time window. For example, the time window for standard lightning impulse voltages and exponential impulse currents ranges from $0.5T_{1\text{min}}$ to $T_{2\text{max}}$. Within the time window, the response of the recorder may not change by more than 1%. The alternative calibration method has

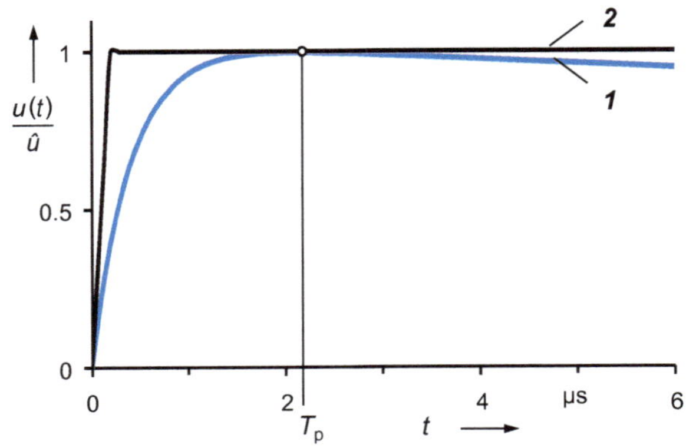

Fig. 10.11 Determination of the peak value of a recorded impulse voltage ***1*** by direct comparison with a step voltage ***2*** at the time to peak T_P

the disadvantage that the evaluation of the step responses requires software other than that developed for the evaluation of test impulses. An additional requirement is that the rise time of the applied step voltage shall be less than 10% of the lower time window limit at $0.5T_{1\min}$. Prior to data evaluation, the noise superimposed on the step response, which occurs during sampling in the AD converter, is largely reduced by filtering.

The example in Fig. 10.11 shows the direct comparison of the peak value of a recorded impulse voltage ***1*** with the amplitude of a step voltage ***2***. Most digital recorders facilitate the determination of the peak value on the screen by means of two horizontal reference lines, which can be set to the zero and peak levels of the impulse. The difference between the two reference lines is given as the peak value of the impulse. This method is also applicable when using analog impulse oscilloscopes. Instead of the step voltage, a DC voltage may also be applied to write the peak reference line. However, the calibration method with the step voltage is preferable because, like the impulse voltage itself, it is a dynamic signal and also indicates the zero level. When noise is superimposed on the recorded impulse voltage, the step voltage is adjusted to the mean curve of the recording.

10.5.3 Calibration with Sinusoidal Voltages

Sinusoidal voltages are usually also suitable for calibrating digital recorders in impulse voltage and current measuring systems. They have the advantage that they are calculable and can be generated very precisely with a sinus calibrator. By comparing the records for the impulse voltage and the sine voltage, not only the measured peak value but also the front time of the impulse voltage can be checked. An example of a 1.2/50 lightning impulse voltage is given in Fig. 10.12. Here, the double sine amplitude is adjusted to the peak value of the impulse voltage. The frequency is selected so that the sine voltage intersects the recorded impulse voltage at 30% and 90% of the peak value, the time difference being equal to the

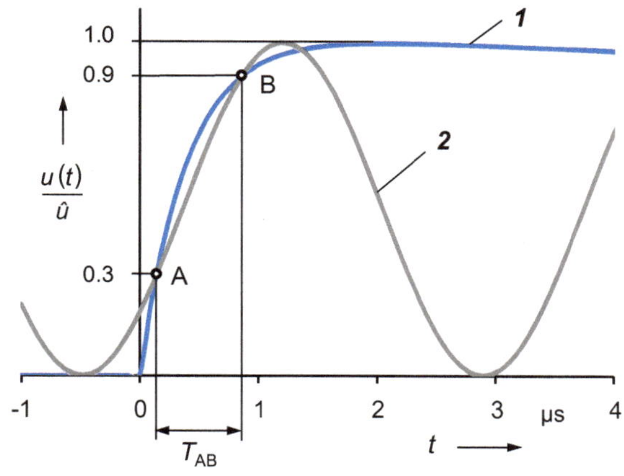

Fig. 10.12 Example of the calibration of a 1.2 lightning impulse voltage ***1*** by comparison with an equally large sine voltage ***2*** with regard to the peak value and the front time T_1

time T_{AB}. For the 1.2/50 lightning impulse voltage, $T_{AB} = 0.6T_1 = 0.72$ μs from which the frequency of the equivalent sine voltage is calculated to be $f = 295.9$ kHz (Ref. [31] of Chap. 7).

References

1. Homepage: www.iec.ch
2. Homepage: www.dke.eu
3. Homepage: www.cigre.org
4. Homepage: www.ptb.de
5. Schon, K., Lucas, W.: Worldwide interlaboratory test comparisons of high voltage impulse dividers. In: Proceeding 2nd ERA Conference on High Voltage Measurements and Calibration, Arnhem, 3.1.1–3.1.9 (1994)
6. Bonamy, A., Bossi, S., Deschamps, F., do Vale, A., Garnacho, F., Hughes, R.C., Lightfoot, H. A., Rizzi, G., Simon, P., Schon, K., Schulte, R., van Boetzelaer, A.W., Vaz, A.: International comparison of HV impulse dividers. In: Proceeding 7th ISH Dresden, paper 61.07 (1991)
7. Hällström, J., Aro, M., Bergman, A., Bovier-Labierre, V., Garnacho, F., Juvik, J.I., Kiseliev, V., Hong, Z.L., Lucas, W., Li, Y., Pykälä, M.-L., Rungis, J., Schon, K., Truong, V.H.: Worldwide comparison of lightning impulse voltage measuring systems at 400 kV level. IEEE Trans. IM **56**, 619–623 (2007)
8. Homepage: www.bipm.org
9. Homepage: www.dakks.eu
10. Schon, K.: Der Deutsche Kalibrierdienst (DKD) auf dem Gebiet der Hochspannungs-Messgrößen. Proc. HIGHVOLT-Kolloquium ’03, Dresden (2003)
11. ISO/IEC 17025: General requirements for the competence of testing and calibration laboratories (2017) German edition: DIN EN ISO/IEC 17025: Allgemeine Anforderungen an die Kompetenz von Prüf- und Kalibrierlaboratorien (2018)
12. ISO 9001: Quality Management Systems - Requirements (2015) German edition: DIN EN ISO 9001: Qualitätssicherungssysteme - Anforderungen (2015)
13. Homepage: www.european-accreditation.org
14. Homepage: www.ilac.org

15. Homepage: www.wto.org
16. Gobbo, R., Pesavento, G.: Analysis of the new procedure of divider qualification according to IEC 60-2. In: Proceeding 9th ISH Graz, paper 4515 (1995)
17. Breilmann, W., Hinrichsen, V.: Two methods of linearity tests of approved measuring systems for LI < 3 MV and SI < 2 MV. In: Proceeding 13th ISH Delft, paper 643 (2003)
18. Suomalainen, E.-P., Hällström, J., Piiroinen, J.: Capacitive divider as a field sensor for voltage linearity measurement on AC dividers. In: Proceeding 13th ISH Delft, paper 418 (2003)
19. Deschamps, F.: Checking linearity of high voltage impulse dividers. In: Proceeding 8th ISH Yokohama, paper 52.04 (1993)
20. Rizzi, G., Tronconi, G., Gobbo, R., Pesavento, G.: Determination of the linearity of impulse divider in the light of the revision of IEC 60: Comparison among several methods. In: Proceeding 8th ISH Yokohama, paper 52.05 (1993)
21. Oliveira, O.B., Junqueira, A.J.S., Chagas, F.A.: Linearity test of HV measuring systems—experimental results. In: Proceeding 8th ISH Yokohama, paper 52.06 (1993)
22. Kind, D., Korff, H., Schon, K.: Abschneidefehler zur Beurteilung von Stoßspannungsteilern. PTB-Bericht PTB-E-28, 22–26 (1986)
23. Kind, D., Korff, H., Schmidt, A., Schon, K.: Chopping errors for characterizing HV impulse dividers. In: Proceeding 5th ISH Braunschweig, paper 71.02(1987)
24. Kind, D., Schon, K., Schulte, R.: The calibration of standard impulse dividers. In: Proceeding 6th ISH New Orleans, paper 41.10 (1989)

Chapter 11
Capacitance and Dissipation Factor

Abstract The optimum performance of high-voltage equipment and apparatus of the electrical energy transmission depends largely on the design and quality of the insulating materials and the error-free execution of the insulation. Solid, liquid or gaseous dielectrics, also in combination, are used as insulation material. Important characteristics of the dielectrics that are exposed to high AC or impulse voltages are the relative permittivity and the dissipation factor (dielectric loss factor). This chapter explains the basics of both measurement quantities and the various analog and digital measurement methods. The basic design of measuring devices such as the Schering bridge with and without Wagner's auxiliary arm, the current comparator bridge and the digital measuring system with A/D converters are discussed. Examples of the calibration of the measuring instruments are given. The properties of compressed gas capacitors according to Schering and Vieweg, which serve as a virtually lossless reference in the measurement of capacitance and dissipation factor, are discussed in detail.

The optimum performance of high-voltage equipment and apparatus of the electrical energy transmission depends largely on the design, the quality of the insulating materials used and the error-free execution of the insulation. Solid, liquid or gaseous dielectrics, also in combination, are used as insulation material. Important characteristics of the dielectrics that are exposed to high AC or impulse voltages are the *relative permittivity* and the *dissipation factor* (*dielectric loss factor*). This chapter explains the basics of both measurement quantities and the various analog and digital measurement methods as well as the calibration of the measuring equipment. The properties of *compressed gas capacitors* according to Schering and Vieweg, serving as almost lossless reference in the measurement of capacitance and dissipation factor, are discussed in detail.

K. Schon, *High Voltage Measurement Techniques*, Power Systems,
https://doi.org/10.1007/978-3-030-21770-9_11

11.1 Basics

Under the influence of an electric field E, the positive and negative charge carriers of the atoms of a dielectric are slightly displaced from their average equilibrium position due to various mechanisms. They form dipoles arranged in the field direction, which lead to *dielectric polarization* and thus to a corresponding electric field that is opposite to the original electric field E in the dielectric. The original field is thus weakened accordingly. When the external electric field becomes zero again, the charge carriers return to their original position within the relaxation time. The effect of the polarization in the dielectric is characterized by the relative permittivity ε_r. Analogous to the magnetic flux density $B = \mu H$ and the electric current density $J = \kappa E$, the *electric flux density* D (*electric displacement*) in the dielectric is defined as:

$$D = \varepsilon E = \varepsilon_0 \varepsilon_r E, \tag{11.1}$$

where $\varepsilon_0 = 8{,}8541878 \times 10^{12}$ F/m is the *electric field constant* and $\varepsilon = \varepsilon_0\ \varepsilon_r$ the *permittivity*. Equation (11.1) applies to isotropic media in which E and D have the same direction. In general notation, D and E are vectors.

Some important solid and liquid insulating materials used in high-voltage engineering, such as mineral oil and thermoplastics, have a relative permittivity that is only slightly greater than 2. For other liquid and solid insulating materials, the relative permittivity is $2 \leq \varepsilon_r \leq 8$. Water that can occur in all insulating materials, however, has a much higher value $\varepsilon_r = 80$. For air and other gases, relative permittivity is $\varepsilon_r \approx 1$. The relative permittivity is not a constant, but depends on various quantities such as temperature, frequency, field strength, humidity, aging, etc.

The relative permittivity, in addition to the dielectric strength, is an important characteristic of insulating materials used in high-voltage engineering. For a capacitive electrode arrangement, this means that the capacitance is correspondingly increased by introducing a dielectric with $\varepsilon_r > 1$. Furthermore, it can be seen that in the case of high-voltage insulation with mixed dielectric, the insulation with the lower relative permittivity is exposed to a higher electric field. This also applies to the case where a small defect in the form of a gas-filled bubble or void exists in liquid or solid insulating dielectrics. The local field strength in the defect is then greater by a factor equal to the relative permittivity ε_r of the surrounding insulating material. Due to the high field strength, then there is a danger that *partial discharges* may occur due to ionization in the gas bubble or void, whereby the life of the insulation could be reduced (see Chap. 12).

In the AC field, the reorientation of the dipoles in the dielectric causes frictional losses, which are called *polarization losses*. Due to the residual ohmic conductivity of insulating material, in addition *conduction losses* occur, which are also present at DC current. The total losses in the dielectric, which occur in the AC field, are characterized by the *dissipation factor* $\tan\delta$. In addition to polarization and conduction losses, *ionization losses* due to partial discharges are also possible. These can start above a certain inception voltage and cause a rapid increase in $\tan\delta$ with increasing voltage.

11.1.1 Dissipation Factor

The dissipation factor is defined as the quotient of the active and reactive components of the power P or, depending on the viewpoint, of the current I or the voltage U in the corresponding equivalent circuit diagram:

$$\boxed{\tan\delta = \frac{P_{\mathrm{w}}}{P_{\mathrm{b}}} = \frac{I_{\mathrm{R}}}{I_{\mathrm{C}}} = \frac{U_{\mathrm{R}}}{U_{\mathrm{C}}}}. \qquad (11.2)$$

Here, only the fundamental oscillation of the AC quantities is considered. As an example, Fig. 11.1a shows a plate capacitor having a lossy dielectric. When an AC voltage U is applied, the current I flowing through the capacitor has two components: a capacitive component I_{C} leading the voltage U by 90°, and a usually much smaller ohmic component I_{R} in phase with U (Fig. 11.1b). The angle between U and I is the *phase angle* φ and that between I and I_{C} is the *loss angle* δ.

For high-voltage insulation, solid and liquid insulating materials with $\tan\delta <$ 0.001 at power frequency are required. Larger $\tan\delta$ values cause heating of the high-voltage insulation, which in turn can further increase the temperature-dependent dissipation factor, thereby inducing *thermal breakdown*. Good solid and liquid high-voltage insulations have dissipation factors in the range of 5×10^{-4} to 1×10^{-5}. For small dissipation factors, we can write $\tan\delta \approx \delta$, and this is almost always true for high-voltage insulations. The dissipation factor depends on a number of influence quantities, such as the magnitude of the applied voltage, ambient temperature, frequency and load duration. Thanks to digital measurement technology, continuous *monitoring* of the dissipation factor of high-voltage apparatus is becoming increasingly common. Therefore, important information about the history of the stressed insulation is always available.

Equation (11.2) refers to the fundamental oscillation of the AC quantities at power frequency, i.e. mainly 50 or 60 Hz. At much higher frequencies, the dipoles generated by the polarization can follow the exciting AC field only delayed due to

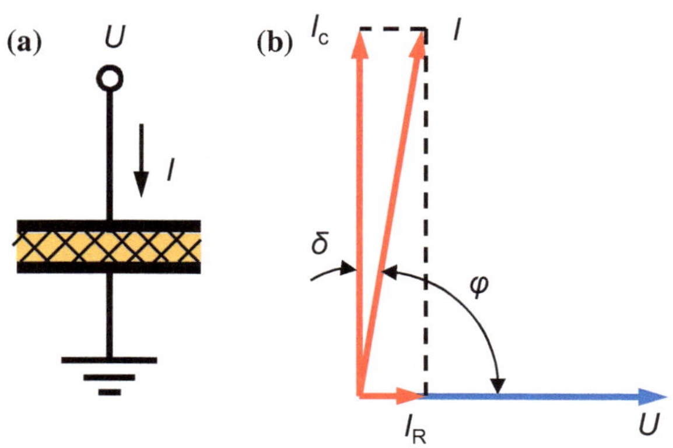

Fig. 11.1 Capacitor with losses and phasor diagram for voltage and current. **a** Capacitor with losses, **b** Vector diagram (parallel equivalent circuit)

the relaxation times, so that a frequency dependence of the relative permittivity occurs. In particular, in theoretical investigations of the processes in the higher frequency range and the occurrence of multiple frequency components, e.g. when excited by a plane wave, it is advantageous to introduce the *complex permittivity*:

$$\boxed{\varepsilon_r^* = \varepsilon_r' - \mathrm{j}\,\varepsilon_r'' \quad \text{mit} \quad \mathrm{j} = \sqrt{-1}}. \tag{11.3}$$

The real part ε_r' here corresponds to the relative permittivity in Eq. (11.1), the imaginary part ε_r'' to the dielectric losses. Analytical expressions for both parts can be derived from the relaxation processes of the dipoles (Ref. [3] of Chap. 1). The dielectric dissipation factor due to the *polarization losses* is:

$$\boxed{\tan \delta_\mathrm{P} = \frac{\varepsilon_r''}{\varepsilon_r'}}, \tag{11.4}$$

and the dissipation factor due to the conduction losses is:

$$\boxed{\tan \delta_\mathrm{L} = \frac{\kappa}{\omega \varepsilon_0 \varepsilon_r'}} \tag{11.5}$$

with the specific conductivity κ.

11.1.2 Equivalent Circuits for Dielectrics with Losses

Lossy dielectrics can be represented by two simple equivalent circuits through which the current I_P or I_S flows (Fig. 11.2). In the one circuit, a resistor and a capacitor are arranged in parallel (index "p"), in the other one in series (index "s"). The resistors R_P and R_S represent the active power P_w due to the losses, the capacitors C_P and C_S the reactive power P_b in Eq. (11.2). The inductive components can be neglected. The dissipation factor results for the *parallel circuit* to (Fig. 11.2a):

$$\boxed{\tan \delta = \frac{1}{\omega R_\mathrm{P} C_\mathrm{P}}} \tag{11.6}$$

and for the series circuit to (Fig. 11.2b):

$$\boxed{\tan \delta = \omega R_\mathrm{S} C_\mathrm{S}}. \tag{11.7}$$

Due to their simplicity, the two equivalent circuit diagrams in Fig. 11.2 have only limited significance. In the parallel connection, the dissipation factor with increasing frequency always seems to be getting smaller, in the series connection

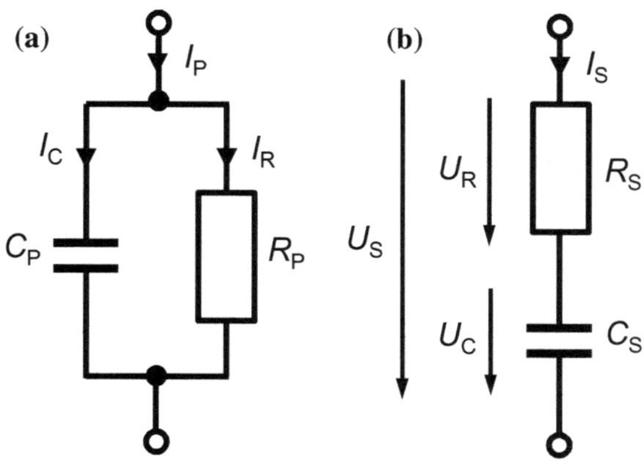

Fig. 11.2 Simple equivalent circuits for a dielectric with losses. **a** Parallel equivalent circuit, **b** Series equivalent circuit

always larger. Of course, both equivalent circuit diagrams should give the same tanδ value when calculated for a given frequency. Under this assumption, the elements of both equivalent circuit diagrams can be converted accordingly by equating the complex impedances (see Sect. 11.3.1). For some insulation arrangements, such as cables and oil insulations, more comprehensive equivalent circuit diagrams are considered in the literature (Ref. [3] of Chap. 1) [1, 2]. The parallel equivalent circuit diagram in Fig. 11.2a can be represented by the phasor diagram with the currents I_C and I_R according to Fig. 11.1. For the series circuit in Fig. 11.2b, a corresponding phasor diagram with the voltages U_R and U_C results.

11.2 Test Arrangements for Solid and Liquid Dielectrics

The relative permittivity ε_r and the *dissipation factor* tanδ—as well as the specific conductivity κ—of solid insulating materials are measured with the test arrangement in Fig. 11.3a. The plate-shaped sample ***1*** is arranged between the voltage electrode ***2*** and the circular measurement electrode ***3***, which is surrounded by the grounded guard ring electrode ***4***. The electrodes ***3*** and ***4*** are connected via a coaxial cable to the measuring instrument M [3]. With the guard ring arrangement, an approximately homogeneous field in the area of the measurement electrode is achieved. In addition, the guard ring ***4*** prevents surface currents from flowing from the voltage electrode ***2*** to the measurement electrode ***3***. The relative permittivity is calculated from the measured capacitance and the geometric dimensions of the sample.

The sample is provided on both sides with adhesive electrodes to ensure good electrical contact with the electrodes. The adhesive electrodes are painted with silver or copper solutions, sprayed with graphite or evaporated in vacuo. However, the solvents may penetrate into the sample, in particular in plastic films, thereby falsifying the measurement. A test arrangement without adhesive electrodes, in

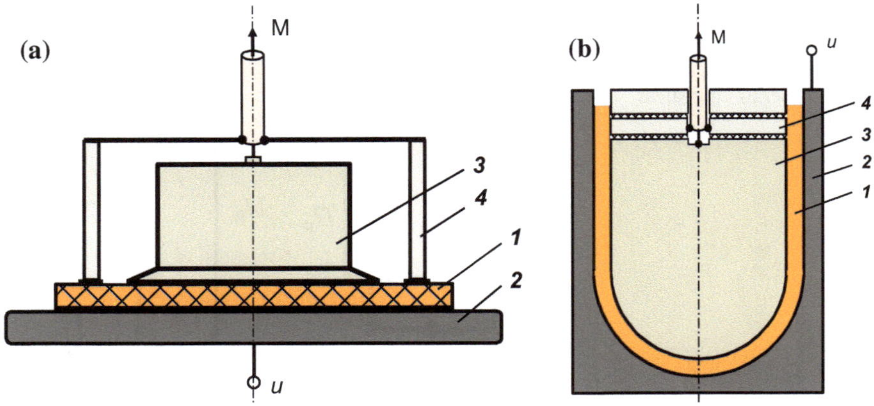

Fig. 11.3 Arrangements for measuring ε_r and $\tan\delta$ of solid and liquid insulating materials. (M: measuring instrument), **a** Plate-shaped insulating sample, **b** Liquid insulating sample, ***1*** insulating sample ***2*** voltage electrode ***3*** measurement electrode ***4*** guard ring electrode

which there is an air gap between the sample and the electrodes at the top and bottom, is investigated theoretically and experimentally in [4]. Due to the positive results, this test arrangement is proposed for inclusion in a new IEC test specification.

Figure 11.3b shows schematically the test vessel for determining the relative permittivity ε_r and dissipation factor $\tan\delta$ of liquid insulating materials. Again, the guard ring principle to eliminate edge effects is clearly visible. First, the air capacitance C_0 of the purified test vessel is measured in the empty state. Subsequently, the insulating liquid is filled into the gap between the voltage and measurement electrode and the capacitance C_X is measured. The relative permittivity results from the two capacitance measurements to $\varepsilon_r = C_X/C_0$. Both test arrangements in Fig. 11.3 allow fundamental investigations of solid or liquid insulating materials with regard to the influence of voltage, temperature, frequency or diffused substances such as water.

There is increasing interest in $\tan\delta$ measurements that can be made directly on apparatus of the electrical power supply, such as cables, bushings and transformers or their replicas [1, 2, 5]. The parameters in the *on-site measurements* are often frequency, voltage or temperature. The aim of the investigations is to obtain reliable information about the long-term behavior and the aging state of the insulation. The rapid development of fast digital technology and data processing meanwhile enables the use of measurement techniques even in *online operation* and thus with significantly shorter evaluation and response times than ever before [6, 7].

Previous $\tan\delta$ measurements relate mainly to approximately sinusoidal voltages in the frequency range from 0.1 to 1000 Hz. According to a new method described in [8], the dielectric response function of the insulation is determined with lightning or switching impulse voltages. The orientation measurements on impregnated paper layers are recorded in the time domain with a digital recorder, and the result is presented in the frequency range of up to 1 MHz. The measurement method is suitable to be applied to transformers and bushings in the high-voltage network, whereby the transient voltages occurring in the network are used for online measurements.

11.3 Measuring Instruments for Capacitance and Tanδ

Basically, a distinction is made between measuring instruments that are used in high-voltage or low-voltage circuits. For both circuits, special bridge circuits are generally used. High-voltage test objects usually have voltage-dependent values of capacitance and dissipation factor and therefore require measurements at power frequency up to the maximum operating voltage. Additional measurements with one of the very accurate low-voltage bridges that normally operate at a measurement frequency of 1000 Hz can be very useful. For example, a low-voltage measurement provides accurate initial values of C and tanδ traceable to SI units on the basis of linearity tests or annual check measurements can be made.

The principle of measuring bridges is to compare the currents flowing through the test object and a virtually lossless standard capacitor in terms of magnitude and phase. About a century ago, Schering presented the classic high-voltage bridge for C and tanδ measurements, which is still used today in several variants [9]. Another classic instrument is the transformer bridge with current comparator, which is also available in self-balancing design. More recent measurement options are provided by electronic C-tanδ measuring devices with A/D converters, with which the currents flowing through the test object and the standard capacitor are separately digitized and evaluated with software.

11.3.1 Schering Bridge

The basic circuit of the *Schering bridge*, which has previously been used almost exclusively for capacitance and tanδ measurements at high voltage, is shown in Fig. 11.4. In the left upper bridge arm, the lossy test object, lying outside the shielded bridge case, is represented as a series connection of the capacitance C_X and the resistance R_X. The standard capacitor C_N, typically a *Schering and Vieweg compressed gas capacitor* (see Sect. 11.5) with a capacitance of 20 pF to 200 pF and a negligible dissipation factor of approximately 5×10^{-6}, lies in the parallel arm outside the bridge case. The AC currents flowing through C_X and C_N generate the voltages u_A and u_B at the bridge points A and B. By setting R_3 and C_4, the bridge circuit can be balanced, i.e. the bridge points A and B are then at the same potential and the null detector NI indicates zero current [9, 10].

The null detector NI is often an oscillographic null indicator. In general, a Lissajous figure can be seen on the screen, which turns into a horizontal straight line

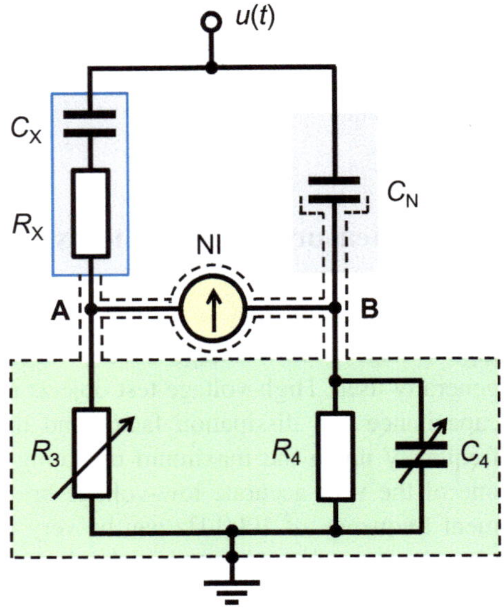

Fig. 11.4 Schering bridge for C and tanδ measurements at high voltage (without protective spark gaps). C_X, R_X: test object with losses, C_N: standard capacitor with negligible dissipation factor, NI: null detector

when the bridge is balanced in phase and magnitude. The balance is achieved only for the fundamental oscillation of the AC voltage $u(t)$, i.e. the higher harmonics of the AC voltage are filtered out. The bridge is protected by spark gaps (not shown in Fig. 11.4) installed at the bridge points A and B. There they limit the voltages in the event of a breakdown or flashover in the high-voltage circuit to safe values. In order to measure a very large capacitance C_X, a low-ohmic resistor can be connected in parallel with R_3, so that the voltage at the bridge point A does not become too large. Variants of the Schering bridge circuit for measuring grounded capacitors are described in (Ref. [5] of Chap. 1) and those with large capacitance in (Ref. [6] of Chap. 1).

According to the Schering bridge in Fig. 11.4, the lossy capacitor or another capacitive test object is represented by C_X and R_X in series in the equivalent circuit diagram. When the bridge is balanced, points A and B have the same potential, resulting in the following equation for the complex impedances and admittances of the bridge elements:

$$\frac{Z_X}{Z_3} = \frac{Z_N}{Z_4} \quad \text{or} \quad Z_X Y_3 = Z_N Y_4, \tag{11.8}$$

and in the complete version:

$$\left(R_X + \frac{1}{j\omega C_X}\right)\frac{1}{R_3} = \frac{1}{j\omega C_N}\left(\frac{1}{R_4} + j\omega C_4\right). \tag{11.9}$$

The separation of Eq. (11.9) into real and imaginary parts supplies equations for the resistance R_X and capacitance C_X in the equivalent circuit diagram:

$$\boxed{R_X = R_3 \frac{C_4}{C_N}} \quad \text{and} \quad \boxed{C_X = C_N \frac{R_4}{R_3}}. \tag{11.10}$$

This results in the expression for the dissipation factor in series connection:

$$\boxed{\tan\delta = \omega\, C_X R_X = \omega\, C_4 R_4}. \tag{11.11}$$

The resistance and capacitance of the series equivalent circuit can be converted into the corresponding elements of the parallel equivalent circuit for a given frequency. By equating the complex impedances $Z_{X,p}$ and $Z_{X,s}$ of the two equivalent circuit diagrams in Fig. 11.2 on the one hand and the two determination equations for tanδ according to Eqs. (11.6) and (11.7) on the other hand, the elements of the parallel equivalent circuit are obtained:

$$\boxed{C_{X,p} = \frac{C_{X,s}}{1 + \tan^2\delta}} \tag{11.12}$$

$$\boxed{R_{X,p} = R_{X,s}\left(1 + \frac{1}{1 + \tan^2\delta}\right)}. \tag{11.13}$$

The Schering bridge is used for measurements at frequencies from a few 10 Hz up to several 100 Hz. The measurement range for C_X is usually between 10 pF and 1 μF and for tanδ between 0.5 and 1×10^{-5}. The dissipation factor of compressed gas capacitors is about 5×10^{-6} (see Sect. 11.5), which should be added to the measured value to achieve the lowest measurement uncertainty. In the case of a test object with small capacitance C_X, the bridge requires a high measurement voltage $u(t) > 10$ kV to achieve the necessary sensitivity of the null indicator. The optimum bridge balance—and thus the lowest measurement uncertainty—is achieved when the capacitances of the two high-voltage capacitors are approximately equal, i.e. $C_X \approx C_N$.

In most cases, the standard capacitor, typically a shielded compressed gas capacitor, and the capacitor under test are located in the high-voltage test hall and are connected via coaxial cables to the bridge in the adjacent control and measurement room. The two coaxial cables should be identical in type and length so that the cable capacitances in both bridge arms are approximately equal. The cable capacitances are parallel to the bridge points A and B lying at voltages on the order of several tens of volts. As a result, leakage currents flow to ground and the bridge balancing is falsified. Also, the stray capacitances of the bridge elements to ground, inductances and resistances of the supply lines and decade switches of the bridge as well as electromagnetic interference coupled into the null detector influence the bridge balance. In some cases, the influence quantities can be taken into account mathematically by correction terms in bridge balancing (Ref. [6] of Chap. 1). If partial discharges occur in the test object, these are reproduced on the screen of the oscillographic null detector as needle pulses superimposed on the Lissajous figure.

11.3.2 Schering Bridge with Wagner's Auxiliary Arm

In the case of bridge balancing, the points A and B of the Schering bridge in Fig. 11.4 are at the same potential but different from zero. Due to the cable capacitances of the two high-voltage capacitors, but also via the stray capacitances of the bridge impedances Z_3 and Z_4, stray currents flow to ground and falsify the bridge balance. This undesirable effect can be avoided by means of Wagner's auxiliary arm [11] (Ref. [4, 6] of Chap. 1). The auxiliary arm consists of the impedance Z_{W1}, usually a high-voltage capacitor, and the adjustable impedance Z_{W2} in the lower part (Fig. 11.5). The connection point E between the impedances Z_{W1} and Z_{W2} is grounded, i.e. the test voltage source must not be grounded and the lower bridge point is protected by a spark gap. The shielding of the Schering bridge (here not drawn), the null detector and the coaxial cables are also connected to point E and thus grounded.

With the switch S, the null detector NI is alternately connected to the bridge point A of the Schering bridge and the grounded point E of Wagner's auxiliary arm. Each time, the two bridges are balanced in magnitude and phase. In general, balancing of each bridge must be repeated several times. After successful balancing, both bridge points A and B are at *virtual zero potential* without being galvanically connected to ground. Since then there is no driving voltage between the conductors and shields of the two coaxial cables, no leakage currents can flow through the cable capacitances to ground. The same applies to the bridge elements connected to A and B.

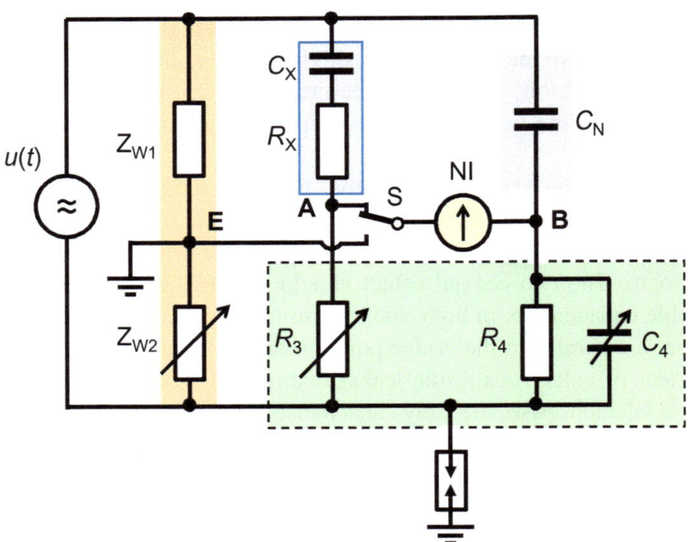

Fig. 11.5 Schering bridge with Wagner's auxiliary arm consisting of Z_{W1} and Z_{W2} (circuit principle)

The manual double adjustment of the Schering bridge and the Wagner auxiliary arm can be tedious and time consuming. Remedy provides an electronic circuit that replaces Wagner's auxiliary arm and allows a semi-automated bridge balancing. Here, the Schering bridge (see Fig. 11.4) is grounded, while the shielded housing and the shielding of the two coaxial cables are connected to the output of an operational amplifier. The amplifier operates as an impedance converter with unity gain and its input is connected to the bridge point B. When manually balancing the Schering bridge, point B—and thus also point A—is raised to a certain potential which is then also applied to the shielding via the operational amplifier. The measuring bridge is thus optimally balanced and since the inner conductors and shields of the long coaxial cables are at the same potential, there are no leakage currents. A further improvement results when using double-shielded cables, with the outer shields grounded (Ref. [6] of Chap. 1).

11.3.3 Measuring Bridge with Current Comparator

The development of very precise *current comparators* with high permeability core and electromagnetic shielding has resulted in a new type of high-voltage bridge for capacitance and dissipation factor measurements. The principle of the *current comparator bridge* or *transformer ratio bridge* is that the currents of the test object and of the standard capacitor flow in opposite directions through two windings of a current comparator, and when the bridge is balanced, the magnetic flux in the core is zero (Fig. 11.6a). The balance for magnitude and phase is indicated by a null detector, which is connected to a third winding of the same magnetic core. The first measuring bridges of this type were still balanced manually by setting the bridge resistance R and the number of turns N_1 and N_2 [12]. Further developments have resulted in transformer ratio bridges with additional windings, via which an electronically generated compensation current is fed into the bridge circuit for balancing (Fig. 11.6b) [13]. Semi-automatic and finally—due to the use of a microprocessor—fully automatic capacitance and tanδ measurements are now possible [14–18]. Thus, continuous monitoring of the insulation of high-voltage capacitors and other equipment can be carried out for long-term studies.

In addition to the automatic balance option, the bridge circuit with current comparator offers further advantages over the Schering bridge. First, the sensitivity of the null detector can be increased by selecting a corresponding number of turns. Secondly, due to the low-ohmic comparator windings, the bridge points A and B are at a lower potential, as a result of which the cable capacitances of the test object and standard capacitor have a negligible influence on the bridge balance. Furthermore, the low bridge potential enables accurate and automated calibration of the bridge by feeding well-defined currents into the bridge points A and B (see Sect. 11.4.1).

In order to be able to measure very large capacitances with correspondingly high charging currents, the test object C_X is connected to the transformer bridge via a

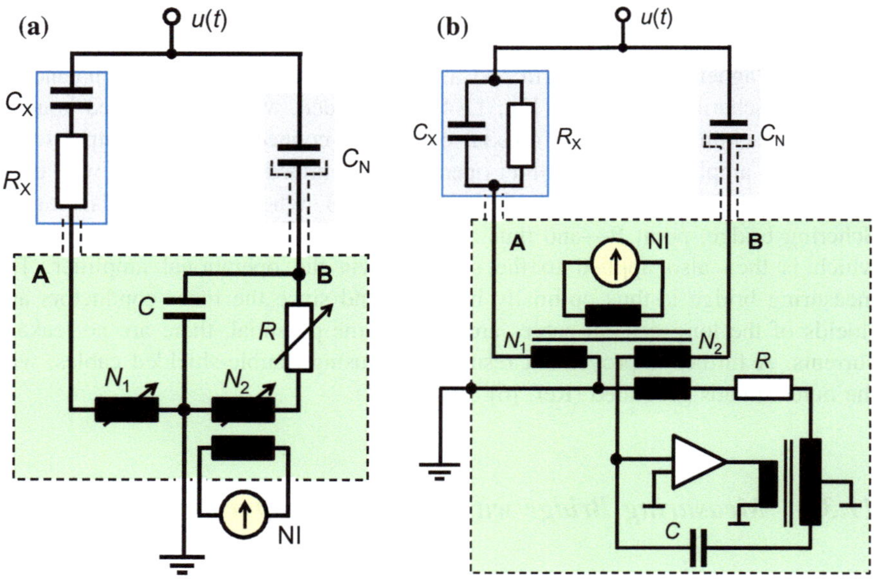

Fig. 11.6 Basic circuits of the current comparator bridge. **a** Bridge with manual balancing, **b** Bridge with automatic balancing

current transformer, which is designed as an electronically error-compensated current comparator [19]. In another study, the standard capacitor C_N is connected to the measuring bridge via a precisely calibrated voltage transformer [20]. Therefore, C_N is at a much lower voltage than the test object C_X. A possibly present voltage dependence of C_N practically does not occur at this reduced voltage. Alternatively, C_N may be a precise low-voltage standard capacitor. However, the test voltage is limited to about 300 kV, because inductive instrument transformers are not usually built for higher voltages.

11.3.4 Digital Capacitance and Tanδ Measurements

As in many other fields of metrology, digital data acquisition with A/D converter and computer-aided evaluation has also found its way into capacitance and dissipation factor measurements. Figure 11.7a shows the basic circuit [21, 22]. The lossy capacitive test object C_X and the standard capacitor C_N are connected to the sensor S_X or S_N on the low voltage side. The capacitive or resistive sensors provide output signals that are directly or after integration proportional to the AC currents I_X and I_N. The two signals are then amplified and digitized with high-resolution A/D converters. Each sensor is either housed with the associated amplifier and A/D converter in a battery-powered unit at the base of C_X or C_N, or is connected

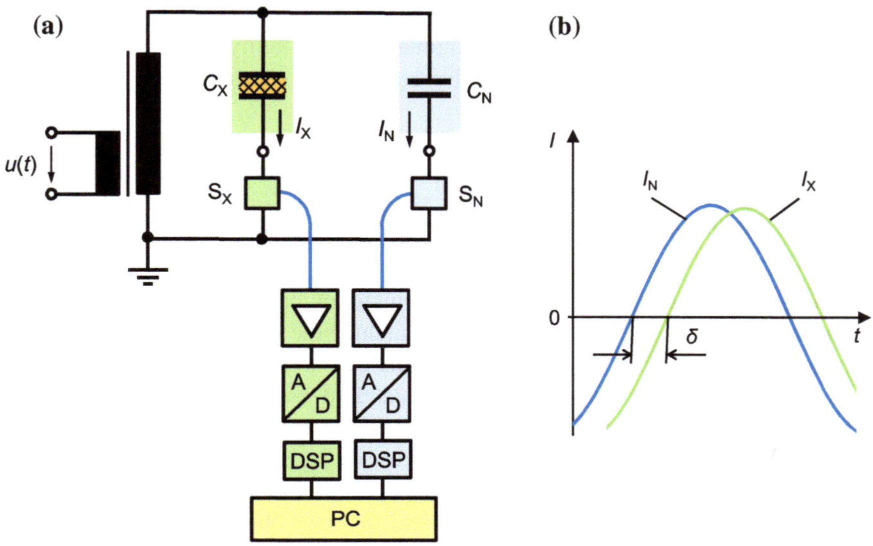

Fig. 11.7 Digital measuring system for capacitance and dissipation factor. **a** Principle of the measuring system, **b** Loss angle δ of the test object C_X

potential-free via optical fiber to the electronic circuit on ground potential. After passing through the digital signal processors DSP, the data of the records are processed with software on the PC. Signal evaluation takes place with the aid of the DFT, wherein the fundamental oscillations of the currents I_X and I_N are evaluated. The capacitance C_X is determined from the amplitude ratio, the loss angle δ from the phase difference of the two currents (Fig. 11.7b). The loss angle δ can also be represented in the phasor diagram (see Fig. 11.1b). The digital measuring system is also suitable for continuous monitoring of capacitive apparatus.

Note The digital measuring system in Fig. 11.7 is often referred to as a "measuring bridge". However, this is not entirely correct, as bridge balancing does not take place and the independent current measurement arms are vectorially evaluated.

11.4 Calibration and Traceability

The dissipation factor is an important characteristic that can provide valuable information about the state of the high-voltage insulation after electrical, thermal and mechanical stress during long-term use. Prerequisite for a correct diagnosis is the reliable measurement of the dissipation factor, which must be verified by a traceable calibration of the measuring instrument. This section explains procedures that can be used to calibrate C and tanδ measuring instruments, standard capacitors and dissipation factor standards. As with all other quantities, the question of traceability of loss factor measurements to national and international units of measurement also arises here. International comparability and recognition of measurement results is demonstrated by comparison measurements on transfer standards.

11.4.1 Calibration of C-Tanδ Measuring Instruments

The achievable uncertainty in the measurement of capacitance and dissipation factor depends on a variety of factors. These include the type, measuring sensitivity and shielding of the measuring instrument, the deviations of the bridge balancing elements from their nominal values, the standard capacitor and the type and capacitance of the test object itself. The previous manual calibration of a Schering bridge by measuring the numerous internal decade resistances and capacitances is extensive and time-consuming. Selective checks of the balanced bridge can be carried out with known loss factor standards as test object C_X. The results of comparison measurements in European laboratories show that Schering bridges and other measuring devices, such as current comparator bridges, do not differ significantly in their C and tanδ values measured on a 100 kV transfer capacitor with artificially enlarged losses, provided the test voltage exceeds 30 kV (see Sect. 11.4.2).

In another accurate calibration method, two calibration currents I_X and I_N are fed directly into the bridge arms, with the bridge connected to neither of the two high-voltage capacitors. The calibration currents are exactly generated in terms of magnitude and phase by means of transformers and inductive dividers from the low-voltage network. I_X and I_N thus simulate the currents flowing through the capacitive test object C_X and the standard capacitor C_N. The working principle of such a calibration device is comparable to that of a computer-controlled measuring system with which the correctness of test sets for current transformers is verified [23]. If, for given values of I_X and I_N, the calibrated bridge appears balanced, the errors of the indicated C and tanδ values can be calculated from the magnitudes and phase angles of the injected calibration currents. The tabulated bridge errors are then used as corrections to obtain accurate C and tanδ values. This calibration procedure is particularly suitable for electronic bridge circuits with current comparator (see Sect. 11.3.3), but also for measuring instruments with vector evaluation (see Sect. 11.3.4). Calibration can then be performed fully automatically, including the printout of the report. Since the comparator windings and the sensors into which the calibration currents are fed are low-ohmic and thus at low potential, stray capacitances to ground have practically no influence.

A further improved calibration method for all types of bridges and other devices for measuring capacitance and dissipation factor is presented in [24]. The sinusoidal calibration currents I_N and I_X to be fed into the measuring bridge ***1*** are generated by means of a digital *two-channel AC voltage source* with D/A converters ***2*** in conjunction with the precision resistors ***3*** according to the equations (Fig. 11.8):

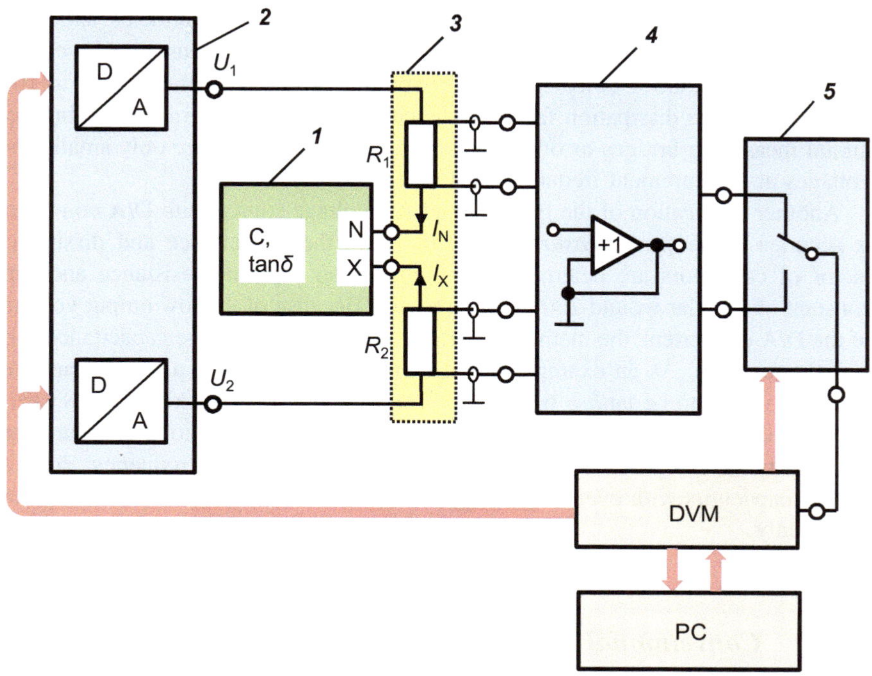

Fig. 11.8 Calibration device for analog and digital C-tanδ measuring systems. *1* C-tanδ measuring system as test object, *2* two-channel AC voltage source with D/A converter, *3* precision resistors R_1 and R_2, *4* buffer amplifiers (fourfold), *5* channel switch

$$\boxed{I_N = \frac{U_1}{R_1}} \quad \text{and} \quad \boxed{I_X = \frac{U_2}{R_2}}. \tag{11.14}$$

The currents generate across the resistors R_1 and R_2 two potential-free voltages which are applied to the channel switch ***5*** via the *buffer amplifiers* ***4***. For each m periods, the two sinusoidal voltages are alternately switched to the precision digital voltmeter DVM, which operates in the sampling mode and digitizes the sinusoidal voltages. The periods immediately before and after the switching points are not taken into account during sampling. The DVM controls the channel switching and also takes over the triggering and control of the two D/A converters, which are thus clocked in exact synchronous operation. The currents, voltages and impedances occurring in the calibration circuit can be expressed as complex quantities in equations with which the capacitance and dissipation factor of the test object are calculated. The evaluation of the measured data and the control of the DVM is done with software on the PC.

In general, the C and tanδ values displayed by the tested bridge differ more or less from those predefined by the calibration device. The differences can be saved and used as corrections for the future use of the bridge. By using high-resolution

D/A converters, precise resistors, a very accurate digital voltmeter and other accurate circuit elements, extremely low uncertainties of measurement can be achieved. These are 1×10^{-6} (relative) for the capacitance and $(1–3) \times 10^{-6}$ (absolute) for the dissipation factor in the calibration of transformer ratio bridges, digital measuring bridges or other measuring devices that require only small input voltages at measurement frequencies of up to 100 Hz.

Another application of the two-channel AC voltage source with D/A converters is accurate *impedance measurement* [25]. Here, the capacitance and dissipation factor of capacitors are determined by comparison with the resistance and time constant of a bifilar wound 200 Ω wire resistor. Because of the low output voltages of the D/A converters, the method is particularly suitable for large capacitances and high frequencies. As an example, the dissipation factor of a 1 μF mica capacitor was determined to be $\tan\delta = 62.1 \times 10^{-6}$ at a frequency of 2π kHz = 6.283 kHz. The detailed consideration of all influence quantities leads to a measurement uncertainty of 2×10^{-6} ($k = 1$). At a reduced measurement frequency, such as 50 Hz, capacitors with even larger capacitances can be measured with similar low uncertainty.

11.4.2 Conventional Dissipation Factor Standards

The accuracy of capacitance and dissipation factor measurements depends on the quality of both the standard capacitor C_N and the measuring bridge. The best low-voltage standard capacitors with 10 pF and 100 pF have a quartz or nitrogen insulation whose dissipation factors are in the range of $(2–4) \times 10^{-6}$. The Schering and Vieweg compressed-gas capacitors (see Sect. 11.5), which are used as C_N at high voltages of up to 1.5 MV, have about twice the dissipation factor. In the case of high accuracy requirements, in particular when measuring other high-voltage standard capacitors, this dissipation factor must be added to the measured $\tan\delta$ value of the test object. To check individual $\tan\delta$ values obtained from a balanced high-voltage bridge, compressed-gas capacitors with series-connected resistors are suitable. However, the dissipation factor of the shielded series circuit cannot simply be calculated using Eq. (11.7), but only taking into account the stray capacitances.

Figure 11.9 shows the simple equivalent circuit diagram of a dissipation factor standard consisting of a shielded capacitor with the main capacitance C in series with a resistor R [26]. Due to the shielding, there are defined stray capacitances of C and R against the grounded housing. They are labeled C_{14}, C_{34} and C_{24}, with C_{24} resulting from the parallel connection of C'_{24} and C''_{24}. The T-network consisting of C, R and C_{24} and the other stray capacitances C_{14} and C_{34} are transformed into an equivalent π-network of the dissipation factor standard. The effective capacitance C_{13} between the input and output terminals ***1*** and ***3*** is calculated from the complex admittances as:

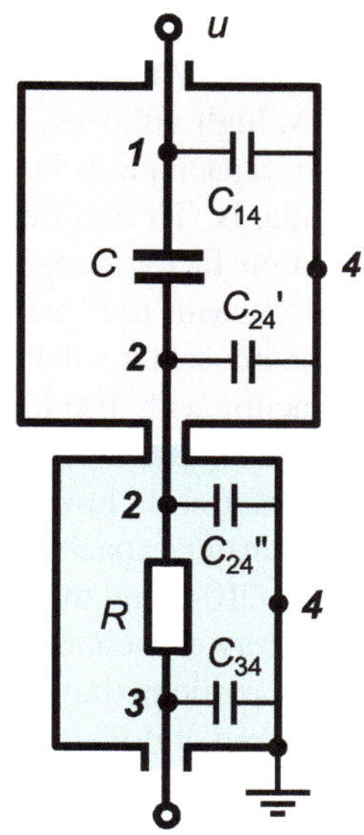

Fig. 11.9 Simple equivalent circuit diagram of a shielded dissipation factor standard for the calibration of C-tanδ measuring systems.
C standard capacitor without losses, R series resistor, C_{14}, C'_{24}, C''_{24}, C_{34} stray capacitances to ground

$$\boxed{C_{13} = \frac{C}{1 + \tan^2 \delta}} \tag{11.15}$$

and the dissipation factor of the standard as:

$$\boxed{\tan \delta = \omega (C + C_{24}) R}. \tag{11.16}$$

The effective capacitance C_{13} according to Eq. (11.15) is lower than the capacitance C actually installed in the dissipation factor standard. In the case $\tan\delta < 10^{-3}$, however, this difference is negligible due to the quadratic contribution $\tan^2\delta$. According to Eq. (11.16) for calculating $\tan\delta$, the capacitance is increased by the stray capacitance C_{24} compared to Eq. (11.7) for the simple series equivalent circuit (see Fig. 11.2b). The other stray capacitances C_{14} and C_{34} have no influence on the effective capacitance and the dissipation factor.

The theoretical results on the dissipation factor standard in [26] were experimentally verified for a gas-filled 1000 pF standard capacitor with nine pluggable resistor boxes connected in series. The dissipation factors at 60 Hz, calculated according to Eq. (11.16), ranged from 5×10^{-6} (for $R = 0$) to 1×10^{-2} (for R = 25671 Ω). For the measurements, the precision bridge presented in [27] was used. The resulting stray capacitance taken for the calculation in Eq. (11.16) was measured to be $C_{24} = 40$ pF. The measured $\tan\delta$ values, which were each determined as the mean of ten individual measurements, were slightly smaller than the corresponding calculated values. For dissipation factors below 2×10^{-4}, the uncertainty of measurement ($k = 2$) is less than 2×10^{-4} (absolute value). Due to the small differences between the corresponding measured and calculated values, the equivalent circuit diagram of the dissipation factor standard shown in Fig. 11.9 is thus fully confirmed.

At high voltages, compressed gas capacitors according to Schering and Vieweg with capacitances between 10 pF and 200 pF are generally used as C and $\tan\delta$ standards. To find out the measurement capability in Europe, capacitance and dissipation factor measurements up to 100 kV were performed in a European comparison with four metrology institutes and three test laboratories [28]. The transfer standard sent to the laboratories was an SF_6-insulated, 120 kV compressed gas capacitor with 100 pF plus two shielded resistors of 250 and 20 kΩ. The resistors were intended to increase the dissipation factor by connecting them in series to the capacitor at its low-voltage output. In the pilot laboratory, the dissipation factor of the transfer capacitor without additional series resistor was measured to be $\tan\delta = 6.2 \times 10^{-6}$ and the stray capacitance to $C_{24} \approx 145$ pF, using a low-voltage bridge. The stray capacitance was thus greater than the rated main capacitance $C = 100$ pF. In the participating laboratories, the C and $\tan\delta$ values of the transfer standard were measured at different temperatures, gas pressures, etc. The values of all participants were related to uniform test conditions and then combined for each of the C or $\tan\delta$ measurement into a common mean. There was a significant deviation of up to 20% between the mean $\tan\delta$ values and the corresponding values calculated according to Eq. (11.16), the cause of which could not be explained.

11.4.3 Cryogenic Dissipation Factor Standard

The basis of every measurement is the proof that the measurand is traceable directly or indirectly with low uncertainty to the units defined in the SI system. The following example illustrates the theoretical and experimental fundamentals to achieve traceability of the dissipation factor to *calorimetric quantities*. In the investigation, the dissipation factor of a 45 μm thick mica capacitor was determined with an elaborate calorimetric method at 4.2 K in the He cryostat [29]. The vapor deposited Pb electrodes and the Pb coated supply lines made of nylon filaments are superconducting at He temperature and therefore lossless. The advantage of the low temperature is further that mica has a very low specific heat capacity and therefore has significant heating when an AC voltage is applied. The measurement principle is that the temperature rise of the mica capacitor due to an applied AC voltage is compared with the temperature rise resulting from the electrical heating by a DC voltage.

The *dielectric power loss* P_w of the mica capacitor is given by Eq. (11.2):

$$P_w = P_b \tan\delta = U^2 \omega C \tan\delta. \tag{11.17}$$

If the AC voltage is applied for the duration Δt_1, the temperature of the lossy mica capacitor increases by ΔT_1. On the other hand, if the capacitor is supplied with a DC heating power I^2R for the duration Δt_2, it heats up by ΔT_2. With an identical temperature increase $\Delta T = \Delta T_1 = \Delta T_2$, the dissipation factor of the mica capacitor is calculated as follows:

$$\boxed{\tan\delta = \frac{I^2R}{U^2\omega C}\frac{\Delta t_2}{\Delta t_1}}. \tag{11.18}$$

In the experiment, at a temperature of 4.2 K and an applied AC voltage $U = 30$ V of the angular frequency $\omega = 10^4\ \mathrm{s}^{-1}$, the dissipation factor of the mica capacitor was determined to be $\tan\delta = 2.23 \times 10^{-6}$ with very low measurement uncertainty. Subsequently, the cryogenic dissipation factor standard was connected to a high-precision bridge at room temperature and several 10 pF and 100 pF low-voltage capacitors were calibrated at frequencies between 53 Hz and 10 kHz.

The dissipation factor of these capacitors, which are subsequently used as reference standards, is thus traceable to the SI units. However, the measurement uncertainty is significantly increased by additional losses of the supply lines from the cryostat to the measuring bridge at room temperature [30, 31]. The uncertainty of the dissipation factor is further increased with decreasing frequency. For example, the uncertainty of $\tan\delta$ that can be achieved with low-voltage measurements at 50 Hz is estimated to be 4.3×10^{-6} (absolute value, $k = 1$). This rather large uncertainty is considered unsatisfactory and will be further increased in the calibration of high voltage capacitors.

11.5 Compressed Gas Capacitors

Compressed gas capacitors of the Schering and Vieweg type with capacitances of typically no more than 100 pF and rated voltages of up to 1.5 MV are often used for accurate measurements in several fields. In contrast to other high-voltage capacitors, compressed gas capacitors are largely shielded by the coaxial arrangement of their cylindrical electrodes and thus protected against electromagnetic interference. Because of other good properties, such as stable capacitance and small dissipation factor, they are mainly used for accurate measurements of AC voltages and as standard capacitors in bridge circuits for capacitance and dissipation factor measurements. The insulating gas used today is mostly SF_6, formerly also N_2 or CO_2. Another type of gas-insulated capacitors has stacked plate electrodes housed in a metal container. They have larger capacitances of up to several nanofarads and also very small dissipation factors. However, the rated voltages are only in the range of (1–30) kV. Gas-insulated capacitors of the parallel-plate type are not dealt with here.

11.5.1 Constructions

Schering and Vieweg compressed gas capacitors have two or more concentric electrodes housed in a gas-filled insulating cylinder [32–34]. In the usual arrangement with two cylinder electrodes for voltages up to a maximum of 1500 kV, the high-voltage electrode ***1*** is arranged concentrically around the measuring electrode ***2***, which is mounted in isolation on the grounded metal tube ***3*** (Fig. 11.10a). The high-voltage plate electrode ***1*** is held by a stable insulating cylinder ***4***, which is covered gas-tight at the top and bottom with metal plate electrodes. The insulating cylinder ***4*** may be made of laminated paper, glass fiber reinforced epoxy resin or polymethylmethacrylate (PMMA, acrylic glass, Perspex, Plexiglas). The surfaces of the electrodes, which are usually made of brass, are polished and often provided with a chrome or nickel layer. This helps to avoid damage to the electrode surfaces by partial discharges, which can occasionally occur during conditioning of the capacitor after filling with not quite pure insulating gas. The connecting line from the measurement electrode ***2*** to the output socket ***5*** is shielded by the metal tube ***3***. Normally, a pressure gauge is installed at the bottom of the capacitor to check the pressure of the insulating gas.

In another embodiment, the measuring electrode ***2*** is divided into two or more sections. While the larger section is used for capacitance and dissipation factor

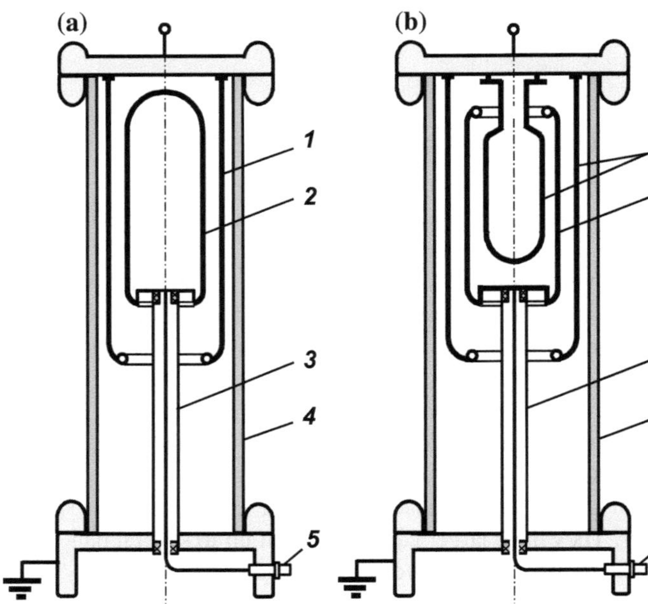

Fig. 11.10 Construction of compressed gas capacitors of the Schering and Vieweg type. **a** Arrangement with two cylinder electrodes **b** Arrangement with three cylinder electrodes, ***1*** high-voltage electrode, ***2*** low-voltage electrode, ***3*** metal tube, ***4*** insulating cylinder, ***5*** output socket

measurements, the other sections are available for voltage measurements or serve as guard ring electrodes to avoid edge effects and the influence of extraneous fields. Figure 11.10b shows a compressed gas capacitor with two high-voltage electrodes ***1***, between which the measurement electrode ***2*** is located. The advantage of this arrangement is in the larger capacitance compared to a capacitor of the same sized in the construction according to Fig. 11.10a.

In both arrangements of Fig. 11.10, the low voltage electrode ***2*** is mounted on a long metal tube ***3***. It therefore takes some effort to position the cylinder electrodes concentrically. In the case of an eccentric electrode arrangement, the capacitance is voltage-dependent since the electrostatic attraction forces increase the eccentricity and thus the capacitance (see Sect. 11.5.5). In addition, this arrangement represents an oscillatory pendulum, which can lead to resonance and thus to changes in capacitance and dissipation factor (see Sect. 11.5.6.2).

In another type of capacitor, the low-voltage and high-voltage electrodes are located in a gas insulated metal vessel provided with a bushing for the high-voltage electrode. The significantly shorter support tube for the low-voltage electrode results in improved mechanical stability and thus in reduced voltage dependence [35].

11.5.2 Influence of Gas Pressure

Compressed gas capacitors are mainly filled with SF6 at a pressure of (2–5) bar. SF_6 has the advantage of higher dielectric strength and inception voltage of partial discharges compared to other insulating gases. Previously used insulating gases were CO_2, N_2 or dry air with significantly higher pressure of up to 15 bar. The relative permittivity ε_r of the insulating gases differs only slightly from one. At 1 bar and 20 °C, the values are $\varepsilon_r = 1.00205$ for SF_6, 1.00095 for CO_2 and 1.00053 for N_2 [36]. The relative permittivity depends on the *gas density*, i.e. the gas pressure p and the temperature T according to (Ref. [4] of Chap. 1):

$$\boxed{\varepsilon_r = 1 + \alpha \frac{273\,p}{100\,T}} \tag{11.19}$$

with p in kPa and T in K. The numerical value α for the gas in question can be calculated from Eq. (11.19) with the above-mentioned relative permittivity.

The pressure dependence of the compressed gas capacitor is given by:

$$\boxed{C(p) = C_0(1 + \alpha \cdot \Delta p)}, \tag{11.20}$$

where C_0 is the capacitance specified for the operation pressure p_0 and Δp is the pressure other than p_0. By measuring the capacitance at different gas pressures, the coefficient α can be determined experimentally. The experimental α values may deviate from the theoretical values given above. The 10 to 20% higher experimental values mentioned in [34, 37] suggest that the examined gases were contaminated, e.g. by moisture with $\varepsilon_r \approx 80$.

The internal gas pressure is usually indicated by a small mechanical pressure gauge with pointer instrument installed at the bottom of the capacitor. However, such a gauge does not allow a sufficiently accurate pressure reading for precision capacitance measurements. The use of a calibrated electronic pressure transducer with sufficiently high sensitivity of 0.1 kPa is reported in [28]. However, the transducer used had a positive drift, so that an approximately equal pressure loss in the capacitor due to a leak could not be detected in time.

Pressure changes caused solely by temperature changes do not affect the capacitance, i.e. the capacitance is corrected only according its temperature coefficient. The reason is that the number of gas molecules—or the *gas density*—present in a gas-filled container remains constant during a temperature change. Although the volume of the gas container may increase slightly when the internal pressure is increased with temperature, the gas density can be assumed to be constant in a first approximation. The very limited influence on capacitance is a few parts per million [33]. The temperature dependence of the capacitance of compressed gas capacitors is mainly caused by the change in the dimensions of the electrodes and their supports (see Sect. 11.5.3).

11.5.3 Temperature Dependence

In high-voltage test halls, temperature gradients of usually more than 0.2 K/m and temporal changes of ±1 K must be expected during a working day. The required height of compressed gas capacitors is up to several meters, depending on the rated voltage. The temperature coefficient for the capacitance of compressed gas capacitors according to Fig. 11.10 is typically in the range of $(2–3) \times 10^{-5}\ \text{K}^{-1}$. For accurate capacitance measurements, it is recommended to use a low-voltage bridge that is more accurate and easier to use than a high-voltage bridge. For temperature measurement, one or two specific locations for the installation of temperature sensors on the capacitor are to be selected. Whether the temperature sensor is to be attached to the bottom or top of the compressed gas capacitor or to both should be determined by orientation measurements. Furthermore, it should be noted that the stationary state is reached only after more than ten hours after a previous change in the air temperature. Instead of correcting the capacitance according to the temperature measurement, it is more accurate to directly determine the capacitance before and after a measurement period using a low-voltage bridge. This check can also help to detect a possible change in capacitance due to a gas leak.

The insulating gas has only minor influence on the temperature dependence of the capacitance. As mentioned above, the insulating cylinder expands with increasing temperature, resulting in a small decrease in the density of the gas molecules. The main cause of the temperature influence is the different thermal expansion of the metal electrodes and the outer insulating cylinder. Here, several parts can be distinguished [33, 34, 38]. The main part comes from the elongation of the low-voltage electrode ***2*** with increasing temperature, which causes an increase in capacitance (see Fig. 11.10). The second part is due to the different elongation of the metal tube **3** and the insulating cylinder **5**. Depending on which elongation prevails, the distance between the hemispherical or dome-shaped head of the low-voltage electrode ***2*** and the plate electrode above becomes larger or smaller, and the partial capacitance changes accordingly. As the temperature rises, the head of the low voltage electrode ***2*** expands and increases the partial capacitance to the plate above. By selecting materials with suitable temperature coefficients, the temperature effect can be partially compensated and the capacitance change limited.

Another situation occurs when there is a sudden temperature change, e.g. when cold or warm air enters the high-voltage test hall through a temporarily opened hall door. When the air flow hits the compressed gas capacitor on one side, the insulating cylinder ***4*** slightly bends. As a result, the high-voltage electrode changes its position relative to the low-voltage electrode and a transient capacitance change occurs. The stationary state will only be reached after several hours. In order to avoid or at least reduce this effect, the capacitor may be surrounded by isolating hard-foam boards.

11.5.4 Eccentricity and Capacitance

The capacitance C of a concentric electrode arrangement compressed gas capacitor (as well as that of a coaxial cable) with the length l, the inner radius r_1 of the outer cylinder and the outer radius r_2 of the inner cylinder is (Ref. [2] of Chap. 9) [35]:

$$\boxed{C = \frac{2\pi\varepsilon l}{\ln(r_1/r_2)}}. \tag{11.21}$$

The prerequisite for this is that no edge effects occur at the ends of the ideal cylinder capacitor. Such edge effects can be largely avoided by two guard ring electrodes. Compared with an eccentric arrangement, the ideal concentric electrode arrangement is characterized in that the capacitance has a minimum and is not voltage-dependent.

The capacitance of cylinder electrodes, whose axes are offset by the eccentricity e, as shown in Fig. 11.11, is given by [35]:

$$\boxed{C = \frac{2\pi\varepsilon l}{\ln\left(y + \sqrt{y^2 - 1}\right)}} \tag{11.22}$$

with:

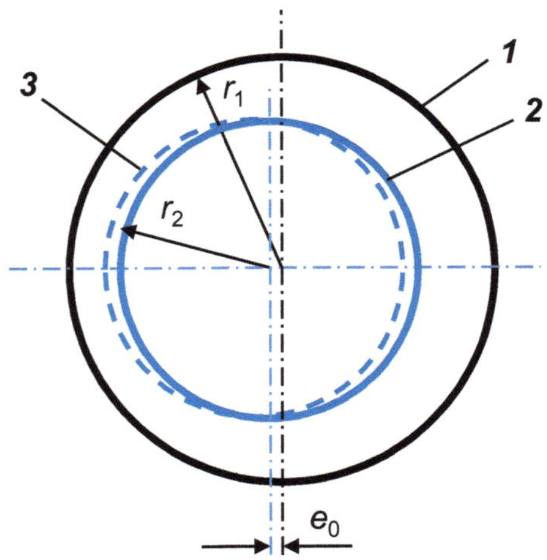

Fig. 11.11 Cross section through a cylindrical compressed gas capacitor with eccentric low-voltage electrode (principle). ***1*** high-voltage electrode with inner radius r_1, ***2*** low-voltage electrode with outer radius r_2, ***3*** position of deflected low-voltage electrode

$$y = \frac{r_1^2 + r_2^2 - e^2}{2r_1r_2}, \quad y > 1. \tag{11.23}$$

Extensive investigations on older compressed gas capacitors with rated voltages between 100 and 800 kV show that the eccentricity of the electrodes can be up to 5 mm.

A simple measurement procedure gives an orienting impression of the eccentricity of the low-voltage electrode, the rigidity of the support tube and the voltage dependence of the capacitance [28]. For this purpose, the compressed gas capacitor is placed horizontally on the ground so that the low-voltage electrode supported by the metal tube tilts down due to gravity. The horizontal compressed gas capacitor is then rotated about its axis by hand and the capacitance is measured as a function of the angle of rotation β with a low-voltage bridge. Figure 11.12 shows the relative change in capacitance versus the angle of rotation β for two compressed gas capacitors with small eccentricity (curve ***1***) or large eccentricity (curve ***2***) of the

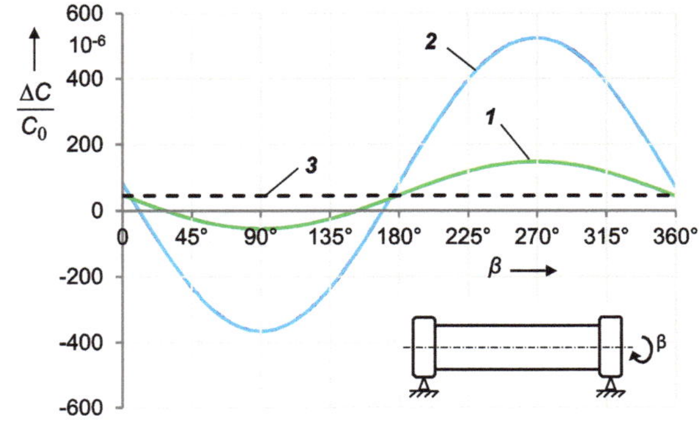

Fig. 11.12 Relative change in the capacitance of horizontal compressed gas capacitors with the angle of rotation β. C_0: Capacitance of the capacitor in the vertical position, ***1*** low eccentricity, *2* large eccentricity, *3* center line of ***1*** and ***2***

low-voltage electrode. Both capacitors tested were of the same type (100 pF, 120 kV) as shown in Fig. 11.10b. Curve ***3*** in Fig. 11.12 represents the center line of the two sinusoidal curves ***1*** and ***2***. The capacitance change $\Delta C(\beta)$ refers to the capacitance C_0 of the capacitor in the normal vertical position. In the ideal case of a precisely concentric electrode arrangement, the inclination of the low-voltage electrode due to gravity would be the same for each angle of rotation and the capacitance independent of the angle of rotation, comparable to curve ***3***.

Capacitance curves similar to the sinusoidal curves shown in Fig. 11.12 are obtained when the compressed gas capacitor is in the usual vertical position and rotated around its axis and a constant force acts laterally on the top of the capacitor. As a result, the insulating cylinder bends and the high voltage electrode attached thereto changes its position. If the force acts exactly in the direction of the eccentricity so that it decreases, the capacitance becomes minimal. The capacitance is again preferably measured at low voltage. Knowing the direction of the eccentricity can be successfully exploited in high-voltage measurements. In order to eliminate or reduce the eccentricity and thus the voltage dependence, the capacitor is tilted accordingly, so that the low-voltage electrode is centered by gravity [33, 37].

11.5.5 Voltage Dependence

The capacitance of a compressed gas capacitor is voltage-dependent, if the cylindrical electrodes ***1*** and ***2*** in Fig. 11.10 are not exactly concentric [39]. When a high voltage is applied, electrostatic forces act on the electrodes. In an eccentric arrangement, the low-voltage electrode attached to the long support tube is deflected so that the *eccentricity* increases. The result is an increase in capacitance. As the voltage increases, the electrostatic forces of attraction and hence the eccentricity and capacitance continue to increase. Relative capacitance changes of up to 200×10^{-6} have been reported in the literature. In addition to the enlarged eccentricity, there is an inclination of the low-voltage electrode, the influence of which will be considered below.

Note Even with concentric arrangement of high and low voltage electrodes, the capacitance is not completely independent of the applied voltage. Slight voltage dependence cannot be excluded due to the attractive forces between the two cylinder electrodes, whereby their distance is minimally reduced and thus the capacitance increased. In the electrode arrangement in Fig. 11.10a, there is even a minimal increase in the partial capacitance between the top of the low-voltage electrode and the high-voltage plate electrode above.

Due to the force acting on the cylinder electrodes, the eccentricity increases by Δe and thus the capacitance by ΔC. In [35], the electrostatic attraction force is compared with the mechanical force ΔF which must be exerted on the low-voltage electrode in the direction of eccentricity in order to achieve an equal increase in Δe and ΔC. Here, it is assumed that electrostatic and mechanical forces cause comparable deflections of the low-voltage electrode. With Eq. (11.22), the calculation gives the following expression for the increase of eccentricity as a function of the applied voltage U:

$$\boxed{\Delta e = \frac{U^2}{2}\frac{\Delta C}{\Delta F}}. \quad (11.24)$$

According to Eq. (11.24), the eccentricity increases with the square of the applied voltage. With the capacitor open, the initial eccentricity e_0 as well as the increase in capacitance ΔC with the mechanical force ΔF are experimentally determined. With the quotient $\Delta C/\Delta F$ being known, Eq. (11.24) allows the calculation of Δe. Since $e = e_0 + \Delta e$, the corresponding capacitance C can be determined for each given voltage U according to Eqs. (11.22) and (11.23).

In most studies on the voltage dependence, the electrodes are assumed to remain exactly parallel to each other. In fact, however, the low-voltage electrode mounted on the support tube makes a small pivotal movement under the effect of the force of attraction. The consideration of this pivotal movement yields an equation for determining the relative change in capacitance, which in addition to the known quadratic term of the voltage also contains a U^4 term [40]. The aim of the investigations was also to extrapolate the capacitance change $\Delta C/C_0 = \mathrm{f}(U)$ measured at a defined voltage to higher voltages by calculation. However, the equation with the U^4 term applied to various compressed gas capacitors did not always reach the target, i.e. the calculated and measured values did not correspond to each other. In studies of other authors, there is also no indication that the voltage dependence deviates from the quadratic one.

In most investigations on the voltage dependence of compressed gas capacitors, the deflection of the low-voltage electrode and the measurement of the capacitance were carried out with the same AC voltage at power frequency. In [41], the capacitance is measured at 190 Hz low voltage while the electrode deflection is done at 50 Hz high voltage. The main problem with this method is the realization of a filter with steep flanks that eliminates the influence of the 50 Hz component and its harmonics on the measurement. Another way is described in [42], in which the electrode is deflected at high DC voltage and the capacitance is measured at 50 Hz. Here, the problem lies in the separation of the high DC voltage from the measuring circuit.

Proven means for reducing the voltage dependence of compressed gas capacitors is the overhaul and readjustment of the electrodes [28, 34]. As an example, Fig. 11.13 shows the voltage-dependent capacitance of a 120 kV compressed gas capacitor in the three-electrode arrangement of Fig. 11.10b. The behavior of this capacitor during the horizontal rotation test is characterized by curve ***2*** in Fig. 11.12. In the original state of the capacitor, the relative capacitance increase is $\Delta C/C_0 = 63 \times 10^{-6}$ at 100 kV (curve ***1*** in Fig. 11.13), where C_0 is the capacitance measured at low voltage. After a simple readjustment of the electrodes in the open capacitor, the capacitance increase is less than half that (curve **2**). After extensive

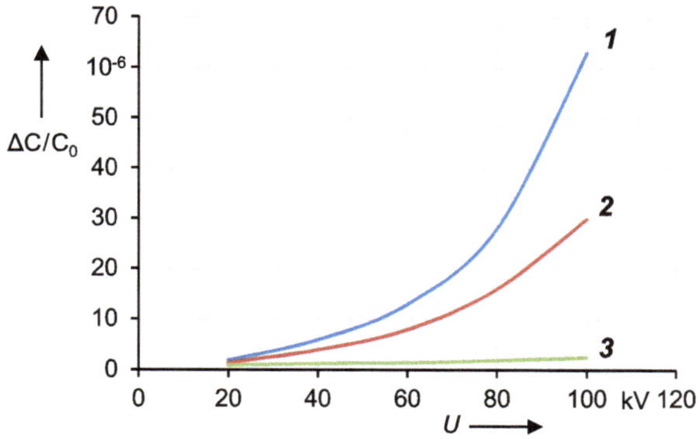

Fig. 11.13 Relative capacitance change $\Delta C/C_0$ of a compressed gas capacitor (120 kV, 100 pF) as a function of the applied voltage *U*. ***1*** original state of the capacitor, ***2*** after readjustment of the electrodes, ***3*** after thorough overhaul of the electrodes and readjustment

smoothing of the electrode surfaces, application of nickel layers and careful readjustment, the relative capacitance increase is only 3×10^{-6} at 100 kV (curve ***3***).

In the typical linearity test, the compressed gas capacitor is compared in a bridge circuit with a standard capacitor, usually also a compressed gas capacitor (see Sect. 11.3). In this case, the standard capacitor should have a much higher rated voltage, so that its capacitance can be considered constant in the voltage range under test.

11.5.6 Mechanical Natural Oscillation, Electrical Resonance

The low-voltage electrode with support tube in a compressed gas capacitor represents an oscillatory system. When a mechanical impulse is applied to the top of the compressed gas capacitor, the impact is passed on to the low-voltage electrode, which is thereby excited to a damped oscillation with the mechanical *natural frequency (eigenfrequency)* f_0. The high-voltage electrode remains almost stationary due to the much greater rigidity of the outer insulating cylinder. From typical oscillograms of the current through the compressed gas capacitor during excitation, important data such as the natural frequency, the eccentricity of the electrode arrangement and the voltage dependence of the capacitance can be derived. Basically, the natural frequency decreases with increasing height of the compressed gas capacitor because of the increasing length of the support tube and the larger mass of the low-voltage electrode. The natural frequency is to be distinguished from the frequency of an AC voltage at which the low-voltage electrode resonates due to the electrostatic force. In the vicinity of the *resonance frequency* corresponding to half the natural frequency, a significant change of the capacitance and the loss factor can be observed.

Note Similar investigations have also been made on the oscillation behavior of gas-insulated standard capacitors with stacked plate electrodes [43]. The tested plate capacitors with rated voltages of up to 1.4 kV also showed mechanical natural frequencies that are several times the power frequency of 50 or 60 Hz. Changes in capacitance and dissipation factor due to resonance again occur near half the natural frequency.

11.5.6.1 Mechanical Natural Oscillation

The oscillation of the low-voltage electrode due to a mechanical impulse acting on the top of a compressed gas capacitor (see Fig. 11.10a), is theoretically and experimentally investigated in [44, 45]. The capacitance C of the cylinder electrode having the initial static eccentricity e_0 is approximately given by the equation:

$$\boxed{C = C_\mathrm{c}\left(1 + b\,e_0^2\right)}. \tag{11.25}$$

Here, C_c is the capacitance of the concentric electrode arrangement according to Eq. (11.21) and b is a constant that depends on the radii r_1 and r_2 of the cylindrical electrodes:

$$b = \frac{1}{\left(r_1^2 - r_2^2\right) \cdot \ln(r_1/r_2)}. \tag{11.26}$$

The natural oscillation, which is excited by the mechanical impulse in the direction of the eccentricity, can be represented as a time-dependent deflection $e(t)$ of the low-voltage electrode:

$$\boxed{e(t) = e_0 - \hat{e}_\mathrm{m} \sin \omega_0 t}, \tag{11.27}$$

which consists of the initial eccentricity e_0 and a superimposed sine oscillation with the amplitude $\hat{e}_\mathrm{m}$ and the angular frequency $\omega_0 = 2\pi f_0$. For the sake of simplicity, the oscillation having the natural frequency f_0 is assumed to be undamped. The oscillating low-voltage electrode causes a characteristic oscillation of the capacitance $C(t)$. From Eq. (11.25), where e_0 is formally replaced by $e(t)$ according to Eq. (11.27), we obtain:

$$\boxed{C(t) = C_\mathrm{c}\left[1 + b\left(e_0 - \hat{e}_\mathrm{m} \sin \omega_0 t\right)^2\right]}. \tag{11.28}$$

The natural oscillation of the low-voltage electrode can be recorded and evaluated using a simple method, called *kinetic method* by the author [44, 45]. In this method, the capacitor is charged by a DC voltage U of several kilovolts and then excited by a mechanical impulse. The current $i(t)$ flowing through the capacitor is given by:

$$\boxed{i(t) = U\frac{\mathrm{d}C(t)}{\mathrm{d}t} = UC_\mathrm{c}e_0^2 b\omega_0\left[\left(\frac{\hat{e}_\mathrm{m}}{e_0}\right)^2 \sin 2\omega_0 t - 2\left(\frac{\hat{e}_\mathrm{m}}{e_0}\right)\cos \omega_0 t\right]}. \tag{11.29}$$

According to Eq. (11.29), $i(t)$ generally consists of two oscillatory components that have single and double natural frequencies and whose amplitudes depend on the ratio $\hat{e}_m/e_0$. Three cases can be distinguished. In the ideal case of concentric cylinder electrodes with $e_0 = 0$, the second term in the square brackets of Eq. (11.29) is zero. This means that the capacitance—and thus the capacitor current—oscillates with $2f_0$, i.e. with twice the natural frequency. For $e_0 \gg e_m$, on the other hand, the first term in the square brackets of Eq. (11.29) is negligible, and C

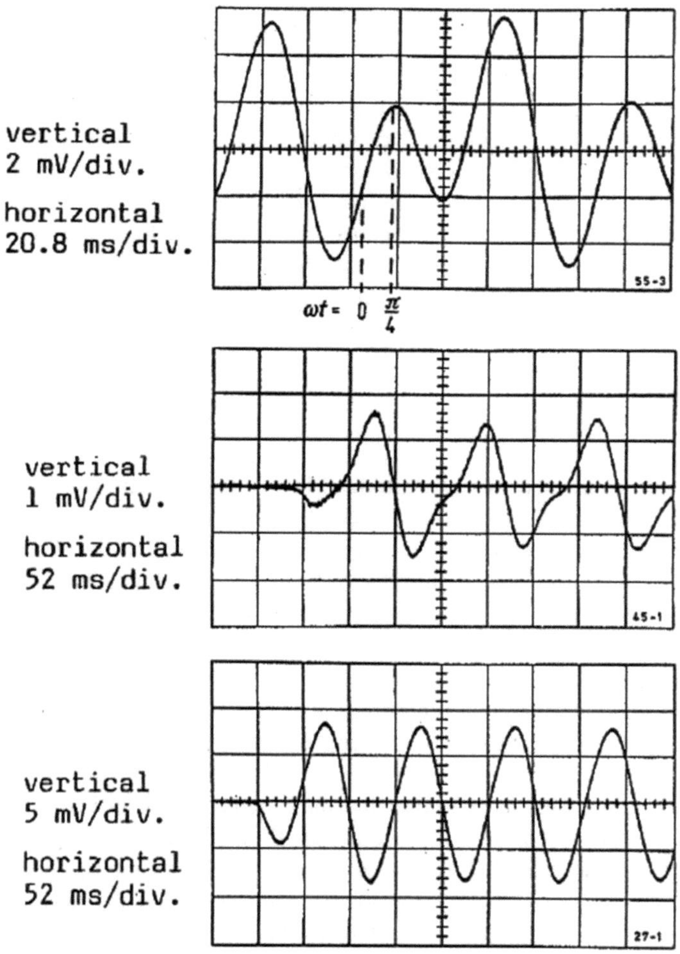

Fig. 11.14 Records of the current $i(t)$ through different compressed gas capacitors with different initial eccentricity after excitation by a mechanical impulse [44, Fig. 11.5], above: $e_0 < \hat{e}_m$ (50 pF, 500 kV) middle: $e_0 > \hat{e}_m$ (68 pF, 800 kV) below: $e_0 \gg \hat{e}_m$ (50 pF, 500 kV)

(t) as well as $i(t)$ oscillate with the simple natural frequency f_0. For all other values of $\hat{e}_m/e_0$, there is a superposition of oscillations with single or double natural frequency.

The theoretical results were confirmed by measurements on two 500 kV and one 800 kV compressed gas capacitors [44, 45]. Typical values of the measured natural frequency ranged from $f_0 = 52$ Hz for a 120 kV capacitor of 50 pF down to $f_0 = 8$ Hz for an 800 kV capacitor of 68 pF. Figure 11.14 shows examples of the recorded current $i(t)$ of the capacitors after being excited by a mechanical impulse. In the upper oscillogram showing the oscillation of the current through a 500 kV capacitor, the two frequency components with f_0 and $2f_0$ are clearly visible. The oscillogram in the middle of Fig. 11.14 shows the oscillating current through an 800 kV capacitor, whereby the $2f_0$ frequency component is only weakly visible. In the lowest oscillogram of Fig. 11.14, the current through another 500 kV capacitor oscillates practically only with the f_0 component. From the recorded currents $i(t)$ in Fig. 11.14, the initial eccentricity e_0 of each compressed gas capacitor can be determined with the values of the current at $\omega t = 0$ and $\omega t = \pi/4$ [45]. For example, the eccentricity of the one 500 kV capacitor, calculated from the current in the upper part of Fig. 11.14, is $e_0 = 4.3$ mm. This eccentricity—and thus the voltage dependence of the capacitance—was thus much larger than that of the other two capacitors investigated.

11.5.6.2 Electrical Resonance

The deflection of the eccentrically arranged low-voltage electrode of a compressed gas capacitor due to the electrostatic force effect of a high AC voltage is described in detail in [46]. For a simple mechanical-electrical model of the compressed gas capacitor, taking into account the electrode mass, damping constant and spring stiffness, a two-dimensional differential equation can be established from the balance of forces. Its solution provides a determination equation for the time-dependent electrode deflection $e(t)$. From this, further equations describing the voltage and frequency dependence of the capacitance and the dissipation factor can be derived. The calculations are supplemented by extensive experimental investigations.

As an example, Fig. 11.15 shows the measured changes in capacitance C and dissipation factor $\tan\delta$ of a 120 kV compressed gas capacitor as a function of the frequency f of an applied AC voltage at 100 kV. Near half the mechanical natural frequency, i.e. at $f = f_0/2 = 18.7$ Hz, the capacitance C and the dissipation factor $\tan\delta$ show a pronounced resonance behavior. The strong increase of $\tan\delta$ results

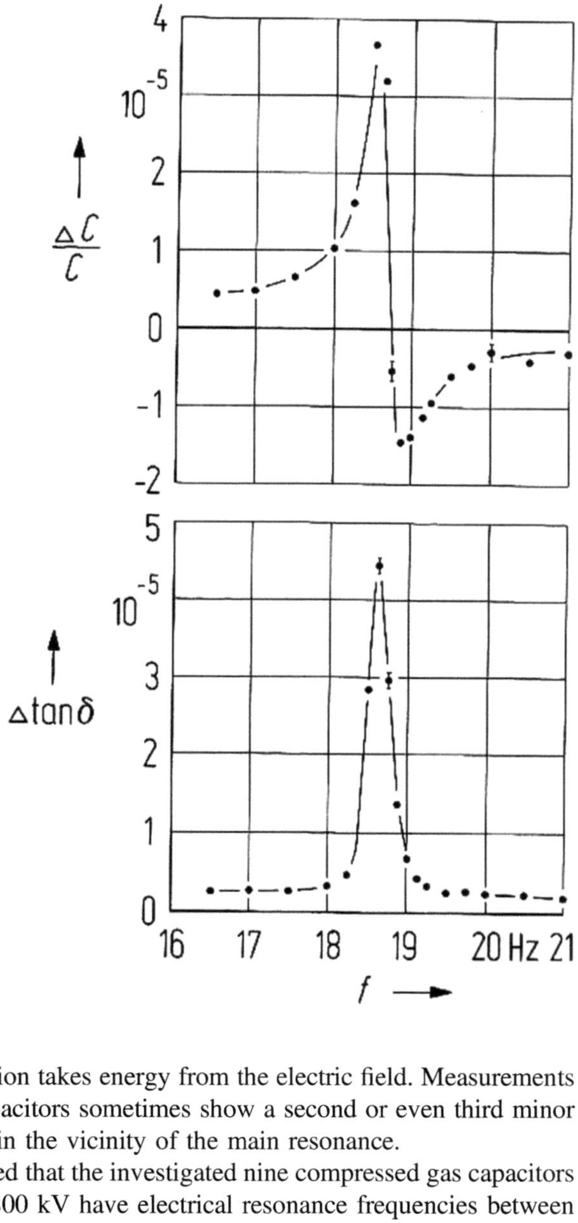

Fig. 11.15 Resonance behavior of a compressed gas capacitor (100 pF, 120 kV) [45, Fig. 11.7a]. Above: relative change in capacitance *C*, Below: relative change in tanδ

from the fact that the oscillation takes energy from the electric field. Measurements on other compressed gas capacitors sometimes show a second or even third minor discontinuity of C and tanδ in the vicinity of the main resonance.

In summary, it can be stated that the investigated nine compressed gas capacitors with rated voltages of 100–800 kV have electrical resonance frequencies between 26 and 4 Hz, which are thus well below the usual power frequencies of 50–60 Hz. Whether a compressed gas capacitor can be used for measurements at lower frequencies, e.g. at the operating frequency of 16.7 Hz of the German railway system, must be checked individually.

References

1. Wimmershoff, R., Wendt, C.: Dielectric diagnostic of cables and cables connected with ring main units. In: Proceeding 10th ISH Montreal, paper 3549 (1997)
2. Hadid, S., Schmidt, U., Schufft, W., Rätzke, S.: Frequency dependence of the dissipation factor of PE/XLPE-insulated medium voltage cables. In: Proceeding 18th ISH Seoul, paper PD-37 (2013)
3. IEC 60250: Recommended methods for the determination of the permittivity and dielectric dissipation factor of electrical insulating materials at power, audio and radio frequencies including meter wavelengths (1969) German edition: DIN IEC 60250 (VDE 0303-4): Bestimmungen für elektrische Prüfungen von Isolierstoffen - Bestimmung der dielektrischen Eigenschaften (1969)
4. Liu, Y., Cao, XL., Li, XM.: Adjustable non-contact electrode and measurements on ε_r and tanδ of insulating materials. In: Proceeding 14th ISH Beijing, paper H-47 (2005)
5. Ohlen, M., Werelius, P., Cheng, J.: Dielectric response measurements in frequency, temperature and time domain. In: Proceeding 18th ISH Seoul (2013), paper OD7-06 (2013)
6. Homagk, C., Leibfried, T.: Insulation diagnosis using dissipation factor measurements. In: Proceeding 14th ISH Beijing, paper G-070 (2005)
7. Reumann, A., Liebschner, M., Küchler, A., Langens, A., Titze, J.: On-line monitoring of capacitance and dissipation factor of HV bushings. In: Proceeding 16th ISH Johannesburg, paper A-43 (2009)
8. Nikjoo, R. et al.: Insulation condition diagnostics of oil impregnated paper by utilizing power system transients. In: Proceeding 18th ISH Seoul, paper OD7-064 (2013)
9. Schering, H.: Brücke für Verlustmessungen. Zeitschr. f. Instrum. **40**, 24 (1920)
10. Braun, A., Schon, K.: Harald Schering, seine Arbeiten und die heutigen Aufgaben der PTB auf den Gebieten Messwandler und Hochspannung. PTB-Mitt. **107**, 227–236 (1997)
11. Wiessner, W.: Beseitigung von Störungen durch Streukapazitäten in Kapazitätsmessbrücken mit Wagnerscher Hilfsschaltung. Zeitschr. f. Instrum. **65**, 139–144 (1957)
12. Baker, W.B.: Recent developments in 50 c/s bridge networks with inductively coupled ratio arms for capacitance and loss-tangent measurements. Proc. IEE Pt. A **109**, 243–247 (1962)
13. Kusters, N.L., Petersons, O.: A transformer-ratio-arm bridge for high-voltage capacitance measurements. IEEE Trans. CE **82**, 606–611 (1963)
14. Petersons, O.: A self-balancing high voltage capacitance bridge. IEEE IM **13**, 216–224 (1964)
15. Zinn, E., Braun, A., Köhler, H.J.: Kapazitäts- und Verlustfaktormesseinrichtung mit selbsttätiger Abgleichung. Techn. Mess. **2**, 924–925 (1977)
16. Seitz, P., Osvath, P.: Microcomputer controlled transformer ratio-arm bridge. In: Proceeding 3rd ISH Mailand, paper 43.11 (1979)
17. Osvath, P., Widmer, S.: Automatische Kapazitäts- und Verlustfaktor tanδ-Messung im industriellen Umfeld. E-wirtschaft **85**, 911–913 (1986)
18. Gourney, P.: Capacitance and dissipation factor measurements under high voltage at BNM-LCIE. CPEM Digest, paper WEP5-5 (2000)
19. Tschirschwitz, T., Seitz, P.: Current transformers with electronic error compensation—An application for precision capacitance and dissipation factor measurements on large capacitive loads. Proc. 5th ISH Braunschweig, paper 73.13 (1987)
20. Braun, A., Richter, H.: Determination of the voltage dependence of the capacitance of high-voltage standard capacitors. In: Proceeding 5th ISH Braunschweig, paper 73.10 (1987)
21. Kaul, G., Plath, R., Kalkner, W.: Development of a computerised loss factor measurement system for different frequencies, including 0.1 Hz and 50/60 Hz. In: Proceeding 8th ISH Yokohama, paper 56.04 (1993)
22. Kornhuber, S., Markalous, S., Muhr, M., Strehl, T., Sumereder, C.: Comparison of methods for the dissipation factor measurement at practical examples. In: Proceeding 16th ISH Johannesburg, paper C-43 (2009)

23. Ramm, G., Roeissle, G., Latzel, H.-G.: Rechnergesteuerte Kalibrierung von Messeinrichtungen für Strom- und Spannungswandler. PTB-Mitt. **108**, 188–200 (1998)
24. Ramm, G., Moser, H.: Calibration of electronic capacitance and dissipation factor bridges. IEEE Trans. IM **52**, 396–399 (2003)
25. Ramm, G., Moser, H.: From the calculable AC resistor to capacitor dissipation factor determination on the basis of time constants. IEEE Trans. IM **50**, 286–289 (2001)
26. Simmon, E.D., FitzPatrick, G.J., Petersons, O.: Calibration of dissipation factor standards. IEEE Trans. IM **48**, 450–452 (1999)
27. Petersons, O.P., Anderson, W.E.: A wide-range high-voltage capacitance bridge with one ppm accuracy. IEEE Trans. IM **24**, 336–344 (1975)
28. Latzel, H.-G., Schon, K.: Internationale Vergleichsmessungen von Kapazität und Verlustfaktor bei Hochspannung. PTB-Mitt. **99**, 227–234 (1989)
29. Thoma, P.: Absolute calorimetric determination of dielectric loss factors at $\omega = 10^4\ s^{-1}$ and 4.2 K and application to the measurement of loss factors of standard capacitors at room temperature. IEEE Trans IM **29**, 328–330 (1980)
30. Hanke, R., Thoma, P.: Messung des Verlustfaktors von Normalkondensatoren. PTB-Jahresbericht, paper 3.2.11 (1980)
31. Thoma, P., Thiemig, M.: Kalibrierung des Verlustfaktors zweier Normalkondensatoren mit Hilfe eines kryokalorimetrischen Verlustfaktornormals zwischen 50 Hz und 10 kHz. PTB-Jahresbericht, paper 2.2.7 (1989)
32. Schering, H., Vieweg, R.: Ein Meßkondensator für Höchstspannungen. Z. f. Techn. Physik **9**, 442–445 (1928)
33. Rungis, J., Brown, D. E.: Experimental study of factors affecting capacitance of high-voltage compressed-gas capacitors. IEE Proc. **128**(Pt. A), 273–277 (1981)
34. Latzel, H.-G., Schon, K.: Precise capacitance measurements of high voltage compressed gas capacitors. IEEE Trans. IM **36**, 381–384 (1987)
35. Hillhouse, D.L., Peterson, A.E.: A 300-kV compressed gas standard capacitor with negligible voltage dependence. IEEE Trans. IM **22**, 408–416 (1973)
36. Ivers-Tiffée, E., von Münch, W.: Werkstoffe der Elektrotechnik. Teubner, Wiesbaden (2007)
37. Anderson, W.E., et al.: An international comparison of high voltage capacitor calibrations. IEEE Trans. PAS **97**, 1217–1223 (1978)
38. Latzel, H.-G.: Temperature-induced transient capacitance change in HV standard capacitors. In: Proceeding 7th ISH Dresden, paper 63.06 (1991)
39. Keller, A.: Konstanz der Kapazität von Preßgaskondenstoren. ETZ-A **80**, 757–761 (1959)
40. Zinkernagel, J.: Modellrechnungen zur Spannungsabhängigkeit der Kapazität von Preßgaskondensatoren. Archiv für Elektrotechnik **60**, 299–305 (1978)
41. Zinkernagel, J.: A double frequency method for the determination of the voltage dependent capacitance variation of compressed gas capacitors. IEEE Trans. PAS **98**, 306–309 (1979)
42. Leren, W., Latzel, H.-G.: Messung der Spannungsabhängigkeit der Kapazität von Druckgaskondensatoren mit dem Gleichspannungsverfahren. PTB-Mitt. **96**, 83–87 (1986)
43. Kusters, N.L., Petersons, O.: The voltage coefficients of precision capacitors. IEEE Trans. CE **60**, 612–621 (1963)
44. Latzel, H.-G.: A new method for detecting voltage dependence of capacitance in compressed gas capacitors. In: Proceeding 5th ISH Braunschweig, paper 73.11 (1987)
45. Latzel, H.-G., Kind, D.: Kinetic method for evaluating the voltage dependence of high-voltage compressed gas standard capacitors. In: Proceeding 6th ISH New Orleans, paper 47.06 (1989)
46. Latzel, H.G.: Frequency dependence of capacitance and dissipation factor in high-voltage compressed gas capacitors due to mechanical resonance. etz-Archiv **12**, 313–319 (1990)

Chapter 12
Basics of Partial Discharge Measurement

Abstract Partial discharge (PD) denotes a small, localized electrical discharge in the insulation between conductors, occurring when the local electric field strength exceeds a critical value. Prolonged exposure to partial discharges is known to degrade the dielectric behavior of the insulation and may even lead to complete breakdown. PD testing of high-voltage equipment and electrical apparatus is one of the most important and difficult tasks in high-voltage testing. This chapter describes the measuring instruments, the calibration methods and the test circuits for measuring the so-called apparent charge. For most apparatus, special PD measurement methods with computerized data processing have been developed, which are constantly being improved. Key words are on-site PD measurements after installation of the equipment, localization of single or multiple PD sources in spatially extended apparatus, synchronous multichannel PD measurement, VHF and UHF PD measurement techniques, permanent PD online monitoring and interference suppression methods. Partial discharges are associated with electromagnetic, acoustic, optical and chemical effects, which are also exploited for their detection and diagnosis, in particular in complex apparatus such as three-phase power transformers and gas-insulated switchgear (GIS). The PD measurement techniques at DC and impulse voltages are briefly discussed.

Partial discharge (PD) denotes a small, localized electrical discharge in the insulation between conductors. PD occurs when the local electric field strength of an insulation arrangement exceeds the critical value for impact ionization and an initial electron is present. Such local discharge is also referred to as incomplete breakdown, since initially the strength of the insulation as a whole remains unaffected. Distinction is made between internal discharges, external discharges and surface discharges. Particularly dangerous are partial discharges within the insulation, as these often remain undetected. Prolonged exposure to partial discharges deteriorates the dielectric behavior of the insulation and may even lead to complete breakdown. External partial discharges, also called *corona*, occur in air or other gases around metal conductors or electrodes with small radii of curvature, e.g. around power transmission lines. Surface discharges appear in arrangements with tangential field distribution along the boundary of two

K. Schon, *High Voltage Measurement Techniques*, Power Systems,
https://doi.org/10.1007/978-3-030-21770-9_12

different insulation materials. A typical example is the shielded high-voltage cable, whose shield is removed at the cable end.

Internal partial discharges and the measurement technology have become more and more important since the second half of the 20th century, when plastic insulation increasingly replaced porcelain and glass in high-voltage engineering. Nowadays, PD testing of high-voltage equipment and electrical apparatus is one of the most important and difficult tasks in high-voltage testing technique. Internal partial discharges are usually not directly accessible for measurements. Measurable are only the pulses occurring as a consequence at the terminals of the test object or are received by a measuring sensor. For most apparatus, special methods of PD measurement with computerized data processing have been developed, which are constantly being improved. Key words here are *on-site PD measurements* after installation of the equipment, *localization* of single or multiple PD sources in spatially extended apparatus, permanent PD *online monitoring* and methods of suppressing interference. Partial discharges are associated with electromagnetic, acoustic, optical and chemical effects, which are also used for their detection and diagnosis, in particular in complex apparatus such as three-phase power transformers and gas-insulated switchgear (GIS).

This chapter describes the basics of measuring instruments and methods for measuring partial discharges at AC voltage, in particular for the determination of the so-called *apparent charge*. The constantly improved electromagnetic, acoustic and optical diagnostic methods are treated in principle. The measurement techniques for DC and impulse voltages will be briefly discussed. Extensive contributions to the development of PD measurements, the various PD forms, equivalent circuit diagrams and further PD measurement techniques can be found in ([1–3]; Refs. [1–6] of Chap. 1).

12.1 Internal Partial Discharges at AC Voltage

The occurrence of partial discharges within high-voltage apparatus is initially a clear indication of the faulty execution of the electrical insulation. For example, in the manufacture of impregnated resin insulation, tiny gas-filled cavities or voids may be formed. Due to the smaller permittivity of the enclosed gas, the electric field strength in the cavity is significantly higher than in the surrounding dielectric. If the field strength in the cavity reaches the value required for the ionization of the gas, and if an initial electron is present, a localized small discharge, i.e. the partial discharge, occurs. The dielectric surrounding the ionized cavity initially does not seem to be affected by the occurrence of partial discharges. However, due to prolonged exposure to partial discharges, the insulation progressively deteriorates by various mechanisms such that finally, sometimes only after several months or years, failure of the apparatus occurs by a complete breakdown. Further examples of the occurrence of internal partial discharges are gas bubbles in oil-insulated transformers and faulty cable fittings. In the case of older XLPE cables, the inclusion of particles in the insulation or irregularities of the conductor smoothing leads to locally increased field strength. Thus, progressive erosion results in tree-like discharge structures that grow to the electrodes and finally initiate electrical breakdown.

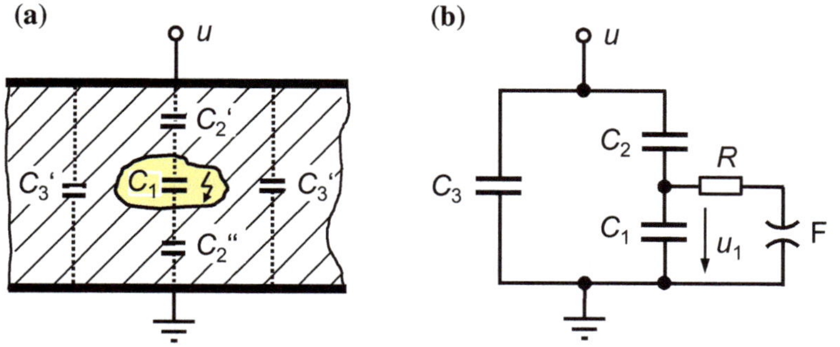

Fig. 12.1 High voltage insulation with gas filled void. **a** Arrangement with partial discharge; **b** equivalent circuit diagram with spark gap F

Figure 12.1a shows the simple model of high-voltage insulation with a gas-filled void as an internal defect, sometimes also located directly in the boundary layer between an electrode and the dielectric. In the simple equivalent circuit diagram in Fig. 12.1b based on Gemant and Philippoff, C_1 is the capacitance of the void, C_2 the capacitance of the insulation in series with the void and C_3 the parallel capacitance of the complete arrangement [4]. Parallel to the void capacitance C_1 is the spark gap F, which fires upon reaching the ionization conditions. The discharge current flows through the (later additionally introduced) resistor R and is thereby limited in duration and height. If need be, additional high-ohmic insulation resistors must be considered in parallel to C_1 and C_2, e.g. for cable testing at very low frequencies.

The processes involved in the occurrence of partial discharges can be described in a simplified manner using the equivalent circuit diagram in Fig. 12.1b. As long as the spark gap F does not fire, the voltage $u_1(t)$ at the spark gap F is proportional to the AC test voltage $u(t)$ according to the capacitive division ratio:

$$\boxed{u_1(t) = u_{10}(t) = \frac{C_2}{C_1 + C_2} u(t)}, \tag{12.1}$$

where $u_{10}(t)$ denotes the (theoretical) voltage at C_1 without partial discharge. When $u_1(t)$ reaches the firing voltage u_Z of the spark gap F, the void capacitance C_1

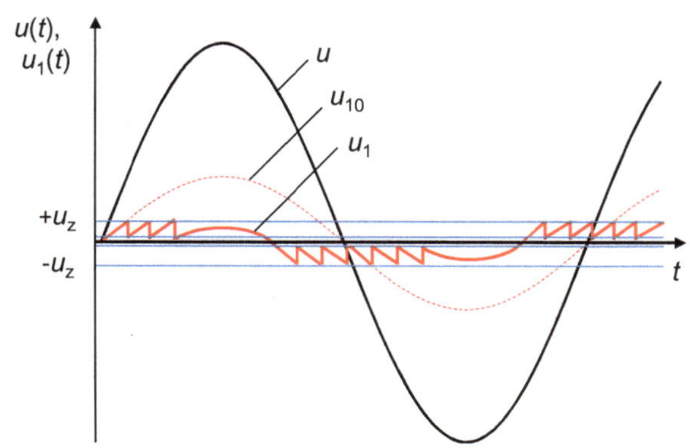

Fig. 12.2 Voltage curves in the equivalent circuit diagram according to Fig. 12.1 for internal partial discharges (principle). u: AC test voltage, u_{10}: voltage at spark gap F without partial discharges, u_1: voltage at spark gap F when partial discharges occur

discharges and $u_1(t)$ collapses to a small residual voltage (Fig. 12.2). As the test voltage $u(t)$ continues to rise, $u_1(t)$ increases again until the next partial breakdown occurs in the void when the firing voltage is reached again, etc. A series of partial discharges is the result. In the negative half of the voltage, the charging and discharging processes are repeated equally. Due to various influences, e.g. lack of initial electrons in the void, formation of space charges and deposition of decomposition products on the void walls, the regularity of the partial discharges may be disturbed.

Note The simple model with the equivalent circuit diagram in Fig. 12.1b is considered by some authors to be unsatisfactory because it does not sufficiently take into account the physics of gas discharges. As an alternative, in (Ref. [5] of Chap. 1) a model is presented in which the partial discharge in the cavity of the insulation is represented by a dipole.

The charging and discharging processes occurring in the cavity are generally not accessible for direct measurements. They cause very small transient changes in current or voltage at the terminals of the test object, which can be measured directly or by means of sensors. However, apart from laboratory investigations on simple samples of known geometry, these measurable pulses do not allow any conclusions about the internal partial discharges and their measurement quantities, e.g. the discharge power. Furthermore, they also show no direct relationship with the gradual deterioration of the insulation. Typical of internal partial discharges is that the PD pulses occur in the region of the zero crossings of the test voltage. Other phase angles and shapes of PD pulses can be observed in external partial discharges and other electrode arrangements. Additional defects in the dielectric cause further partial discharges at different voltage levels, which further complicate the phenomena of PD pulses.

During the PD test of high-voltage apparatus, the test voltage is increased from a low initial value to the voltage below the permissible limit, at which partial discharges are measured for the first time. At this *PD inception voltage*, usually one PD pulse occurs in each positive and negative half of the voltage. If the test voltage is increased further up to the permissible limit, the number of PD pulses per cycle increases. In addition, the *PD magnitude* may increase with further defects in the insulation. As the test voltage gradually decreases again, the partial discharges cease to occur at the *PD extinction voltage*, which is below the PD inception voltage. An important requirement for high-voltage equipment is that not only the PD inception voltage but also the PD extinction voltage shall be above the operating voltage. This ensures that partial discharges resulting from a transient overvoltage no longer occur when the operating voltage is restored [5, 6].

Note According to IEC 60270 [5], "PD magnitude" is the correct term, not "PD intensity". The term "intensity" is related to a surface or an area. The same applies to "current magnitude" and "current intensity".

12.2 Characteristics of PD Pulses

Internal partial discharges are very short electrical pulses whose rise time and pulse duration can be in the nanosecond range [7, 8]. Partial discharges in gas insulation have extremely short rise times, i.e. much less than 1 ns. Minimum rise times of

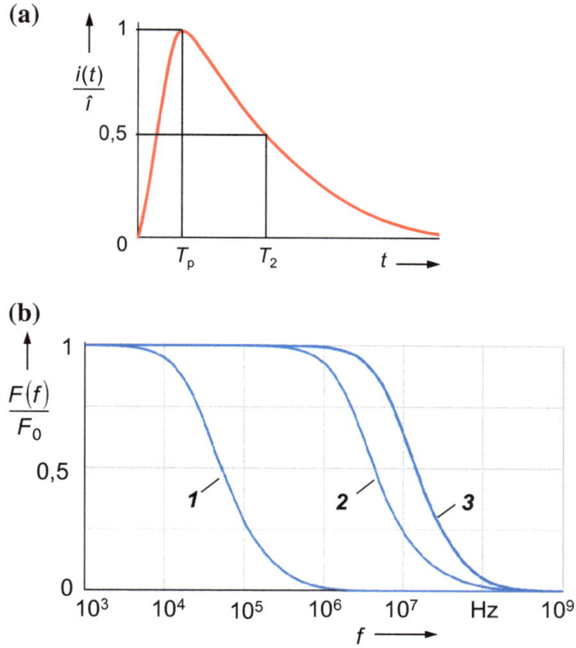

Fig. 12.3 Idealized current pulse $i(t)$ in the time and frequency domain. **a** Current pulse $i(t)/\hat{\imath}$ with the time to maximum T_p and time to half-value T_2; **b** amplitude density $F(f)/F_0$ of the pulse for different values of T_p and T_2. ***1*** $T_p = 1$ µs, $T_2 = 5$ µs, ***2*** $T_p = 5$ ns, $T_2 = 50$ ns, ***3*** $T_p = 5$ ns, $T_2 = 20$ ns

only 35 and 22.3 ps were measured in a point-to-plane electrode configuration in an SF_6-filled pressure chamber, with the bandwidth of the digital oscilloscope used being 32 GHz [9, 10]. As already mentioned above, partial discharges within the insulating material are generally not measurable. They cause current or voltage pulses at the terminals of the apparatus, which are accessible to measurements. These pulses are generally understood to be the PD pulses, but differ significantly from the "true" partial discharges occurring within the dielectric.

Formally, a PD pulse can be represented as a current pulse $i(t)$ by superposition of two exponential functions [11]. To characterize the PD pulse, here T_p is referred to as the time to maximum i_{max} and T_2 as the time to half-value at which $i(t)$ has fallen to $0.5i_{max}$ (Fig. 12.3a). The spectrum of these pulses can be calculated by means of the Fourier integral or the Laplace transform (see Chap. 8). For three different value pairs of T_p and T_2, Fig. 12.3b shows the amplitude density $F(f)$ relative to the respective value $F_0 = F(f = 0)$. For extremely short PD pulses with rise times of a few 10 ps, the spectrum extends into the GHz range.

Note 1 The infinitely narrow Dirac pulse is known to have a spectrum that remains constant up to infinitely high frequencies.

Note 2 Partial discharges are not always pulse-shaped. Investigations on point-to-plane arrangements show that under certain experimental conditions a transition of the PD pulses into a more continuous form, the so-called pulseless discharges in gases, takes place [12].

The aim of decades of research was and is to find the PD measurement quantity that enables a well-founded statement about the lifetime reduction of high-voltage apparatus. This goal has not yet been fully and satisfactorily achieved for all high-voltage insulations. It was not possible to derive a meaningful measurable quantity, neither from the PD pulses measured at the terminals of the test object nor from the "true" partial discharges inside the insulating material itself, which, incidentally, are not accessible for the measurement with the exception of simple test specimens. As a compromise, there is an international agreement to determine the charge q of the PD pulse:

$$\boxed{q = \int_0^\infty i(t)\mathrm{d}t} \tag{12.2}$$

as the decisive measurement quantity in PD tests. The charge of the PD pulse has the metrological advantage of being invariant when the pulse changes its shape as it passes through the high voltage test and measurement circuitry.

Examples of the *invariance of the pulse charge* can be seen in Fig. 12.4. It shows the calculated time courses of a current pulse in its original state (curve ***1***) and after deformation by a low-pass filter with (curve **2**) and without resonance point (curve ***3***). The integration of the three current pulses according to Eq. (12.2) always gives the same charge $q = q_1 = q_2 = q_3$, even taking into account the oscillation in curve **2** with negative charge components. This invariance of the pulse charge is the basic prerequisite for PD measurements in spatially extended high-voltage circuits, which are generally not designed for the transmission of high frequency signals. However, due to stray capacitances in the test and measurement circuit, some of the pulse charge can flow to ground and is not displayed by the PD instrument. This charge fraction is detected by calibrating the PD instrument in the complete test and measurement circuit (see Sect. 12.7.2).

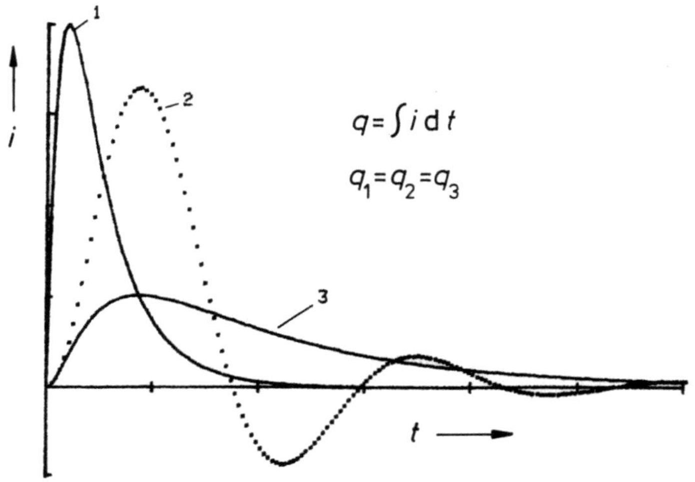

Fig. 12.4 Example of the invariance of the pulse charge. ***1*** Original pulse, **2** deformation by oscillating circuit, ***3*** deformation by low-pass filter

The pulse charge q has a further specific characteristic. By comparing Eq. (12.2) with the general equation for the complex spectrum $F(\mathrm{j}\omega)$ of a current pulse $i(t)$ [13]:

$$F(\mathrm{j}\omega) = F\{i(t)\} = \int_{-\infty}^{\infty} i(t)\mathrm{e}^{-\mathrm{j}\omega t}\mathrm{d}t \tag{12.3}$$

we obtain for the positive time range and $\omega = 2\pi f = 0$ the identity:

$$\boxed{q = \int_{0}^{\infty} i(t)\mathrm{d}t = |F(\mathrm{j}\omega = 0)| = F(0)}, \tag{12.4}$$

i.e. the pulse charge q corresponds to the DC component $F(0)$ of the amplitude density. However, this identity between q and $F(0)$ is not directly usable for the practice of PD measurements. The PD pulses are usually coupled out of the test circuit with a high-voltage capacitor, the so-called *coupling capacitor*, whereby the DC component of the pulse spectrum is lost. As can be seen in Fig. 12.3b, however, the amplitude density of the PD pulse remains approximately equal to $F(0)$, even at higher frequencies. For this frequency range in which $F(f) \approx F(0)$ holds, the important finding is that the charge q is approximately equal to the amplitude density $F(f)$:

$$\boxed{q = F(0) \approx F(f)}. \tag{12.5}$$

Equations (12.2) and (12.5) form the basis for the different operating principles of conventional PD charge measuring instruments (see Sect. 12.5).

12.3 Apparent Charge

The charge of a PD pulse which is coupled out on the high-voltage side of a test object is called *apparent charge q*. In the International Standard IEC 60270, the apparent charge of a PD pulse is defined as "that charge which, if injected within a very short time between the terminals of the test object in a specified test circuit, would give the same reading on the measuring instrument as the PD current pulse itself" [5]. The apparent charge of the PD pulses is usually expressed in picocoulomb (pC). Depending on the high-voltage apparatus and measurement circuit, values of less than 1 pC up to several 1000 pC are typical. However, the "true" charge actually involved in the discharge within the insulation of an apparatus cannot be deduced from the apparent charge.

Despite the knowledge that neither the true nor the apparent charge can give a well-founded statement about the lifetime reduction of insulation, the PD measurement nonetheless represents an important non-destructive test method for detecting defects in insulation. Based on decades of experience, the test specifications lay down limits for the apparent charge for each apparatus of the electrical power supply, which must not be exceeded during acceptance tests before the first use. If the requirement is met, failure of the equipment due to partial discharges within the scheduled service life is not considered very likely.

Determining the apparent charge as the most important PD measurement quantity is particularly advantageous from a metrological point of view. Although the high-frequency PD pulses are more or less deformed on their way from the test object to the PD measuring instrument, the apparent charge of the PD pulses is maintained (see Sect. 12.2). However, part of the apparent charge can flow to ground via stray capacitances and therefore will not be displayed by the PD instrument. This loss of charge is determined by means of calibration pulses of known charge, which are injected into the complete test and measurement circuit (see Sect. 12.7.2). From the calibration, the *scale factor k* is determined by which the reading of the instrument in the PD test is to be multiplied in order to obtain the actual PD magnitude.

12.4 Derived PD Quantities

In some countries, other PD quantities derived from the charge q are used. These quantities are generally approved by IEC 270 for PD testing and it is the responsibility of each National Technical Committee to admit them for PD tests on the relevant power apparatus. In addition to the *radio disturbance voltage* U_r, which is dealt with separately in Sect. 12.5.1.3, the following quantities are accepted [5, 14]:

average discharge current I:

$$I = \frac{1}{T}(|q_1| + |q_2| + \cdots |q_m|), \tag{12.6}$$

discharge power P:

$$P = \frac{1}{T}(q_1 u_1 + q_2 u_2 + \cdots q_m u_m), \tag{12.7}$$

quadratic rate D:

$$D = \frac{1}{T}(q_1^2 + q_2^2 + \cdots q_m^2). \tag{12.8}$$

In the equations for I, P and D, the terms $q_1 \ldots q_m$ denote the individual charges in a given time interval T, e.g. in a period of the AC test voltage. In Eq. (12.7), $u_1 \ldots u_m$ are the instantaneous values of the test voltage at the times of occurrence of the individual charges $q_1 \ldots q_m$. Occasionally, the sum of the PD charges $Q_s = |q_1| + |q_2| + \cdots |q_m|$ is also determined within a given time interval.

12.5 PD Measuring Instruments for the Apparent Charge

PD measuring instruments for the apparent charge q can be subdivided into two main groups according to their working principle. In one group, integration is performed by band-limited processing of the frequency response of the PD pulses according to Eq. (12.5), which is also called *quasi-integration*. This integration method forms the basis for the measuring instruments used mainly in the early days of PD measurement technology. In detail, a distinction is made here between narrow-band instruments with bandwidths of 4–10 kHz and wide-band instruments with bandwidths of 100 kHz to 1 MHz (formerly: 500 kHz). As electronic development progressed, it became possible to determine the charge of PD pulses by two "real" integration methods according to Eq. (12.2). Integration of the PD pulse is achieved either analogously using an integration amplifier or numerically with an algorithm applied to the digitally recorded pulse. Combinations of analog and digital integration techniques are also possible. Knowledge of the functioning of the measuring instrument used is helpful for correct measurement and calibration (see Sect. 12.8).

The various PD measuring instruments have advantages and disadvantages, so that they are not equally suitable for all measurement tasks [15, 16]. Digital PD measuring instruments with computer-aided evaluation have the decisive advantage that they enable comprehensive representation and diagnosis of the measurement results by means of software. For *on-site tests* after installation of the equipment and during *online monitoring* in practice, the determination of the apparent charge is often not possible or impractical. Then, completely different evaluation methods are used (see Sect. 12.9). For PD tests on low-voltage devices such as isolating transformers and opto-couplers, basically the same PD measuring instruments are used as for PD tests at high-voltage [17].

12.5.1 Quasi-Integration Through Band Limitation

A PD measuring instrument with quasi-integration is characterized in that it has a limited bandwidth compared to the PD spectrum. This bandwidth must be within the frequency range of the approximately constant spectral amplitude density of the PD pulse (see Fig. 12.3b). In this frequency range, the relation holds: $F(f) \approx F(0) = q$, i.e. the charge q is approximately equal to the amplitude density $F(f)$ according to Eq. (12.5). If the bandwidth of the measuring instrument extends into the higher-frequency range in which the amplitude density decreases more and more, the condition for the charge-proportional display would no longer be satisfied. The resulting error in the charge is called *integration error* [5, 18].

To determine the amplitude density $F(f)$ and thus the charge q, the use of narrow-band frequency analyzers is basically possible. The conventional PD measuring instruments, however, use a different measurement principle, which was feasible at the beginning of the PD measurement technique with the technical possibilities available at that time. As a result, the measurement of $F(f)$ is replaced by a quantity measured in the time domain. To understand this measurement principle, a measuring system is considered which has a constant transfer factor $A(f) = A_0$ within the limit frequencies f_1 and f_2. The ideal bandpass filter has the mid-band frequency f_0 and the bandwidth Δf as follows:

$$f_0 = \frac{f_1 + f_2}{2} \tag{12.9}$$

$$\Delta f = f_2 - f_1. \tag{12.10}$$

If a *Dirac pulse* is applied to the input of the measuring system, at the output the response $u(t)$ is given by [13]:

$$u(t) = 2A_0 F(f)\{f_2 \mathrm{si}[2\pi f_2 (t - t_0)] - f_1 \mathrm{si}[2\pi f_1 (t - t_0)]\}. \tag{12.11}$$

Here $\mathrm{si}(x) = \sin(x)/x$ is the si function and t_0 is the transit time of the pulse through the measuring system. The response is thus a transient oscillation, which will be discussed later in two examples. The maximum value of the oscillating pulse response is:

$$\boxed{u_{\max} = 2A_0 F(f)\Delta f}. \tag{12.12}$$

The amplitude density $F(f)$ of the Dirac pulse is constant up to infinitely high frequencies. Therefore, the same maximum value $u_{\max}$ always results for any arbitrary mid-band frequencies f_0 and constant bandwidth Δf of the measuring system.

The result derived for the Dirac pulse can be applied to PD pulses with finite pulse width, as long as their amplitude density remains approximately constant at least up to the upper limit frequency f_2 of the measuring system. Because of $F(f) \approx q$ according to Eq. (12.5), Eq. (12.12) provides the important information that the maximum value of the pulse response is proportional to the PD pulse charge:

$$\boxed{u_{\max} \sim q}. \tag{12.13}$$

The results obtained for ideal band-limited systems are generally transferable to PD measuring instruments. However, due to their non-ideal band filter characteristic (finite slope of the filter, etc.), small deviations in the response may occur. Depending on the bandwidth of the PD measuring instrument, two characteristic pulse responses can be distinguished as follows.

12.5.1.1 Wide-Band PD Instruments

The first example of quasi-integration deals with a PD instrument with ideal wide-band filter characteristic $A(f) = A_0$ within the limit frequencies $f_1 = 10$ kHz and $f_2 = 100$ kHz (Fig. 12.5). A PD input pulse having a constant amplitude density $F \approx F(0)$ until at least the limit frequency f_2 generates at the filter output the response $u(t)$ calculated according to Eq. (12.11). The maximum value u_{max} is detected by a peak detector and displayed as the charge q according to Eq. (12.13). Alternatively, the output pulses of the bandpass system can be evaluated with a pulse height analyzer. Characteristic of the band-limited pulse processing is also the widening of the pulse response relative to the original PD pulse at the input. For the limit case $f_1 = 0$ (ideal low-pass behavior), the *pulse duration* at half the amplitude of the response is calculated from the bandwidth Δf as follows [13]:

$$\boxed{\tau = \frac{1}{2f_2} = \frac{1}{2\Delta f}}. \tag{12.14}$$

In Fig. 12.5, for example, the duration of the pulse response is $\tau \approx 5$ μs, which is at least 100 times greater than the duration of the PD input pulse. Due to their large pulse duration, successive pulse responses can overlap, resulting in *superposition errors* and thus incorrect PD readings. In order to avoid superposition of pulse responses, the time interval between two consecutive PD pulses and to reflected pulses, as may occur in long cables, must be greater than τ. With a PD measuring instrument having a larger bandwidth, a smaller pulse width of the pulse response and thus a higher pulse resolution can be achieved.

Older wide-band PD measuring instruments display the measured PD value with a pointer instrument, possibly supplemented by an oscillographic display of the PD pulses on an elliptic time base representing the AC test voltage [3]. As a result, the magnitude, phase and repetition rate of the PD pulses are visible. In newer PD measuring instruments, the pulse response of the bandpass filter according to Fig. 12.5 is sampled with an A/D converter. Figure 12.6 shows the principle of the

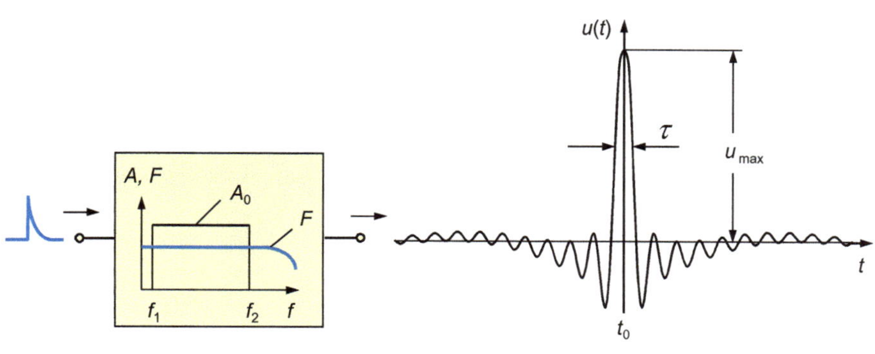

Fig. 12.5 Pulse response $u(t)$ of an ideal wide-band system with constant transfer factor A_0 within the limit frequencies $f_1 = 10$ kHz and $f_2 = 100$ kHz

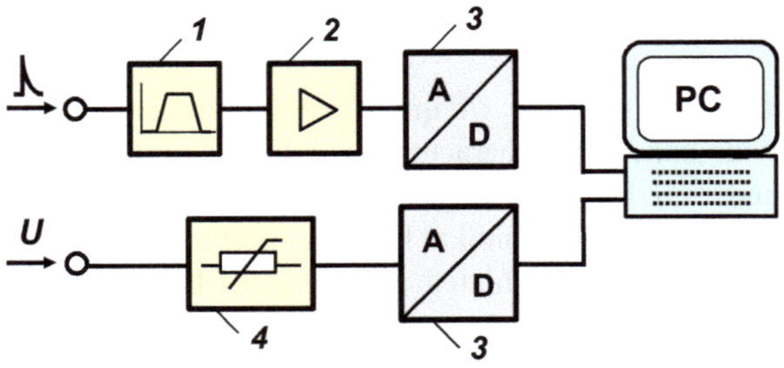

Fig. 12.6 Circuit principle of a wide-band measuring instrument with quasi-integration and digital post-processing of the recorded data. ***1*** Wide-band filter ($\Delta f \leq 1$ MHz), ***2*** amplifier, ***3*** A/D converter, ***4*** voltage divider

measuring circuit. The PD pulses decoupled from the high-voltage circuit enter the upper channel and first pass through a bandpass filter ***1*** having a bandwidth of usually not more than 1 MHz. As a result, quasi-integration of the PD pulses takes place in the same way as in the purely analog wide-band instrument. The analog output pulses are amplified in ***2*** and then digitized using an A/D converter ***3*** (*post-processing*). The maximum values u_{max}, which according to Eq. (12.13) are proportional to the charge, are stored for one or more periods in the PC. The second A/D converter in the lower channel of Fig. 12.6 digitizes the test voltage. With the PC, the pulse charges and the phases related to the test voltage are determined for a preselected number of periods and displayed graphically.

These instruments are often referred to as "digital" PD instruments although the analog measurement principle based on quasi-integration with bandpass filter is used. Since the charge of the PD pulse is stored, these instruments can also be used for PD measurements with DC voltage.

12.5.1.2 Narrow-Band PD Instruments

As the second example of quasi-integration, Fig. 12.7 shows the pulse response of a narrow-band system with a bandwidth $\Delta f = 9$ kHz and a mid-band frequency

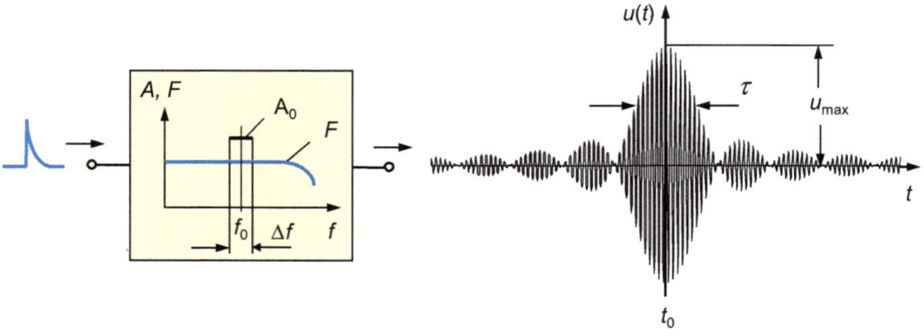

Fig. 12.7 Pulse response $u(t)$ of an ideal narrow-band system with constant transfer factor A_0 within the bandwidth $\Delta f = 10$ kHz (mid-band frequency $f_0 = 100$ kHz)

$f_0 = 100$ kHz, calculated again with Eq. (12.11). The PD pulse at the input of the narrowband system has a constant amplitude density $F \approx F(0)$ to at least the frequency $f_0 + \Delta f/2$, i.e. the mid-band frequency f_0 is thus in the frequency range of the approximately constant amplitude density of the PD pulse. The pulse response oscillates at the mid-band frequency and is significantly smaller and wider than that of the wide-band PD instrument in Fig. 12.5. The charge q of the PD pulse again results from the maximum value u_{max} of the pulse response (envelope curve) according to Eq. (12.13). A statement about the polarity is not possible, since positive and negative PD pulses generate the same pulse responses.

The pulse duration τ at half the amplitude of the pulse response is quite large because of the low bandwidth of the PD measuring instrument. In order to avoid superimposition of pulse responses, the interval between two successive PD pulses must not fall below a corresponding minimum value. In some PD measuring instruments, the mid-band frequency f_0 can be tuned in a wide frequency range up to the MHz range. This has the advantage that interference can be avoided by choosing an optimal mid-band frequency. Possible causes of such disturbances are, on the one hand, radio transmitters, thyristors in the low-voltage circuit and fluorescent lamps, on the other hand, resonance points in the test and measurement circuit, which lead to an increase in amplitude in the pulse spectrum.

12.5.1.3 Radio Disturbance Meters

A special type of the narrow-band measuring instrument with quasi-integration is the *radio disturbance meter* (also known as *radio interference meter*). It was used worldwide in the early days of PD measurement technique and is still widely used in North America for standard-conforming PD measurements. The radio disturbance meter has a tunable mid-band frequency, which can be from 150 kHz to 30 MHz. The frequency is to be chosen so that it lies within the range of the constant amplitude density of the PD pulse. With a 6 dB bandwidth of 9 kHz, radio disturbance meters basically correspond to narrowband charge measuring instruments. However, the display depends not only on the charge q, but also on the pulse repetition rate N according to a *weighting function* f(N) standardized in CISPRE 16-1 [19] (see Fig. 12.10, curve ***4***). The display for PD pulses is the *disturbance voltage* U_r in microvolts:

$$U_r \sim q \cdot \mathrm{f}(N) \cdot \Delta f \cdot R_m, \tag{12.15}$$

where Δf denotes the bandwidth and R_m the measuring resistor at the input of the radio disturbance meter. In individual cases, a conversion of the disturbance voltage U_r into a corresponding apparent charge q is possible. An example is given for the PD inception voltage with an equal PD pulse in each positive and negative half cycle. For a 50 Hz test voltage, the repetition rate is $N = 100\ \mathrm{s}^{-1}$, and in this case 1 μV = 2.6 pC for a measuring resistor $R_m = 60\ \Omega$ [5, 20–22].

Note The weighting function f(N) in Eq. (12.15) originally characterizes the loudness perceived by the human ear when there are crackling noises in the radio receiver. The function f(N) also appears to be acceptable for PD measurements, as an increased repetition rate of PD pulses is often accompanied by accelerated damage to the insulation.

12.5.2 Integration of the PD Pulses in the Time Domain

In this measurement principle, the apparent charge is determined by integrating the time course of the PD pulses. Here, the PD measuring instruments use two variants. In one variant, integration is performed by means of an electronic analog circuit. In commercial PD instruments, however, this principle was only applied for a limited time. In the other, increasingly used variant, the individual PD pulses are digitized with a fast A/D converter, and integration is done by software with a numerical algorithm. As an alternative to numerical integration, digital filtering of the data set is also possible, which effects quasi-integration.

12.5.2.1 Analog Integration Circuit

PD instruments with analog integration circuit have been produced for a limited time. The measurement principle, however, is still used for the calibration of pulse calibrators in performance tests. The current pulses $i(t)$ are integrated in the simplest case with an RC element as a low-pass filter or, much better, with an operational amplifier with capacitive feedback [23]. A current pulse at the input of the integrator leads to the pulse voltage $u(t)$ at the output:

$$u(t) \sim \frac{1}{RC} \int_{0}^{\infty} i(t)\mathrm{d}t, \tag{12.16}$$

which theoretically reaches its final value u_∞ after an infinite time. The comparison of Eq. (12.16) with Eq. (12.2) shows that the final value of the pulse response, strictly speaking at $t = \infty$, is proportional to the charge:

$$\boxed{u(t = \infty) = u_\infty \sim q}. \tag{12.17}$$

As an example, Fig. 12.8a shows on the left side the idealized current pulse $i(t)$ at the input of the integrator and on the right side its output voltage $u(t)$, which asymptotically approaches the final value u_∞. According to Eq. (12.17), u_∞ is proportional to the charge q of the current pulse. In an ideal integrator, u_∞ remains constant and can be measured with a peak detector, an oscilloscope or an A/D converter. In the practical implementation of the integrator, the pulse response $u(t)$ is set to zero after a few microseconds, so that the PD instrument is then ready to measure the next PD pulse. Otherwise, if the integrator is not set to zero, each

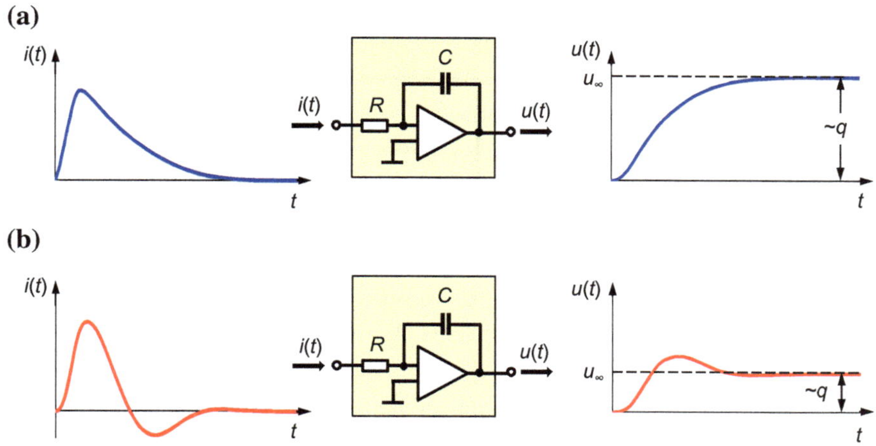

Fig. 12.8 Principle of charge measurement with an integration amplifier using the example of two current pulses *i*(*t*) with and without undershoot ($q \sim u_\infty$). **a** Pulse without oscillation; **b** pulse with negative oscillation in the tail

additional PD pulse leads to a corresponding increase in the output voltage. The result is a staircase curve whose final value represents the sum of the charges within the selected time, e.g. a period of the AC test voltage. The measurement principle with analog integrator is therefore also suitable for PD measurements at DC voltage within a defined measurement time.

It has already been pointed out that the PD pulses are often deformed by the high-voltage test and measurement circuit and therefore may be superimposed by oscillations. Also, calibration pulses may have oscillations due to inductances in the circuit of the calibrator. As an example, Fig. 12.8b on the left shows a current pulse *i*(*t*) with an oscillation in the tail, which thus causes a negative charge portion. After integration of this pulse by the operational amplifier, the output voltage *u*(*t*) overshoots before reaching the final value u_∞. The charge *q* is again proportional to the final value u_∞, not to the maximum of *u*(*t*).

12.5.2.2 Digital Data Acquisition, Numerical Integration

Advances in the hardware of fast A/D converters, data storage and computers now enable the use of digital measuring instruments that directly digitalize each PD pulse with high bandwidth and sampling frequency, store it as a digital data set and process it by software. The first PD measuring instruments of this type were digital oscilloscopes with PC and user-developed software, whereby the determination of the pulse charge was not necessarily in the foreground [24–27]. The principle of a very complex PD measuring instrument with digital acquisition of PD pulses and comprehensive data processing is shown in Fig. 12.9 [28]. In the upper input

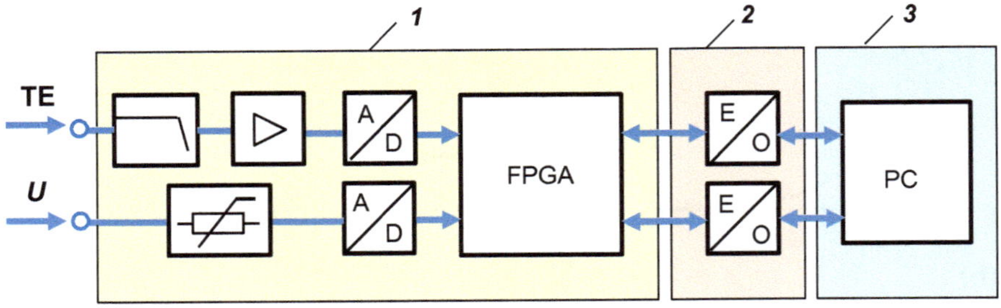

Fig. 12.9 Block diagram of a digital PD measuring instrument *1*, which is connected potential-free via optocouplers *2* and optical fibers to the PC *3*

channel, the PD pulses first pass through an *anti-aliasing filter* with a bandwidth of 20 MHz, are then amplified and subsequently digitized by an A/D converter (14 bits, 64 MS/s). In the lower channel, the attenuated test voltage is digitized with a second A/D converter (24 bits, 100 kS/s). Both data sets are sent to the FPGA (Field Programmable Gate Array) and processed in real time in online mode with up to 1.4×10^6 PD pulses per second. The measured data and control commands are sent via optocouplers *2* and optical fibers to the PC *3*.

The large bandwidth combined with the high sampling rate of the digital measuring device in Fig. 12.9 allows in many cases a largely true-to-original recording of the entire time course of PD pulses. Even fast successive pulses are reliably resolved in time. Thanks to the large bandwidth, effective suppression of interference is possible. The PD pulse recorded in the time domain can be displayed in the frequency domain with its spectrum using FFT. The fast evaluation of PD measurements is particularly advantageous in the continuous online monitoring of high-voltage apparatus. Changes in the PD pulse shape, which can be an early indication of a defect, are detected in good time, so that the apparatus can be shut down immediately if necessary. The simultaneous use of several of these battery-powered PD meters enables discrimination and location of PD sources that occur in various defects of a complex test object, such as a power transformer or cable (*synchronous multi-channel measurement*, see Sect. 12.8.2). With these digital devices, a comprehensive evaluation of PD pulses is possible even at DC or impulse voltages.

There are two ways to determine the charge of PD pulses with these measuring devices. In the first method, the charge is calculated by numerical integration of the digitally stored pulse according to Eq. (12.2). In the second, faster method, the recorded data passes through a digital filter and the charge results from the maximum value of the filtered pulse, as in analog quasi-integration (see Sect. 12.5.1). The mid-band frequency and the bandwidth of the digital filter can be freely selected within a wide range.

PD measuring instruments with digital data acquisition and processing are now included in IEC 60270 [5, 6]. The digital measurement principle with numerical integration has long been used to verify the charge of pulse calibrators needed to calibrate PD meters and test circuits [11]. The calibration pulses generated by the calibrator are recorded with a digital recorder and their charge is determined using numerical algorithms (see Sect. 12.7.1). In a sense, the predecessor of the digital wideband PD meter is the high bandwidth analog oscilloscope previously used in research and development for PD measurements with manual ("digital") counting of the area under the pulse shape [7, 14].

12.5.3 Pulse Train Response

PD pulses show a stochastic pattern and may originate from different defects in the insulation. In AC voltage tests, PD measuring instruments must display the *largest repeatedly occurring PD magnitude* [5]. Occasionally occurring disturbance pulses and smaller pulse charges resulting from other defects shall not contribute to the display. Older analog PD measuring devices with mechanical pointer fulfill this requirement for uniformly distributed pulses with *pulse repetition frequencies* $N \geq 100\ \text{s}^{-1}$. At lower pulse repetition frequencies, a pulsating value is indicated due to the mechanical inertia of the pointer, resulting in a smaller average value of the charge.

Note Some first generation PD measuring instruments with digital data processing behave differently. Their display remains constant even if the pulse rate falls well below $100\ \text{s}^{-1}$. Therefore, they indicate greater PD charges than analogue PD measuring instruments with pointer [29].

For all types of analog and digital PD measuring instruments, the display is now standardized as a function of the pulse repetition frequency N [5]. Figure 12.10 shows the gray tolerance band ***1***, in which the *pulse train response* of a PD instrument shall lie. Here, $N = 100\ \text{s}^{-1}$ means that one PD pulse each occurs in the positive and negative half cycles of the 50 Hz test voltage. The tolerance band ***1*** is chosen so that a large number of older analog PD measuring instruments with pointer meet this requirement, as the example of curve ***2*** in Fig. 12.10 shows. Curve ***3*** applies to the digital PD measuring device with numerical data processing shown in Fig. 12.9, which also complies with the IEC specification [28]. For comparison, curve ***4*** shows the characteristic of a radio disturbance meter according to CISPRE 16-1 [19], which lies outside the permitted tolerance band ***1***.

12.6 PD Test Circuits According to IEC 60270

The small PD pulses appear at the high-voltage terminals of the apparatus under test. They are superimposed on the high AC test voltage and must be coupled out of the test circuit for measurement. Figure 12.11 shows three basic circuits for the detection and measurement of partial discharges according to IEC 60270, from

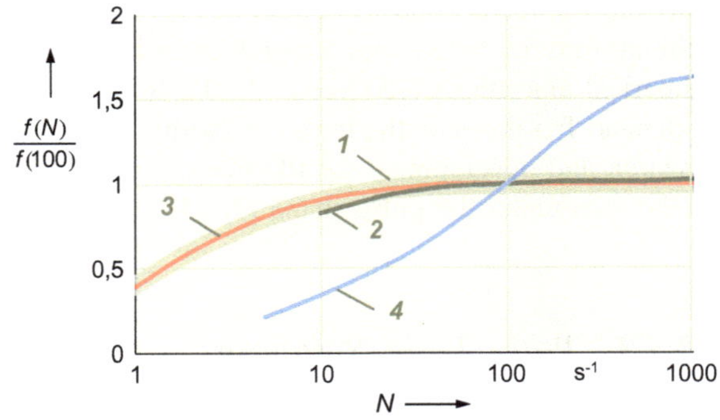

Fig. 12.10 Pulse train response f(*N*) of PD measuring instruments versus pulse repetition frequency *N*. ***1*** Specified tolerance band [5], ***2*** older wide-band meter with pointer instrument, ***3*** ultra-wide-band digital instrument with numerical data evaluation [28], ***4*** radio disturbance meter according to CISPRE 16-1 [19]

which further PD test circuits are derived [5, 30]. In each PD test circuit, the test object is shown simplified as a capacitor C_a with internal partial discharges. In the test circuit shown in Fig. 12.11a, the PD pulses reach the *coupling device* CD via the high-voltage *coupling capacitor* C_k, which should have a low inductance and itself no partial discharges or a sufficiently low PD magnitude. The coupling device is connected via a more or less long coaxial cable to the PD measuring instrument M outside the high-voltage area.

The coupling capacitor C_k is intended, on the one hand, to couple out the PD pulses of the high-voltage circuit and, on the other hand, to keep away the high AC test voltage, including the harmonics, from the PD measuring circuit. The capacitance of C_k should be large compared to the stray capacitance of the test circuit in order to achieve high measurement sensitivity and signal-to-noise ratio [8]. Since C_k acts as a high pass filter, preferably the higher frequency components of the PD spectrum can pass. When using a narrow-band PD measuring instrument according to Sect. 12.5.1, therefore, the choice of a high mid-band frequency f_0 is advantageous. On the other hand, f_0 must still lie in the frequency range of the approximately constant amplitude density of the PD pulse to be measured. Wide-band PD measuring instruments for IEC tests shall have a bandwidth of no more than 1 MHz (formerly 500 kHz).

The coupling device CD, the coaxial cable and the PD measuring instrument M represent a unit with regard to the transfer behavior. The coupling device is usually shielded and, in conjunction with the input circuit of M, forms the measuring impedance Z_m. In many cases, $Z_m = R_m$ is a resistance equal to the wave impedance of the coaxial cable in order to avoid reflection phenomena. In parallel to Z_m, there are usually a surge arrester and an inductor to short-circuit the remaining AC current component, including its harmonics, to ground. The coupling device may be equipped with a preamplifier. In conjunction with a narrow-band PD instrument, the coupling device used may be a resonant circuit whose resonance frequency is matched to the mid-band frequency of the PD instrument. Alternatively, the coupling device may be mounted on the high-voltage side of C_k, which can then be grounded directly. In this case, the PD pulses are transmitted to the PD instrument via a potential-free optical data link.

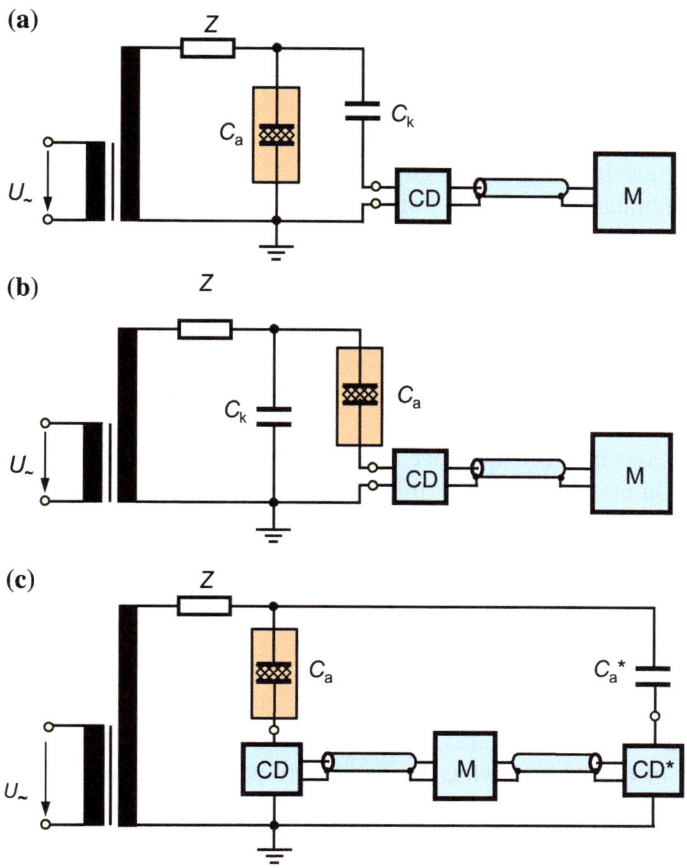

Fig. 12.11 Basic PD test and measurement circuits according to IEC 60270 [5]. **a** Coupling device CD in series with the coupling capacitor C_k; **b** coupling device CD in the ground connection of the test object C_a; **c** bridge circuit with parallel arm C_a* and CD*

Figure 12.11b shows a PD measurement circuit with the coupling device CD in the ground connection of the test object C_a. This presupposes that C_a can be operated without direct grounding and that CD can carry the high AC load current through C_a. With direct grounding of C_a, the connection to CD can be achieved using a Rogowski coil. Special constructions of the Rogowski coil can be opened and conveniently placed around the ground line of C_a. The electrical path of the PD pulses in turn leads via the coupling capacitor C_k. In the case that the stray capacitances of the test circuit to ground are large compared to the capacitance C_a of the test object, the coupling capacitor C_k may be omitted.

PD tests usually take place in a shielded high-voltage hall. External electromagnetic interference, e.g. caused by radio stations, can then not superimpose the PD pulses and distort the PD measurement. If the shielding has been carried out carefully, especially in the area of the hall gates, doors and windows, a noise level corresponding to a PD magnitude of less than 1 pC can be achieved. If a shielded hall is not available or if internal disturbance sources such as fluorescent lamps, thyristor circuits or floating metal parts with corona discharges are active, the bridge circuit in Fig. 12.11c can reduce the effects of interference. For optimal bridge balance, the parallel bridge arm with C_a* and CD* should be as identical as possible to the main bridge arm with the test object C_a and, of course, must not itself have any significant partial discharges. If necessary, the stray capacitances and series inductances must also be taken into account, as a result of which the bridge balance becomes frequency-dependent. When the bridge is ideally balanced, the PD measuring instrument M is connected potential-free between CD and CD* and displays the PD pulses at C_a without interference. The differential measurement technique with the bridge circuit can also be used for PD measurements at DC voltage (see Sect. 12.10).

The low pass filter or impedance Z in the three PD test circuits of Fig. 12.11 has two tasks. On the one hand, high-frequency disturbances that might come from the high-voltage supply and interfere with the PD measurement is blocked or at least attenuated. On the other hand, Z prevents the PD pulses from being bypassed to ground through the parallel stray capacitances of the transformer winding, resulting in greater measurement sensitivity. However, in the circuit of Fig. 12.11b, it may sometimes be more favorable to omit the filter Z, i.e. C_k is then parallel to the test transformer. In this case, the stray capacitances of the transformer to ground are fully effective, so that C_k and hence the measurement sensitivity are increased.

For transformers with capacitance-graded bushing, Fig. 12.12 shows a variant of the circuit in Fig. 12.11a, which is particularly preferred for on-site tests. In this case, the capacitance of the outermost grading element of the bushing is used as the coupling capacitor for coupling out the PD pulses. A separate high-voltage coupling capacitor is thus not required.

Another solution for measuring partial discharges in transformers is described in [31]. In the patented measurement arrangement, a metal strip is mounted on the

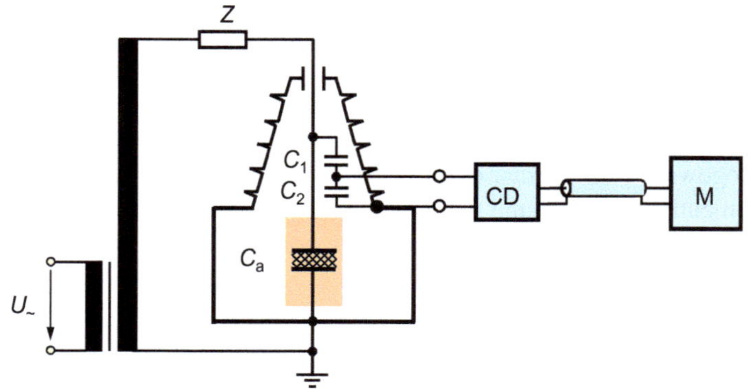

Fig. 12.12 PD test circuit for transformers with the bushing capacitance C_1 as coupling capacitor

bottom of the bushing and connected by a short lead to the coupling device CD, which is located on the transformer tank. The easy-to-install metal strip serves as a capacitive sensor for decoupling the PD pulses. After broadband amplification in the frequency range of 10 kHz to 10 MHz, the PD pulse is transmitted via a fiber optic link to the digital oscilloscope. The digital data is then evaluated by the PC, including sophisticated digital filtering to separate the PD pulses from continuous and pulsed noise.

12.7 Calibration of PD Instruments and Test Circuits

To ensure reliable PD measurements, comprehensive calibration is required. On the one hand, the PD measuring instrument is calibrated together with the coupling device CD in order to determine the basic properties and correctness of the displayed charge in all measurement ranges. On the other hand, the complete high-voltage test and measurement circuit is calibrated to determine the *scale factor k*. It takes into account the proportion of the pulse charge that flows through the stray capacitances to ground and therefore is not displayed by the PD measuring instrument. In both cases, calibration is accomplished by injecting current pulses of known charge and repetition frequency into the test object. These calibration pulses can be generated within specified uncertainties by means of a *pulse calibrator* with a relatively simple circuit. However, to determine the characteristics of a PD instrument in detail, a more sophisticated circuit of the calibrator is required. The bandpass characteristic of the PD measuring instrument including the coupling device CD shall be calibrated with sinusoidal currents and the result given as the *transfer impedance Z(f)*. In a comparison test among 13 European metrology institutes and calibration laboratories, different calibration procedures and the achievable measurement uncertainties were investigated.

12.7.1 Calibration Pulses

The form of the calibration pulses and the method for their generation are specified in detail in IEC 60270 [5, 6]. The standardized calibrator essentially consists of a rectangular or step voltage generator with the open-circuit output voltage U_0 and a series capacitor C_0 (Fig. 12.13). When the calibrator is short-circuited, the charge q_0 of the generated current pulse is formally given by:

$$\boxed{q_0 = U_0 C_0}, \qquad (12.18)$$

where C_0 is usually in the range of 1–100 pF, but not more than 200 pF. The rectangular or step voltage shall have a rise time $t_r \leq 60$ ns. Then the calibration pulse generated in conjunction with C_0 is sufficiently steep to produce a nearly

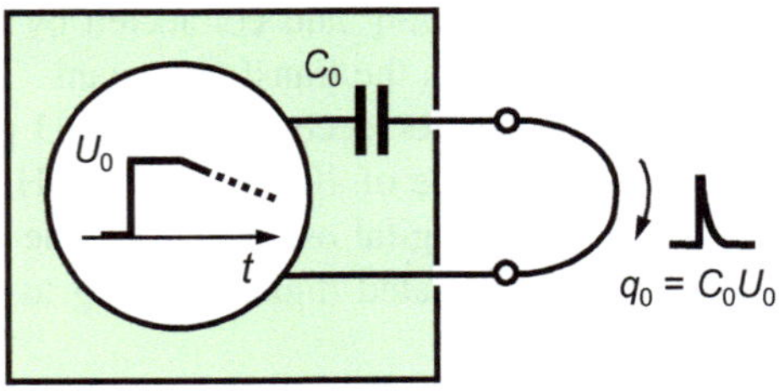

Fig. 12.13 Principle of a PD calibrator that generates positive pulses with the charge $q_0 = C_0U_0$ (output short-circuited)

constant amplitude density up to high frequencies. For the calibration of a very wide-band PD instrument with an upper limit frequency $f_2 > 500$ kHz (see Sect. 12.5.1.1), an even shorter rise time $t_r < 0.03/f_2$ is required.

Further requirements for the characteristics of the rectangular or step voltage are given in Annex A1 to IEC 60270, which is now included in the "Consolidated version" of the standard [5, 6]. Accordingly, the amplitude of the step voltage shall be approximately constant for at least 5 μs and then slowly decrease to $0.1U_0$ after not shorter than 500 μs. This, in conjunction with C_0, ensures that no negative pulse with appreciable amplitude is generated. At the beginning of the generated step voltage, oscillations are usually superimposed. After a maximum time of 200 ns, any deviation ΔU from the step voltage amplitude U_0, e.g. caused by superimposed oscillations, must not exceed 3% of U_0.

Note It is unusual that in the main part of an IEC Standard a mandatory construction manual is specified as it exists for the calibrator in IEC 60270. Not the calibration pulse itself, but the rectangular or step voltage is standardized. Furthermore, most specifications concerning the pulse shape of the step voltages and spectrum of the calibration pulses are only relevant to PD measuring instruments with quasi-integration (see Sect. 12.5.1).

Equation (12.18) also applies if the calibration pulse is injected into a resistor R_m, since the pulse charge q_0 is theoretically independent of R_m and the internal resistance of the calibrator. However, with very large resistances, part of the charge can be lost through parallel and stray capacitances. If the calibration pulse is injected into a capacitor C, only the pulse charge:

$$\boxed{q_0* = q_0\frac{C}{C_0 + C}} \tag{12.19}$$

is transferred. Therefore, a large capacitance ratio $C/C_0 \gg 1$ is required so that as much as possible of the entire calibration charge is injected into the test object C and approximately $q_0* \approx q_0$ applies.

In measurement practice, small battery-powered calibrators are used which generate unipolar calibration pulses with a pulse frequency of $N = 100\ \text{s}^{-1}$ and a multi-stage adjustable pulse charge. They serve to quickly check the display of PD measuring instruments and to determine the *scale factor k* of the complete test and measurement circuit. More detailed examination and calibration of PD instruments require a multifunctional calibrator that generates pulses with different pulse shapes, continuously adjustable charges and repetition rates. Instead of the simple step voltage generator in Fig. 12.13, a fast D/A converter with an amplitude resolution of 12–16 bits and a maximum sampling rate of 50–200 MHz may be used [32–34]. With the programmable calibrator, arbitrary pulse shapes and pulse frequencies are achievable in conjunction with the capacitor C_0. Furthermore, oscillating pulses, pulses with stochastically distributed charges and double pulses required for calibrating PD measuring instruments for cable tests can be generated. This enables a comprehensive investigation of analogue and digital PD measuring instruments in terms of linearity, dynamic behavior and phase-resolved PD pattern (see Sect. 12.8.1).

Occasionally, the capacitor C_0 in Fig. 12.13 is a high-voltage capacitor located outside the step generator. The calibration can then be performed with the test voltage applied to the test object. In another circuit variant, the calibration device called *online calibrator* is operated at high-voltage potential [35]. The electronics with D/A converter (for generating U_0) and A/D converter (for data acquisition) is housed between two metal hollow sphere halves, which are galvanically insulated. The lower half of the hollow sphere is attached via a short metal rod directly to the high-voltage electrode of the test object. The upper half of the hollow sphere is capacitively coupled to ground potential via its stray capacitance, which represents the capacitance C_0. An optical fiber link connects the electronics to a PC at ground potential for transmission of measurement data and control signals. Thus, online calibrations of the test object in high-voltage operation are thus possible at any time.

The pulse calibrator is decisive for the accuracy of the PD measuring instrument used and must therefore be regularly verified by *performance tests* and *performance checks*. There are several methods to prove the correctness of the generated calibration charge [5, 36–38]. Provided that the circuit of the calibrator allows this, the step voltage U_0 is measured with a digital oscilloscope and the capacitance C_0 is measured with a bridge. According to Eq. (12.18), the calibration charge is $q_0 = U_0 C_0$. The calibration charge can also be determined by two further methods. Here, the calibration pulses are either injected into a measuring resistor R_m or capacitor C_m and the output voltage is digitally recorded for further data processing (Fig. 12.14). In both measuring circuits, the pulse charge of the calibrator can also be determined by comparison with a *reference calibrator* traceable to national standards. The digital oscilloscope alternately records the pulses from both calibrators, and since it operates in both cases in the same input range, it does not need to be calibrated.

In the measuring circuit of Fig. 12.14a, the resistance R_m at the recorder input is typically 50–200 Ω. The larger values are preferred for measuring very small charges or for attenuating oscillations superimposed on the recorded pulse voltage. From the area A of the recorded pulse, the calibration charge is calculated numerically according to Eq. (12.2). In the second measuring circuit shown in Fig. 12.14b, the capacitor C_0 of the calibrator and the measuring capacitor C_m form a voltage divider. For calibration, C_m should be about 10 nF, i.e. $C_m \gg C_0$, so that

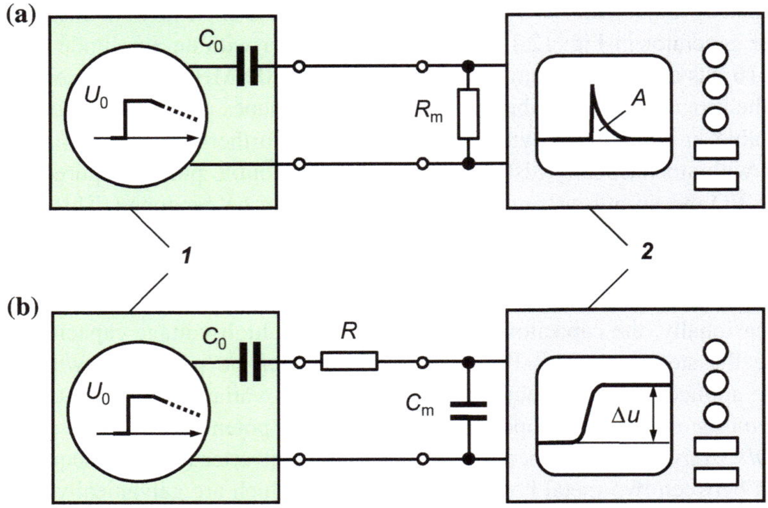

Fig. 12.14 Two measurement circuits for verifying impulses charges. *1* Pulse calibrator, *2* digital recorder. **a** Injection of the calibration pulse into a measuring resistor R_m; **b** injection of the calibration pulse into a measuring capacitor C_m

approximately the entire calibration charge q_0 is injected into C_m and displayed by the oscilloscope. In the case of $R = 0$, the voltage at C_m recorded by the digital oscilloscope is directly proportional to the internal step voltage of the calibrator. The series resistor R of the order of 100 Ω is inserted to attenuate superimposed oscillations, but results in a slower voltage rise. After the decay of the initial oscillations, the voltage Δu is proportional to the calibration charge q according to Eq. (12.18). When measuring charges smaller than 50 pC, Δu becomes quite small. It is then advantageous to replace C_m by an operational amplifier with capacitive feedback.

As an example, Fig. 12.15a shows the time course of a calibration pulse recorded in the circuit according to Fig. 12.14a with the resistance $R_m = 50\ \Omega$. The charge calculated by numerical integration of the pulse is $q_0 = 48$ pC. The amplitude density of the calibration pulse calculated with FFT is approximately constant up to 1 MHz (Fig. 12.15b). The recorded pulse is superimposed by an oscillation caused by line inductances in the circuitry of the calibrator. The oscillation disappears when the calibration pulse is injected into a larger resistance $R_m = 200\ \Omega$. The charge is then maintained according to the invariance principle, but the spectrum changes and the range of approximately constant amplitude density is shortened.

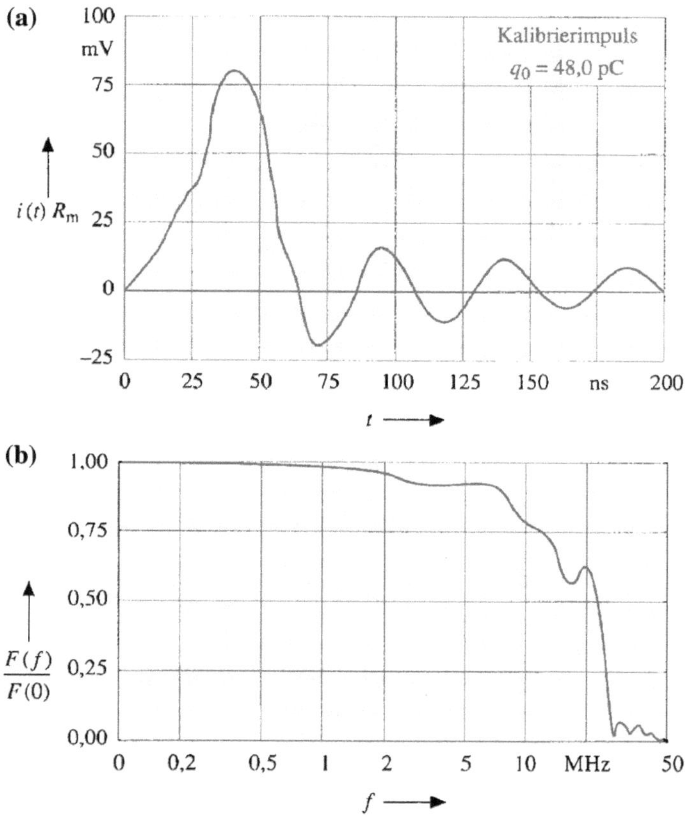

Fig. 12.15 Calibration pulse recorded with a digital recorder according to Fig. 12.14a. **a** Time course $i(t)R_m$ (R_m = 50 Ω); **b** normalized amplitude density $F(f)/F(0)$

12.7.2 Calibrating the Complete Test and Measurement Circuit

Due to the stray capacitances of the spatially extended high-voltage test and measurement circuit, the PD pulses lose part of their apparent charge q on their way from the test object to the PD measuring instrument. As a result of the charge losses, only the portion q/k is indicated by the PD meter. The *scale factor* k is determined by calibrating the complete test and measurement circuit with the high-voltage generator connected but usually de-energized (Fig. 12.16). The calibration pulses with the charge q_0 are injected into the terminals of the test object C_a and undergo on their way to the PD instrument M a comparable charge loss as the PD pulses. The indicated charge q_M is therefore smaller than the injected charge q_0. The scale factor k of the PD test circuit with the test object C_a is defined as:

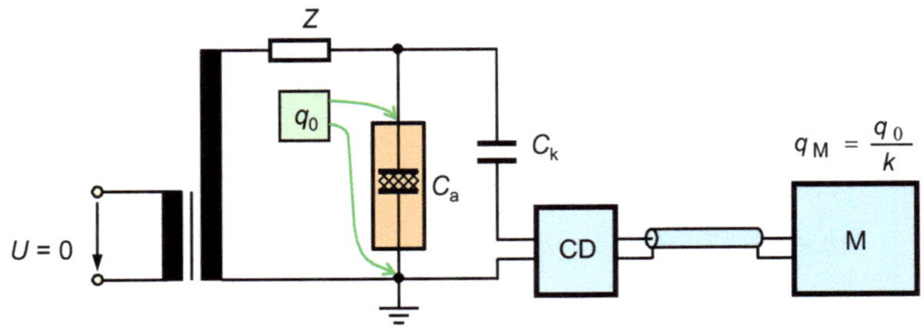

Fig. 12.16 Basic arrangement for calibrating the complete PD test circuit by injecting the calibration charge q_0 into the test object C_a and reading the charge q_M indicated by the PD meter M

$$k = \frac{q_0}{q_M}. \tag{12.20}$$

The pulse calibrator should be placed as close as possible to the high-voltage terminal of the test object in order to avoid loss of charge due to the stray capacitance of the lead. The actually injected charge also depends on the capacitance C_a of the test object and the coupling capacitance C_k. If the condition $C_0 < 0.1C_a$ is satisfied, the injected charge can be considered equal to the calibration charge q_0 [5]. Unlike the commonly used calibrators, the calibrator with external high-voltage capacitor and the online calibrator already presented in Sect. 12.7.1 can even be used with the test object at high-voltage [35].

In the subsequent PD test, the display of the PD instrument must be multiplied by the scale factor k to obtain the correct value of the apparent charge q. In order to achieve high measurement sensitivity, k should not be much greater than 1, i.e. the charge losses in the test and measuring circuit must be kept small. This is achieved with a coupling capacitance C_k that is much larger than the stray capacitances. The scale factor applies only to the same arrangement of the test and measurement circuit as in the calibration. Any change of the test object or in the test and measurement circuit requires a new determination of the scale factor.

12.7.3 *Accuracy Requirements*

IEC 60270 specifies the accuracy requirements of PD measuring instruments and pulse calibrators [5]. The uncertainty of pulse calibrators, which represents the basis for the accuracy of PD measuring instruments, must not exceed 5% of the set charge (or 1 pC for small charges). Furthermore, the impulse charge must be traceable to national or international standards of measurement (Refs. [9, 10] of Chap. 10). A calibrated PD measuring instrument conformal to the provisions of the standard is considered to have an uncertainty of not more than 10% or 1 pC, whichever is greater. Real PD measurements are, however, influenced in practice by various parameters, including electrical disturbances or background noise. A sound estimate of the uncertainty of the measured apparent charge with the test object in high-voltage operation can thus hardly be given.

In order to increase the mutual acceptance of pulse charge measurements, a ring comparison was carried out with thirteen European laboratories, which were metrology institutes, calibration and test laboratories. The task was to measure the charge of different pulses generated by an arbitrary waveform generator with a series capacitor. The participating laboratories used various methods and circuits for charge measurement, inter alia the two circuits shown in Fig. 12.14. The results show that the charges of the various calibration pulses were correctly measured by the majority of participants within (2–3)%. The ring comparison also included a check of the PD measuring instruments used in the participating laboratories. For some pulse shapes, the PD meters partly showed impermissibly large deviations of the measured charge values [39].

12.8 Visualization of PD Pulses

At the beginning of the PD measurement technology, a comprehensive evaluation of PD pulses with the analog measuring instruments available at that time was not possible. Due to the inertia of the moving coil meter, only the largest repeatedly occurring PD magnitude was displayed. An important advance in the field of PD measurement technique was the assignment of the individual PD pulses to the phase of the AC test voltage. The PD pulses were displayed on the screen of an analog dual-beam oscilloscope, with the voltage waveform being a sine or ellipse. The phase-resolved representation of PD pulses was later adopted and refined with the help of digital measuring instruments. For each insulation arrangement, characteristic distribution patterns of the PD pulses are now available in dependence of the phase of the AC voltage, which are compiled in comprehensive databases. In simple cases, such as when only a single PD source exists, the comparison with the database allows a statement about the nature of the measured partial discharges and the degree of risk of the insulation. Further basic principles of today's digital PD measurement technique are the *synchronous multi-channel* and *multi-frequency measurement techniques* that have already been implemented with analog measuring instruments.

12.8.1 Phase Resolved PD Pattern

Digital PD measuring instruments have been in use for more than three decades. Thanks to computer-aided data processing, they enable fast evaluation and comprehensive visualization of the measurement results. In first-generation digital PD meters, the A/D converter is connected downstream of the analog band filter. This means that not the actual PD pulse, but the much wider response of the band filter is digitized (see Fig. 12.5). Its peak value is determined, which is proportional to the charge due to the quasi-integration of the PD pulse. This evaluation method is called *post-processing*. The charge values digitally stored for several periods can be displayed differently [40–44]. For example, a simple graphical representation shows the frequency distribution $H(q)$ of the charge q determined with a pulse height discriminator. A more complex representation of the PD measurement is provided by the extended three-dimensional φ, q, n graph, where the phase φ of the AC voltage, the charge q and the number n of PD pulses are the axes.

In another fairly clear representation, the pulse charges are given as *phase-resolved partial discharge pattern* (*PRPD pattern*) over the AC test voltage. Each PD pulse is represented by a small dot, and the number n of PD pulses is indicated by different gray levels or colors. As an example, Fig. 12.17a shows the φ, q, n pattern of a 10 kV induction motor in operation having multiple defects with a critical average PD charge greater than 10 nC [45]. The φ, q, n pattern in Fig. 12.17b represents the state of polymeric material immediately after fabrication with at least six gas-filled cavities of different sizes.

Today, ultra-wideband recording devices with very high sampling rates are available so that they can directly record the true time course of PD pulses. The maximum value u of each PD pulse is then evaluated and represented in a φ, u, n pattern. That is, the charge q is not evaluated because the knowledge of the charge magnitude for the diagnosis is often not needed in on-site tests and online measurements.

The digital dataset in combination with the appropriate software offers a wide range of possibilities for distinguishing PD pulses from disturbances [46–50]. The pattern of phase-resolved PD pulses is typical for each test object and each PD pulse shape and provides information on the nature of the partial discharges. It can also be compared with the stored PD patterns of a large reference database [51, 52]. It should be noted, however, that the PD pattern may be affected by various quantities, such as temperature, harmonic content and frequency of the test voltage [53, 54].

PD measurements with digital data processing are performed not only during acceptance tests in the high-voltage hall or during on-site inspections, but also increasingly for continuous online monitoring of high-voltage or power apparatus due to the fast and automated data evaluation. This makes it possible at any time to compare the actual phase-resolved PD patterns with previously recorded data. Changes in the PD behavior indicating the state of aging of the insulation are thus readily apparent. An important goal is the definition of critical PD threshold values that, if exceeded, would lead to the automatic deactivation of the endangered equipment.

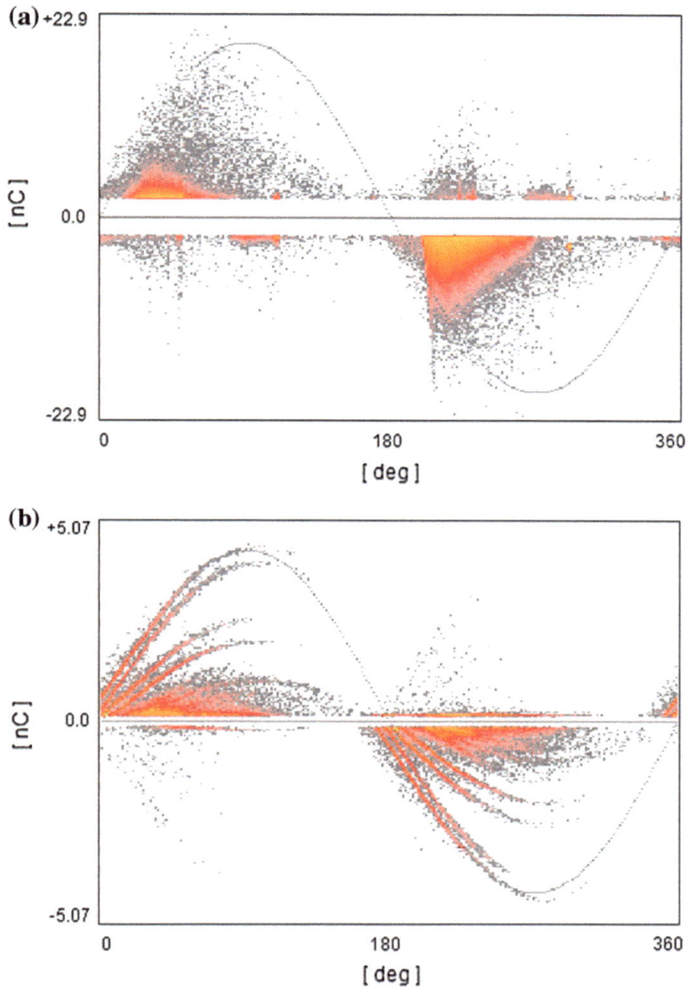

Fig. 12.17 Examples of phase-resolved PD patterns of defects. **a** φ, q, n pattern of a 10 kV induction motor ([45], Fig. 4); **b** φ, q, n pattern of at least six spherical gas inclusions in polymeric material (Power Diagnostix Systems)

12.8.2 Synchronous Multichannel PD Measurement

Complex test objects with large dimensions such as three-phase power transformers may have multiple defects. Partial discharges in the vicinity of one phase are coupled in an attenuated form into the other two phases and are to be regarded as interference pulses there. The separation of the pulses and localization of the defect by measuring at only one point is difficult or even impossible. For this reason, investigations on power transformers were carried out at an early stage with the aim of measuring PD pulses at the three bushings as simultaneously as possible. Initially, only one PD meter—first analog, later digital—was used, which alternatively was connected via a fast multiplexer to the three high-voltage bushings and also to the ground connection if applicable [55–58]. However, the separation and localization of defects was not very accurate due to the time delay in switching and the large pulse widths due to quasi-integration. In addition to these *multi-terminal measurements*, *multi-frequency measurements* with a frequency analyzer are also mentioned in [55].

The direct digitization of PD pulses with very fast ultra-wideband A/D converters and data processing software represent a significant advance in PD measurement technique. The combination of several of these PD measuring devices with the corresponding coupling units or sensors enables synchronous multi-channel measurement of PD pulses in complex equipment. Here, two variants can be distinguished. The PD pulses are simultaneously detected either with ultra-wide bandwidth at different, usually three measurement terminals (*synchronous multi-terminal measurement*) or with narrow bandwidth at one terminal with different center frequencies (*synchronous multi-frequency measurement*). The PD pulses of each defect in the insulation are characteristically affected on their way through the test object to the measurement terminal. Mathematically speaking, this influence corresponds to the convolution of the pulse by the transfer function of this particular propagation path. With both measurement variants, PD pulses from different defects can be detected and separated from external interference pulses. A combination of both measurement methods is also possible.

The principle of synchronous PD multi-terminal measurement and the visualization of the results will be explained using a simple example. In the case of a three-phase test object with an internal PD source, differently sized PD pulses occur at the measurement terminals of the three phases L1, L2 and L3. When the PD source is closest to the phase L1, the PD pulse measured there has the largest amplitude (Fig. 12.18a). Due to the attenuation during the longer signal propagation, the pulse amplitude measured at L2 is smaller and even smaller at L3. The three measured PD amplitudes are drawn into the triaxial *multi-terminal star diagram* in a two-dimensional representation and added vectorially (Fig. 12.18b). The end point indicates the position of this single discharge in the star diagram. With longer recording time, a large number of PD pulses are detected whose amplitudes are plotted in the same way in the star diagram. Due to the known spread of partial discharges and depending on the operating status of the test object (e.g. voltage instabilities, oil temperature changes and magnetostriction in a transformer), the points cover a more or less large area in the star diagram (Fig. 12.18c). These result in a characteristic pattern of partial discharges from this PD source, referred to as ***3-Phase Amplitude Relation Diagram*** (*3PARD*). The different frequency of occurrence of the PD pulses with approximately the same amplitude is indicated by different colors or gray levels [59–61].

PD pulses from other PD sources in the test object also produce characteristic patterns elsewhere in the star diagram. This representation also has the advantage that each individual PD pattern in the star diagram can be converted into a

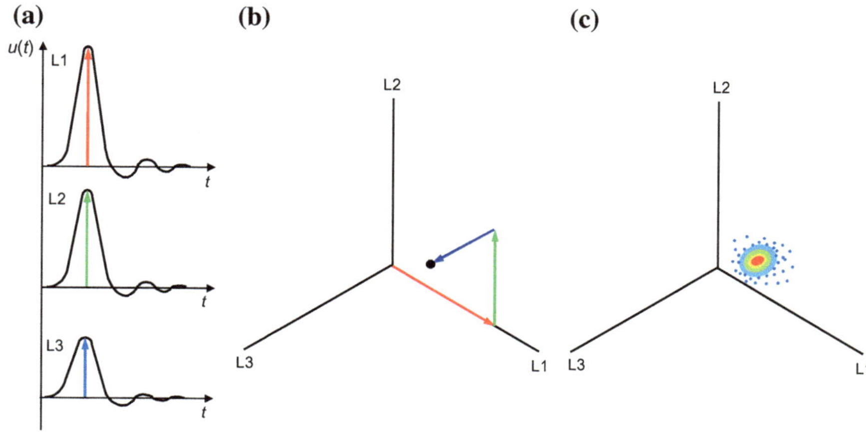

Fig. 12.18 Explanation of the two-dimensional multi-terminal star diagram (3PARD). **a** Signals of a single PD pulse measured at the three phases L1, L2 and L3 of a transformer; **b** vectorial addition of the amplitudes of the three measured PD signals in the star diagram; **c** visualization of the PD signals measured over a longer time period

corresponding φ, q, n pattern (see Sect. 12.8.1). Thus, the PD pulses from each individual PD source can be clearly assigned to the phase of the test voltage and interpreted based on decades of experience. (Compared with that, a complete φ, q, n pattern with superimposed PD pulses from all existing PD sources is very complex and would be difficult to evaluate.) Disturbance signals from outside the test object usually have the same effect on all measurement terminals and are therefore mapped in a small area around the origin of the diagram axes.

A variant of the multi-terminal star diagram is the ***3-Phase Time Relation Diagram*** *(3PTRD)* which shows the time differences between the occurrence of the PD signals measured in the three phases [60, 62]. Alternatively, or in addition to the synchronous multi-terminal measurement, the *synchronous multi-frequency measurement* is performed. The advantage is that the measurement is performed only at a single terminal of the test object. For a three-phase apparatus under test, three narrow-band PD measuring instruments with different center frequencies are connected to the terminal. Due to the band limitation, there is a quasi-integration of the PD signals, so that the evaluated pulse amplitudes are, in principle, proportional to the charge. The measurement results are shown in the ***3-Center Frequency Relation Diagram*** *(3CFRD)* [63].

Another advancement of the synchronous multi-terminal and multi-frequency measurement methods is that they can also be used in PD tests with DC voltage, which in turn allows different PD sources to be distinguished (see Sect. 12.10).

12.9 Special Measurement and Detection Methods

The conventional PD measurement technique according to IEC 60270 for determining the apparent charge usually requires a coupling capacitor to decouple the PD pulses from the test circuit. Such capacitor becomes larger and more expensive with increasing operating voltage. Three-phase high-voltage apparatus with large dimensions such as power transformers and GIS require multiple coupling capacitors to provide meaningful measurement results. On the other hand, the electromagnetic and acoustic processes associated with partial discharges allow the use of completely different PD measurement techniques. In conjunction with appropriate hardware and software, these techniques can be used advantageously in particular for on-site and online PD measurements. Calibration with charge pulses to determine the apparent charge is often not possible and not required. The PD pulses are detected with small electrical or acoustic sensors positioned inside or outside the test object and measured with ultra-fast digital oscilloscopes in the time domain or with frequency analyzers in the frequency domain. An IEC publication on electromagnetic and acoustic PD measurements is available as a *Technical Specification* for information [64]. A special role in these measurement techniques is played by the software, with which different PD sources can be distinguished and separated from interference, e.g. caused by external discharges. The occurrence of partial discharges is also associated with optical and chemical processes that are also used for diagnosis.

12.9.1 VHF and UHF Measurement Methods

Partial discharges in gas-filled voids have spectral components of more than 20 GHz and rise times of less than 50 ps [9, 10]. The very fast gas discharges generate electromagnetic waves whose propagation in a homogeneous medium is described by Maxwell's equations. Velocity and attenuation of wave propagation depend on the properties of the insulating material and the construction of the test object. Characteristic of the VHF and UHF measuring method is that the electromagnetic field generated by partial discharges is measured and evaluated in a frequency range of a few 100 MHz (VHF range) to some gigahertz (UHF range). This frequency range is thus considerably larger than that of conventional PD measurements according to IEC 60270 ($f_2 \leq 1$ MHz). Various sensors inside or outside the test object are used for VHF or UHF measurements. In particular, the UHF measurement technology is now thanks to high-performance sensors and measuring instruments a very meaningful diagnostic tool for partial discharges in GIS, GIL, transformers and cables. With on-site tests, the correct assembly of the individual components is checked and with increasing online monitoring, changes in the PD behavior can be identified as hazard criteria.

12.9.1.1 UHF PD Measurements in GIS

The reasons of the occurrence of partial discharges in GIS are manifold. During manufacture and assembly, small metal particles may remain in the GIS and be charged in the strong 50 Hz field. Depending on the particle size, the AC field may be high enough to lift the particles so that they can move freely by hopping within the GIS. When approaching the metal wall, the charged particle causes a small electric discharge [65, 66]. Electric discharges also occur due to microtips on the current conductor. In both cases, there is a risk of an electrical flashover when overvoltages occur. Partial discharges may also occur in case of defects in cast resin GIS spacers that support the current conductor. Furthermore, partial discharges lead to decomposition of the insulating gas (SF6), whose fission products also serve to detect partial discharges (see Sect. 12.9.5).

In a gas-filled coaxial conductor, TEM waves and, with corresponding dimensions, even higher TE and TM modes propagate at near the speed of light. Wave propagation in GIS is affected by attenuation and reflection phenomena, which are the subject of numerous theoretical and experimental investigations [67–73]. The typical mean attenuation of wave propagation in gaseous material is considered to be 2 dB/m. A particular advantage of UHF measurement technique in GIS lies in the low influence of electromagnetic interference on PD pulse measurements. External sources of interference are largely shielded and internal interferences only have a small share in the UHF range. For noise suppression, it is often sufficient to average the recorded PD pulses over a few minutes. With the UHF measurement method, the perfect assembly of the GIS components is checked during on-site tests. The method is also increasingly used in online measurements, as changes in PD behavior are immediately recognizable. The long-term goal of PD pulse monitoring is to send automatic hazard information to personnel or even to switch off the operating voltage when a certain PD pulse magnitude is exceeded.

The results of studies on various very *wide-band sensors* for PD measurements in GIS can be found in [74–76]. Basically, a distinction is made between internal and external sensors. Internal sensors are, for example, disc-shaped or cone-shaped. They are installed in one or more inspection openings of the grounded outer conductor in such a way that the symmetrical structure of the GIS remains as undisturbed as possible. The PD signal is capacitively detected by the sensor and fed to the measuring instrument outside the GIS (Fig. 12.19). Another measurement option is provided by sensors in antenna or ring form embedded in prefabricated GIS spacers made of cast resin [77]. External sensors are mounted on *dielectric windows* in the outer conductor and detect the field that leaks to the outside. Alternatively, the high-frequency stray field of the PD pulses is detected by horn or other antennas mounted externally at the position of built-in spacers [78, 79].

Internal sensors are generally more sensitive than external sensors or antennas. The latter are also susceptible to external disturbances and must therefore be screened. The sensitivity of internal sensors corresponds to an apparent charge of less than 1 pC. The decisive characteristic of a sensor or an antenna is the *effective length* (or *effective height*), expressed as the quotient of the output voltage of the

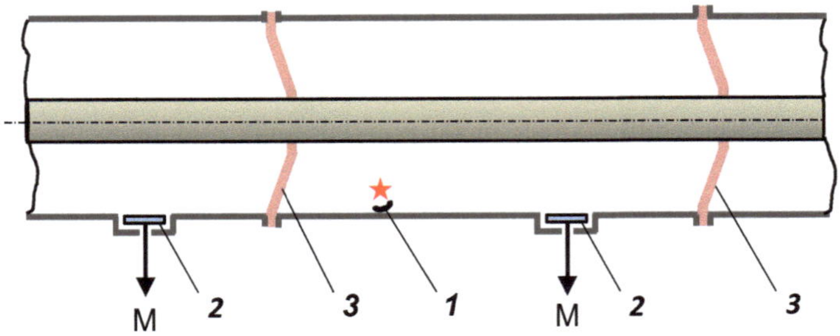

Fig. 12.19 PD diagnosis in GIS with UHF sensors (example). *1* Metal particle as PD source, *2* UHF sensor with instrument M, *3* spacer

sensor and the surrounding electric or magnetic field. The reciprocal is called the *antenna factor*. The basic properties and characteristic values of UHF sensors can be determined in the field of a GTEM cell or of a cone-shaped antenna placed on a metal plate [80–82].

There are two ways to process the PD pulses received from a sensor. In the first method, the PD pulse is evaluated in the time domain with a wide-band (>1 GHz) fast digital oscilloscope (>1 GS/s). The second method is to evaluate the pulse in the frequency domain with a narrow-band spectrum analyzer with center frequencies of up to several gigahertzes. The UHF measurement method has the advantage that disturbances in the GHz range have largely subsided. The time representation of the PD pulses as a function of the phase of the AC voltage again provides characteristic distributions (PRPD, see Sect. 12.8.1), which give valuable information about the type and cause of the partial discharges [70, 79]. Since the propagation of the PD impulses in GIS is damped, several sensors are mounted at intervals. By measuring the transit times of the PD pulses between two or more sensors, it is possible to locate the PD source.

A disadvantage of the UHF measurement is that the apparent charge of the PD pulses cannot be determined directly as with the IEC method. There are several suggestions for calibrating the UHF measuring instrument including the sensor. The measured quantity in the time domain is the maximum value of the PD pulse recorded with the digital oscilloscope, usually expressed in millivolts. Two methods are used to establish an approximate relationship between the display in mV and the apparent charge in pC. In one method, partial discharges are generated by an artificial defect within the GIS and measured simultaneously with the UHF and IEC methods with separate sensors. Or electrical pulses with rise times less than 1 ns and known charge are fed via an internal sensor into the GIS, which are then received by a second sensor and evaluated with the UHF meter [83–85].

Partial discharges in GIS emit not only electromagnetic waves but also sound waves. The acoustic measurement methods are discussed in more detail in Sect. 12.9.3, which deals with PD measurements in transformers. They are occasionally used in combination with the electric UHF measurement technique. Acoustic sensors have a frequency range from 50 to 200 kHz. For PD measurements, they are fixed individually or in groups outside the GIS with paste or glue. The electric output signals are measured with a digital oscilloscope or a comparable digital measuring device. The results of the investigation show that not only the partial discharges caused by bouncing metal particles but also the partial discharges caused by metal tips on the inner and outer conductors can be detected well with acoustic sensors [86, 87].

12.9.1.2 UHF PD Measurements in Transformers

Partial discharges in oil-insulated transformers often originate from gas bubbles in the insulating oil or in the winding insulation. The rise times and spectra of the PD pulses in oil transformers are therefore quite comparable to those in GIS. However, the propagation of electromagnetic waves in insulating oil is only about 200 m/μs, i.e. two-thirds of the speed of light. The higher spectral components are more attenuated than in GIS, so that the evaluation of the measurements usually only covers the frequency range up to 1 GHz. The application of the GIS-proven UHF measurement technique to PD measurements in power transformers is the goal of intensive investigations. External disturbances such as corona discharges, which can penetrate through the bushings into the transformer tank, are usually in the frequency range below 300 MHz or even below 200 MHz and can therefore easily be distinguished from the high-frequency internal partial discharges.

Two versions of capacitive sensors have proven to be particularly suitable for on-site measurements using the UHF method (Ref. [5] of Chap. 1; [1, 88–90]). Internal sensors can be inserted directly into the transformer tank using the standard oil filling valves (Fig. 12.20a). These sensors are referred to as *drain valve sensors*. The same UHF sensor can also be mounted as an external sensor on a *dielectric window* in the tank wall and is referred to as *top hatch sensor* (Fig. 12.20b). Often, the UHF sensor is directly equipped with a monopole antenna that can emit electric calibration pulses into the oil tank. This allows direct function and sensitivity tests of the UHF sensor before the PD measurement. The suitability of further sensors for UHF measurements in model arrangements is discussed in [91–93].

A monitored power transformer is generally equipped with UHF sensors at several points (Fig. 12.21). From the differences in the measured transit times and amplitudes, the position of the PD source in the transformer can be determined to a limited extent. However, due to multiple reflection and attenuation phenomena, accurate PD localization is possible only in special cases. By combining the UHF method with acoustic measurement methods, the location of PD sources in transformers appear to be more effective (see Sect. 12.9.3). For UHF PD measurements in transformers, comparable digital oscilloscopes or frequency analyzers are used as for PD measurements in GIS.

As with GIS, UHF measurements of partial discharges in transformers have the disadvantage that the apparent charge of the PD pulses cannot be determined directly. Sensitivity measurements are reported in detail in [94]. Defined pulses are injected via a monopole antenna into an oil-filled tubular test chamber and the

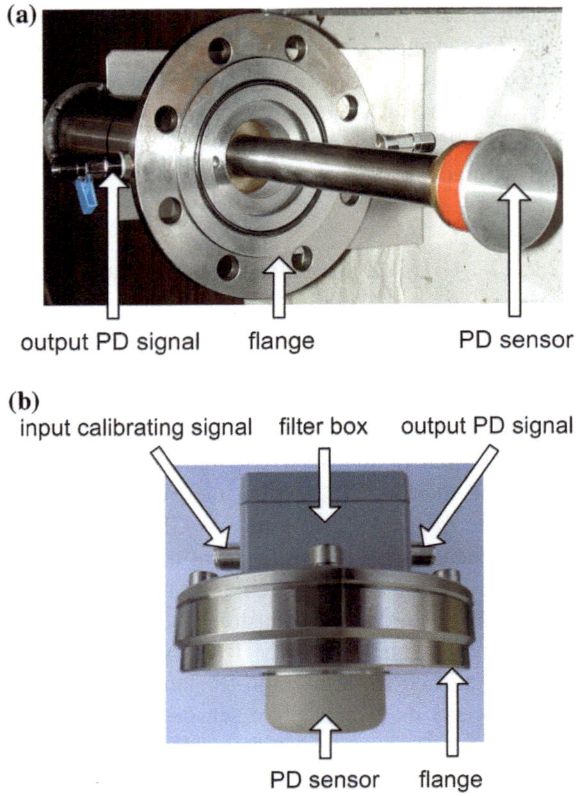

Fig. 12.20 UHF sensors for PD measurements in power transformers (Ref. [5] of Chap. 1; Fig. 10.35). **a** Internal UHF sensor as drain valve sensor; **b** external UHF top hatch sensor for use on dielectric window

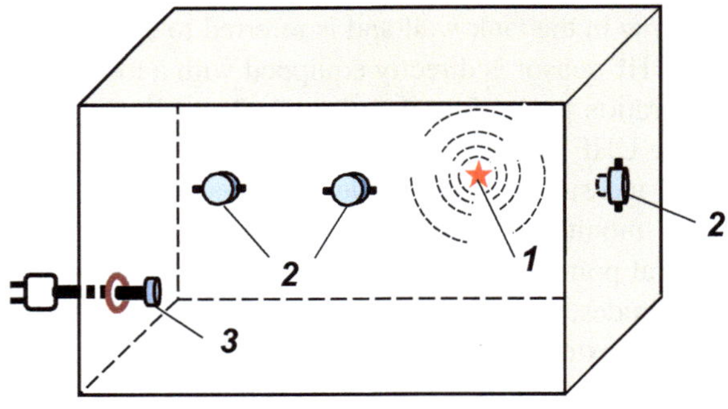

Fig. 12.21 Possible positions of UHF sensors for PD measurement in the oil tank (schematic). *1* PD source with electromagnetic radiation, *2* UHF sensors on dielectric windows, *3* UHF sensors as drain valve sensor

radiated electromagnetic signal is measured with a UHF sensor. The injected pulses with a half-width of less than 0.5 ns are generated electronically by means of a bipolar square wave generator and a fast *step recovery diode*. The injected pulse, which simulates a PD pulse, and its UHF signal are recorded with a two-channel digital oscilloscope and evaluated with the PC. Further investigations are considered necessary to obtain a quantitative statement about the injected pulse charge

q and the displayed value of the UHV sensor in mV. A theoretical study in [95] presents the dimensioning of an oil-filled GTEM cell that would be suitable for the basic calibration of UHF sensors.

12.9.1.3 PD Probe for Fast on-Site Measurement

The high-frequency electromagnetic field generated by partial discharges in GIS, transformers, cable terminations, etc., can generally leak out due to small cracks and gaps in the metallic housing and can therefore be detected outside. Figure 12.22 shows a battery-operated PD probe that enables fast potential-free measurements of this high-frequency field on site (Ref. [5] of Chap. 1; [96]). The probe is equipped on the front with two capacitive sensors, which are connected in a bridge circuit in order to achieve a large signal-to-noise ratio. The highly sensitive PD probe can be manually guided at a distance of a few centimeters along the test object without touching it. The high-frequency field of the partial discharges is detected in the range of 300–800 MHz. After processing the measured PD signal, the PD probe displays the maximum value on a pointer instrument. In addition, a digital oscilloscope or PC for further data processing can be connected via an optical data link. Instead of capacitive sensors, the PD probe can also be operated with inductive sensors.

Fig. 12.22 Portable PD probe with two capacitive sensors arranged in a bridge circuit for fast on-site measurements (Ref. [5] of Chap. 1; Fig. 4.60c)

12.9.2 PD Measurements in Cables and Accessories

The insulation of newly manufactured high-voltage cables consists mainly of *cross-linked polyethylene* (XLPE). Older cables with oil-insulated paper as insulation or mass-insulated cables will continue to be used for decades. Partial discharges can be caused by gas bubbles or contamination in the cable insulation, and by damage to the semiconducting layers on the inner or outer conductor of XLPE cables. With older plastic cables, there is a risk that the lifetime will be impaired by the formation of *water trees*. Newer cables with production lengths of up to several 1000 m are subjected to a conventional PD test according to IEC 60270 in the test laboratory. If partial discharges occur in the cable, the PD pulses travel at a speed of 140–200 m/µs in both directions away from the defect (Fig. 12.23). Since the cable end B is usually not terminated during the test, the PD pulses arriving there are reflected and return to point A at the beginning of the cable. The pulses coming directly from the PD source and the reflected pulses are more or less attenuated depending on the distance traveled through the cable. At point A, the pulses are coupled out via the capacitor C_k and measured with an oscilloscope as voltage pulses at the measuring resistor R_m [97–99].

Note The first version of the so-called Lemke probe still worked according to the integration principle given in Sect. 12.5.2.1. The capacitively received PD pulse passed through an input amplifier with a bandwidth of 10 MHz to an integration amplifier, so that the charge of the PD pulse was displayed.

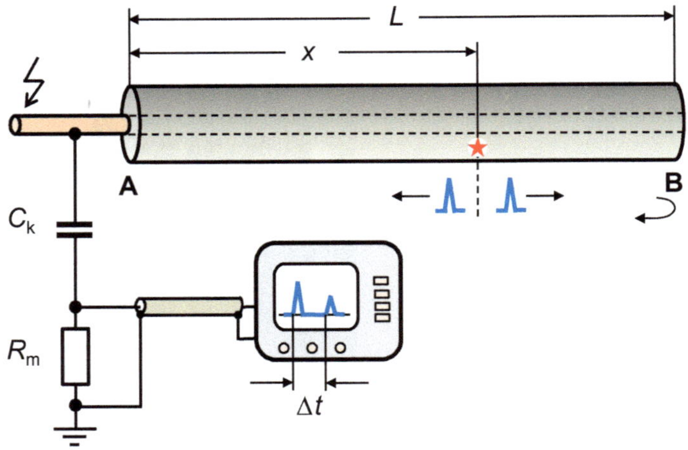

Fig. 12.23 Basic circuit for localizing partial discharges in a cable (high-voltage power supply and cable accessories not shown)

The time difference Δt between the direct PD pulse and the reflected PD pulse when they arrive at point A is:

$$\Delta t = \frac{2(L - x)}{v}, \tag{12.21}$$

where L is the cable length, x is the distance of the PD source from point A at the beginning of the cable and v is the propagation velocity of the PD pulse in the cable. With Eq. (12.21), it is possible to calculate the distance x and thus the location of the PD source. In general, the measurement is repeated, the measuring system then being connected to the cable end B. The mean of the distance values x_A and x_B provides a more accurate value x for the location of the PD source. The propagation velocity v in the cable of length L is determined by measuring the transit time of a calibration pulse fed into the cable end B.

This basically simple location method will usually fail if the PD source is near the cable ends A or B, or if there are multiple PD sources in the cable. For very long cables, the reflected pulse is no longer measurable due to the strong signal attenuation. In these cases, especially for on-site testing, the following two methods can often be used successfully. In the first method, the PD test is performed with a sinusoidal voltage of significantly reduced frequency of 0.1 Hz due to the large capacitive load of the coaxial cable. In the second PD measurement method, which is also well-suited for on-site testing, the test is performed with a decaying sinusoidal oscillation, resulting from the sudden discharge of a charged DC cable [100, 101].

For an external 5.6 km long three-phase gas pressure cable system, an improved method for locating PD pulses is described in [102] wherein the synchronous multi-terminal measurement technique is used (see Sect. 12.8.2). During the test, only one phase is connected to high voltage. The pulses from a possible PD source propagate directly to both phase ends, where they are decoupled and evaluated in real time by two ultra-wideband digital PD measuring systems of the type shown in Fig. 12.9. The main part of the publication deals with a new technique to synchronize the two PD measuring systems that are several kilometers apart. For this purpose, a carrier signal is fed into the end of one of the two other cables, which activates the generation of steep trigger pulses at both cable ends. Taking into account the signal propagation time in the cable, the two PD measuring systems thus have exactly the same time base for each recording, i.e. they are synchronized. From the measured time differences of the PD pulses arriving at the PD measuring systems, the PD source can be located. The location of multiple PD sources and the separation of external disturbances are also possible. According to the results of the investigation with artificial defects at both cable ends, the uncertainty of locating the PD source is given as 0.2% of the cable length.

Defects in new cables are usually detected in PD tests by the manufacturer before delivery. Problems with partial discharges therefore occur mainly during cable assembly on site, especially in *cable joints* or *cable terminations*. In particular in cable joints (or cable terminations), internal partial discharges can occur for various reasons. In order to detect or exclude assembly errors—or a previously

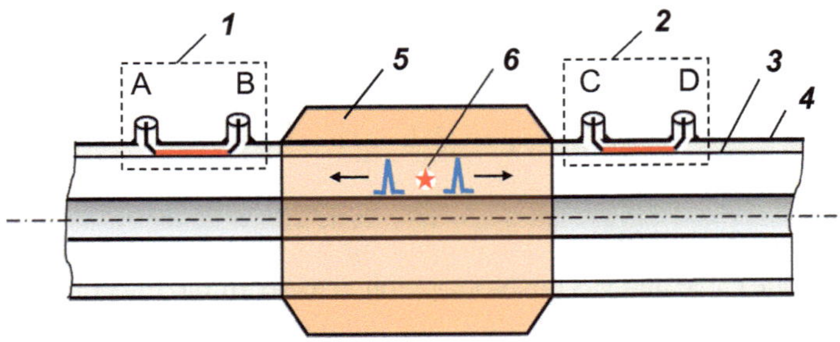

Fig. 12.24 Principle of directional coupler technology for locating partial discharges in a cable joint. *1*, *2* Directional coupler, *3* outer semi-conductive layer, *4* cable sheath, *5* cable joint, *6* PD source in the cable joint

unrecognized defect in the cable—*directional couplers* are installed on both sides of the cable joint [103, 104]. The sensors of the two directional couplers ***1*** and ***2*** are installed between the outer semi-conductive layer ***3*** and the metal cable sheath ***4*** (Fig. 12.24). Each directional coupler has two ports denoted A-B and C-D, which are connected to the PD analyzing system. Alternatively, the sensors can be installed directly in the joint ***5***. Optimal directivity and sensitivity are achieved when the sensors are mounted directly on the cable shield ***4***.

If a PD source is active in the cable system, a TEM wave will propagate to both directional couplers and cause voltage pulses at the sensor ports A to D due to capacitive and inductive interactions. Depending on the direction from which the TEM wave hits the sensor, an addition or subtraction takes place, in the ideal case even a complete extinction of these pulses. If, for example, the PD source ***6*** is located in the cable joint, much higher pulse voltages occur at the ports B and C than at the ports A and D. On the other hand, when the PD source is in the left cable, the pulse voltages at the ports A and C are higher than at the ports B and D. With directional coupler technology, multiple PD sources in the cable system can be distinguished and separated from external interference.

PD pulses may also be decoupled inductively by means of an HF current transformer arranged around the ground line of the cable at its termination, as discussed in several publications. Preliminary studies on several de-energized three-phase cables show that calibration pulses injected into the beginning of a single phase can be inductively coupled out and detected at each of the three ground connections at the other end of the short-circuited cable [105]. However, this measurement technique is very susceptible to noise and therefore not suitable for on-site PD tests. The signal-to-noise ratio is significantly improved by a special arrangement of the ground connections through HF ring transformers [106]. Hereby, the three phases are short-circuited at the measurement station and grounded, which, of course, requires a powerful AC voltage supply. In this PD test, the ground connections are led in pairs through the ferrite cores of three HF current transformers as shown in Fig. 12.25. For example, the left core with the currents $+i_1$ and $-i_2$ detects PD pulses in phases ***1*** and ***2***, but with different polarity. Due to the

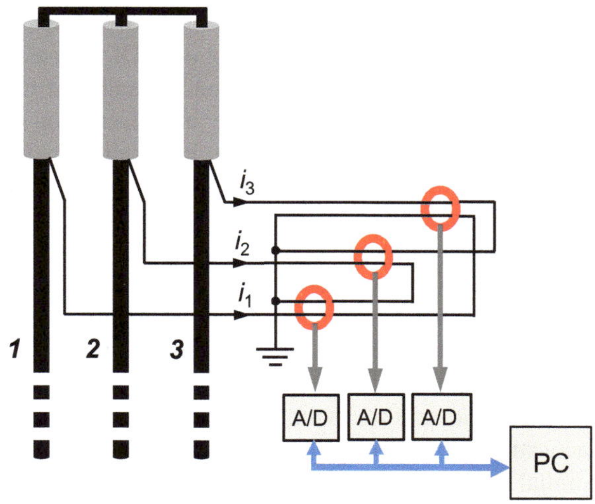

Fig. 12.25 Inductive decoupling of PD pulses from the ground lines of a short-circuited three-conductor cable in conjunction with synchronous multi-channel PD measurement technique

opposite current flow in each ferrite core, common-mode noise will be subtracted so that external disturbances largely cancel each other out. The output signals of the three HF ring transformers are recorded repeatedly using a potential-free three-channel ultra-wideband PD measuring system with A/D converters. Using the example of calibration pulses injected into the cable, the recorded output signals are evaluated by software and displayed in various ways, e.g. in a star diagram (3PARD, see Sect. 12.8.2). The PD measurement principle can also be used in such a way that, instead of the ground connections, the high-voltage conductors are led through the ferrite cores.

The propagation of electromagnetic fields generated by partial discharges in cables and joints is the subject of numerous theoretical and experimental investigations. Due to the high spectral components of the PD pulses in the GHz range, the shielding of cables and joints is not perfect, so that the fields are partially emitted. This is shown by field calculations for three-dimensional model arrangements using different software solutions of Maxwell's field equations. The experimental investigations carried out in this context largely confirm the theoretical results. In [107], the use of small air coils is described to measure the external magnetic fields generated by PD pulses in an artificial void in a short test cable. Another paper deals with small planar wide-band antennas for measuring the UHF field of PD pulses emitted by a defect in cable terminations [108]. In this context, reference is also made to the PD measuring probe in Sect. 12.9.1.3, which is capable of detecting the emitted field of partial discharges in shielded test objects.

12.9.3 Acoustic PD Measurement

In addition to electromagnetic waves, partial discharges also generate ultrasonic waves which propagate in gases, liquids and solids at different velocities. External partial discharges in air, such as on high-voltage lines or fittings, can be easily located with a directional microphone. Of course, the corona source must be in direct line of sight, and its location is facilitated by using the microphone with a laser pointer. If the sound wave is reflected on walls or larger objects in the environment of the corona source, its location may be erroneous.

The acoustic measurement of partial discharges in oil-filled power transformers has gained particular significance. The velocity of undisturbed propagation of acoustic waves in oil is about 1340 m/s. However, the shape and propagation of the acoustic pulses generated by partial discharges in the oil-filled transformer are affected by attenuation and superposition of reflected pulses. When the pulses impinge on the walls of the transformer tank, they also propagate along the metal walls. The ultrasonic waves impinging on the inner tank wall can be detected with *piezoelectric sensors* mounted on the outside of the metal tank. They convert the acoustic signal into an electric signal that is recorded by a digital oscilloscope or digital PD measuring system and evaluated on the PC using special software [16, 109–112]. The main spectral component of the oscillating signals is between 50 and 200 kHz.

The advantage of the acoustic PD measurement method is that electric noise, e.g. caused by corona, has no effect, provided that the electric part of the sensor unit, the supply line and the measuring instrument are sufficiently shielded. Mechanical vibrations of the transformer, e.g. caused by magnetostriction, occur well below the above frequency range. By averaging multiple records of the acoustic signals, noise and disturbances are largely reduced. Acoustic PD measurements are often combined with methods for the detection of decomposition products (see Sect. 12.9.5).

With several ultrasonic sensors mounted on the outside of the transformer tank, it is possible to locate a PD source *1* from the propagation time differences of the detected acoustic pulses (Fig. 12.26a). However, the sensors *2* receive not only the directly incoming acoustic wave, but also the components that are reflected from the tank wall to the sensor or that propagate along the wall. Due to the

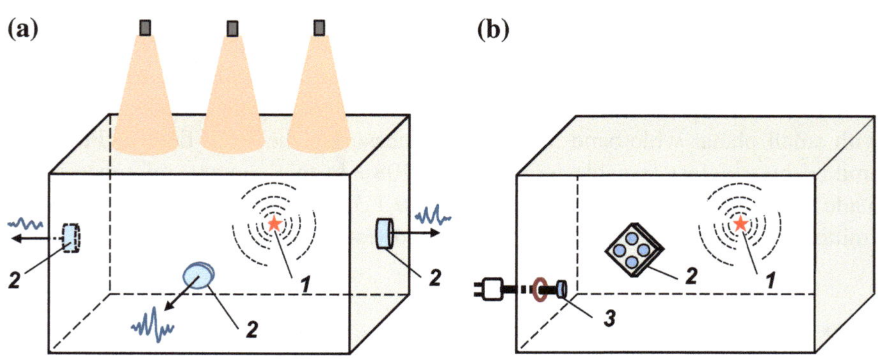

Fig. 12.26 Acoustic detection of partial discharges in oil-insulated transformers (principle). **a** Locating a PD source *1* in the oil tank with three ultrasonic sensors *2*; **b** locating a PD source *1* in the oil tank with an acoustic sensor array *2* and an UHF sensor *3*

superimpositions, the beginning of the acoustic pulse and thus the location of the PD source cannot be determined exactly. By changing the position of the individual sensors, the measurement conditions can be improved if necessary. Specially designed ultrasonic sensors can also be positioned within the oil tank for PD measurements. In this position, the sensors commonly referred to as hydrophones have the advantage of higher sensitivity, but can only be used in areas of low field strength.

The accuracy of the acoustic localization of PD sources in large power transformers can be significantly improved by combination with the UHF method [113]. The high-frequency PD signal rapidly propagating in the oil is measured with a UHF sensor, which is inserted into the oil tank through a drain valve (Fig. 12.26b). The UHF signal serves as a trigger signal for the recorder channels for measuring the acoustic pulses. It thus represents the exact zero point for the evaluation of the acoustic transit time measurements. Further, the suitability of an *acoustic sensor array* ***2*** for locating a PD source ***1*** is investigated in [114]. The array attached to the outer wall of the tank consists of two orthogonal sensor pairs. The advantage is that the position of the sensor array can be changed more easily than that of four individual acoustic sensors. The results of the study show that in combination with the UHF sensor ***3*** even multiple PD sources can be localized very precisely at the same time.

In [115], the prototype of a combined sensor for acoustic and UHF measurements of partial discharges as well as for transformer vibrations due to magnetostriction is presented. The sensor can be easily inserted into the oil tank of the laboratory test setup through a drain valve. As an alternative to the trigger signal derived from the UHF signal, the electrical PD signal can be coupled out with a coupling capacitor or HF current transformer [116]. The different transit times of the electric and acoustic signals from the PD source in oil to the measuring system as well as the influencing parameters were specifically investigated in order to obtain a reliable trigger signal, which is indispensable for locating the PD source.

With ultrasonic measurements, as well as with UHF measurements, there is no direct correlation between the measured quantity and the apparent charge of the PD pulses. In [117], two methods for the generation of defined ultrasonic waves are descried. In one method, a small rod-plane gap is used as an artificial PD source to calibrate ultrasonic sensors when installed. The miniaturized calibration source is fixed at the end of a coaxial cable that can be immersed from above at different depths into the oil-insulated transformer tank. Depending on the voltage applied to the cable, partial discharges of 10–2000 pC can be generated. The spectrum of the ultrasonic pulses generated by the artificial PD source is comparable to that of partial discharges in oil, i.e. the main part lies in the frequency range between 50 and 200 kHz. The charge of the thus generated PD pulses was measured according to IEC 60270 with a separate coupling capacitor and coupling unit or by utilizing the capacitance of the coaxial cable to the rod-plane arrangement.

The second method described in [117] deals with the injection of defined mechanical pulses from the outside into the metal wall of a tank via an acoustic transmitter. With the electronically controlled pulse calibrator, the tank wall is made to vibrate, so that acoustic waves propagate along the tank wall. The vibration is received by an acoustic sensor mounted at some distance on the tank wall. From the excited and the received vibrations, the location-dependent *acoustic transfer function* of the tank wall between the points of injection and measurement is obtained, the knowledge of which is useful for theoretical investigations.

The acoustic PD measurement technique in model arrangements of transformers and bushings with epoxy resin insulation is basically studied in [118, 119]. The main part of the acoustic PD spectrum lies in the frequency range from 100 kHz to a maximum of 1 MHz. The attenuation of the acoustic PD signal in the epoxy resin insulation is between 62 and 140 dB/m, depending on the direction of propagation and type of the test object. With the acoustic sensor, therefore, only very strong partial discharges can be detected within 0.5 m. Acoustic disturbances, e.g. by corona, are eliminated quite well by averaging multiple records.

Other investigations are concerned with the influence of partial discharges on a polarized light beam which is coupled by a He–Ne laser via a polarizer into an optical fiber. The pressure wave of a partial discharge causes a temporal change in the refractive index of the optical fiber and thus a phase modulation of the light beam, which is then converted into an electrical signal and evaluated. In [120], a model with a point-plane electrode in a cylindrical glass chamber filled with SF6 is investigated, around which a single mode fiber of only 5 μm thickness is wound in several turns. The pressure wave propagating from the PD source causes phase modulation of the optical beam in the fiber. The detection sensitivity of the experimental set-up intended to simulate part of a GIS corresponds to a PD magnitude of at least 20 pC. The *acousto-optic sensor* has the advantage that it is non-invasive and insensitive to electromagnetic interference and can also be used in areas of high field strengths.

Another acousto-optic sensor described in [121] consists of a tightly wound optical fiber coil embedded in silicone elastomer. During the test, the sensor is attached to one side of a cubic sample of transparent silicone elastomer having a cast-in point-plate electrode. The thus generated partial discharges cause acoustic pulses which propagate to the acousto-optic sensor and modulate the phase of the light beam in the optical fiber. The investigation results show that the sensor sensitivity is 6 pC for AC voltage and better than 1 pC for DC voltage. The acousto-optic sensor system is intended for use in high-voltage cable accessories. However, the measurements at DC and AC voltages caused partly, especially at positive DC voltage, anomalous display of the acousto-optic sensor system.

12.9.4 Optical PD Measurement

In addition to electric and acoustic measurement methods, partial discharges can also be detected by optical methods. The spectrum of the light emitted by partial discharges due to ionization depends on the surrounding medium, ranging from the infrared to the ultraviolet range. External partial discharges (corona) on high-voltage power lines, fittings, etc. can be easily located in the dark with the naked eye or residual light amplifiers. In daylight, special UV camera systems are successfully used, which capture an image or sequence of external discharges together with the high-voltage circuit. By using optical filters in the wavelength range λ = 240–280 nm, the ambient daylight is sufficiently well suppressed [122, 123].

The detection of light emission from internal partial discharges is the goal of a series of fundamental investigations. Several papers deal mainly with two optical measurement methods. Of course, both methods require that the PD source be surrounded by a more or less translucent insulation, e.g. insulating gas, transformer oil, silicone, etc. and that no external light disturbs the measurements. Advantages of these methods are that the optical sensor practically does not disturb the electric field, is immune to electromagnetic interference and causes only limited additional partial discharges.

In preliminary studies, a conventional optical fiber is used as a sensor for the partial discharges of a point-plane electrode arrangement ***1*** in a test vessel ***2*** filled with air or oil (Fig. 12.27). For maximum sensitivity, one end of the optical fiber ***3*** is directed at a distance of less than 1 cm directly onto the point electrode. A greater distance of up to 5 cm is possible when placing a lens in front of the optical fiber, thereby increasing the angle of incidence. The other end of the optical fiber is connected to a photodiode ***4*** which, in conjunction with a current-voltage transducer, converts the optical signal into an electrical voltage for further evaluation. The optical PD measuring system in Fig. 12.27 has a sensitivity of 35 pC. The displayed pulse height is proportional to the apparent charge [124].

In the second method of measurement, optical fibers with a fluorescent additive are used to measure the luminous appearance of partial discharges. The particular characteristic of the *fluorescent optical fiber* ***1*** (FOF) is to absorb the light rays ***3*** emanating from a PD source ***2*** and to emit photons ***4*** (Fig. 12.28a). By multiple reflections at the boundary layer between the core and the coating of the fiber, the photons propagate along the FOF. At its end, a conventional optical fiber is connected, which forwards the light to a photomultiplier for conversion into an

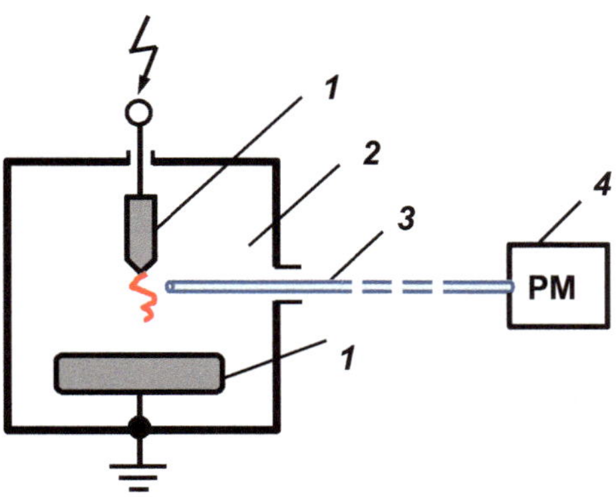

Fig. 12.27 Experimental setup for PD measurement with optical fiber. ***1*** Point-plane electrode arrangement, ***2*** test cell filled with air or oil, ***3*** optical fiber, ***4*** photo diode with current-voltage transducer

electrical signal. The angle of incidence at which the excitation beams of the partial discharge strike the fluorescent fiber has only a small influence on the emitted photocurrent.

In the experimental setup in Fig. 12.28b, a fluorescent optical fiber ***1*** (FOF) is positioned in the continuous groove of a two-part sample ***2*** of transformer board. The sample is placed between two plate electrodes ***3*** in an oil-filled test container ***4*** [125]. The partial discharges occurring in the groove activate the optical processes in the FOF with their excitation radiation. These processes take place very quickly, including the subsequent conversion into electrical pulses with a photomultiplier. The rise time of these pulses is in the range of only a few nanoseconds, whereas the PD pulses measured with a conventional PD measuring system according to IEC 60270 have strong oscillations for more than 1 μs. PD measurements with and without FOF show that the number of PD pulses is only slightly increased by the presence of the FOF. Due to the positive results of the investigation, the authors recommend the future use of the optical measuring system for PD measurement in transformers.

In another experimental setup, the FOF is wound in multiple turns around a darkened cylindrical sample of transparent silicone in which a point-plane electrode arrangement generates partial discharges [126]. A conventional optical fiber connects the FOF end via an optoelectronic converter to the channel of a digital PD measuring system according to Sect. 12.5.2.2. In addition, the PD pulses are conventionally coupled out via a coupling capacitor and measuring impedance and passed directly into a second channel of the PD measuring system. Comparison of the PRPD patterns obtained with optical and electrical PD measurement methods (see Sect. 12.8.1) shows good agreement, although small differences are visible. Due to the overall positive experimental results, the use of fluorescent optical fibers for PD monitoring in cable terminations is suggested. Further details on the same topic are also presented in [127].

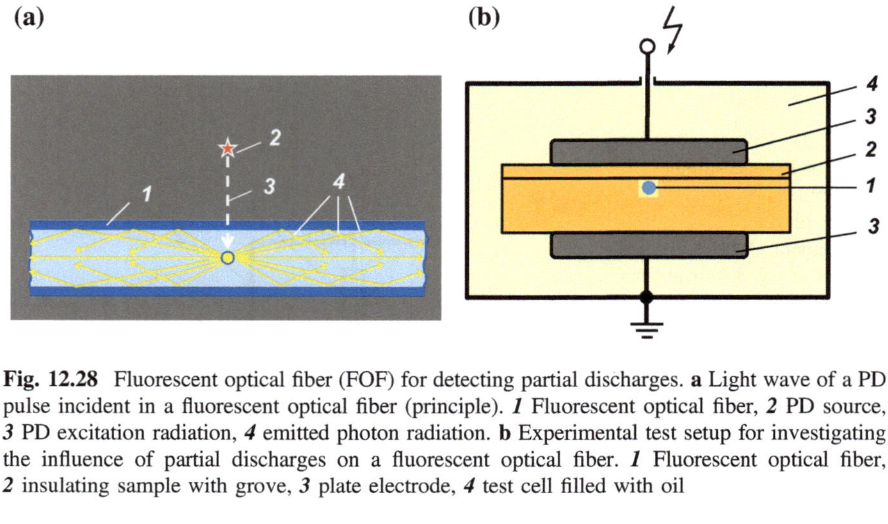

Fig. 12.28 Fluorescent optical fiber (FOF) for detecting partial discharges. **a** Light wave of a PD pulse incident in a fluorescent optical fiber (principle). ***1*** Fluorescent optical fiber, ***2*** PD source, ***3*** PD excitation radiation, ***4*** emitted photon radiation. **b** Experimental test setup for investigating the influence of partial discharges on a fluorescent optical fiber. ***1*** Fluorescent optical fiber, ***2*** insulating sample with grove, ***3*** plate electrode, ***4*** test cell filled with oil

12.9.5 Chemical PD Detection Methods

Partial discharges cause by various influences a gradual decomposition of the surrounding insulating material. Here, a distinction can be made between the direct effects on the insulating material and secondary effects. UV radiation and temperature rise due to partial discharges act directly on the insulating material. Secondary effects of partial discharges are caused by decomposition products such as ozone, hydrogen, nitrogen oxides and—in the presence of moisture—nitric acid, which then attack the surrounding insulating material. For example, the *Dissolved Gas Analysis* (*DGA*) has long been a proven and reliable method of detecting the condition of oil-filled transformers after prolonged exposure to temperature and partial discharges. For this purpose, oil samples are taken at regular intervals and analyzed with a *gas chromatograph* with regard to the individual gas fractions. Based on decades of experience, typical decomposition gases in oil-paper insulations can be detected after PD exposure. According to IEC 60599, the concentration ratios of certain gases such as CH_4/H_2 and C_2H_4/C_2H_6 clearly show that partial discharges have occurred in the oil-insulated system [128]. Permanent online monitoring is increasingly being carried out with gas sensors mounted directly on the transformer. Doing this, the total pressure of some typical gases dissolved in the oil is continuously recorded and displayed. The connection can be clearly shown by simultaneous measurements of the partial discharges and the gas pressure in the transformer [129–131].

Partial discharges in GIS also lead to a decomposition of the insulating gas, usually SF_6. In [132], a novel gas sensor based on the *single-walled carbon nanotube* (*SWNT*) is investigated. With this sensor, the decomposition products of a gas mixture of SF_6, O_2 and H_2O in a GIS-like test chamber can be detected. The gas molecules of the decomposition products are deposited on the gas sensor and cause an integral increase in the electrical conductivity, which is digitally recorded and evaluated in online operation. Synchronous measurements with the novel gas sensor system, a UHF measuring system and a conventional PD measuring system at partial discharges of 10 and 20 pC show consistent results with regard to the PD behavior of the test sample. Advantages of the novel gas sensor are its small size, fast response and high sensitivity compared to other gas sensors. Also advantageous are the simple evaluation and the very low data volume of the measurements.

12.10 Partial Discharge Measurement at DC Voltage

The occurrence and measurement of partial discharges at DC voltage in HVDC systems, X-ray equipment, cathode ray tubes, accelerator systems and radar systems has long been a well-known problem. The growing demand for power transmission systems operating at very high DC voltages has led to an increasing interest in the behavior of insulation under DC stress. In principle, partial discharges occur at DC voltage as well as in the case of AC voltage when the insulation fails due to locally increased field strength. Partial discharges within equipment with solid, liquid or gaseous insulation are also particularly dangerous as they may shorten the service time [133–135].

Figure 12.29 shows a simple equivalent circuit diagram for internal partial discharges at DC voltage. The defect in the insulation is characterized by R_1, C_1 and the spark gap F. The processes in the generation of partial discharges at DC voltage differ from those at AC voltage, since the capacitances in the circuit have little influence. After firing of the spark gap F and the collapse of the voltage u_1, the recharging process in the defect via the high-ohmic resistance R_2 of the insulation takes place only very slowly. Compared to the charging process with AC voltage, therefore, so much more time passes until a new partial discharge occurs at the defect. Due to various influences, such as the formation of space charges in the insulation or the instability of the applied DC voltage, the magnitudes and in particular the time intervals between successive PD pulses are subject to greater scattering. This scattering is also noticeable in the inception and extinction voltages of partial discharges.

The characteristic measurement quantity of PD pulses at DC voltage is again the apparent charge. Further quantities are the pulse shape and the spectrum, the time between successive pulses and the pulse frequency in a defined interval. The diagnosis is not easy, since a clear relationship of the PD pulses to the DC voltage, as it is the phase of the AC voltage, does not exist. Due to the random nature of PD pulses, a distinction between partial discharges and interference pulses is made more difficult. In some cases, the ripple of a rectified DC test voltage can establish a phase dependent correlation with the PD pulses. Some information and suggestions for standardized PD measurements at DC voltage can be found in IEC 60270 [6]. In addition to the measurement of the individual pulse charges, in particular the *accumulated apparent charge* is of interest. It represents the sum of the apparent charges of all individual PD pulses above a specified threshold level within a predetermined time interval (Fig. 12.30). An additional criterion for passing the PD test of a test object at DC voltage may be, for example, that the number of PD pulses during a given time is less than a specified total.

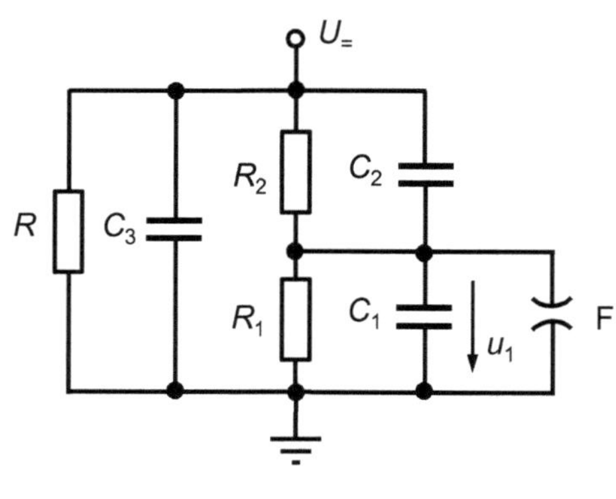

Fig. 12.29 Simple equivalent circuit diagram for internal partial discharges of an insulating arrangement at DC voltage

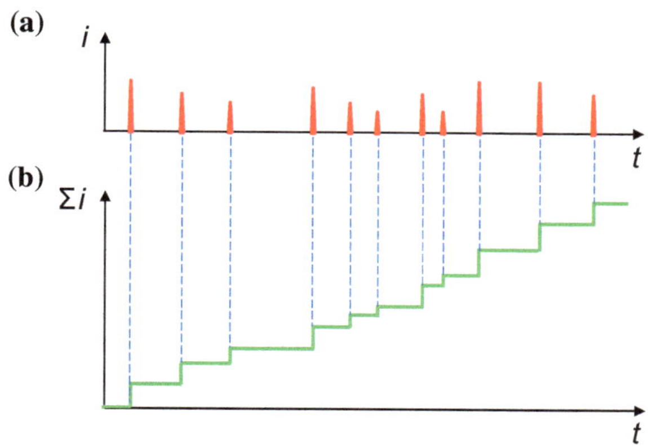

Fig. 12.30 Measurement of partial discharges at DC voltage (principle methods). **a** PD pulses versus time; **b** accumulated apparent charge of the individual PD pulses

Basically, the digital measuring instruments used for PD tests with AC voltage are also suitable for DC voltage tests if they can display individual PD pulses and the accumulated charge. The same applies to the different measurement and evaluation methods, e.g. the conventional test method with coupling capacitor according to IEC 60270, the UHF method or the acoustic and optical diagnosis. Particularly advantageous are wide-band digital PD measuring instruments that record the PD pulses as true to the original as possible and enable comprehensive computer-assisted evaluation. The two synchronous PD multichannel measurement methods 3PARD and 3CFRD (see Sect. 12.8.2) are also applicable for the detection and separation of partial discharges in PD tests with DC voltage [136]. Particularly advantageous here is the ability to distinguish internal PD pulses of the test object from external disturbances. These can be, for example, corona from the DC power supply (e.g. Greinacher cascade) or discharges of metal pieces that are charged by influence. A long recording time of one or several hours is a prerequisite for evaluating a sufficiently large number of partial discharges.

The problem of distinguishing internal partial discharges from external interference pulses can also be solved by means of a bridge circuit with differential signal evaluation [137]. The computer-controlled bridge in conjunction with a digital PD measuring system can be used in both AC and DC voltage tests. The basic measurement arrangement corresponds in principle to the conventional analog bridge circuit in Fig. 12.11c, but no bridge balancing is performed. Instead, the difference signal and the individual signals of the two bridge arms are digitally recorded and multiplied by weighting factors in order to achieve a *virtual bridge balance* by calculation. The weighting factors for the bridge arms result from the calibration with known pulse charges. The measurement of the PD pulses and determination of the charge is done either with wide-band devices in the time domain or with narrow-band devices in the frequency domain. The common mode rejection for external noise is given as 25 dB. The paper describes application examples for the computer-controlled measuring bridge with differential signal evaluation in order to measure PD pulses at AC and DC test voltage. The results show that partial discharges in a test object at 400 kV DC voltage can be clearly distinguished from corona discharges of the DC voltage supply.

12.11 Partial Discharge Measurement at Impulse Voltage

Partial discharges generated by impulse voltages are attracting increasing interest. Here, two areas of interest can be distinguished. On the one hand, impulse voltages are the better choice for certain test tasks than AC voltages. On the other hand, high-voltage equipment is constantly exposed to impulse voltages during normal operation. For example, due to the large cable capacitance, PD testing on AC cables requires a powerful AC voltage supply at mains frequency, which in particular is a transport problem for on-site testing. PD tests to be performed at AC voltage with reduced frequency down to 0.1 Hz or at impulse voltage are under discussion, with the impulse voltage generator having the advantage that it can be transported in parts and assembled on site for the PD test. The 250/2500 switching impulse voltage has a rise time which is approximately 20 times smaller than that of the 50 Hz AC voltage, the upper 3 dB limit frequency being about 50 Hz (see Sect. 8.1.4). Comparative PD tests provide a factor of about 2 by which the switching impulse voltage must be greater than the AC test voltage to detect partial discharges of the same defect (Ref. [5] of Chap. 1). It should be noted here that partial discharges can occur first on the impulse back, i.e. at a lower voltage than the peak value. This is due to the statistical time delay in the formation of the electron avalanche required for the partial breakdown. Examples of devices whose normal operation is characterized by permanent impulse stress and thus the risk of partial discharges are rotating machines and their converters, transformers, cables etc., as well as devices in power electronics, especially in converters for wind turbines and photovoltaic systems.

Information on electrical PD measurements at impulse voltages is given in the *Technical Specification* IEC TS 61934 [138]. In the PD test with impulse voltages, the basic measurement problem is that the spectrum of the test voltage at higher frequencies may be close to the spectrum of the PD impulses coupled out by the coupling capacitor. A simple high-pass filter, as is common in AC tests, is therefore not sufficient to separate the small PD signal components from the frequency components of the test voltage. Therefore, an active high-pass filter of higher order is required. If the test generator is a conventional impulse voltage generator with open spark gaps, strong electromagnetic interference is generated with spectral components in the upper frequency range. The usable frequency range for PD measurements is then above 100 MHz. Extremely wide-band PD measuring instruments with direct digital data acquisition are preferred (see Sect. 12.5.2.2).

Orienting investigations deal with a method in which a PD measuring system with an acoustic sensor (see Sect. 12.9.3) detects and evaluates the ultrasonic waves originating from partial discharges that are generated by 1.2/50 impulse voltages of negative polarity [139]. The partial discharge source is a needle-plane electrode arrangement in oil, with a 1.6 mm thick pressboard insulation placed on the plane electrode. The emitted ultrasonic signal propagates in the test chamber at a speed of about 1400 m/s in oil and hits on the acoustic sensor mounted outside the chamber wall only after several 100 μs. After this time delay, the electromagnetic interference caused by the firing of the impulse generator has already subsided. The measurement method is proposed as suitable for PD detection and diagnosis in impulse voltage tests of power transformers.

References

1. Temmen, K., Plath, R. (eds.): Teilentladungen in hochbeanspruchten elektrischen Isolierungen. VDE-Verlag (in preparation)
2. König, D., Rao, Y.N. (eds.): Partial Discharges in Electrical Power Apparatus. VDE-Verlag GMBH, Berlin Offenbach (1993)
3. Kreuger, F.H.: Discharge Detection in High Voltage Equipment. Heywood, London (1964)
4. Gemant, A., Philippoff, W.: Die Funkenstrecke mit Vorkondensator. Z. Tech. Phys. **13**, 425–430 (1932)
5. IEC 60270:2000 + AMD1:2015 CSV: High-Voltage Test Techniques—Partial Discharge Measurements (2015) [German edition: DIN EN 60270 (VDE 0434): Hochspannungs-Prüftechnik, Teilentladungsmessungen (2016)]
6. IEC 60270 Amendment A1: High-Voltage Test Techniques—Partial Discharge Measurements (2015)
7. Kind, D.: Grundlagen der Messeinrichtungen für Korona-Isolationsprüfungen. ETZ-A **84**, 781–787 (1963)
8. Beyer, M.: Möglichkeiten und Grenzen der Teilentladungsmessungen und -ortung, Grundlagen und Messeinrichtungen. etz-a **99**, 96–99, 128–131 (1978)
9. Reid, A.J., Judd, M.D.: High bandwidth measurement of partial discharge pulses in SF_6. In: Proceedings of 14th ISH, Beijing, Paper G-012 (2005)
10. Ohtsuka, S., Fukuda, K., Sogabe, A.: Measurement of PD current waveform in SF_6 gas with a super high frequency wide band measurement system. In: Proceedings of 17th ISH, Hannover, Paper D-074 (2011)
11. Schon, K.: Konzept der Impulsladungsmessung bei Teilentladungsprüfungen. etzArchiv **8**, 319–324 (1986)
12. Schwab, A., Zentner, R.: Der Übergang von der impulsförmigen in die impulslose Koronaentladung. ETZ-A **89**, 402–407 (1968)
13. Küpfmüller, K.: Die Systemtheorie der elektrischen Nachrichtenübertragung. S. Hirzel Verlag, Stuttgart (1974)
14. König, D.: Erfassung von Teilentladungen in Hohlräumen von Epoxydharzplatten zur Beurteilung des Alterungsverhaltens bei Wechselspannung. Thesis, TU Braunschweig (1967)
15. Szaloky, G.: Schmal- und breitbandige Teilentladungsmessungen. Bull. ASE/UCS **76**, 1144–1148 (1985)
16. Kachler, A.J., Nieschwietz, H.: Broad and narrow band pd measurements on power transformers. In: Proceedings of 5th ISH, Braunschweig, Paper 41.09 (1987)
17. Facklam, Th, Pfeiffer, W.: Teilentladungsprüfung an Bauelementen der Niederspan nungstechnik. etz **109**, 440–447 (1988)

18. Zaengl, W.S., Osvath, P., Weber, H.J.: Correlation between the bandwidth of PD detectors and its inherent integration errors. In: IEEE International Symposium on Electrical Insulation, Washington D.C. (1986)
19. CISPRE 16-1: Specifications for Radio Disturbance and Immunity Measuring Apparatus and Methods—Part 1: Radio Disturbance and Immunity Measuring Apparatus (1993)
20. Debinski, E.M., Douglas, J.L.: Calibration and comparison of partial-discharge and radio-interference measuring circuits. IEE Proc. **115**, 1332–1340 (1968)
21. Harrold, R.T., Dakin, T.W.: The relationship between the picocoulomb and microvolt for corona measurements on HV transformers and other apparatus. IEEE Trans. PAS **91**, 187–198 (1972)
22. Vaillancourt, G.H., Dechamplain, A., Malewski, R.A.: Simultaneous measurement of partial discharge and radio-interference voltage. IEEE Trans. IM **31**, 49–52 (1982)
23. Lemke, E.: Ein neues Verfahren zur breitbandigen Messung von Teilentladungen. ELEKTRIE **23**, 468–469 (1969)
24. Hartje, M.: Primary results with a partial discharge (pd) computer aided measuring system on power transformers. In: Proceedings of 5th ISH, Braunschweig, Paper 41.10 (1987)
25. Borsi, H., Hartje, M.: A new system for computer aided automation of different commercially available partial discharge (PD) detectors. In: Proceedings of 6th ISH, New Orleans, Paper 22.18 (1989)
26. Kurrat, M., Peier, D.: Wideband measurement of partial discharges for fundamental diagnostics. In: Proceedings of 7th ISH, Dresden, Paper 72.02 (1991)
27. Buchalla, H., Koch, H., Pfeiffer, W., Stietzel, U.: A new adaptive partial discharge measuring system using digital signal processing. In: Proceedings of 8th ISH, Yokohama, Paper 62.09 (1993)
28. Holle, R., Plath, R., Schon, K., Lucas, W.: Typprüfung eines digitalen TE-Messsystems nach IEC 60270. VDE-ETG-Fachveranstaltung, Kassel (2006)
29. Zondervan, J.P., Gulski, E., Brooks, R.: Digital reading procedure for apparent partial discharge magnitudes. In: Proceedings of 10th ISH, Montreal, Paper 3440 (1997)
30. CIGRE D1.33: Guide for Electrical Partial Discharge Measurements in Compliance with IEC 60270. ELECTRA Technical Brochure 366, vol. 60, p. 241 (2008)
31. Werle, P., Akbari, A., Gockenbach, E., Borsi, H.: An enhanced system for partial discharge diagnostic on power transformers. In: Proceedings of 13th ISH, Delft, Paper 430 (2003)
32. Schon, K., Valentini, H.-D.: Programmable impulse charge generator for calibrating PD instruments. In: Proceedings of 10th ISH, Montreal, 33–36 (1997)
33. Brzostek, E., Schon, K.: Rechnergesteuerter Impulsladungsgenerator zur Kalibrierung von Teilentladungsmessgeräten. Report PTB-E-79, Wirtschaftsverlag NW, Braunschweig (2002)
34. Hu, Y., Chiampi, M., Crotti, G., Sardi, A.: Development of a set-up for the evaluation of advanced partial discharge measuring instruments. In: Proceedings of 16th ISH, Johannesburg, Paper D-13 (2009)
35. Seifert, S., Kessler, O., Plath, R., Emanuel, H., Boschert, V.: Influences of parasitic effects on online calibration of partial discharge measurements. In: Proceedings of 16th ISH, Johannesburg, Paper D-5 (2009)
36. Crotti, G., D'Emilio, S., Giorgi, P.A., Saracco, O.: Systems for the calibration of partial discharge calibrators. In: Proceedings of 9th ISH, Graz, 4511-1–4511-3 (1995)
37. Lucas, W., Schon, K., Lemke, E., Elze, H.: Comparison of two techniques for calibrating PD calibrators. In: Proceedings of 10th ISH, Montreal, 67–70 (1997)
38. Cherbaucich, C., Rizzi, G., Gobbo, R., Pesavento, G., Scroccaro, A.: Evaluation of the characteristics of calibrators and PD measuring systems according to IEC 60270. In: Proceedings of 12th ISH, Bangalore, Paper 6-71 (2001)
39. Schon, K., Lucas, W., et al.: Intercomparison on PD calibrators and PD instruments. In: Proceedings of 11th ISH, London, 1.5.S1–1.6.S1 (1999)
40. Vaillancourt, G., Malewski, R.: Digital acquisition and processing of partial discharges during acceptance test of HV transformers. In: IEEE/PES Winter Meeting, New York, Paper 88WM 049-9 (1988)

41. Fruth, B., Niemeyer, L., Hässig, M., Fuhr, J., Dunz, Th.: Phase resolved partial discharge measurements and computer aided partial discharge analysis performed on different high voltage apparatus. In: Proceedings of 6th ISH, New Orleans, Paper 15.03 (1989)
42. Ward, B.H.: Digital Techniques for Partial Discharge Measurements. 91 SM 355-8 PWRD, with comprehensive reference list (1991)
43. Gulski, E.: Digital analysis of partial discharges. IEEE Trans. DEI **2**, 822–837 (1995)
44. Florkowska, B., Zydron, P.: Interpretation of partial discharge patterns for insulation diagnostic aims. In: Proceedings of 10th ISH, Montreal, Paper 3436 (1997)
45. Gross, D.W., Söller, M.: Partial discharge diagnosis of large rotating machines. In: Proceedings of 14th ISH, Beijing, Paper G-044 (2005)
46. Wang, Y., Li, Y.M., Qiu, Y.: Application of wavelet analysis to the detection of transformer winding deformation. In: Proceedings of 10th ISH, Montreal, Paper 3271 (1997)
47. Zhou, C., Ma, X., Kemp, I.J.: Optimal algorithm for wavelet-based denoising in PD measurement. In: Proceedings of 13th ISH, Delft, Paper 168 (2003)
48. Quak, B., Gulski, E., Smit, J.J., Seitz, P.P.: Numeric data processing to enable PD site (2005) location by time-domain reflectometry in distribution power cable system. In: Proceedings of 14th ISH, Beijing, Paper G-08 (2005)
49. Kubota, H., Furihata, H., Inui, A., Kawaguchi, Y.: High sensitive distinction of partial discharge signal from noisy signal by wavelet transform. In: Proceedings of 14th ISH, Beijing, Paper G-097 (2005)
50. Zhang, H., Blackburn, T.R., Phung, B.T., Shuying Kang, S., Naderi, M.S., Nam, H.O.: Comparison of wavelet transform de-noising methods for partial discharge detection in power cables. In: Proceedings of 14th ISH, Beijing, Paper G-103 (2005)
51. Lapp, A., Kranz, H.-G.: The influence of the quality of the reference data base on PD pattern recognition results. In: Proceedings of 10th ISH, Montreal, Paper 3065 (1997)
52. Hücker, T.: UHF partial discharge expert system diagnosis. In: Proceedings of 10th ISH, Montreal, Paper 3186 (1997)
53. Florkowska, B., Florkowski, M., Zydron, P.: Influence of testing voltage frequency on partial discharge phase-resolved images. In: Proceedings of 13th ISH, Delft, Paper 309 (2003)
54. Florkowski, M.: Distortion of partial discharge images caused by high voltage harmonics. In: Proceedings of 10th ISH, Montreal, Paper 3080 (1997)
55. Nieschwitz, H., Stein, W.: Teilentladungsmessungen an Hochspannungstransformatoren als Mittel der Qualitätskontrolle. etz-a **97**, 657–663 (1976)
56. Burger, H.P., Gulski, E., Smit, J.J., Brooks, R.: Digital tools for PD analysis in three-phase transformers. In: Proceedings of 10th ISH, Montreal, Paper 3510 (1997)
57. Lortie, R., Aubin, J., Vaillancourt, G.H., Su, Q.: Partial discharge detection on power transformers using a multi-terminal measurement method. In: Proceedings of 10th ISH, Montreal, Paper 3210 (1997)
58. Werle, P., Wasserberg, V., Borsi, H., Gockenbach, E.: A sensor system for partial discharge detection and localisation on dry type transformers. In: Proceedings of 12th ISH, Bangalore, Paper 6-14 (2001)
59. Plath, K.-D., Plath, R., Emanuel, H., Kalkner, W.: Synchrone dreiphasige Teilentladungsmessung an Leistungstransformatoren vor Ort und im Labor. In: ETG-Fachtagung "Diagnostik elektrischer Betriebsmittel", Berlin, Paper O-11 (2002)
60. Obralic, A., Kalkner, W., Plath, R.: Separation and individual evaluation of PD sources in stator winding of rotating machines by synchronous multi-channel measurement. In: Proceedings of 15th ISH, Ljubljana, Paper T10-512 (2007)
61. Johnson Hio Nam, O., Blackburn, T.R., Phung, B.T.: PD characteristics and defect patterns in three phase power cables. In: Proceedings of 16th ISH, Johannesburg, Paper D-17 (2009)
62. Schaper, S., Obralic, A., Kalkner, W., Plath, R.: Synchronous multi-terminal on-site PD measurements on power transformers with an enhanced time differential evaluation method. In: Proceedings of 15th ISH, Ljubljana, Paper T7-535 (2007)

63. Balkon, C., Kalkner, W., Obralic, A., Plath, R., Rethmeier, K.: Potential of multispectral PD measurement for differentiation of interfering impulses and multiple PD sources. In: Proceedings of 16th ISH, Johannesburg, Paper D-30 (2009)
64. IEC TS 62478: High Voltage Test Techniques—Measurement of Partial Discharges by Electromagnetic and Acoustic Methods (2016)
65. Wanninger, G.: Discharge currents of free mowing particles in GIS. In: Proceedings of 10th ISH, Montreal, Paper 3147 (1997)
66. Park, K., et al.: Chaotic behaviour of a free moving particle in the 362 kV gas-insulated system. In: Proceedings of 13th ISH, Delft, Paper 396 (2003)
67. Judd, M.D., Farish, O.: FDTD simulation of UHF signals in GIS. In: Proceedings of 10th ISH, Montreal, Paper 3129 (1997)
68. Hampton, B.F., Meats, R.J.: Diagnostic measurements at UHF in gas insulated substations. IEE Proc. **135**(Pt. C), 137–144 (1988)
69. Kurrer, R., Feser, K.: The application of ultra-high frequency partial discharge measurements to gas-insulated substations. IEEE Trans. PD **13**, 103–106 (1998)
70. Meijer, S., Gulski, E., Rutgers, W.R.: Evaluation of partial discharge measurements in SF_6 gas insulated systems. In: Proceedings of 10th ISH, Montreal, Paper 3507 (1997)
71. Zoetmulder, R.G.A., Meijer, S., Smit, J.J.: Condition based maintenance based on on-line partial discharge measurements of HV switchgear systems. In: Proceedings of 13th ISH, Delft, Paper 794 (2003)
72. Hoeck, S.M., Koch, M., Heindl, M.: Distribution and propagation mechanisms of PD pulses for UHF and traditional electrical measurements. In: Proceedings of 17th ISH, Hannover, Paper D-075 (2011)
73. Suzuki, K., Yoshida, M., Kojima, H., Hayakawa, N., Hanai, M., Okubo, H.: Correlation between UHF electromagnetic waveforms and partial discharge current waveforms in GIS. In: Proceedings of 17th ISH, Hannover, Paper F-045 (2011)
74. Hoshino, T., Nojima, K., Matsumoto, S.: Investigation into frequency characteristics of UHF sensor. In: Proceedings of 12th ISH, Bangalore, Paper 7-20 (2001)
75. Neuhold, S., et al.: Experiences with UHF PD detection in GIS using external capacitive sensors on windows and disk-insulators. In: Proceedings of 15th ISH, Ljubljana, Paper T7-480 (2007)
76. Hoshino, T., et al.: Sensitivity of UHF coupler and loop electrode using UHF method and comparison for detecting a partial discharge in GIS (2). In: Proceedings of 16th ISH, Johannesburg, Paper C-37 (2009)
77. Kim, Y.-H., et al.: A study of UHF sensors embedded into GIS spacer for PD detection. In: Proceedings of 15th ISH, Ljubljana, Paper T10-630 (2007)
78. Liu, W., Huang, Y., Wang, J., Qian, J.: UHF sensing and locating of partial discharges in GIS. In: Proceedings of 10th ISH, Montreal, Paper 3202 (1997)
79. Miyashita, M., et al.: Novel microwave techniques for PD diagnosis in existing gas insulated equipment. In: Proceedings of 16th ISH, Johannesburg, Paper D-48 (2009)
80. Judd, M.D., Farish, O., Coventry, P.F.: UHF couplers for GIS—sensitivity and specification. In: Proceedings of 10th ISH, Montreal, Paper 3130 (1997)
81. Fan, C.-L., Liang, W., Chung, M., Judd, M.: Comparison of UHF partial discharge sensor calibration system. In: Proceedings of 17th ISH, Hannover, Paper D-028 (2011)
82. Gautschi, D., Bertholet, P.: Design and testing of a novel calibration system for UHF sensors for GIS. In: Proceedings of 17th ISH, Hannover, Paper D-045 (2011)
83. Hoshino, T., Koyama, H., Maruyama, S., Hanai, M.: Comparison of sensitivity between UHF method and IEC 60270 standard for on-site calibration in various GIS. In: Proceedings of 15th ISH, Ljubljana, Paper T7-264 (2007)
84. Okabe, S., et al.: A new verification method of the UHF PD detection technique. In: Proceedings of 16th ISH, Johannesburg, Paper D-41 (2009)
85. Troeger, A., Riechert, U.: Influence of different parameters on sensitivity verification for UHF PD measurement. In: Proceedings of 16th ISH, Johannesburg, Paper B-33 (2009)

86. Schichler, U., Reuter, M., Gorablenkow, J.: Partial discharge diagnostics on GIS using UHF and acoustic method. In: Proceedings of 16th ISH, Johannesburg, Paper D-9 (2009)
87. Twittmann, J., et al.: Measuring techniques and setup for synchronous acoustic and electric multi-site PD analysis. In: Proceedings of 16th ISH, Johannesburg, Paper D-57 (2009)
88. Siegel, M., Kornhuber, S., Beltle, M., Müller, A., Tenbohlen, S.: Monitoring von Teilentladungen in Leistungstransformatoren. In: Proceedings of Hochspannungs Symposium, Stuttgart
89. Judd, M.D., Cleary, G.P., Meijer, S.: Testing UHF partial discharge detection on a laboratory based power transformer. In: Proceedings of 13th ISH, Delft, Paper 229 (2003)
90. Meijer, S., Agoris, P.D., Smit, J.J.: UHF PD sensitivity check on power transformers. In: Proceedings of 14th ISH, Beijing, Paper F-08 (2005)
91. Rutgers, W.R., Fu, Y.H.: UHF PD detection in a power transformer. In: Proceedings of 10th ISH, Montreal, Paper 3120 (1997)
92. Sinaga, H.H., Phung, B.T., Ao, A.P., Blackburn, T.R.: UHF sensors sensitivity in partial discharge sources in a transformer. In: Proceedings of 17th ISH, Hannover, Paper D-069 (2011)
93. Mirzaei, H.R., et al.: New antenna design for UHF monitoring of power transformers. In: Proceedings of 18th ISH, Seoul, Paper PF-36 (2013)
94. Judd, M.D., Gray, D., Reid, A.J.: Using a step recovery diode to generate repeatable pulses for partial discharge simulation. In: Proceedings of 14th ISH, Beijing, Paper J-39 (2005)
95. Tenbohlen, S., Siegel, M., Beltle, M., Reuter, M.: Suitability of ultra-high frequency partial discharge measurement for quality assurance and testing of power transformers. In: CIGRE SC A2&C4 Joint Colloquium, Zurich (2013)
96. Lemke, E., Elze, H., Weissenberg, W.: Experience in PD diagnosis tests of HV cable terminations in service using the ultra-wide band PD probing. In: Proceedings of 8th ISH, Delft, Paper 404 (2003)
97. Mole, G.: Measurement of the magnitude of internal corona in cables. IEEE Trans. PAS **89**, 204–212 (1970)
98. Lukaschewitch, A., Puff, E.: Messung von Teilentladungen (TE) an langen Kabeln. Z. prakt. Energietechnik **28**, 32–39 (1976)
99. FGH: Zustandsdiagnose von Papiermasse-Kabelanlagen in Verteilungsnetzen. Technischer Bericht 300 (2006)
100. Muhr, M., Sumereder, C., Woschitz, R.: The use of the 0,1 Hz cable testing method as substitution to 50 Hz measurement and the application for PD measuring and cable fault location. In: Proceedings of 12th ISH, Bangalore, Paper 6-38 (2001)
101. Boltze, M., Lemke, E., Strehl, T.: Comparative PD measurements under damped and continuous AC energising voltages with respect to preventive on-line PD diagnosis tests of medium voltage power cables. In: Proceedings of 13th ISH, Delft, Paper 406 (2003)
102. Kumm, T., Rethmeier, K., Kalkner, W., Zinburg, E.: A new approach of synchronous PD measurement at both ends of a H.V. gas pressure cable system. In: Proceedings of 14th ISH, Beijing, Paper F-11 (2005)
103. Pommerenke, D., Strehl, T., Kalkner, W.: Directional coupler sensor for partial discharge recognition on high voltage cable systems. In: Proceedings of 10. ISH, Montreal (1997), Paper 3447
104. Heinrich, R., Kalkner, W., Plath, R., Obst, D.: On-site application of directional coupler sensors for sensitive PD measurement and location on a 380 kV cable line. In: Proceedings of 12th ISH, Bangalore, Paper 6-34 (2001)
105. Gulski, E., van Vliet, M.G.A., Smit, J.J., de Vries, F.: Practical experiences on on-line PD detection on distribution power cables. In: Proceedings of 14th ISH, Beijing, Paper F-24 (2005)
106. Plath, R., Vaterrodt, K., Habel, M.: Symmetric 3-channel inductive PD detection with optimized SNR for extruded power cables. In: Proceedings of 17th ISH, Hannover, Paper D-016 (2011)

107. Ghani, A.B.A., Chacrabarty, C.K.: Investigation of the transient magnetic field components propagation characteristics due to partial discharge in XLPE cable. In: Proceedings of 16th ISH, Johannesburg, Paper B-18 (2009)
108. Denissov, D., Köhler, W., Tenbohlen, S., Grund, R., Klein, T.: Optimization of UHF sensor geometry for on-line partial discharge detection in cable terminations. In: Proceedings of 16th ISH, Johannesburg, Paper D-24 (2009)
109. Howells, E., Norton, E.: Detection of partial discharges in transformers using acoustic emission techniques. IEEE Trans. PAS **97**, 1538–1549 (1978)
110. Bengtsson, T., Kols, H., Jönsson, B.: Transformer PD diagnosis using acoustic emission techniques. In: Proceedings of 10th ISH, Montreal, Paper 3057 (1997)
111. Phung, B.T., Liu, Z., Blackburn, T.R., James, R.E.: Signal characteristics of partial discharge acoustic emissions. In: Proceedings of 12th ISH, Bangalore, Paper 6-59 (2001)
112. Markalous, S.M., Grossmann, E., Feser, K.: Online acoustic PD-measurements of oil/paper-insulated transformers—methods and results. In: Proceedings of 13th ISH, Delft, Paper 164 (2003)
113. Markalous, S.M., Tenbohlen, S., Feser, K.: Improvement of acoustic detection and localization accuracy by sensitive electromagnetic PD measurements under oil in the UHF range. In: Proceedings of 14th ISH, Beijing, Paper G-039 (2005)
114. Siegel, M., Tenbohlen, S., Kornhuber, S.: Neue Methoden zur Ortung mehrerer TE-Quellen mittels akustischem Sensorarray. VDE ETG Fachtagung, Fulda (2012)
115. Beltle, M., Siegel, M., Tenbohlen, S.: Investigations of in-oil methods for PD detection and vibration measurement. In: Proceedings of 18th ISH, Seoul, Paper OD5-04 (2013)
116. Broniecki, U., et al.: Localization of partial discharges in power transformers by combined acoustic and electric measurements. In: Proceedings of 17th ISH, Hannover, Paper D-059 (2011)
117. Grossmann, E., Feser, K.: New calibrators for acoustic PD-measurements. In: Proceedings of 12th ISH, Bangalore, Paper 6-61 (2001)
118. Beyer, M., Borsi, H., Cachay, O.: Some aspects about basic investigations of acoustic partial discharge (PD) detection and location in cast epoxy resin coils. In: Proceedings of 6th ISH, New Orleans, Paper 22.20 (1989)
119. Hartje, M., Krump, R., Lange, H., Schulenberg, J.: Acoustic emission qualities of partial discharges (PD) in bushing materials. In: Proceedings of 15th ISH, Ljubljana, Paper TC-638 (2007)
120. Zargari, A., Blackburn, T.R.: Partial discharge detection in SF_6 GIS using optical fibre techniques. In: Proceedings of 10th ISH, Montreal, Paper 3518 (1997)
121. Rohwetter, P., Habel, W.R., Heidmann, G., Pepper, D.: Fibre-optic acoustic detection of damage processes in elastomeric insulation under AC and DC stress. In: Proceedings of 18th ISH, Seoul, Paper OD1-03 (2013)
122. Borneburg, D.: Über den praktischen Einsatz einer UV-Kamera zur Detektion, Lokalisierung und Echtzeitdarstellung von Korona-Entladungen an elektrischen Betriebsmitteln. In: ETG Fachbericht Nr. 104, Diagnostik elektrischer Betriebsmittel, Köln, 45–50 (2006)
123. Cardoso, J.A.A., Oliveira Filho, O., de Mello, D.R.: Use of UV cameras for corona tests in high voltage. In: Proceedings of 16th ISH, Johannesburg, Paper A-17 (2009)
124. Schwarz, R., Muhr, M., Pack, S.: Partial discharge detection and localisation for application in transformers. In: Proceedings of 13th ISH, Rotterdam, Paper 183 (2003)
125. Muhr, M., Schwarz, R.: Partial discharge behaviour of oil board arrangements by the installation of fibre-optic technology for monitoring. In: Proceedings of 15th ISH, Ljubljana, Paper T10-325 (2007)
126. Behrend, S., Kalkner, W., Heidmann, G., Emanuel, H., Plath, R.: Synchronous optical and electrical PD measurements. In: Proceedings of 17th ISH, Hannover, Paper D-048 (2011)
127. Pepper, D., et al.: Sensitive optical detection of partial discharges with fluorescent fibers. In: Proceedings of 19th ISH, Pilsen, Paper 598 (2015)

128. IEC 60599: Mineral Oil-Impregnated Electrical Equipment in Service—Guide to the Interpretation of Dissolved and Free Gases Analysis (1999 + 2007) [German edition: DIN EN 60599, VDE 0370 Teil 7: Im Betrieb befindliche, mit Mineralöl imprägnierte elektrische Geräte – Leitfaden zur Interpretation der Analyse gelöster und freier Gase (2008)]
129. Werle, P., et. al.: Comparison of different DGA (Dissolved Gas Analysis) methods for the conditioning assessment of power transformers. In: Proceedings of 12th ISH, Bangalore, Paper F-58 (2001)
130. Tenbohlen, S., et al.: Investigation on sampling, measurement and interpretation of gas-in-oil analysis for power transformers. In: CIGRE Colloquium, Paris, Paper D1-204 (2008)
131. Müller, A., Beltle, M., Coenen, S., Tenbohlen, S.: Correlation of DGA, UHF PD measurement and vibration data for power transformer monitoring. In: Proceedings of 17th ISH, Hannover, Paper F-071 (2011)
132. Chang, Y., et al.: Partial discharge detection of GIS using SWNT-UHF fusion sensor. In: Proceedings of 16th ISH, Johannesburg, Paper C-25 (2009)
133. Fromm, U.: Interpretation of partial discharges at dc voltages. IEEE Trans. Dielectr. Electr. Insul. **2** (1995)
134. Morshuis, P.H.M., Smit, J.J.: Partial discharges at dc voltage: their mechanism, detection and analysis. IEEE Trans. Dielectr. Electr. Insul. **12**, 328–340 (2005)
135. Schichler, U., Kuschel, M., Gorablenkow, J.: Partial discharge measurement on gas-insulated HVDC equipment. In: Proceedings of 18th ISH, Seoul, Paper OH1-01 (2013)
136. Rethmeier, K., Küchler, A., Liebschner, M., Krause, Ch., Kraetge, A., Krüger, M.: Enhanced partial discharge evaluation methods for DC PD measurements using fully digital PD analysing systems. In: Proceedings of 16th ISH, Cape Town, Paper A-13 (2009)
137. Kästner, B., Hoek, S.M., Plath, R., Rethmeier, K.: A modern approach to differential partial discharge diagnosis. In: Proceedings of 19th ISH, Pilsen, Paper 573 (2015)
138. IEC TS 61934: Electrical Insulation Materials and Systems—Electrical Measurement of Partial Discharges (PD) Under Short Rise Time and Repetitive Voltage Impulses (2011)
139. Umemoto, T., Kainaga, S., Ishikura, T., Yoshimura, M., Tsurimoto, T.: Partial discharge detection under impulse voltage application by acoustic emission sensors in oil/pressboard composite insulation system. In: Proceedings of 19th ISH, Pilsen, Paper 206 (2015)

Chapter 13
Evaluation of Uncertainties of Measurement

Abstract Each measurement is not perfect and therefore cannot give the "true" value of the quantity to be measured, but only a more or less accurate approximate value, called estimate (of the value). Even if the measurement is repeated under seemingly identical conditions, the measuring instrument with sufficiently high resolution will display values usually differing from one another. The imperfection, or, positively considered, the quality of a measurement is expressed quantitatively by a numerical value, the uncertainty of measurement. The result of a measurement is the more reliable, the smaller the uncertainty. If the specified measurement uncertainty limits are not met during tests and calibrations, the test object will not pass the acceptance test. This chapter describes the definitions and the concept of how the uncertainty of a measurement can be determined. Key words are model function of the measurement, Type A and Type B evaluation methods, standard uncertainties, expanded uncertainty, uncertainty budget and statement of uncertainties. Several examples are given in Appendix B.

Each measurement is imperfect and therefore cannot give the "true" value of the quantity to be measured, but only a more or less accurate approximate value, called *estimate* (*of the value*). Even if the measurement on a test object is repeated under seemingly identical conditions, the measuring instrument with sufficiently high resolution displays values usually differing from one another. The imperfection, or, positively considered, the quality of a measurement is expressed quantitatively by a numerical value, the *uncertainty of measurement*. As defined in the International Vocabulary of Metrology (VIM), it is a "parameter, associated with the result of a measurement that characterizes the spread of the values that could reasonably be attributed to the measurand" [1].

The knowledge about the uncertainty (of measurement) and its evaluation according to uniform specifications is of great economic importance in the international trade in goods. The result of a measurement is the more reliable, the smaller the uncertainty. If a measuring system consisting of several components is to be improved, it makes sense to first replace the component with the greatest uncertainty. The comparability of measurements at different locations or at different times is meaningful

K. Schon, *High Voltage Measurement Techniques*, Power Systems,
https://doi.org/10.1007/978-3-030-21770-9_13

only with the statement of the uncertainty. This also applies to the *traceability* of a measured quantity to national or international measurement standards (see Sect. 10.1). If the specified limit values of the measurement uncertainty are not met during tests and calibrations, the test object concerned will not pass the acceptance test.

13.1 The GUM

The idea of making a statement about the accuracy of measurement in addition to the measured value is already very old. In this context, reference is made to the classical Gaussian error calculation, which, however, only takes into account measurement deviations based on statistical effects. Deviations due to non-statistical effects, insofar they were not known exactly, were not included in uncertainty calculations in the past. Increasing globalization of the world economy and the rising demands on the accuracy of products and services require uniform international rules for the determination of measurement uncertainties, taking into account statistical and non-statistical influence quantities. As a result of the cooperation of the most important bodies and organizations in this field, under the leadership of the *Bureau International des Poids et Mesures* (*BIPM*), a guideline for the estimation of measurement uncertainties was presented. A revised version of the guideline was published as an ISO-Guide in 1995 and supplemented in 2008 [2]. This guide with more than 100 pages, referred to as *GUM* or *Guide*, is a detailed instruction for the evaluation of uncertainties in all fields of measurement practice. Besides a main part in general terms, the GUM contains several appendices with practical hints and recommendations for many measurement tasks.

The GUM is binding on the *National Metrology Institutes* (*NMI*) as well as accredited test and calibration laboratories all over the world. A shortened version of the GUM applies to the European laboratories [3, 4]. The GUM is also the basis for the specification of uncertainties in test standards in various areas, including high-voltage and high-current testing. The GUM replaces all previous methods for determining uncertainties, including the old terminology. The basic concept of the GUM is presented here in simple form and illustrated by examples from the areas of calibration and testing [5–7].

13.1.1 Basic Concept of the GUM

At the beginning of this chapter, the imperfection of each measurement has already been pointed out. This means that, if a measurement is repeated with the utmost care, a more or less deviating value usually results. Possible reasons for the deviations are the inconstancy of the measuring instruments used, the instability of the test object itself and the not exactly reproducible measurement and ambient conditions. In the GUM, the *standard uncertainty* (*of measurement*), which is defined for influence quantities of statistical as well as non-statistical nature, is of fundamental importance. It serves to characterize the range of values within which the unknown "true" value of a measured quantity can be expected. This range and the probability distribution of

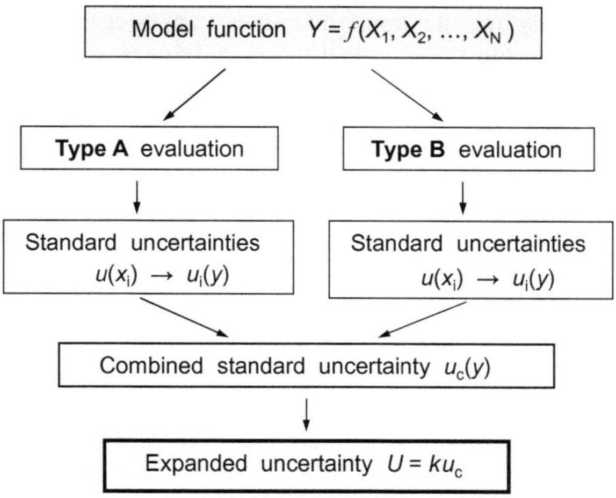

Fig. 13.1 Concept for the estimation of uncertainties according to the GUM (schematic)

the values of an input quantity result either from the measurements themselves or must be estimated on the basis of reliable information. Frequently, a *normal* or *rectangular distribution* of the possible values can be assumed.

The individual steps for evaluating the *measurand* and its measurement uncertainty are shown schematically in Fig. 13.1 and are described in more detail in the following sub-sections. The equations and examples given apply to *uncorrelated input quantities*, as is the case in high-voltage and high-current testing. In the first step, the *model function* of the measurement is set up, which describes the functional dependence of the desired measurand Y on all conceivable input variables X_i. Each of the N input quantities X_i is associated with a standard uncertainty $u(x_i)$, which results either directly from repeated measurements using the *Type A method* or from an estimate of reliable data using the *Type B method*. From the individual contributions $u(x_i)$, the corresponding standard uncertainties $u_i(y)$ of the measurand Y are calculated with the help of the model function and combined as *combined standard uncertainty* $u_c(y)$. After multiplication by the *coverage factor* k, the expanded uncertainty $U = ku_c(y)$ is given in industrial metrology, which characterizes the range of the possible values of Y with a *coverage probability* of at least 95%.

The estimation of the standard uncertainties $u(x_i)$ of the input quantities that are not determined by repeated measurements, requires great expertise and generally the greatest effort in the evaluation of uncertainty. A certain degree of subjectivity in evaluating a measurement by different experts cannot be ruled out, so that slightly different approaches and values may result for the same measurement. Further steps for the evaluation of uncertainty up to the statement of the expanded uncertainty U are rather to be carried out formally applying the given formulas.

13.1.2 Model Function of a Measurement

As a rule, a measurand Y (also called *result quantity* or *output quantity*) results from the combination of N different *input quantities* X_i. The dependence of the output quantity on the input quantities can be generally expressed by the functional relationship *f*, also called the *model function of the measurement*:

$$\boxed{Y = f(X_1, X_2, \ldots, X_i, \ldots, X_N)}. \tag{13.1}$$

In this case, the input quantities X_i themselves may depend on other variables, such as ambient temperature, air pressure, etc., or are subject to corrections for systematic deviations. Each input quantity X_i in the model function has not only a value x_i but also a standard uncertainty $u(x_i)$. With the model function according to Eq. (13.1), then not only the result value *y*, but also the combined standard uncertainty $u_c(y)$ is calculated according to the rules of the GUM. In a complex measurement task with a large number of input quantities, the model function is also very complex. It is generally set up as a single equation or multiple analytic expressions, but it can also be a numerical arithmetic algorithm or an experimentally determined data table. In any case, the model function shall comprise every input quantity X_i, including all corrections and correction factors that lead to a significant contribution to the result value and its uncertainty.

As a simple example of a model function, the measurement of a temperature-dependent resistor *R* is considered. For this example, the model function is:

$$R = f(V, I, T_k, \Theta) = \frac{V}{I}[1 + T_K(\Theta - 20\ °C)]$$

where *V* is the applied voltage, *I* the current magnitude, T_K the temperature coefficient and θ the ambient temperature. Usually, V, I and θ are measured while T_K is taken from a data sheet. The corresponding standard uncertainties result either directly from the measurements or from the data sheets of the measuring instruments. With the model function, both the resistance *R* and its expanded uncertainty are then calculated.

In measurements with smallest uncertainty, it will be indispensable to determine the influence of each input quantity very accurately by measurements. However, the GUM offers the possibility of determining the uncertainty contribution of an input quantity by means of a reliable estimate based on experience and knowledge. The accuracy of the estimate may be slightly less than the result of precise measurements, but which is often negligible in the final result. According to the GUM, an uncertainty contribution that is determined by a reliable estimate or by a measurement is considered to be equivalent. The estimate has the advantage over the accurate measurement that time and effort for this task are saved and thus costs are reduced.

13.1.3 Type A Evaluation Method

The *Type A method* for evaluating standard uncertainties is applied to quantities that result from the statistical evaluation of a series of individual measurements under identical test conditions. This refers in particular to the comparison measurement between the measuring system to be calibrated and the reference system for determining the scale factor and the measurement deviations of the time parameters.

For an infinitely large number of repetition measurements, the individual measured values x have dispersion according to the *normal (Gaussian) distribution* $p(x)$:

$$p(x) = \frac{1}{\sigma\sqrt{2\pi}} \exp\left[\frac{-(x-\mu)^2}{2\sigma^2}\right], \tag{13.2}$$

where μ denotes the expected value of the measured quantity with the greatest probability of occurrence and σ the standard deviation (Fig. 13.2). Corresponding to the bell-shaped normal distribution $p(x)$, the probability of occurrence of a certain value x is the lower the further it deviates from the expectation value. The distribution at p = 68.3% of the maximum value occurs at the values $\mu + \sigma$ and $\mu - \sigma$.

For a limited number n of repetition measurements, the best estimate for the expectation value μ is given by the *arithmetic mean*:

$$\bar{x} = \frac{1}{n}\sum_{k=1}^{n} x_k, \tag{13.3}$$

and the best estimate for the standard deviation σ is given by the empirical standard deviation s of the individual measurements:

$$s(x) = \sqrt{\frac{1}{n-1}\sum_{k=1}^{n}(x_k - \bar{x})^2}. \tag{13.4a}$$

The square value $s^2(x)$ is called *empirical variance*. The *empirical standard deviation of the mean* is:

$$s(\bar{x}) = \frac{s(x)}{\sqrt{n}}. \tag{13.4b}$$

It states how well $\bar{x}$ meets the expectation value of X. By comparing the results of the statistical evaluation with the Type A evaluation method, the best estimate for the input variable X_i is obtained as its arithmetic mean:

Fig. 13.2 Gaussian normal distribution of measured values

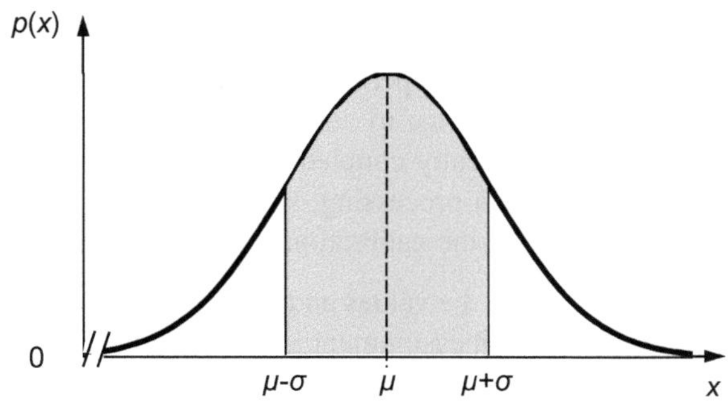

$$\mu = x_i = \bar{X}_i, \tag{13.5}$$

and the best estimate for the standard uncertainty $u(x_i)$ is the empirical standard deviation of the mean:

$$\boxed{u(x_i) = s(\bar{X}_i) = \frac{s(x_i)}{\sqrt{n}}}. \tag{13.6}$$

The number of measurements should be $n \geq 10$, otherwise the reliability of the Type A evaluation of uncertainty according to Eq. (13.6) must be verified on the basis of the *effective degrees of freedom* (see Sect. 13.1.7). If an empirical standard deviation s_p of the individual measurement is already known from previous measurements under perfect statistical conditions, it is recommended that in a comparable measurement series with small number n, i.e. $n = 1, 2, 3, \ldots$ etc., $s(x_i)$ in Eq. (13.6) is replaced by s_p.

13.1.4 Type B Evaluation Method

The *Type B method* for the evaluation of standard uncertainties is to be applied whenever the influence of an input quantity on the measurand does not result from the statistical evaluation of a measurement series. In principle, the influence of an input quantity can always be determined by statistical evaluation of a series of measurement, which, however, means a great experimental effort. The time-consuming statistical evaluation is avoided, especially if it is a secondary input quantity, by determining the standard uncertainty by the Type B method. Its application appears simple, but requires extensive knowledge and experience of the metrological and physical relationships between the desired measurand Y and the input quantities X_i. When used competently, the Type B method is just as reliable as the Type A method.

Contributions to the Type B standard uncertainty originate from:

- Nonlinearities of voltage dividers and measuring instruments
- Dynamic behavior of the measuring system for different impulse forms
- Resolution of digital instruments, reading errors with analog display
- Short-term stability, self-heating
- Long-term stability, drift
- Dependence on temperature, humidity and pressure
- Proximity effect due to neighboring objects
- Electromagnetically coupled or line-conducted interferences
- Method of data processing, software
- Uncertainty in the calibration of the measuring system and its components.

Information on the values and uncertainties of input quantities can be taken from current and past measurement results, calibration certificates, manufacturer data, data from manuals and test specifications, or they are based on empirical data and general knowledge about material and measuring instrument properties. The following cases are distinguished:

(a) There is only a single value for the input quantity X_i, e.g. a single measured value, a correction value or a reference value from technical literature. This value is then used as the input value x_i with the given standard uncertainty $u(x_i)$. If $u(x_i)$ is not known, the value shall be determined from the given reliable data or empirically estimated.

(b) The input quantity X_i is recorded with a measuring instrument whose expanded uncertainty $U = ku_c$ is stated in a calibration certificate or in a data book of the manufacturer (see Sect. 13.1.6). As a rule, a normal (Gaussian) distribution can be assumed, so that the coverage factor $k = 2$. The standard uncertainty then results as follows:

$$\boxed{u(x_i) = \frac{U}{k}}. \tag{13.7}$$

(c) For the possible values of the input quantity X_i, there is no special knowledge about the probability distribution, but only the upper and lower limits a_+ and a_- can be estimated. A rectangular distribution is then assumed in which all possible values of X_i within the interval boundaries are equally probable and the outside values are equal to zero (Fig. 13.3). On the assumption of a rectangular distribution, the best estimate of the input quantity is given by the mean:

$$x_i = \frac{a_+ + a_-}{2} \tag{13.8}$$

and its standard uncertainty by:

$$\boxed{u(x_i) = \frac{a}{\sqrt{3}}}, \tag{13.9}$$

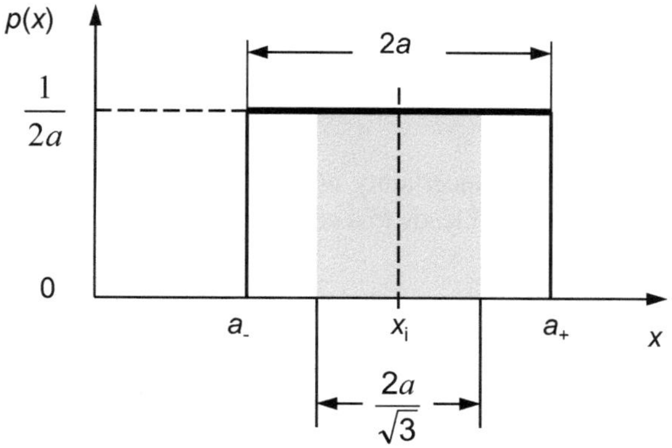

Fig. 13.3 Rectangular distribution of measured values

where a is half the interval width:

$$a = \frac{a_+ - a_-}{2}. \tag{13.10}$$

The rectangular distribution is often assumed because of its simplicity, unless more detailed information about the distribution of the values is not available. However, the discontinuity of the rectangular distribution at the boundaries is often not physically justified, so that the assumption of other distributions such as triangle, trapezoidal or normal distribution seems more appropriate. The standard uncertainty is $u(x_i) = a/\sqrt{6}$ for the triangular distribution and $u(x_i) = \sigma$ for the normal distribution. The rectangle distribution thus provides the largest uncertainty contribution, so that one is on the safe side of the estimate.

The GUM explicitly points out that an uncertainty contribution that has already been taken into account as a Type A contribution shall not be re-entered in full size as a Type B contribution. This applies, for example, to the use of digital recorders when calibrating the scale factor. The spread of the n measured values observed in the multiple measurements, which gives a Type A uncertainty contribution, can be attributed to the limited resolution and internal noise of the digital recorder (see Chap. 7). This spread no longer needs to be taken into account as a Type B uncertainty contribution, or only to a limited extent. However, if the digital recorder is used for a single measurement, the limited resolution corresponding to the stated number of bits and the internal noise must be considered as a Type B uncertainty contribution in its full magnitude.

Often, the value of an input quantity must be adjusted or corrected in order to eliminate the systematic influence of a parameter, e.g. a significant temperature or voltage dependency. Since such a correction can never be absolutely accurate, a residual uncertainty contribution is to be reliably estimated and taken into account in the uncertainty budget. Altogether, the uncertainty is to be determined realistically and on the basis of standard uncertainties. Special safety factors for achieving uncertainties greater than those determined by the GUM method are not permitted, with the exception of the coverage factor k.

13.1.5 Combined Standard Uncertainty

The standard uncertainty $u(x_i)$ of an input quantity X_i evaluated by the Type A or Type B method leads to a corresponding uncertainty contribution $u_i(y)$ of the output quantity Y:

$$\boxed{u_i(y) = c_i u(x_i)}, \tag{13.11}$$

where c_i is the *sensitivity coefficient*. It describes the dependence of the output quantity Y on the input quantity X_i and can be determined analytically or numerically by partial derivation of the model function f with respect to X_i:

$$\boxed{c_i = \frac{\partial f}{\partial X_i}\bigg|_{X_i = x_i} = \frac{\partial f}{\partial x_i}}. \tag{13.12}$$

Example of a sensitivity coefficient is the temperature coefficient of a resistor, which has a dimension and can be positive or negative. Under the assumption of uncorrelated input quantities, the sign of the sensitivity coefficient is without any effect, as only the square of the standard uncertainty is used in the further calculation. If the model function f in Eq. (13.1) is very complex and its derivation corresponding to Eq. (13.12) is not possible, the sensitivity coefficient c_i is determined numerically. For this purpose, the model function is calculated for different input values x_i, and c_i is determined as the difference quotient $\Delta f/\Delta x_i$. This numerical solution is also practiced in software for the programmed calculation of measurement uncertainties.

The N uncertainty contributions $u_i(y)$ of all input quantities assumed to be uncorrelated result in the *combined standard uncertainty* $u_c(y)$ of the output quantity according to the calculation rule:

$$\boxed{u_c(y) = \sqrt{u_1^2(y) + u_2^2(y) + \ldots u_N^2(y)} = \sqrt{\sum_{i=1}^{N} u_i^2(y)} = \sqrt{\sum_{i=1}^{N} [c_i\, u(x_i)]^2}}. \tag{13.13}$$

The form of the equation recalls the quadratic error propagation law of Gauss. The combined standard uncertainty $u_c(y)$ characterizes the dispersion of the values which can be "reasonably" attributed to the output quantity Y. These values are approximately normally distributed if at least three Type B uncertainties of about the same size and defined probability distribution, e.g. rectangular, triangle or normal distribution, contribute to the combined standard uncertainty $u_c(y)$, and the Type A standard uncertainty results from at least $n = 10$ repetition measurements. The combined standard uncertainty according to Eq. (13.13) then covers 68.3% of the possible values of the output quantity Y.

Note The standard uncertainty $u_c(y)$ attributed to the output quantity is referred to as "combined standard uncertainty" in the GUM and is therefore given the index "c". In the European brief version of the GUM for accredited European test and calibration laboratories, published later by the EA, the index "c" is not given [3]. For the sake of clarity, the index is retained here.

If two or more input quantities are correlated, also linear terms exist in Eq. (13.13) and the sign of the sensitivity coefficients is important. For example, correlation exists if the same measuring instrument is used to measure several input quantities. Complicated uncertainty calculations can be avoided by taking into account additional input quantities with the calibration values and uncertainties of the measuring instrument in the model function. If there is a correlation, the combined uncertainty may be even smaller than for uncorrelated quantities. Consideration of correlations is indispensable for precise uncertainty analyzes with very small uncertainty contributions.

If the output quantity is a product or quotient of several input quantities:

$$Y = cX_1^{p_1} X_2^{p_2} \ldots X_N^{p_N} = c \prod_{i=1}^{N} X_i^{p_i}, \tag{13.14}$$

we obtain an expression comparable to Eq. (A2.13):

$$\boxed{w_c(y) = \sqrt{\sum_{i=1}^{N} \left[p_i w(x_i) \right]^2}}, \tag{13.15}$$

where $w(x_i)$ and $w_c(y)$ are the relative standard uncertainties:

$$w(x_i) = \frac{u(x_i)}{|x_i|} \quad \text{and} \quad w_c(y) = \frac{u_c(y)}{|y|}.$$

The quadratic error propagation law is therefore valid for both types of expressing uncertainties with uncorrelated input variables.

13.1.6 Expanded Uncertainty

In many areas of industrial measurement practice, a coverage probability of $p = 68.3\%$ is considered to be too low. It has therefore been agreed worldwide to specify an *expanded uncertainty U* covering a larger proportion of the possible values of Y. Therefore, a coverage probability of not less than $p = 95\%$ is required, which also applies to high-voltage and high-current test techniques. The corresponding expanded uncertainty U is simply obtained by multiplying the combined standard uncertainty $u_c(y)$ by the *coverage factor k*:

$$\boxed{U = k \cdot u_c(y)}. \tag{13.16}$$

If a normal distribution can be assigned to the possible values of the output quantity and the combined standard uncertainty $u_c(y)$ is sufficiently reliable, then $k = 2$. Reliability is judged on the basis of the effective degrees of freedom (see Sect. 13.1.7). If necessary, $k > 2$ must be set to achieve the required coverage probability of not less than 95%.

The expanded uncertainty U is positive like all other uncertainties and is given without sign. If, however, the relevant uncertainty interval is meant within its limits and is linked to the measured value y, the information is given as $y \pm U$.

The term "overall uncertainty" is often found in older test regulations and other sources. This statement can usually be interpreted as expanded uncertainty U with the coverage factor $k = 2$.

13.1.7 Effective Degrees of Freedom

Prerequisite for the assumption $k = 2$ in Eq. (13.16) corresponding to a coverage probability $p \geq 95\%$ is that the possible values of the output quantity Y are normally distributed. The correctness of this assumption is verified by means of the *effective degrees of freedom* ν_{eff} according to the equation:

$$\boxed{\nu_{\mathrm{eff}} = \frac{u_c^4(y)}{\sum_{i=1}^{N} \frac{u_i^4(y)}{\nu_i}}}, \tag{13.17}$$

where ν_i is the number of degrees of freedom of the corresponding uncertainty contribution $u_i(y)$ according to Eq. (13.11) for $i = 1, 2, 3, \ldots, N$. If the assumption of a normal distribution is not justified, a value $k > 2$ has to be determined in order to achieve a coverage probability of approximately 95%.

The degree of freedom ν_i is a measure of the reliability of the estimation of the respective standard uncertainty. Generally accepted values of ν_i for the various uncertainty contributions are:

- $\nu_i = n - 1$ for a Type A standard uncertainty for n observations,
- $\nu_i = \infty$ for a Type B contribution with rectangular distribution,
- $\nu_i \geq 50$ for an uncertainty statement in a calibration certificate with a coverage probability of approximately 95%.

With these values for ν_i, the effective degrees of freedom ν_{eff} can be calculated according to Eq. (13.17). At first glance, the equation looks more complicated than it actually is. However, it is often simplified since in the sum in the denominator of Eq. (13.17), due to $\nu_i = \infty$, the Type B contributions $u_i(y)$ with rectangular distribution become zero. For the calculated value of ν_{eff}, the corresponding coverage factor k for a coverage probability $p = 95.45\%$ is given in Table 13.1. For $\nu_{\mathrm{eff}} < 50$, the coverage factor is $k > 2$.

The coverage factor k can alternatively be calculated using the approximation formula given in IEC 60060-2:

$$\boxed{k = 1.96 + \frac{2.374}{\nu_{\mathrm{eff}}} + \frac{2.818}{\nu_{\mathrm{eff}}^2} + \frac{2.547}{\nu_{\mathrm{eff}}^3}}. \tag{13.18}$$

Table 13.1 Coverage factor k dependent on the effective degrees of freedom ν_{eff} for a coverage probability of $p = 95.45\%$

ν_{eff}	1	2	3	4	5	6	7	8	10	20	50	∞
k	13.97	4.53	3.31	2.87	2.65	2.52	2.43	2.37	2.28	2.13	2.05	2.00

13.1.8 Uncertainty Budget

In the *uncertainty budget*, all essential data of the measurement and evaluation according to the model function in Eq. (13.1) are summarized. It is usually presented in the form of a table, as exemplified in Table 13.2. It is recommended to enter all input values and standard uncertainties as absolute values, including their units. When using special software, the uncertainty calculation and the generation of the uncertainty budget are performed automatically after entering the model function and all input data into the computer [7, 8]. The last line in Table 13.2 shows the result value y calculated from the measured quantities and, if necessary, from corrections. In addition, the combined standard uncertainty $u_c(y)$ and the effective degree of freedom ν_{eff} is given.

13.1.9 Statement of the Complete Result of a Measurement

In calibration and test certificates, the complete result of a measurement is to be given in the form "$y \pm U$", where y is the measurand including each correction and U is the expanded uncertainty for a coverage probability (or confidence level) of approximately 95%. In other words, 95% of the possible values y lie within the interval $(y - U) \leq Y \leq (y + U)$. The numerical value of U is rounded to no more than two significant digits. If the numerical value of the uncertainty due to rounding decreases by more than 5%, the rounded-up value is to be given. The measurand itself shall be rounded corresponding to the least significant digit of the expanded uncertainty. Examples for the recommended statement of a measured voltage value with expanded uncertainty are:

(227.2 ± 2.4) kV,
227.2 (1 ± 0.011) kV,
227.2 (1 ± 1.1 × 10^{-2}) kV.

Table 13.2 Scheme of a measurement uncertainty budget

Quantity X_i	Estimate x_i	Standard uncertainty $u(x_i)$	Degrees of freedom ν_i (ν_{eff})	Sensitivity coefficient c_i	Uncertainty contribution $u_i(y)$
X_1	x_1	$u(x_1)$	ν_1	c_1	$u_1(y)$
X_2	x_2	$u(x_2)$	ν_2	c_2	$u_2(y)$
⋮	⋮	⋮	⋮	⋮	⋮
X_N	x_N	$u(x_N)$	ν_N	c_N	$u_N(y)$
Y	y	$u_c(y)$	ν_{eff}		

The frequently chosen statement "227.2 kV ± 1.1%" is therefore not acceptable. Furthermore, the coverage probability p and the coverage factor k shall be stated (usually $p \approx 95\%$ and $k = 2$ for normal distribution).

13.2 Concluding Remarks

The evaluation of measurement uncertainties is an important tool in quality management and is part of the work of the personnel in test and calibration laboratories that meet the requirements of ISO IEC 17025. Occasionally, the question arises as to whether the personnel in test laboratories have to deal with the topic "uncertainty of measurement" at all, since during tests they use the high-voltage systems calibrated by accredited calibration laboratories within the specified measurement uncertainties. The answer is "yes" for high-voltage and high-current test laboratories, especially if the calibration was performed not in the test site but in the calibration laboratory. When using the measuring system in the test laboratory, further uncertainties are then to be taken into account because of the differing measuring and ambient conditions. This also applies if the calibration is incomplete because not all influence quantities can be determined, e.g. if in the calibration laboratory the linearity test cannot be performed up to the full rated voltage of the high-voltage measuring system.

If the calibration of the measuring system is carried out in the test laboratory, it can be assumed that the effects of temperature, proximity effect, grounding conditions, shape of the test voltage, etc., will be measured completely and correctly and will enter into the uncertainty budget. The uncertainty stated in the calibration certificate for the measuring system can then usually be taken over simply when measuring test voltages or currents (see examples in Annex B). If need be, another uncertainty contribution for the long-term behavior of the measuring system has to be considered.

It is the responsibility of the personnel in the test laboratory to verify the data in the calibration certificate of the measuring system and, if necessary, to add missing uncertainty contributions. The subsequent consideration of uncertainty contributions is formally quite simple. In the first step, the standard uncertainty U/k of the measuring system is determined by dividing the expanded uncertainty U by the coverage factor k given in the calibration certificate (see Sect. 13.1.4, case b). Then, the value of U/k is combined with the standard uncertainties of the additional input quantities in accordance with Eq. (13.13). Contributions that arise in the test during the voltage or current measurement are treated in the same way and are included in the uncertainty budget. For example, by connecting the test object to the test circuit, the impulse test voltage may be superimposed by peak oscillations. Finally, the combined standard uncertainty $u_c(y)$ is calculated with the additional uncertainty contributions and by multiplying by the coverage factor k, the expanded uncertainty $U = ku_c(y)$ of the measurement of the test voltage or test current is obtained.

References

1. ISO/IEC Guide 99: International vocabulary of metrology—basic and general concepts and associated terms (VIM) (2007)
2. ISO/IEC Guide 98-3: Uncertainty of measurement—part 3: guide to the expression of uncertainty in measurement (GUM: 1995) (2008)
3. EA-4/02M: Evaluation of the uncertainty of measurement in calibration (2013)
4. DIN 1319-3: Grundlagen der Messtechnik – Teil 3: Auswertung von Messungen einer einzelnen Messgröße, Messunsicherheit (1996)
5. PTB-Mitt. 111 (Special Print): Themenschwerpunkt Messunsicherheit. Wirtschaftsverlag NM, Bremerhaven (2001)
6. Schon, K.: What is new in the future IEC 60060-2: Uncertainty of measurement and convolution. HIGHVOLT Kolloquium '07, Dresden, paper 1.2 (2007)
7. Li, Y., Schon, K., Mohaupt, P.: Determinations of measurement uncertainty of the atmospheric correction factor for high-voltage testing. In: Proceedings of 14th ISH Ljubljana, paper T10-500 (2007)
8. Metrodata: GUM Workbench, Homepage: www.metrodata.de

Appendix A

Fourier and Laplace Transforms

The *Fourier transform* and the *Laplace transform* are integral transforms which have great practical significance in science and technology. They are applied very successfully for solving many mathematical problems. In most applications, a real continuous function in the time domain is transformed into a complex function in the *complex variable domain* (*s-domain*, *spectral domain*, *frequency domain*). The Laplace transform is considered a generalization of the Fourier transform. A direct application of both transforms provides the *complex transfer function* of a linear system in the frequency domain from which the amplitude response and phase response are obtained. With the Laplace transform, complicated calculations in the time domain such as differentiation and integration can be replaced by simple algebraic operations in the s-domain. The function obtained in the s-domain is then transformed back into the time domain and expressed as a time function. For a number of functions, there are appropriate correspondences listed in tables. The Laplace transform can also be successfully applied to solve the convolution integral. Analogous to the Laplace transform is the *z-transform*, which will not be discussed further here and by which a time-discrete function is transformed into a complex discrete function in the z-domain.

A.1 Fourier Transform

An impulse is a one-time temporal event that can be represented by the superposition of an infinite number of sinusoidal partial oscillations with different amplitudes and phase angles. Mathematically, the decomposition of a time function $f(t)$ into partial oscillations is described by the complex *Fourier integral*:

K. Schon, *High Voltage Measurement Techniques*, Power Systems,
https://doi.org/10.1007/978-3-030-21770-9

$$\boxed{F(\mathrm{j}\omega) = F(\omega)\, e^{\mathrm{j}\,\varphi(\omega)} = \int_{-\infty}^{\infty} f(t)\, e^{-\mathrm{j}\omega t}\mathrm{d}t} \tag{A.1}$$

where $\omega = 2\pi f$ is the angular frequency and $F(\mathrm{j}\omega)$ the spectral function or Fourier transform of the time function $f(t)$. In contrast to a periodic signal known to have a discrete spectrum and which can be represented by an infinite Fourier series, an impulse has a continuous spectrum. The amplitude of a partial oscillation is hereby related to an infinitesimal frequency interval $d\omega$. The absolute value of the complex spectral function $F(\mathrm{j}\omega)$ is the *amplitude density* $F(\omega)$, resulting from the square root of the sum of squares of the real and imaginary parts of $F(\mathrm{j}\omega)$:

$$\boxed{F(\omega) = |F(\mathrm{j}\omega)| = \sqrt{\mathrm{Re}\{F(\mathrm{j}\omega)\}^2 + \mathrm{Im}\{F(\mathrm{j}\omega)\}^2}.} \tag{A.2}$$

The individual partial oscillations of the signal with the amplitude density according to Eq. (A.2) have a phase displacement expressed in its entirety as the *phase response* $\varphi(\omega)$ of the signal:

$$\boxed{\varphi(\omega) = \arctan\frac{\mathrm{Im}\{F(\mathrm{j}\omega)\}}{\mathrm{Re}\{F(\mathrm{j}\omega)\}}.} \tag{A.3}$$

The knowledge of $\varphi(\omega)$ is generally of secondary importance in high-voltage and high-current impulse measurement technique.

On the other hand, if the complex spectrum $F(\mathrm{j}\omega)$ of a signal is known, the corresponding time function $f(t)$ is obtained by the *inverse Fourier transform* into the time domain:

$$\boxed{f(t) = \frac{1}{2\pi}\int_{-\infty}^{\infty} F(\mathrm{j}\omega)\, \mathrm{e}^{\mathrm{j}\omega t}\mathrm{d}\omega} \tag{A.4}$$

Apart from determining the spectrum of an impulse according to Eq. (A.1), the transformation of a time function into a spectral function has the advantage that certain arithmetic operations in the frequency domain can be performed better or easier than in the time domain. The resulting new spectral function is then transformed back into the time domain according to Eq. (A.4) and yields the appropriate time function. However, this arithmetic operation is preferably performed with the Laplace transform (see Appendix A.2).

The complex Fourier integral according to Eq. (A.1) and the inverse Fourier transform according to Eq. (A.4) can only be solved for some analytically defined impulse shapes and spectra. In practice, therefore, the real presentation of the time signal and Fourier integral is of greater importance. The time function is represented in real form by the Fourier series:

$$f(t) = \int_0^{\infty} a(\omega)\sin\omega t\,\mathrm{d}\omega + \int_0^{\infty} b(\omega)\cos\omega t\,\mathrm{d}\omega \tag{A.5}$$

with the two spectral functions:

$$a(\omega) = \frac{1}{\pi}\int_{-\infty}^{\infty} f(t)\sin\omega t\,\mathrm{d}t \quad \text{and} \quad b(\omega) = \frac{1}{\pi}\int_{-\infty}^{\infty} f(t)\cos\omega t\,\mathrm{d}t. \tag{A.6}$$

The amplitude density of a signal is given by:

$$F(\omega) = \sqrt{a^2(\omega) + b^2(\omega)}. \tag{A.7}$$

The real form is advantageous for numerical calculations, e.g. when $f(t)$ is available as data set of a digital recorder. The integrals in Eqs. (A.5) and (A.6) are then replaced by series. In order to reduce the computation time required for large amounts of data, the *Fast Fourier Transform* (*FFT*) is used instead of the Discrete Fourier transform (DFT) in Eqs. (A.5) and (A.6). The number of samples must be a power of 2. The advantage of speed, however, is achieved at the expense of accuracy of calculation, which is acceptable for many practical applications (Refs. [1, 2, 4] of Chap. 9).

A.2 Laplace Transform

The Laplace transform of a time function into the complex variable domain provides another possibility for signal analysis. It was originally developed to solve linear differential equations by means of transforming them into algebraic equations. The Laplace transform generally applies to any function, but is commonly applied to time functions. Besides signal analysis, another important field of application of the Laplace transform is the simple calculation of switching operations in electrical circuits. Under the assumption that the time function $f(t)$ for $t < 0$ is equal to zero, the Laplace transform of $f(t)$ reads:

$$L\{f(t)\} = F(s) = \int_0^{\infty} f(t)\,\mathrm{e}^{-st}\mathrm{d}t, \tag{A.8}$$

where $s = \sigma + \mathrm{j}\omega$ is a complex number. In order to determine the spectrum of a time function, $s = \mathrm{j}\omega$ is inserted.

In analogy to Eq. (A.2), the amplitude density $F(\omega)$ of a signal is given by the absolute value of the Laplace transform:

$$F(\omega) = |F(s)| = \sqrt{\text{Re}\{F(s)\}^2 + \text{Im}\{F(s)\}^2}. \tag{A.9}$$

Apart from calculating the spectrum of a time signal, the practical advantage of the Laplace transform lies in the fact that calculations in the s-domain can be done much more easily than in the time domain. The resulting new image function F(s) is then transformed back into the time domain and yields the associated time function. The inverse Laplace transform of the image function F(s) into the time domain formally reads:

$$f(t) = \frac{1}{2\pi \text{j}} \int_{\sigma - \text{j}\infty}^{\sigma + \text{j}\infty} F(s)\, \text{e}^{st} \text{d}s. \tag{A.10}$$

The application of the Laplace transform to electric circuits requires that all energy sources be uncharged at time $t = 0$. If this condition is satisfied, the Laplace transform can be developed directly in the s-domain with the operator s from the impedances of the circuit. For a circuit with energy sources, first the differential equation for the circuit must be set up, from which the Laplace transform is then formed.

For the Laplace transform and inverse transform, there are general calculation rules that can be advantageously applied. Thus, differentiation and integration of a function as well as the convolution integral in the time domain are replaced by simple algebraic operations with the operator s in the complex variable domain. For a large number of functions of the Laplace transform, the two integrals in Eqs. (A.8) and (A.10) have already been evaluated and are available in the literature in table

Table A.1 Some calculation rules of the Laplace transform

Rule No.	Laplace F(s)	Function $f(t \geq 0)$	Remark
1	$\frac{1}{s}F(s)$	$\int_0^t f(t)\, dt$	Integration for $t \geq 0$
2	$\frac{dF(s)}{ds}$	$-tf(t)$	Multiplication
3	$sF(s) - f(0)$	$\frac{df(t)}{dt}$	First derivative
4	$e^{-as}F(s)$	$f(t + a)$	Time delay by $t = a$
5	$F_1(s) \cdot F_2(s)$	$\int_0^t f_1(x) \cdot f_2(t - x)\, dx$	Convolution
6	$\frac{1}{a}F\left(\frac{s}{a}\right)$	$f(at)$	Time scaling (a > 0)
7	$a_1F_1(s) + a_2F_2(s)$	$a_1f_1(t) + a_2f_2(t)$	Linearity

Table A.2 Some correspondences of the Laplace transform

Example No.	Laplace $F(s)$	Function $f(t \geq 0)$	Remark
1	$\frac{1}{s}$	$u_s(t) = 1$	Unit step function, $u_s = 0$ for $t < 0$
2	1	$\delta(t)$	Dirac impulse
3	$\frac{a}{s^2}$	$a\,t$	Ramp function
4	$\frac{1}{s+a}$	e^{-at}	Exponential function
5	$\frac{1}{s\,(s+a)}$	$1 - e^{-at}$	Mirrored exponential function
6	$\frac{1}{s^2(s+a)}$	$t - a(1 - e^{-at})$	
7	$\frac{1}{(1+as)\,(1+bs)}$	$\frac{1}{a-b}\left(e^{-t/a} - e^{-t/b}\right)$	Double exponential function
8	$\frac{1}{s^2+as+b}$	$\begin{cases} \frac{1}{\omega_0} \sin \omega_0 t \cdot e^{-at/2} \\ \frac{1}{\omega_0^*} \sinh \omega_0^* t \cdot e^{-at/2} \end{cases}$	$\omega_0 = \sqrt{b - \frac{a^2}{4}}$ for $b > \frac{a^2}{4}$ $\omega_0^* = \sqrt{\frac{a^2}{4} - b}$ for $b < \frac{a^2}{4}$

form (Refs. [1, 2, 4] of Chap. 9). A small selection of the calculation rules and correspondences that are used in Chaps. 8 and 9 is compiled in Tables A.1 and A.2.

Appendix B

Examples of Estimating Uncertainties

By means of three examples, the procedure for estimating uncertainties according to the GUM is shown for non-correlated input quantities. The first example deals with the calibration of a high-voltage impulse measuring system by comparison with a reference system. After creating the model equation for the scale factor of the measuring system, the individual uncertainty contributions are estimated. From the combined uncertainty, the expanded uncertainty of the scale factor is determined. The second and third examples deal with the later use of the calibrated measuring system in the high-voltage test of a power apparatus, with the impulse voltage superimposed with and without peak oscillations. The uncertainty of the voltage measurement is composed of the combined uncertainty of the scale factor and additional uncertainty contributions that occur during the voltage test and were not taken into account during the previous calibration. Another example in (Ref. [7] of Chap. 7) deals with various methods for estimating the uncertainty of the atmospheric correction factor.

B.1 Scale Factor of an Impulse Voltage Measuring System

The scale factor of a high-voltage impulse measuring system X is determined by comparison with a reference system N in the measurement setup according to Fig. 10.2. Each measuring system consists of a 1 MV voltage divider with digital recorder for recording the divider output voltage. The comparison measurement is carried out at a temperature of 15 °C in the high-voltage hall, in which the measuring system X is also used for tests on power apparatuses. For the reference system N, a scale factor $F_N = 1.015$ at 20 °C with expanded uncertainty $U = 0.8\%$

K. Schon, *High Voltage Measurement Techniques*, Power Systems,
https://doi.org/10.1007/978-3-030-21770-9

(k = 2) is stated in the calibration certificate. The calibration certificate issued 11 months ago contains no uncertainty contribution for long-term stability.

B.1.1 Setting up the Model Function

The first step consists of the analysis of the measurement procedure with the aim of setting up the model function according to Eq. (13.1). In the ideal case, both the measuring system X and the reference system N provide the correct peak value $\hat{u}$ of the applied impulse voltage during the comparison measurement. Then we can write (see Fig. 10.3):

$$\hat{u} = \hat{u}_X F_X = \hat{u}_N F_N,$$

where F_X and F_N are the scale factors and $\hat{u}_X$ and $\hat{u}_N$ are the peak values measured at the low-voltage side of the measuring system X or reference system N. The above equation yields the basic form of the model function for the scale factor of the measuring system X:

$$\boxed{F_X = \frac{\hat{u}_N}{\hat{u}_X} F_N.} \tag{B.1}$$

Other basic forms of the model function are quite conceivable, but will not be discussed here.

The two measuring systems X and N are subject to various influences, which are quantified by measurement or reliable estimation and taken into account in the model function for the scale factor F_X. In the case of the measuring system X, these are the measured values for the quotient $\hat{u}_N/\hat{u}_X$ and the standard deviation, the dependence on the peak value $\hat{u}$ and the front time T_1 of the test voltage, the clearance L to neighboring objects and the short-term stability S_k. The scale factor F_N of the reference system is affected by the long-term stability S_L and the ambient temperature Θ. The complete model function for the scale factor F_X accordingly has the general form:

$$F_x = f(\hat{u}_N/\hat{u}_X, F_N, S_L, \Theta, \hat{u}, T_1, L, S_k). \tag{B.2}$$

The exact functional dependence between the output quantity F_X and most input quantities in Eq. (B.2) is not known in detail. To avoid extensive investigations, the model function is set up in a simplified form. The influence of the input quantities on F_N and F_X is directly expressed, i.e. implicitly the sensitivity coefficients c_i (see Sect. 13.1.5), by the deviations $\Delta F_{N,i}$ and $\Delta F_{X,k}$ and reliably estimated. Each of these deviations is formally given a negative sign and put to the respective scale factor on the appropriate side of the model function.

After inserting these deviations in Eq. (B.1) and solving for the scale factor F_X, the complete model function for the scale factor reads:

$$F_X = \frac{\hat{u}_N}{\hat{u}_X}\left(F_N - \sum_{i=1}^{2}\Delta F_{N,i}\right) + \sum_{k=1}^{4}\Delta F_{X,k} \tag{B.3}$$

with the numbered deviations $\Delta F_{N,i}$ and $\Delta F_{X,k}$:

- $\Delta F_{N,1}$ influence of the ambient temperature Θ on the reference system N
- $\Delta F_{N,2}$ long-term stability S_L of the reference system N
- $\Delta F_{X,1}$ voltage dependence of the measuring system X
- $\Delta F_{X,2}$ influence of the front time on the measuring system X
- $\Delta F_{X,3}$ proximity effect of the measuring system X due to the wall distance L
- $\Delta F_{X,4}$ short-term stability S_K of the measuring system X.

Each of the listed deviations basically consists of a numerical value for the deviation itself and the corresponding standard uncertainty. The numerical value of a deviation may even be zero within the specified uncertainty. The long-term behavior of the measuring system X is not taken into account here. A corresponding uncertainty contribution is therefore to be determined by the test laboratory itself from subsequent performance checks.

B.1.2 Estimation of Standard Uncertainties

According to the GUM, there are two methods for estimating standard uncertainties (see Sects. 13.1.3 and 13.1.4). The Type A method is applied to quantities resulting from the statistical evaluation of a series of individual measurements under identical test conditions. The standard uncertainty is then simply deduced from the standard deviation s and the number of measurements according to Eq. (13.6). The Type B method is applied to quantities that do not result from the statistical evaluation of a measurement series. If the exact distribution of the measured value is not known, it then often makes sense and is practical to assume a rectangular distribution with symmetrical interval limits a_+ und a_-, in which the respective deviations $\Delta F_{N,i}$ and $\Delta F_{X,k}$ have uniform probability of occurrence (see Fig. 13.3). From the half-width a of the rectangular distribution, the standard uncertainty of the deviation results in $a/\sqrt{3}$ according to Eq. (13.9). Since the uncertainty contributions of ΔF_{X1} to ΔF_{X4} in Eq. (B.3) are directly related to the scale factor F_X, the model function has a relatively simple form.

In the following steps, the input quantities, deviations, corrections and standard uncertainties are determined by measurements or obtained from reliable data sources. First, the scale factor F_N of the reference system determined at 20 °C is calculated for the ambient temperature of 15 °C during the comparison measurement. According to the manufacturer's data for the temperature coefficient, the scale factor has to be corrected by −0.3% so that the actual value at 15 °C is $F_N = 1.012$. However, corrections are always associated with uncertainties. Therefore, the last decimal place of F_N is considered uncertain and formally a residual deviation within

±0.001 is inserted, assuming a rectangular distribution according to Fig. 13.3. The corresponding Type B standard uncertainty thus results in $u_1(F_N) = 0.001/\sqrt{3} = 0.000577$ (absolute value) according to Eq. (13.9). The long-term stability of the reference system within one year is ±0.5%, as stated by the manufacturer. This results in a further uncertainty contribution $u_2(F_N) = 0.005 \cdot 1.012/\sqrt{3} = 0.00292$.

The comparison measurement between the measuring system X with the rated voltage $U_0 = 1$ MV and the reference system N is carried out at five voltage levels between 20 and 100% of U_0. The impulse voltage generator is set to a front time $T_1 = 1.1$ µs, which is approximately in the middle of the permissible tolerance limits, and to the maximum time to half-value $T_{2max} = 60$ µs. The impulse voltages are smooth and have no peak oscillation or overshoot. For each of the five voltage levels, $n = 10$ pairs of the peak values $\hat{u}_N$ and $\hat{u}_X$ are measured. As an example, Table B.1 shows the values $\hat{u}_N$ and $\hat{u}_X$ measured at 20% of U_0. The simultaneous measurement of both peak values eliminates the effect of instability of the impulse voltage generator. The spread of the values for the quotient is thus smaller than for the individually measured peak values and correspondingly also for the Type A standard uncertainty.

In the same way, the peak values $\hat{u}_N$, $\hat{u}_X$ and the quotients $\hat{u}_N/\hat{u}_X$ are determined at 40–100% of the rated voltage U_0. The result of the comparison measurement at the five voltage levels between 0.2 and 1 MV is summarized in Table B.2. For each voltage value, the mean quotient $\hat{u}_N/\hat{u}_X$ and the standard deviation $s(\hat{u}_N/\hat{u}_X)$ are given. With increasing impulse voltage, an increase of $\hat{u}_N/\hat{u}_X$ is visible. According to the model function, the scale factor F_X also increases. For calculating the Type A standard uncertainty according to Eq. (13.6) with $n = 10$, the maximum value s_{max} of the standard deviations is taken.

For the later employment of the measuring system X during tests, it is expedient to use a mean scale factor $F = F_{Xm}$ for the entire voltage range (see Fig. 10.4).

Table B.1 Measured peak values $\hat{u}_N$ of the reference system N and $\hat{u}_X$ of the measuring system X and their quotients $\hat{u}_N/\hat{u}_X$ at approximately 20% of $U_0 = 1$ MV

Measurement No.	$\hat{u}_N$ kV	$\hat{u}_X$ V	$\hat{u}_N/\hat{u}_X$
1	208.0	103.6	2007.7
2	208.2	103.6	2009.7
3	207.1	102.9	2012.6
4	205.9	102.3	2012.9
5	207.3	102.3	2026.4
6	207.7	103.1	2014.5
7	207.8	103.3	2011.6
8	207.7	103.3	2010.6
9	206.8	102.9	2009.7
10	207.8	103.5	2007.7
Mean quotient $\hat{u}_N/\hat{u}_X$			2012.3
Standard deviation $s(\hat{u}_N/\hat{u}_X)$			5.4

Table B.2 Result of the comparison measurement at five impulse voltage levels

Voltage $\hat{u}$ MV	Quotient $\hat{u}_N/\hat{u}_X$	Standard deviation $s(\hat{u}_N/\hat{u}_X)$
0.2	2012.3	5.4
0.4	2011.0	5.6
0.6	2015.2	6.2
0.8	2019.9	6.1
1	2025.7	6.9 (=s_{max})
Mean quotient $\hat{u}_N/\hat{u}_X$	2016.7	
Maximum deviation from mean	9	

Therefore, the mean value of the quotients $\hat{u}_N/\hat{u}_X$ from the five measurement series is inserted into Eq. (B.3). The voltage dependence of the scale factor is taken into account by an uncertainty contribution at $\Delta F_{X,1}$. This contribution is obtained from half the interval width a_1 of a rectangular distribution, where a_1 denotes the maximum deviation of the individual quotients $\hat{u}_N/\hat{u}_X$ from their mean multiplied by the scale factor F_N of the reference system (see Table B.2).

In further comparison measurements with the reference system N, the dynamic behavior of the measuring system X is investigated. The front time T_1 of the lightning impulse voltage is varied and the effect on the scale factor is determined. The measurements show that within the permissible tolerance range T_1 = 1.2 µs ± 30%, the scale factor F_X changes within $\pm a_2$ = 0.5%. The proximity effect during the comparison measurements is taken into account with a proportion $\pm a_3$ = 0.2%, depending on the position of the test object. The short-term stability test leads to a deviation of F_X within $\pm a_4$ = 0.2%.

With the deviations a_1–a_4, the corresponding Type B standard uncertainties are calculated according to Eq. (13.9). During the interference test, disturbances of less than 1% occur in the initial part of the recorded impulse voltage. However, the disturbances do not affect the determination of the impulse parameters and are therefore not taken into account in the uncertainty budget.

B.1.3 Uncertainty Budget and Result of the Calibration

The values of the input quantities and uncertainty contributions are summarized in the uncertainty budget (Table B.3). The numerical evaluation of the model function in Eq. (B.3) is conveniently performed using validated software (Ref. [8] of Chap. 13). As a result of the calculation, the mean scale factor F_{Xm}, the combined standard uncertainty F_{Xm} and the number of the effective degrees of freedom ν_{eff} is given in the last line of Table B.3. The relatively large value ν_{eff} = 370 means that there is a normal distribution with the coverage factor k = 2 for a coverage probability of at least 95%.

Table B.3 Uncertainty budget for the mean scale factor F_{Xm}

Quantity X_i	Value x_i	Standard deviation $u(x_i)$	Degrees of freedom ν_i (ν_{eff})	Sensitivity coefficient c_i	Uncertainty contribution $u_i(F_{Xm})$
$\hat{u}_N/\hat{u}_X$	2016.7	2.18[a]	9	1.0	2.2
F_N	1.0150	0.00400[a]	50	2000	8.1
ΔF_{N1}	0.003036	0.000577	∞	−2000	−1.2
ΔF_{N2}	0.0	0.00292	∞	−2000	−5.9
ΔF_{X1}	0.0	5.25	∞	1.0	5.3
ΔF_{X2}	0.0	5.89	∞	1.0	5.9
ΔF_{X3}	0.0	2.36	∞	1.0	2.4
ΔF_{X4}	0.0	2.36	∞	1.0	2.4
F_{Xm}	2041	u_c = 13.4	ν_{eff} = 370		

[a]Normal distribution (all other input quantities: rectangular distribution)

The complete result of the calibration is finally stated in the calibration certificate for the scale factor of the measuring system X in the form:

$$F = F_{Xm} = 2041 \pm 27 \text{ with a coverage probability of } p \geq 95\% \ (k = 2).$$

The specified expanded uncertainty of the assigned scale factor $F = F_{Xm}$ corresponds to a relative value of 1.3%.

Finally, it should be pointed out once again that the uncertainty of the assigned scale factor as a result of a calibration is not necessarily identical with the uncertainty resulting from the use of the measuring system X for voltage measurements during a test. In addition to the long-term stability of the measuring system, further input quantities may also have to be considered in the uncertainty budget for voltage measurements. This will be discussed in the following two examples in which impulse voltages with and without superimposed peak oscillations are measured.

B.2 Uncertainty of Voltage Measurement During a Test

The measuring system X calibrated in Appendix B.1 is used to measure the test voltage value of a single lightning impulse when testing a power apparatus. In the first example, the lightning impulse voltage is measured without and in the second example with superimposed oscillation. The scale factor F_{Xm} = 2041 with the expanded uncertainty $U(F_{Xm})$ = 27 determined by the calibration at 15 °C is to be adapted to the actual test conditions. The following additional input quantities and uncertainty contributions are taken into account. The impulse voltage test takes place at a temperature of 21 °C. Thereby, the scale factor F_{Xm} referred to 15 °C increases by ΔF_1 = 0.3% corresponding to the temperature coefficient of the measuring system. Performance checks show that the scale factor F_{Xm} has increased by $\Delta F_2 = 0.4\%$ as a result of the long-term drift. Since the two corrections ΔF_1 and ΔF_2 represent only approximate values, a rectangular distribution with half the interval width of 0.1% each, referred to F_{Xm}, is assumed for the possible values of the corrected scale factor. With all input quantities and standard uncertainties, the numerical evaluation of the model function will be done again with software (Ref. [8] of Chap. 13).

B.2.1 Impulse Voltage Without Peak Oscillation

In this example, the impulse voltage to be measured has a time course without peak oscillation. The digital recorder displays a peak value of $U_{rec} = 324.5$V. Since the deflection of the 10-bit recorder is only 80%, a deviation of ±0.2% within a rectangular distribution is set for U_{rec}, which also takes into account the contribution of superimposed noise. For the desired value U_t of the test voltage, the following model equation can be set up:

$$\boxed{U_t = U_{rec} \cdot F = U_{rec}(F_{Xm} + \Delta F_1 + \Delta F_2).} \tag{B.4}$$

The given values and uncertainties are summarized in the uncertainty budget (Table B.4). The number of degrees of freedom F_{Xm} is set to $\nu = 50$ because the value $\nu_{eff} = 370$ determined during calibration and indicated in Table B.3 is generally not included in the calibration certificate and is thus unknown (see Sect. 13.1.7). The last line of Table B.3 gives the result for the test voltage value U_t. The number of effective degrees of freedom is 57, so that there is a normal distribution of the result values and the coverage factor is $k = 2$ (coverage probability $p \geq 95\%$). The result for the test voltage value is:

$$U_t = 666.9\,\text{kV} \pm 9.1\,\text{kV} \quad (p \geq 95\%,\ k = 2).$$

Table B.4 Uncertainty budget for the test voltage value U_t of an impulse voltage without superimposed peak oscillation

Quantity X_i	Value x_i	Standard deviation $u(x_i)$	Degrees of freedom ν_i (ν_{eff})	Sensitivity coefficient c_i	Uncertainty contribution $u_i(U_t)$ kV
F_{Xm}	2041	13.5[a]	50	320 V	4.4
ΔF_1	6.12	2.04	∞	320 V	0.66
ΔF_2	8.16	2.04	∞	320 V	0.66
U_{rec}	324.5 V	0.375 V	∞	2100	0.77
U_t	666.9 kV	4.55 kV	57		

[a]Normal distribution (all other input quantities: rectangular distribution)

The expanded uncertainty of the test voltage value U_t is relatively 1.6%. It is thus only slightly greater than the uncertainty of 1.3% for the scale factor, as determined in the first example for the calibration of the impulse voltage measuring system.

B.2.2 *Impulse Voltage with Peak Oscillation*

In this example, the lightning impulse test voltage has overshoot or oscillation in the peak region. The test voltage value U_t that is decisive for the stress of the insulation of the power apparatus, must therefore be determined by means of the test voltage function $k(f)$ (see Sect. 4.1.1.2). The recorded impulse voltage with superimposed peak oscillation has the extreme value $U_e = U_{rec} = 324.5$ V. To determine the test voltage value, the alternative method based on manual evaluation of the recorded impulse is used. A double exponential base voltage according to Eq. (8.8) is fitted to the oscillating lightning impulse voltage. This method gives a peak value of $U_b = 299.3$ V, for which a standard uncertainty of 0.2% is estimated, assuming a rectangular distribution. The difference between the oscillating lightning impulse voltage and the exponential voltage yields the oscillating residual curve with the amplitude $\beta = U_{rec} - U_b$. The oscillation frequency, which is the reciprocal of twice the duration of the half-period in the time region of the peak, is $f = 0.3$ MHz. For this frequency, the value of the test voltage function is $k(f) = 0.835$ according to Eq. (4.4). The amplitude β of the residual curve is multiplied by $k(f)$ and superimposed on the peak value of the double exponential base curve. The result is the desired value U_t of the test voltage.

With the same input data for the scale factor (see example in Sect. B.2.1), the following model equation for the desired test voltage value U_t can be set up:

$$U_t = [U_b + k(f) \cdot \beta] \cdot F = [U_b + k(f) \cdot (U_{rec} - U_b)] \cdot (F_{Xm} \cdot + \Delta F_1 + \Delta F_2). \tag{B.5}$$

For the test voltage function $k(f)$ according to Eq. (4.4), no uncertainty is given in (Ref. [1] of Chap. 2). However, the result of the investigation in (Ref. [2] of Chap. 4) shows that the experimentally determined k-values deviate more or less from the function values of $k(f)$ calculated according to Eq. (4.4). The typical spread of the values is within ± 0.2 (absolute). For the test voltage function $k(f)$, a Type B standard uncertainty $u(k) = 0.2/\sqrt{3} = 0.115$ is assumed, which also includes the uncertainty of the frequency determination.

The values and uncertainties are summarized in the uncertainty budget (Table B.5). The last line in Table B.5 contains the result for the test voltage value U_t. The large number of effective degrees of freedom of 440 indicates that there is a normal distribution of the result values with the coverage factor $k = 2$ for a coverage probability $p \geq 95\%$.

Table B.5 Uncertainty budget for the test voltage value U_t of a lightning impulse voltage with peak oscillation taking into account the test voltage function $k(f)$ according to Eq. (4.4)

Quantity X_i	Value x_i	Standard uncertainty $u(x_i)$	Degrees of freedom ν_i (ν_{eff})	Sensitivity coefficient c_i	Uncertainty contribution $u_i(U_t)$ kV
F_{Xm}	2041	13.5[a]	50	320 V	4.3
ΔF_1	6.12	2.04	∞	320 V	0.65
ΔF_2	8.16	2.04	∞	320 V	0.65
U_{rec}	324.5 V	0.375 V	∞	1700	0.64
U_b	299.3 V	0.346 V	∞	340	0.12
$k(f)$	0.835	0.115	∞	52×10^3 V	6.0
U_t	658.4 kV	7.5 kV	440		

[a]Normal distribution (all other input quantities: rectangular distribution)

The result of the voltage measurement during the test can finally be stated in the form:

$$U_t = 658 \text{ kV} \pm 15 \text{ kV} \quad (p \geq 95\%, k = 2).$$

The expanded uncertainty of the test voltage value U_t for the lightning impulse voltage with superimposed peak oscillation is relatively 2.3%. It is thus greater than the uncertainty of 1.6%, which is estimated for the lightning impulse voltage without peak oscillation, but is still below the permissible limit value of 3%.

Index

K. Schon, *High Voltage Measurement Techniques*, Power Systems,
https://doi.org/10.1007/978-3-030-21770-9

T

MIX
Papier aus verantwortungsvollen Quellen
Paper from responsible sources
FSC® C105338

If you have any concerns about our products,
you can contact us on
ProductSafety@springernature.com

In case Publisher is established outside the EU,
the EU authorized representative is:
Springer Nature Customer Service Center GmbH
Europaplatz 3, 69115 Heidelberg, Germany

Printed by Libri Plureos GmbH
in Hamburg, Germany